9/17/93

PRINCIPLES OF ENGINEERING GEOLOGY

PRINCIPLES OF ENGINEERING GEOLOGY

Robert B. Johnson
Colorado State University

Jerome V. DeGraff
U.S. Forest Service

JOHN WILEY & SONS
New York Chichester Brisbane Toronto Singapore

Cover photo shows the toe of the Muddy Creek Landslide restricting Muddy Creek north of Paonia Reservoir, west-central Colorado. (*Photo by Robert Johnson.*)

Library of Congress Cataloging in Publication Data:

Johnson, Robert B. (Robert Britten)
Principles of engineering geology.

1. Engineering geology. I. DeGraff, Jerome V. II. Title.
TA705.J65 1988 624.1′51 88-5510
ISBN 0-471-03436-3

Printed in the United States of America

Printed and bound by the Hamilton Printing Company.

10 9 8 7 6 5 4 3

To our wives, Garnet and Sandy, for their forebearance, encouragement, assistance, and continuing ability to envision the light at the end of the tunnel.

Preface

Engineering geology is the application of geologic fundamentals to civil engineering. The intent of this text is to provide a balanced, comprehensive view of the subject. Because this area of study deals with applied science, every effort is made to present both relevant geologic principles and the basics of engineering geologic practice.

The text is designed for the upper-division undergraduate or the beginning graduate student in both geology and civil engineering. Geology students can expect to learn theory and applied aspects of engineering geology; rather than becoming practicing engineers, they will be prepared to function as important contributors to civil engineering work. Civil engineering students with geotechnical interests should gain an understanding of the impact geology has on civil engineering planning, design, construction, and monitoring. No attempt is made to provide the engineering student with a first course in geology. Other than a brief introductory review of geologic fundamentals, the material presented is understandable to students in both disciplines. The senior author's success in using this format for a number of years in classes composed of geology and civil engineering students has demonstrated its validity.

Topics such as applicable geophysical methods, investigation fundamentals, use of aggregate materials, site instrumentation, and remote-sensing applications often receive only cursory treatment in engineering-geologic course work. This text gives more breadth in these topics and offers many examples of practical application. The use of Standard International (SI) units in much of the data is intended to increase understanding and usage of these units and their relationship to the more common English units.

The text is liberally illustrated with material drawn from the literature and contains a wealth of references. This is a deliberate effort to impress on students the availability of a large reservoir of information they can utilize in later professional practice. A student completing an undergraduate program or beginning graduate study should understand that engineering geology is a dynamic field. It requires a continuing effort to learn new methods and approaches that are useful in dealing with the great variety of projects and conditions that may be encountered. In addition, the broad spectrum of references in each chapter will assist faculty who have not had an opportunity to encounter the extensive geologic and geotechnical literature that is helpful in understanding and presenting the subject matter of engineering geology.

Engineering geology provides a unique opportunity for the productive, cooperative interaction between two disciplines to solve many problems facing us today. It is hoped that this text illustrates both the basis as well as the need for such interaction. If we help to provide better-trained individuals to meet the exciting challenges provided by this subject, which bridges geology and civil engineering, we will have succeeded in achieving our purpose.

ROBERT B. JOHNSON
JEROME V. DEGRAFF

Contents

1

Engineering Geology: An Overview

Engineering geology is an applied discipline of geology that relies heavily on knowledge of geologic principles and processes. As its name implies, it is an interdisciplinary profession in which the engineering geologist works closely with, and must understand and respond to the needs of, the civil engineer. To do so one must be proficient with the properties and uses of earth materials outside those commonly encountered by the practicing geologist. Opportunities arise to apply basic geologic knowledge in fresh ways as well as to be resourceful in using what is learned for engineering purposes in such disciplines as geophysics, remote sensing, and instrumentation.

It is important to recognize early on how the knowledge gained in an undergraduate geology curriculum has immediate application to engineering problems. Indeed, this knowledge is coupled with the additional need for engineering geologists to become involved in mechanics, which bridges structural geology and civil engineering practice, as well as to gain a full understanding of those disciplines that are specific to civil engineering.

The following overview of basic geology and mechanics is intended to provide insight into the aspects of geology of immediate value in engineering geology and to highlight some of the common ground as well as differences that exist—and will continue to exist—between geologists and civil engineers. In the following chapters, these subject areas will be developed more fully in the context of engineering geology.

GEOLOGIC FUNDAMENTALS

Rocks, Minerals, and Soils

The activities of the engineering geologist invariably are directly or indirectly associated with rocks and rock-forming minerals. Knowledge of rock types and the environments in which they formed as well as their responses to weathering, erosion, and tectonic processes are useful in making estimates of site conditions and in formulating site investigation programs. For most engineering geologic applications, it is of greater value to know the physical properties that characterize the three major rock groups and their common rock types than to use detailed mineralogic and/or petrologic analyses and the accompanying specialized rock names. In addition to the physical properties of the individual, or intact, rock sample, the mass properties of the rock also have engineering importance. For instance, all rocks are discontinuous as a result of partings such as joints, bedding, foliation, and faults. *Discontinuity* is a nongenetic and useful name for such a parting.

Except for clay minerals, the engineering geologist is less concerned with minerals per se than the physical properties of the rocks and soils that they compose. The engineering importance of clays is discussed in chapter 3.

The terms *rock* and *soil* have already been used. It is important to be aware of the various meanings of these commonly used terms among different disciplines. For example, what may be a mappable rock unit to the geologist may be a soil to the civil engineer. Highly weathered granite would appear as granite on a geologic map but would have the physical properties of an engineering soil. This, in turn, creates a terminology problem with *soil* as we find significantly different definitions among geologists, soil scientists, and civil engineers.

The problems have arisen because of the needs of the individual disciplines. The engineering definitions are governed by the need for quantifiable physical properties that may group materials together without regard to the more specialized classifications of the geologist or soil scientist.

The engineering properties and resultant classifications of soils and rocks are addressed in detail in chapters 3 and 4. At this point, a summary of definitions will alert the reader to examine with care the context of how *rock* and *soil* are used in publications and reports. It should be emphasized that there are no universally accepted definitions or classifications of rock and soil even within a given discipline.

Rock is defined geologically as a naturally occurring consolidated or unconsolidated material composed of one or more minerals (Gary et al., 1972). Although the definition is useful to the geologist, it is apparent that the definition includes materials with physical properties that the engineer would consider to be engineering soils, that is, unconsolidated materials. As a result, a commonly used engineering definition of rock is that of a hard, compact, naturally occurring aggregate of minerals (Krynine and Judd, 1957). Variations of this definition are found in Terzaghi and Peck (1967), Geological Society (1972), Geological Society (1977), IAEG (1981), and ASTM (1983). The physical state and the accompanying implied behavior of the material, which are of importance to the engineer, are the bases for the engineering definition of rock.

The "soil" of the civil engineer is an aggregate of mineral grains that can be separated by gentle means such as agitation in water (Terzaghi and Peck, 1967). Again, physical characteristics supercede the geological and pedological definitions of soil that restrict the term to surficial materials that support growth of land plants. Gradations exist between the engineering definitions of soil and rock (as in compact clay-rich soils) and rocks (e.g., some glacial tills and shales) that have physical properties falling within a gray zone between the "typical" soil and rock. Some attach ease or difficulty of excavation to engineering soil and rock definitions to further delineate materials when working in this gray zone. To the civil engineer physical properties are of practical value and a terminology has evolved to satisfy engineering requirements.

Before examining some of the engineering-related characteristics of the main rock groups, it is of value to consider some general relationships. The engineering properties of rocks are uniquely related to rock type. This is particularly true for unbroken or intact specimens. Ideally, the name of a rock should provide information useful for engineering applications. This is the case with most igneous and metamorphic rocks in which mineralogy, texture, crystal (grain) size, and structure are implicit in the name, as are the conditions prevailing at the time of the rock's origin. Classification schemes for these major rock groups are based logically on these variables.

Special care must be taken when interpreting sedimentary rock names or applying names to given specimens or exposures. The complex interaction of sedimentary environment, parent material, the detrital and/or soluble products of weathering, transporting mechanism, lithification, and postdepositional changes preclude a classification system that can account for all factors involved. The correct application of descriptive terms to sedimentary rock masses has special importance to one who has to estimate engineering properties from names and descriptions of sedimentary rocks.

Knowledge of the regional geologic history of an area is of additional value. It permits the engineering geologist to broaden the interpretation of an area to include the rock mass characteristics and possible field associations as well as variations of rocks native to the area. Review of the geologic history of several diverse regions in the United States such as the Gulf Coastal Plain, Central Interior, and Southern Rockies is recommended to illustrate the close ties among rock types, physical properties, structure, and geographic distribution with geologic history.

Igneous Rocks

Igneous rocks have silicate mineral compositions and interlocking textures. The latter characteristic does not apply to pyroclastic rocks. Engineering geologic classifications of igneous rocks are based primarily on composition of crystal (grain) size as shown in Table 1.1. For the typical fresh igneous rock, mineralogy and texture combine to cause high strength and excellent elastic deformation characteristics. Crystal size inversely affects strength.

The emplacement mode of intrusive igneous rocks has engineering significance. Massive intrusive bodies such as stocks and batholiths tend to have relatively homogeneous compositions and textures that are three-dimensional throughout (Figure 1.1). Knowing the boundary limits and rock type of such an intrusive, one may predict a variety of physical properties that may affect tunneling, mining, quarrying operations, slope stability,

Table 1.1 Classification of Igneous Rocks

Pyroclastic	Igneous				Genetic Group	
Massive					Usual Structure	
At least 50% of grains are of igneous rock	Quartz, feldspars, micas, dark minerals		Feldspar; dark minerals	Dark minerals	Composition	
	Acid	Intermediate	Basic	Ultrabasic		Predominant Grain Size (mm)
Rounded grains AGGLOMERATE	PEGMATITE	PEGMATITE	PEGMATITE	Pyroxenite Peridotite	Very Coarse-grained	
						60
Angular grains VOLCANIC BRECCIA	GRANITE	DIORITE	GABBRO	Pyroxenite Peridotite	Coarse-grained	
						2
TUFF	GRANITE	DIORITE	DOLERITE	Pyroxenite Peridotite	Medium-grained	
						0.06
Fine-grained TUFF	RHYOLITE	ANDESITE	BASALT		Fine-grained	
						0.002
Very fine-grained TUFF	RHYOLITE	ANDESITE	BASALT		Very fine-grained	
	VOLCANIC GLASSES	VOLCANIC GLASSES	VOLCANIC GLASSES		GLASSY AMORPHOUS	

Source: Reprinted with permission from Bull. Int. Assoc. Eng. Geol., No. 24, Rock and Soil Description and Classification for Engineering Geological Mapping Report by the IAEG Commission on Engineering Geological Mapping, 1981.

Figure 1.1 Yosemite Falls in Yosemite National Park, California, is a striking scenic feature in part owing to the homogeneity of the rock composing this section of the Sierra Nevada batholith.

and the rock's use as construction material. Mapping of large homogeneous rock masses reveals jointing patterns resulting from crystallization and relief of overburden and tectonic stresses.

Tabular intrusive bodies such as dikes and sills may create more construction or rock-utilization problems than massive intrusives because of the inherent lack of the three-dimensional continuity that is found in massive intrusives. This is especially the case with tabular intrusives in rocks of markedly different physical properties such as shales. The typically sharp contact of intrusives with surrounding country rock may create stability problems where relatively planar contacts intersect tunnels and rock slopes in addition to normally occurring jointing. Rock type heterogeneity and closely spaced contacts are illustrated by Figure 1.2, in which thin, tabular intrusives have been controlled by foliation in the metamorphic country rock.

The cooling histories of tabular intrusives that have different thicknesses introduce variations in crystal size and resultant differences in crushing strengths. Jointing perpendicular to the contact surfaces may result from cooling and be additional to jointing from other causes. All things considered, tabular intrusives present more problems in mapping and in engineering construction and rock use than the massive intrusives.

Extrusive igneous rocks generally either crystallize from lava flows with finely crystalline textures or they crystallize during the explosive eruptive phases of volcanism to form tuffs, welded tuffs, and volcanic breccias and agglomerates with clastic textures. Although extrusives may be compositionally similar, their origins greatly influence their engineering properties, as illustrated by the physical differences between welded or ash-flow tuffs

Figure 1.2 Granitic intrusions along near-vertical foliations in metamorphic rock. Cache la Poudre Canyon, Colorado.

and tuffs. In either the flow or explosive origins, extrusive igneous rocks have some attributes of sedimentary rocks. These are rocks that, in a sense, have been "deposited" on the earth's surface. As such they conform to that surface and may be layered as a result of multiple eruptions (Figures 1.3 and 1.4).

Closely spaced columnar jointing (Figure 1.3) characterizes many lava flows, whereas more widely spaced, sometimes crudely columnar jointing occurs in the more heterogeneous pyroclastic rock masses. Pyroclastic rocks exhibit a wide range of physical properties. For example, ash beds may be soft, highly altered, and easily eroded, whereas welded tuffs may be resistant, cliff-forming units. Resistance to weathering is exhibited by the welded ash-flow tuff shown in Figure 1.4.

Sedimentary Rocks

Sedimentary rocks present many challenges to the engineering geologist. As products of numerous marine, freshwater, and terrestrial environments, they exhibit a wide range of physical properties, lateral extents, and thicknesses (Figure 1.5). Classification of sedimentary rocks is complicated by the fact that grain size separates the rocks composed of detrital materials and composition separates those of chemical and organic origin (Table 1.2).

Stratification characterizes all sedimentary rocks. Primary sedimentary structures such as bedding surfaces and cross bedding (Figure 1.6) create discontinuities in addition to those formed by secondary structures such as joints and faults. Primary and secondary structures reduce rock-mass strength and may contribute to slope instability.

Sandstones typify the variability of sedimentary rocks. Beginning ideally with sand-sized material ($1/16$ mm to 2 mm), many possible differences in composition and size exist. The predominantly sand-sized material often is composed of quartz. However, in arkosic sandstone, the sand is composed

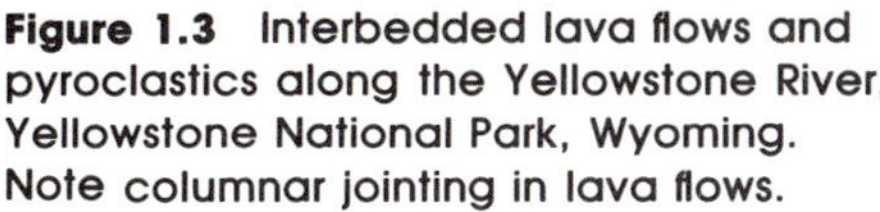

Figure 1.3 Interbedded lava flows and pyroclastics along the Yellowstone River, Yellowstone National Park, Wyoming. Note columnar jointing in lava flows.

Figure 1.4 Cliff-forming welded ash-flow tuff near Gunnison, Colorado. Note layering in this view.

largely of one of the feldspars, with attendant problems of postdepositional alteration of the feldspars into clay minerals, and thus with resultant loss of strength. A sandstone may have silt- and clay-sized fractions as well as different kinds and amounts of cementing material. All affect the strength and weathering characteristics of a sandstone. It is critical in engineering geologic applications that—in addition to such descriptive adjectives as

Figure 1.5 Vertical and lateral variations in thickness in sandstone strata. Near McGraw, New York.

Table 1.2 Classification of Sedimentary Rocks

Genetic Group	Detrital Sedimentary				Chemical/ Organic	Genetic Group	
Usual Structure	Bedded					Usual Structure	
Composition	Grains of rock, quartz, feldspar and clay minerals		At least 50% of grains are of carbonate		Salts, carbonates, silica, carbonaceous	Composition	
Argillaceous or Lutaceous	Grains are of rock fragments	Rounded grains: CONGLOMERATE Angular grains: BRECCIA	LIMESTONE (undifferentiated)	Calci- rudite	SALINE ROCKS Halite Anhydrite Gypsum	Very coarse- grained	Predominant Grain Size (mm)
							60
						Coarse- grained	
							2
Aren- aceous	Grains are mainly mineral fragments	SANDSTONE Grains are mainly mineral fragments	LIMESTONE (undifferentiated)	Calc- arenite	CALCAREOUS ROCKS	Medium- grained	
							0.06
Rudaceous	MUDSTONE SHALE: fissile mudstone	SILTSTONE 50% fine-grained particles	Marlstone / LIMESTONE (undifferentiated)	Calci- siltite	LIMESTONE	Fine- grained	
				CHALK	DOLOMITE		0.002
		Claystone 50% very fine grained particles	Marlstone / LIMESTONE (undifferentiated)	Calci- lutite		Very fine- grained	
					SILICEOUS ROCKS Chert Flint		
					CARBONACEOUS ROCKS LIGNITE COAL	GLASSY AMORPHOUS	

Source: Reprinted with permission from Bull. Int. Assoc. Eng. Geol., No. 24, Rock and Soil Description and Classification for Engineering Geological Mapping Report by the IAEG Commission on Engineering Geological Mapping, 1981.

Figure 1.6 Discontinuities in sandstone caused by cross bedding. Zion National Park, Utah.

thinly bedded or cross-bedded—terms such as *silty, calcareous,* and *arkosic* should be used to describe a sandstone.

As in the example of sandstone, the physical properties of shales and limestones are similarly influenced by differences in compaction, composition, grain-size range, texture, and the kind and amount of cementing material. Vertical and horizontal gradation into other sedimentary rock types as well as variations of the same rock type may be expected. A calcareous shale may grade into a shaly or argillaceous limestone or a sandy or silty limestone may grade into a calcareous sandstone or siltstone as a result of variations in sedimentary environments and sediment sources. Engineering investigations must take such potential changes into consideration. For any given area, knowledge of the regional tectonic history aids in understanding environments and areal distribution of rock formations.

Metamorphic Rocks

Metamorphism causes textural, structural, and often mineralogic changes in the original rock, modifying its physical properties. The modifications may improve some engineering properties, whereas other changes may result in reductions in strength, slope stability, and abrasion resistance. Metamorphic rock classification is based primarily on the presence or absence of foliation (Table 1.3). Massive, nonfoliated quartzite and marble from pure quartzose sandstone, and pure limestone are characterized by high strength because of textural changes.

Slates, phyllites, schists, and gneisses by comparison progressively exhibit mineral reorientation and generation of crystalline textures from clay-rich clastic sedimentary rocks. During regional metamorphism, mineral reorientation and recrystallization in these rocks create foliation that reduces any improvement in engineering properties gained by recrystallization (Figure 1.7). Rock strength differs with orientation of applied stess (compressive, tensile, or shear) to foliation. The least strength anisotropy

Table 1.3 Classification of Metamorphic Rocks

Metamorphic		Genetic Group		
Foliated	Massive	Usual Structure		
Quartz, feldspars, micas, dark minerals	Quartz, feldspars, micas, dark minerals, carbonates	Composition		
Tectonic breccia		Very coarse-grained		Predominant Grain Size (mm)
MIGMATITE			60	
GNEISS	HORNFELS Marble Granulite QUARTZITE	Coarse-grained		
			2	
SCHIST Phyllite Amphibolite		Medium-grained		
			0.06	
SLATE		Fine-grained		
			0.002	
Mylonite		Very fine-grained		
		GLASSY AMORPHOUS		

Source: Reprinted with permission from Bull. Intl. Assoc. Eng. Geol., No. 24, Rock and Soil Description and Classification for Engineering Geological Mapping Report by the IAEG Commission on Engineering Geological Mapping, 1981.

Figure 1.7 Jointed, foliated metadiorite requiring rock bolts to maintain cut slope stability. Crystal Dam, Black Canyon of Gunnison River, Colorado.

occurs in gneiss because the higher degree of recrystallization reduces the weakening influence of foliation.

Weathering Processes

Weathering processes are responsible for the development of sediments that may lithify into sedimentary rocks or occur as engineering soils. Apart from the role of weathering in the development of engineering soils (chapter 3), it plays a significant part in altering the engineering properties of intact rock and rock masses (chapter 4). Weathering may influence all rock types, the degree being dependent on the rock type, the kind of weathering process, the new environment to which the rock is subjected, the climate, and time.

Weathering processes are divided into physical (mechanical) disintegration and chemical decomposition. From the engineering geologic viewpoint, it is important to recognize the part played by weathering processes in the performance of rock and soil in civil engineering applications.

Physical Weathering

Physical weathering processes break down a rock mechanically, that is, without any chemical changes. Frost wedging and heaving from ice formation in discontinuities (such as joints and bedding surfaces) and exfoliation and spheroidal jointing from rock expansion are the common types of physical weathering. The most notable products of physical weathering of importance to engineering geology are talus slopes and rock falls that originate from frost wedging and gravity movement of jointed rock masses (Figure 1.8). Both result in construction and maintenance problems and in hazards along transportation corridors. On a smaller scale, but of no less importance in engineering applications, is the mechanical breakdown of crushed rock into smaller pieces that change a prescribed gradation or disrupt a structure.

In many cases, physical weathering is the precursor to chemical weathering. For instance, exfoliation joints or fractures permit entry of water that may further mechanically disrupt the rock mass by freezing and thawing and thus expose fresh rock surfaces to chemical decomposition by water. The resulting progressive reduction in rock mass strength is described in chapter 4.

Chemical Weathering

Chemical weathering processes include oxidation, solution, and hydrolysis. Of these, solution and hydrolysis have the greatest importance in engineering geology. Widening of joints in limestone (Figure 1.9) and ultimate development of caves that have the potential of surface collapse are both products of solution of calcium carbonate in water with dissolved carbon dioxide. The chemical weathering of certain silicate minerals is typified by the decomposition of feldspars into insoluble clay minerals, soluble salts of cations (such as potassium, sodium, and calcium), and silica in solution. Oxidation may be of importance, as in cases where pyrite or marcasite

Figure 1.8 Massive talus deposit extending downward into a rock glacier. Near Cumberland Pass, Colorado.

oxidizes in the presence of water to iron oxide and sulfuric acid. The acid will dissolve matrix carbonate minerals, thereby weakening the rock. Iron oxide structurally weakens the rock and is cosmetically distracting when exposed, as in Portland cement concrete structures.

The importance of the combination of multiple joint sets and joint-controlled chemical decomposition on rock-mass strengths is readily apparent (Figure 1.10). Chemical weathering affects rock masses proportionally to the number and persistence of discontinuities. It must be remembered that a weathered rock mass is composed of intact rock blocks between the

Figure 1.9 Solution widening of joint in cross-bedded, clastic limestone. Cave In Rock, Illinois.

Figure 1.10 Conversion of granite into soil by chemical weathering along joints. Near Buena Vista, Colorado.

discontinuities and that the mineralogy and texture of the intact rock controls, in part, the rate at which weathering progresses in the rock mass. The more comprehensive rock mass classifications for civil engineering applications include both jointing and chemical weathering in rating schemes for slope and tunnel stability and construction practice (chapter 4).

Chemical weathering tends to establish equilibrium between rocks of different compositions and surface or near-surface environments. It is to be expected, therefore, that the rock-forming minerals that exhibit different degrees of weathering stability will control the weathering performance of a given rock. Table 1.4 lists common silicates and oxides that are found in igneous, sedimentary, and metamorphic rocks. The presence of detrital quartz grains in many sedimentary rocks, other than pure quartzose sandstones, is testimony to the chemical stability of quartz. The abundant clay minerals, which are insoluble weathering products from the hydrolysis of aluminosilicates (such as feldspars), are in equilibrium with surface conditions and thus are almost as stable as quartz. In addition to the insoluble

Table 1.4 Chemical Weathering Stability of Selected Minerals

Most stable	Fe-oxides
	Quartz
	Clay minerals
	Muscovite
	Orthoclase (K-feldspar)
	Biotite
	Albite (Na-feldspar)
	Amphibole
	Pyroxene
	Anorthite (Ca-feldspar)
Least stable	Olivine

clay minerals, the minerals in Table 1.4 decompose to various oxides, soluble salts, and soluble silica.

The rock-forming and cementing carbonate minerals, calcite and dolomite, dissolve completely in water that contains dissolved carbon dioxide. Karst topography, with its associated internal drainage and collapse features, characterizes areas underlain by limestone and, to a lesser degree, dolomite. Site exploration in such terranes must be designed to locate areas of potential subsidence, irregular bedrock surfaces, or leakage problems if the place cannot be avoided (Figure 1.11). A sandstone composed of quartz grains cemented by a carbonate cementing agent such as calcite may be converted into a sandy soil by chemical weathering.

Chemical weathering is not restricted to near-surface or surficial materials. Extensive cavern systems such as Mammoth Cave, Kentucky, and Carlsbad Caverns, New Mexico, illustrate this for carbonate rocks. Faults may conduct water to considerable depths, where weathering takes place just as at the surface. Faults with breccia and gouge zones are especially susceptible to chemical decomposition from increased surface areas (Figure 1.12).

Decomposition of rock also results from the migration of chemically active solutions through a rock mass from a source such as a nearby magma chamber or volcanic vent. This hydrothermal alteration typically is more pervasive, as it may progress through the solid rock at depths where discontinuities are not present in large numbers. The end result of alteration of this sort or that from chemical weathering is essentially the same from the engineering standpoint, that is, a weakening of the rock mass.

Recognition of the two possible ways in which rocks may chemically decompose is of value to the engineering geologist in predicting conditions at depth and in designing a subsurface exploration program. Chemical weathering progresses downward from the surface, decreases with depth,

Figure 1.11 Irregular bedrock surface developed on limestone by differential solution. Overlying soil has been hydraulically removed to expose bedrock surface. George Dam, Fort Gaines, Georgia.

Figure 1.12 Weathering along fault zone in sediments. Northern Big Horn Mountains, Wyoming.

and is controlled to a large degree by the extent and character of discontinuities. By comparison, hydrothermal alteration typically either increases with depth or, at least, is pervasive outward from a source with no reference to the surface.

Geomorphology

Erosion and transport agents combine with weathering processes to sculpture the earth's surface, forming familiar landforms such as valleys, ridges, moraines, and dunes. The responsible agents are water, ice, wind, and mass wasting. Preliminary investigations of rock and soil types, geologic structure, and soil thickness may be made for engineering sites by examining landforms and related drainage patterns.

Figure 1.13 illustrates the controls exerted on topography and drainage patterns by dipping sedimentary strata and nonstratified igneous and metamorphic rocks. The hogbacks and strike valleys graphically outline stratified rock units of differing resistances to weathering and erosion in contrast to the absence of similar topography in the nonstratified crystalline rocks.

Drainage Patterns and Density

Given similar climatic conditions, drainage density is a useful indicator of subsurface conditions. For instance, lower density or more widely spaced drainage channels indicate subsurface drainage through thick, permeable soils or through karstic limestone terrane when compared with greater drainage density over less permeable soils and bedrock units.

Drainage patterns also tend to define rock types and geologic structure. Relatively flat-lying limestone bedrock may be indicated by the even distribution of lakes and depressions that result from limestone solution, that is, karst topography. By comparison, the disrupted drainage patterns and

Figure 1.13 Air view showing irregular topography and modified dendritic drainage in metamorphic/igneous rock anticline core (in right half of photo) and linear, controlled topography—hogbacks and strike valleys—on limb of anticline (in left half of photo). Near Fort Collins, Colorado. Courtesy of J. A. Campbell.

irregular topography in areas of continental glacial deposition may show little influence from bedrock or structure.

Changes in slope angle or the presence of topographic linearity, or possibly, discontinuous linear features are the result of rock type and structure interaction (Figure 1.13). Where exposed to view by erosion and transport of materials, they provide geologic insight of an area. Changes in stream gradient, pattern, and density as well as associated flood-plain width and interdrainage-divide characteristics are useful indicators of soil and/or rock control of drainage. All such features should be evaluated carefully in light of the known geologic history of an area. The interpretation of erosional and depositional landforms and drainage patterns for engineering purposes is dealt with in greater detail in chapter 7.

Mass Wasting

Mass wasting or downslope movement of soil and/or rock under the action of gravity is of concern, as it may affect manmade structures as well as natural features such as hill slopes and drainages. Transportation corridors may be interrupted and buildings destroyed. Many factors enter into mass wasting or landsliding.

Given a particular site, thickness of soil over bedrock, steepness of slope, smoothness of the bedrock surface, amount of soil moisture, physical properties of the soil, and natural or manmade vibrations may all contribute to either slow or catastrophically rapid downslope movement of material. In rock masses, inclined planar discontinuities such as bedding surfaces and joints may combine with natural or manmade slopes, water, and vibrations as factors that contribute to mass wasting. Certain soil-like rocks such as shale may creep downslope or fail rapidly in the absence of well-defined discontinuities. Indications of earlier mass wasting such as hummocky topography, cirquelike scars above such areas, and well-defined boundaries in flowlike masses of material should alert the engineering geologist to the possibility of new or renewed movement in an area. The influence of soil and rock properties on mass wasting and landslides as specific hazards are dealt with in chapter 9.

Correct interpretation of landforms is of considerable importance in defining the geology of an area for engineering purposes. Site investigations are dependent on these preliminary interpretations.

Structural Geology

Structural geology is the study of rock responses to compressive, tensile, and shear forces with the earth. Folds, faults, some joints, and foliation are products of structural deformation. Each has relatively planar, three-dimensional features that affect rock mass strengths and that may permit measurement and mapping for engineering purposes. Recognition and interpretation of rock deformation is enhanced by the presence of primary structures such as the discontinuities formed by bedding surfaces in sedimentary rocks.

The aspects of structural geology that have engineering geologic importance will be treated here. Review of the concepts of structural domain and structural style are of special value in engineering geology. A structural domain is a mapped area that is structurally homogeneous. Examples are areas characterized by multiple plunging anticlines and synclines or by numerous parallel or intersecting normal faults. In either case, uniformity of structure simplifies engineering design and construction procedures within the domain. Structural style commonly refers to fold structures in an architectural sense. Style is morphological when used in describing the shape of a fold as isoclinal or cylindrical. It may include secondary metamorphic characteristics. The usefulness of style is found in the reproducibility or predictability of structural features when similar styles are encountered that involve the same rock type.

The scale of an engineering project often is a conditioning factor in determining the scale at which structural features are mapped by the engineering geologist (chapter 2). The design and construction of a highway through an area of folded sedimentary rocks, as in the Appalachian Fold Belt, may serve as an example of small- and large-scale applications of structural geologic mapping to engineering geology. On a scale of 1 : 24,000 used for U.S. Geological Survey (USGS) 7½-minute geologic maps, the various rock units and related topography to be encountered during construction of grades, road cuts, and tunnels may be obtained for the proposed alignment. Construction procedures, equipment types, and time and cost estimates may be made from such data.

At a larger scale, the dip and strike of all discontinuities that occur in road cuts control the design of cut slopes that will not be susceptible to failure along inclined surfaces (Figure 1.14). Detailed mapping of orientation and frequency of occurrence of bedding surfaces, joints, and faults is a necessity as such features may differ from location to location at a construction site, depending on the rock type, degree of deformation, and

Figure 1.14 Steeply dipping planar discontinuities intersect tunnel portal, creating hazardous conditions stabilized by rock bolts. Crystal Dam, Black Canyon of Gunnison River, Colorado.

orientation of deforming forces. Presence of localized differential movement along bedding surfaces and joints that occurred within competent rocks during folding must be recognized. Project design, construction techniques, construction equipment, and design of rock-support systems all are dependent on such detailed, large-scale mapping.

In areas of folded rock units, orientation and frequency of joints as well as the number of joint sets are for the most part dependent on the orientation and magnitude of the compressive stresses that caused the folding and the competence of the involved rock types. In stratified rocks, these secondary discontinuities are accompanied by primary bedding surfaces that have various frequencies of occurrence as well as orientations that are dependent on position within the fold. As a result, rock mass strength will differ within the same rock unit as a function of variations in primary rock structures, position in the fold, and the style of the fold.

Localized differential movements along discontinuities within competent rock masses result in slickensided surfaces that enhance later tendencies for movement (Figure 1.15).

In areas that have undergone fault displacements, the simplest of cases one might encounter is the same rock type on either side of a fault where there is minimal displacement or where a thick, homogeneous rock mass is involved. A fault in such cases may be a simple discontinuity. However, in most cases faulting introduces a number of conditions that may have negative impacts on the design and construction of engineering projects. Among these are differing rock types on either side of the fault, rotation of blocks accompanying relative block movements, variations in fault-surface orientation and curvature laterally and at depth, and the presence of

Figure 1.15 Localized cross-cutting failure surface resulting from folding-induced stresses in massive sandstone. Friction-reducing slickensides on the surface are visible. Near Fort Collins, Colorado.

closely spaced fault surfaces or fault gouge and/or breccia in fault and shear zones. With such a plethora of possibilities, careful mapping and exploration drilling of faulted rock masses are of critical importance in engineering projects. Implicit in such exploration is correct interpretation of relative block movements, that is, normal, reverse, low-angle thrust, or strike slip.

Fault zones introduce radial changes in rock mass strength when encountered either at the surface or in the subsurface. In addition, they provide access by water to exposed rock surfaces at considerable depths. The dual construction problems of encountering "rock" with the properties of weak soil as well as large flows of water have plagued tunnel construction contractors over the years. The influence of these fault-related construction problems on progress made in construction of part of the Roberts Tunnel in Colorado is illustrated by Figure 1.16.

The association of folding and faulting is important to the engineering geologist in areas affected by tectonic forces. Knowledge of the regional tectonic history, local rock types, and perception of structural possibilities should guide the mapping and subsurface exploration of a site. It is necessary to determine as many of the structural problems as possible within the limits imposed by funding. Examination of published and open-file geologic maps of an area should precede any fieldwork or design of an exploration program.

Although of relatively surficial value in some areas, geologic maps should be carefully evaluated. Map legends, outcrop patterns, dip and strike symbols, mapped folds and faults, and topography provide valuable engineering information. These factors emphasize the inherent interaction of rock type, geomorphology, structure, and geologic history that must be considered if engineering geology is to properly service the civil engineering profession.

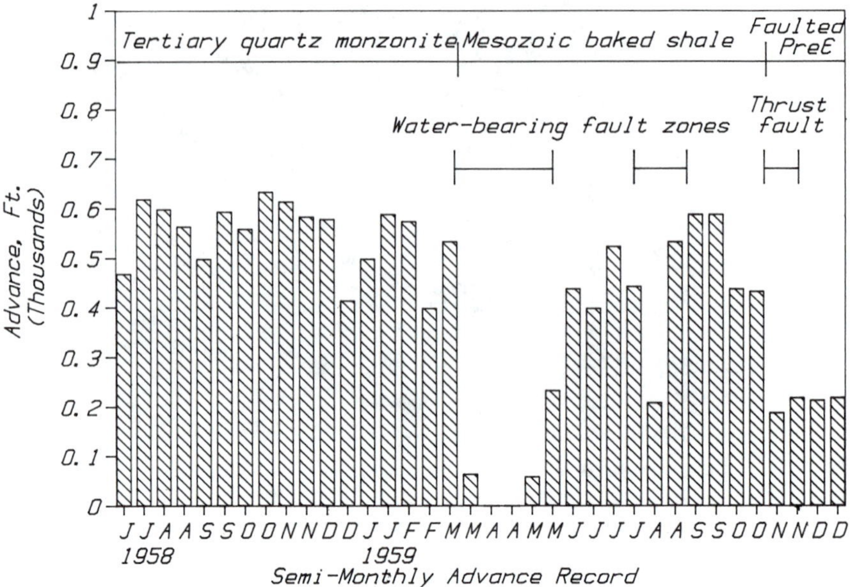

Figure 1.16 Influence of faulting and fault-related rock weathering on tunneling progress in a section of the Harold D. Roberts Tunnel, Colorado. (Modified from Wahlstrom, 1981.)

MECHANICS FUNDAMENTALS

In addition to the geologic fundamentals just presented, there are basic aspects of mechanics that are common, for the most part, to geology and civil engineering. In geology, mechanics has a role primarily in structural geology, whereas soil mechanics, rock mechanics, and structural engineering most commonly utilize principles of mechanics in civil engineering. Structural engineering applications are not of primary interest in the context of engineering geology. The concepts of mechanics are pervasive in much of the subject matter addressed in this text. Although there are few specific applications made in the following chapters, principles of stress and strain and related material responses provide a basis for understanding qualitative and quantitative aspects of engineering geology that involve soil and rock materials.

Stress

Natural forces that act on soil and rock masses and forces that act on materials as a result of construction as well as those that are applied to samples during laboratory testing are examples of forces that cause deformation and ultimately failure of soil and rock. Force, however, is not a useful measure because the force required for a given amount of deformation of a material varies proportionally with the surface area to which the force is applied. Stress is a more useful parameter because it is not dependent on differences in area. Stress is defined as force per unit area and is synonymous with pressure. Force and stress may be given in English, CGS (centimeter-gram-second), and SI (Standard International) units. Common units of force are pound-force (lbf), dyne, and newton (N), respectively. Associated stress units are lbf/in^2 (psi), dynes/cm^2, and N/m^2 or pascals (Pa). In some geologic literature, stress is given in bars, which equal 0.9869 atmospheres (14.503 psi, 10^6 dynes/cm^2, 10^5 Pa). Throughout this text, SI units are used. Unit conversions are given in table form on p. 491.

Stress is a vector quantity and its analysis is independent of material physical properties. A stress vector acting on a surface typically is resolved into normal and parallel (shear) components relative to the surface. For an elemental cube of unit dimensions in an equilibrium state, the sum of forces acting on the surfaces and their moments equal zero. Figure 1.17 illustrates the stresses acting on a two-dimensional portion of an elemental cube lying in the X-Y plane and having negligible thickness in the Z direction (plane state of stress). By convention in engineering literature and practice, the normal and shear stresses on a given surface are positive as drawn. Thus, by definition, positive normal stresses are compressive. Geologic notations for normal and shear stress components are opposite in direction. Care must be taken to adhere to sign conventions in the analysis of state of stress in a body. Subscript notations for normal and shear stresses are useful in positioning the stresses correctly in the stress field. In the equilibrium state where moments are zero, τ_{xy} must equal τ_{yx}.

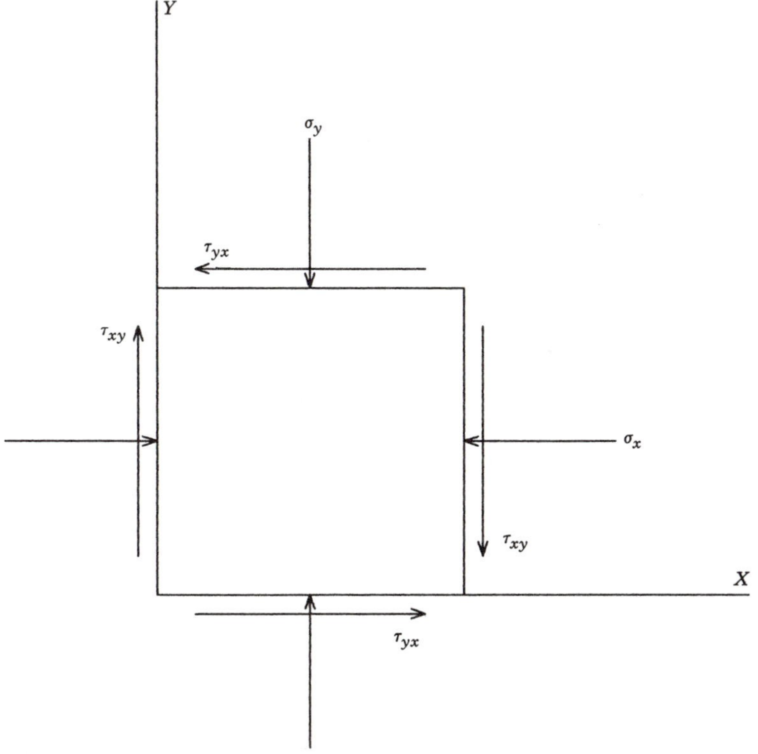

Figure 1.17 Compressive and shear stresses acting in the X-Y plane of an elemental cube. (See text for notation.)

Analysis of Stress

The analysis of stress within a body has many practical applications in the analysis of soil and rock strengths. Although stresses normally act three-dimensionally on a given point within a body, analysis of the state of stress at a point in a plane is less rigorous but remains conceptually sound. An elementary but useful illustration of the analysis of the two-dimensional state of stress at a point can be achieved in an element of negligible thickness (Figure 1.18). The stress field is that shown in Figure 1.17. For simplicity, the stress vector, $\boldsymbol{P_n}$, is drawn normal to the surface and inclined at an angle $\theta°$ to the X axis. The stresses σ_x, σ_y, τ_{xy}, and τ_{yx} are known.

In Figure 1.19, the stress vector, $\boldsymbol{P_n}$, can be resolved into vectors $\boldsymbol{P_x}$ and $\boldsymbol{P_y}$ acting parallel to the X and Y axes, respectively. For a static equilibrium state, the forces acting on the triangular element must be equal in the X and Y directions. These forces can be expressed by

$$\text{AB } \boldsymbol{P_x} = \text{OB } \sigma_x + \text{OA } \tau_{yx} \quad (1.1)$$

$$\text{AB } \boldsymbol{P_y} = \text{OA } \sigma_y + \text{OB } \tau_{xy} \quad (1.2)$$

wherein each force component is obtained by multiplying each stress by the surface on which it is acting; recall that each surface is of negligible thickness in the Z direction. From Figure 1.18,

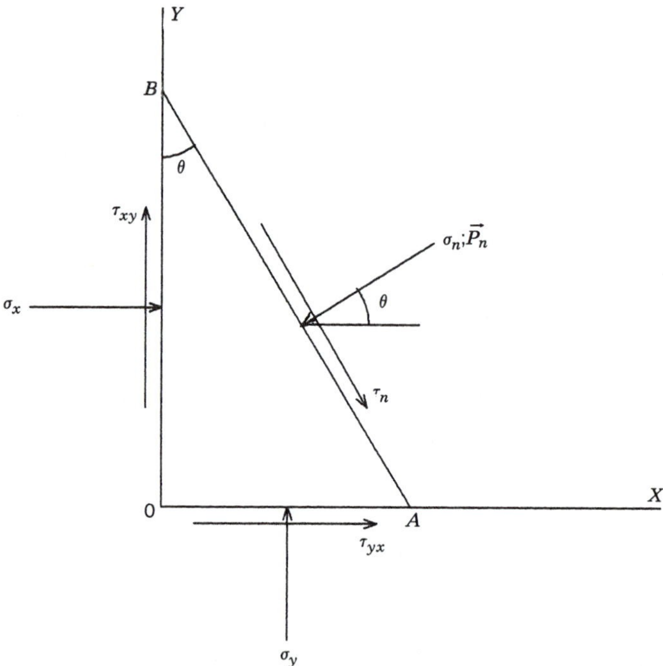

Figure 1.18 Stress field acting at a point on an inclined plane in an elemental body of negligible thickness. (See text for notation.)

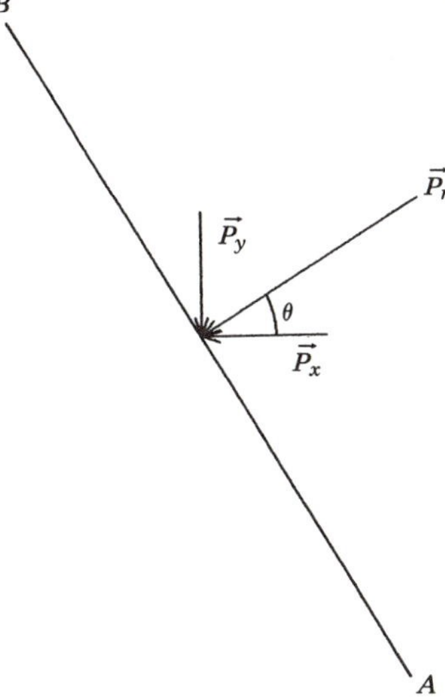

Figure 1.19 Resolution of normal stress vector on an inclined plane into X and Y axis components.

$$\mathrm{OA} = \mathrm{AB} \sin \theta \tag{1.3}$$

$$\mathrm{OB} = \mathrm{AB} \cos \theta \tag{1.4}$$

Substituting equations 1.3 and 1.4 into equations 1.1 and 1.2, we obtain

$$\boldsymbol{P}_x = \sigma_x \cos \theta + \tau_{yx} \sin \theta \tag{1.5}$$

$$\boldsymbol{P}_y = \sigma_y \sin \theta + \tau_{xy} \cos \theta \tag{1.6}$$

An expression for σ_{n} in terms of σ_x, σ_y, τ_{xy}, and τ_{yx} may be obtained by substituting equations 1.5 and 1.6 into the following equation for σ_{n} based on Figures 1.18 and 1.19:

$$\sigma_n = \mathrm{P}_x \cos \theta + \boldsymbol{P}_y \sin \theta \tag{1.7}$$

$$\sigma_n = \sigma_x \cos^2 \theta + 2\tau_{xy} \sin \theta \cos \theta + \sigma_y \sin^2 \theta \tag{1.8}$$

where $\tau_{xy} = \tau_{yx}$ for an equilibrium state.

In similar fashion, the shear stress, τ_n, on the surface AB is obtained from substitution of equations 1.5 and 1.6 into the following equation for τ_n, where signs are governed by the convention stated earlier:

$$\tau_n = \boldsymbol{P}_y \cos \theta - \boldsymbol{P}_x \sin \theta \tag{1.9}$$

$$\tau_n = \tfrac{1}{2} (\sigma_y - \sigma_x) \sin 2\theta + \tau_{xy} \cos 2\theta \tag{1.10}$$

where the identities $\cos 2\theta = \cos^2 \theta - \sin^2 \theta$ and $\sin 2\theta = 2 \sin \theta \cos \theta$ are used.

In stress fields such as the one illustrated here, there will be unique and orthogonal orientations of normal stresses where the shear stresses will be zero. The stress directions resulting in zero shear components are the principal stress axes and the normal stresses will be the principal stresses. In a two-dimensional case, the principal stresses are σ_1 and σ_2 and $\sigma_1 > \sigma_2$ by convention. The angle θ (Figure 1.18) that results from rotating the normal stress vector to the principal stress axis is found by setting equation 1.10 to zero. From this operation, the following equations are obtained:

$$\tan 2\theta = 2\,\tau_{xy}/(\sigma_x - \sigma_y) \qquad (1.11)$$

$$\theta = \tfrac{1}{2}\tan^{-1}[2\,\tau_{xy}/(\sigma_x - \sigma_y)] + n90^\circ \qquad (1.12)$$

Where:

$$n = 0,1,2,3.$$

The values of σ_1 and σ_2 can be calculated from the following equations:

$$\sigma_1 = \tfrac{1}{2}(\sigma_x + \sigma_y) + \tfrac{1}{2}[(\sigma_x - \sigma_y)^2 + 4\tau_{xy}{}^2]^{1/2} \qquad (1.13)$$

$$\sigma_2 = \tfrac{1}{2}(\sigma_x + \sigma_y) - \tfrac{1}{2}[(\sigma_x - \sigma_y)^2 + 4\tau_{xy}{}^2]^{1/2} \qquad (1.14)$$

Often it is convenient in stress analysis to set the X and Y axes in the directions σ_1 and σ_2, respectively. When this is done, equations 1.8 and 1.10 become

$$\sigma_n = \sigma_1 \cos^2\theta + \sigma_2 \sin^2\theta = \tfrac{1}{2}(\sigma_1 + \sigma_2) + \tfrac{1}{2}(\sigma_1 - \sigma_2)\cos 2\,\theta \qquad (1.15)$$

Where:

$$\cos^2\theta = \tfrac{1}{2}(1 + \cos 2\theta) \text{ and } \sin^2\theta = \tfrac{1}{2}(1 - \cos 2\theta) \text{ and}$$

$$\tau_n = -\tfrac{1}{2}(\sigma_1 - \sigma_2)\sin 2\theta \qquad (1.16)$$

Mohr Circle of Stress. The graphical representation of the state of stress at a point in a body is known as Mohr's circle or Mohr circle of stress. Axes for construction of the circle (Figure 1.20) are the stress components σ and τ are referred to as the principal stress axes. For two-dimensional or plane state of stress analysis, the principal axes are σ_1 and σ_2. The circle is the locus of coordinate values for σ_n and τ_n for points on planes defined by all values of θ. The two limiting cases of σ_n for τ_n equals zero are σ_1 and σ_2, which fall on the axis by definition. The center and radius of the circle are defined by $(\sigma_1 + \sigma_2)/2$ and $(\sigma_1 - \sigma_2)/2$ in equations 1.15 and 1.16.

The relationship between a stress field illustrated by a Mohr circle construction and the state of stress at a point developed earlier in our discussion is illustrated by Figure 1.21. Points E and D are principal stresses σ_1 and σ_2, respectively, where $\sigma_1 > \sigma_2$ by convention. Point C at the center is the mean stress amount from which all normal stresses deviate from different values of θ. In the Mohr circle construction used here, the rotation about the circle is clockwise from point E. In Figure 1.21, the angle BCA measured clockwise from E is equal to angle 2θ. The value of OB obtained from the intersection of a perpendicular to OE from point A is σ_n acting at θ° from σ_1 on a plane as in Figure 1.22*a*. The following

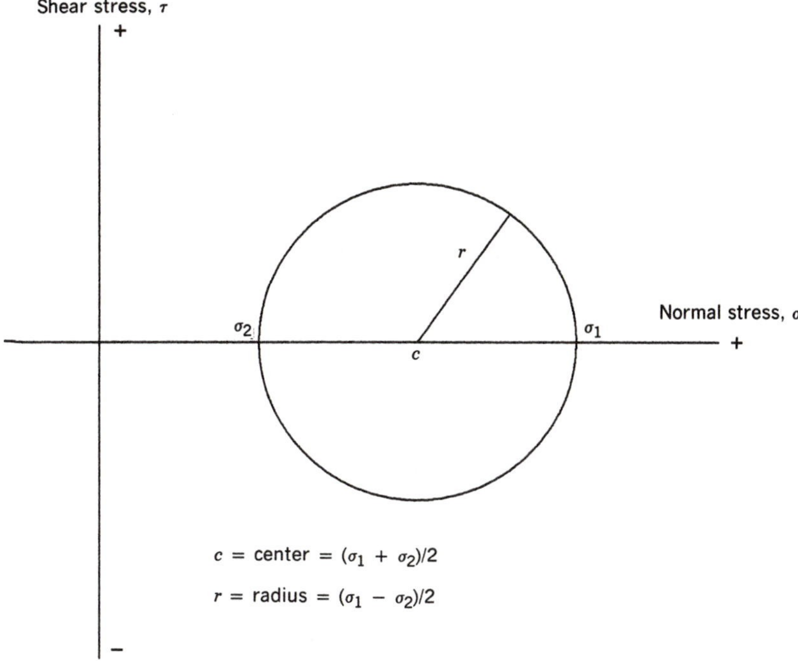

Figure 1.20 Fundamental elements of a Mohr circle of stress.

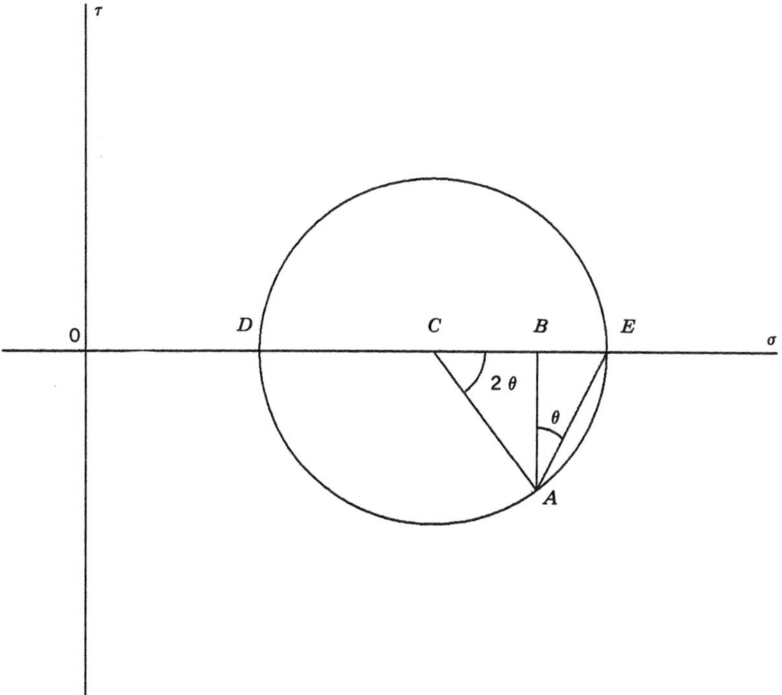

Figure 1.21 Mohr circle representation of state of stress at point A for a limiting case of Figure 1.18 where $\sigma_x = \sigma_1$ and $\sigma_y = \sigma_2$. (See text for notation.)

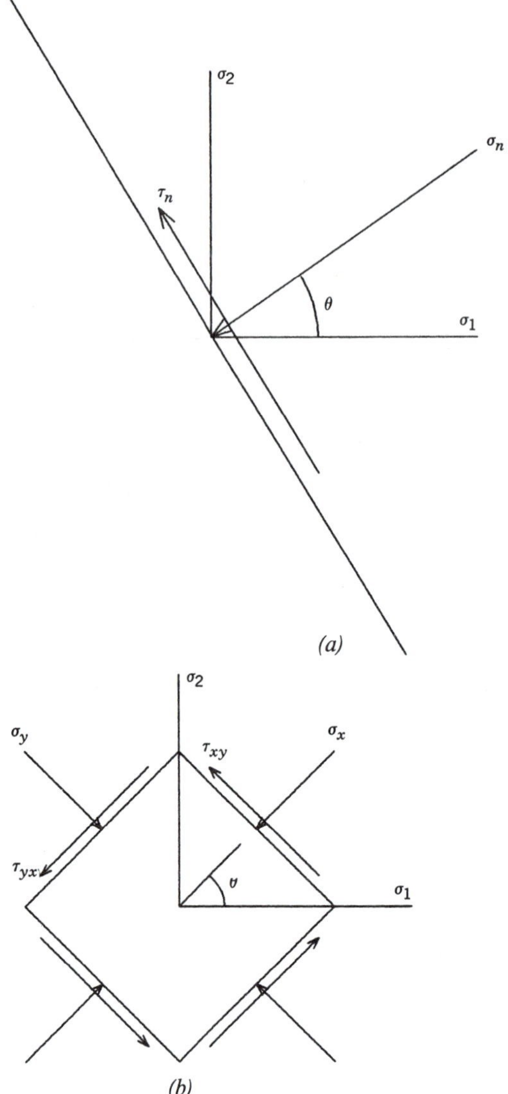

Figure 1.22 (*a*) State of stress at a point equivalent to A on Figure 1.21. (*b*) General case for Figure 1.22*a*, using a square element to show all stresses.

solution from Figures 1.20 and 1.21 is identical to equation 1.15 for σ_n:

$$OB = \sigma_x = \sigma_n = OC + CB = (\sigma_1 + \sigma_2)/2 + [(\sigma_1 - \sigma_2)/2] \cos 2\theta \quad (1.17)$$

The shear stress, τ_n, on the same plane at point A is −AB, which is the same value and sign as in equation 1.16 as follows:

$$-AB = \tau_{xy} = \tau_n = -[(\sigma_1 - \sigma_2)/2] \sin 2\theta \quad (1.18)$$

Figure 1.21 can be expanded with simplified notation to Figure 1.23 to include all normal and shear stresses acting on a point on the square element in Figure 1.22*b*.

Using the Mohr circle (case 1), it is possible to determine the magnitude and direction of the principal stresses, given the orthogonal stresses σ_x and σ_y and associated shear stress, τ_{xy}, using equations 1.12, 1.13, and 1.14 and Figure 1.23. In similar fashion (case 2), the state of stress at a point on any plane can be obtained by construction, knowing the principal stresses σ_1 and σ_2. The example given in Module 1.1 illustrates case 1. Case 2 merely requires reversing the procedure shown, following construction of the σ_1, σ_2 circle and selection of the plane orientation desired.

The Mohr circle is a useful means of illustrating a variety of states of stress. The examples that have been examined up to here involve positive or compressive principal stresses, as in Figure 1.24*a*. Figure 1.24*b* illustrates a uniaxial compression case where $\sigma_2 = 0$. A limiting case in which σ_1 is compressive stress and σ_2 is tensile stress is shown in Figure 1.24*c*. In this instance, the two principal stresses are equal and opposite in magnitude, resulting in a pure shear stress state. Figure 1.24*d* is a uniaxial tension case where $\sigma_1 = 0$. Where both principal stresses are tensile, Figure 1.24*e* defines the possible stress states.

Shear stresses are possible only when principal stresses are not equal. The difference in stress is known as deviatoric stress. *In situ* and laboratory-controlled states of stress may exist where σ_1 and σ_2 are equal. This is the hydrostatic state where no shear stresses act on a material regardless of the orientation of a plane relative to the principal stress axes. There will be

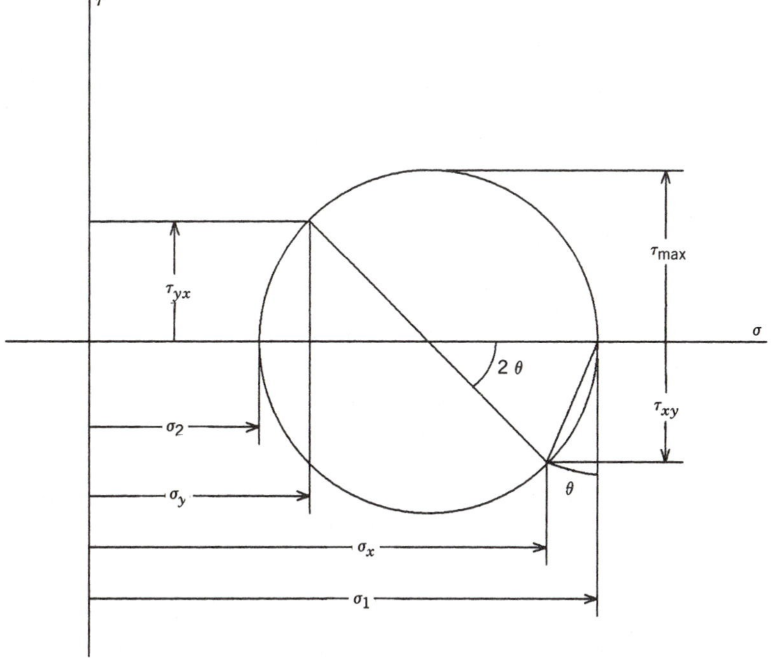

Figure 1.23 Mohr circle representation of stresses shown in Figure 1.22*b*.

MODULE 1.1

Mohr circle solution of problem to find values of σ_1, σ_2, and θ when given σ_x, σ_y, and τ_{xy}.

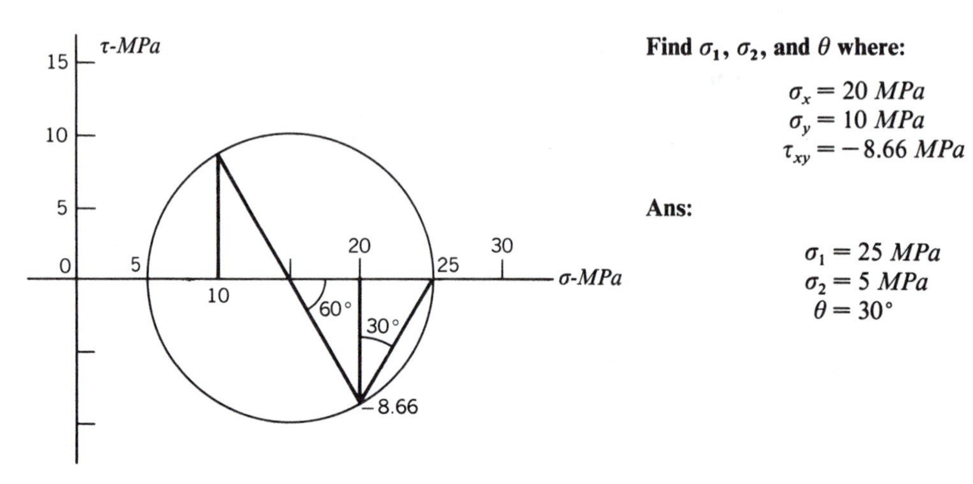

no Mohr circle for such cases, as the compressive and tension states will each appear as a single point to the right or left of the origin, respectively, on the σ axis.

Although analysis of the three-dimensional or triaxial state of stress at a point will not be pursued, it is of value to note that the Mohr circle can be used for such cases. However, Mohr circle determination of normal and shear stresses for various planes oriented three-dimensionally in a material introduces graphical techniques beyond the needs of this text. A simple, two-dimensional representation of the three unequal principal stresses is shown in Figure 1.25.

Mohr-Coulomb Failure Envelope. Planar shear failure in soil or rock has been found to occur through the interaction of normal stress on the failure plane and shear stress acting along the plane. The state of stress on a sample may be uniaxial or triaxial. Figure 1.26 illustrates a two-dimensional view of a triaxial case where σ_1 is the maximum principal stress and σ_3 the minimum principal stress. For this two-dimensional case, the intermediate principal stress, σ_2, is ignored. During most triaxial testing $\sigma_2 = \sigma_3$ so that the two-dimensional representation of soil and rock testing is acceptable.

The shear failure plane in Figure 1.26 lies at an angle β to the major principal stress plane. The same angle is formed by the intersection of the major principal stress axis, σ_1, and the stress normal to the failure plane, σ_n. The angle β is used in this section to specifically relate it to the angle

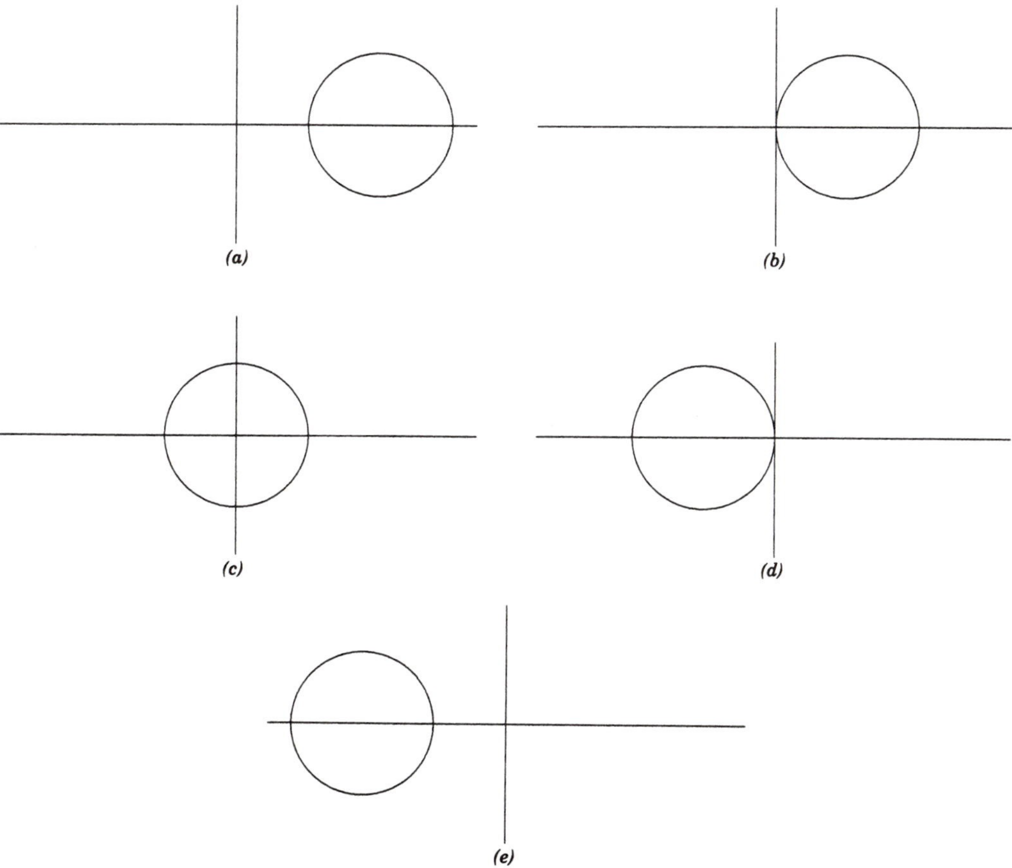

Figure 1.24 Idealized Mohr circle for the following states of stress: (*a*) Triaxial compressive. (*b*) Uniaxial compressive. (*c*) Pure shear. (*d*) Uniaxial tensile. (*e*) Triaxial tensile.

between the major principal stress and the normal stress rather than to stresses other than principal stresses such as σ_x, σ_y, and so on.

The criterion of planar stress failure was introduced by Coulomb. The criterion is a linear relationship that involves shear stress at failure acting along the plane, τ_f, cohesion of the material, c, and by a constant, tan ϕ, multiplied by the normal stress plane, σ_n, as shown by the following equation:

$$\tau_f = c + \sigma_n \tan \phi \tag{1.19}$$

The expression $\sigma_n \tan \phi$ is the coefficient of internal friction of the material; the angle ϕ is the angle of internal friction.

The stresses acting during shear failure may be illustrated by the use of the Mohr circle. The line defined by Coulomb's criterion, equation 1.19, may be shown on a Mohr circle diagram (Figure 1.27), which gives rise to the commonly used term, *Mohr-Coulomb failure envelope.* Both the Mohr circle and the failure, or rupture, line are symmetrical about the normal

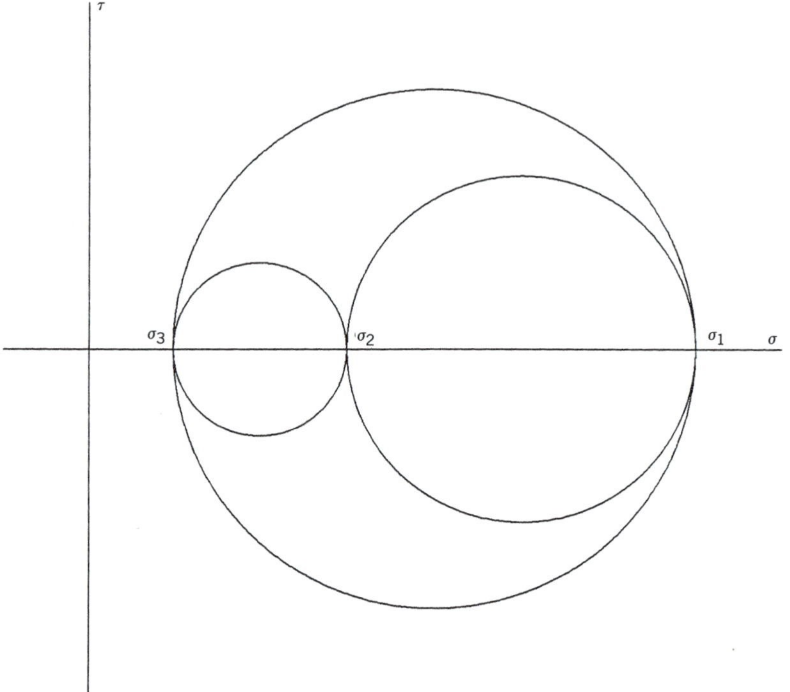

Figure 1.25 Mohr circle diagram for triaxial stress state where $\sigma_1 > \sigma_2 > \sigma_3$.

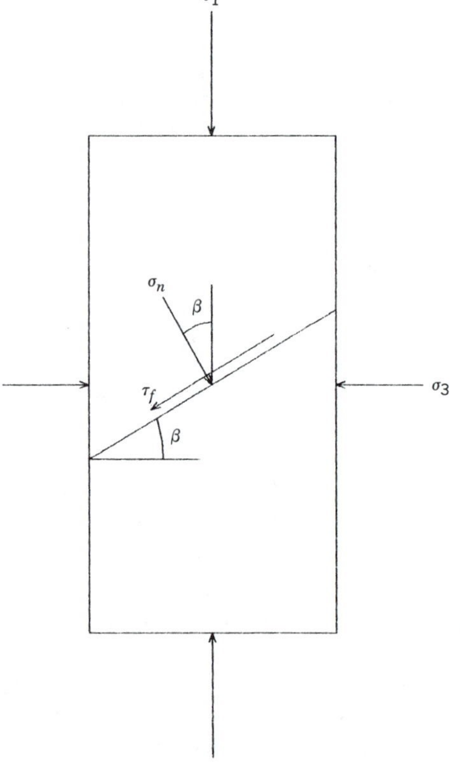

Figure 1.26 Stresses acting to cause development of a shear-failure plane in a sample under compressive stresses.

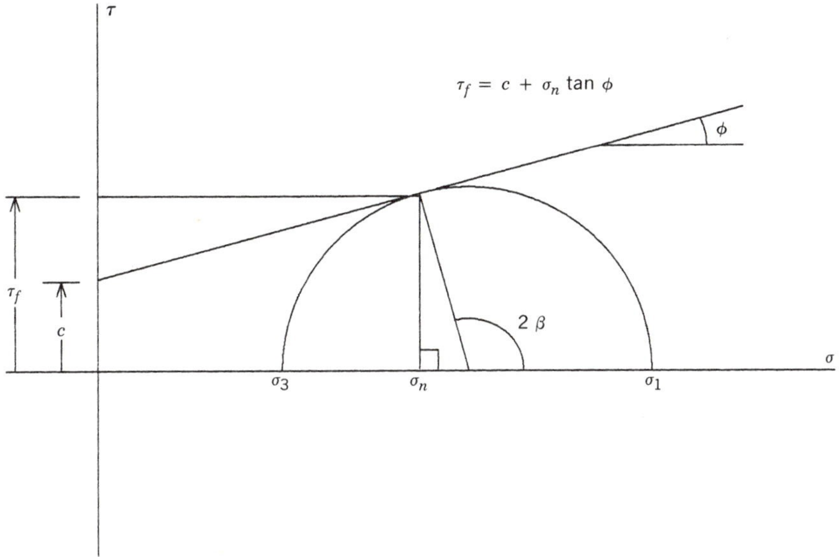

Figure 1.27 Components of Mohr-Coulomb failure envelope at sample failure. (See text for notation.)

stress axis. Therefore, only the upper half of the circle and the upper line need to be constructed. Contrary to what might be expected, neither the shear stress, τ_f, at failure nor the normal stress, σ_n, at failure is at a maximum value. The angle 2β is plotted counterclockwise because the orientation of the shear stress, τ_f, in Figure 1.26 is opposite to that used earlier in our discussion of the Mohr circle of stress. The sign of τ only shows the direction of shear stress acting on a surface and does not affect its magnitude.

It is seen in Figure 1.27 that the angle 2β equals $90° + \theta$ or $\beta = 45° + \theta/2$, which relates the Mohr diagram to the ideal physical orientation of the shear failure plane obtained by testing, as shown in Figure 1.26. Shear failure of a sample having defined values for cohesion and internal angle of friction will only occur at the unique combination of σ_1, σ_3, and σ_n, which causes the circle to become tangent to the rupture line. Thus, in Figure 1.27, the Mohr circle is unique and is often referred to as the rupture, or failure, circle. The graphical values for σ_1 and σ_n are the same as those that may be calculated by use of equations 1.15 and 1.16 where $\tau_f = \tau_n$ and $\beta = \theta$.

The values of cohesion, c, normal stress, σ_n, and angle of internal friction, ϕ, necessary for Coulomb's equation are obtained from uniaxial and/or triaxial testing of samples to failure. In Module 1.2, typical data from uniaxial and triaxial tests on a granite have been used to construct two Mohr circles for respective values of σ_1 and σ_3 obtained at the moment of failure in each test. The line drawn tangent to these two rupture circles defines the rupture or failure line of Coulomb (Mohr-Coulomb failure envelope). Note that although cohesion for a given material remains constant, the compressive, normal, and shear strengths increase with addi-

MODULE 1.2

Use of test data to illustrate construction of the Mohr-Coulomb failure envelope and kinds of information derived from it.

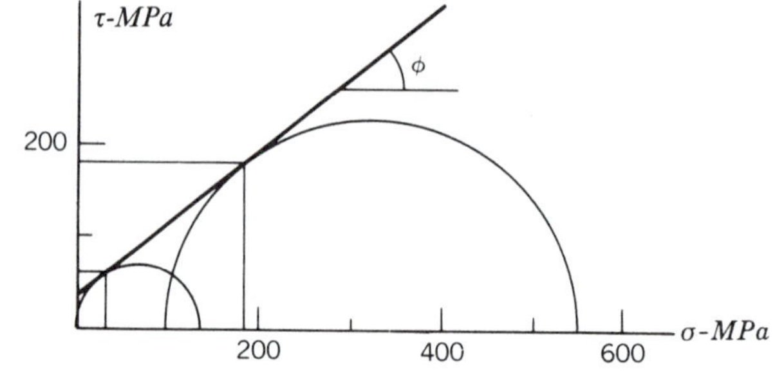

Given data for typical granite:

Test 1—uniaxial

$\sigma_1 = 140\ MPa$
$\sigma_3 = 0$

Test 2—triaxial

$\sigma_1 = 550\ MPa$
$\sigma_3 = 100\ MPa$

Graph-derived data:
Cohesion, *C*-35 *MPa*
Angle of internal friction, ϕ-37°
Shear strengths-τ_f
Test 1— 58 *MPa*
Test 2—180 *MPa*
Normal stresses-σ_n
Test 1— 30 *MPa*
Test 2—190 *MPa*

Calculated shear strengths
(using $\tau_f = \sigma_n \tan\phi + c$):
Test 1— 57.6 *MPa*
Test 2—178.2 *MPa*

tional confining stress, σ_3. The angle ϕ defines the relationship between compressive strength, σ_1, and shear strength, τ_f, for different confining stresses.

In conclusion, during testing that involves more than two different tests to failure, the tangent to the multiple rupture circles is curved convex upward rather than being linear as defined by Coulomb's criterion. This physical relationship was recognized by Mohr, who simply related shear stress at failure to normal stress at failure by the expression

$$\tau_f = f(\sigma_n) \qquad (1.20)$$

Common practice is to fit a linear rupture line to multiple Mohr circles for the sake of simplicity and uniformity.

Strain

Strain or deformation of a material is intimately related to the foregoing discussion on stress. Without the application of forces on a body, there would be no strain. Whereas stresses are analyzed at a point at a given instant, strains are defined as changes in relative positions of particles in a body at two different times. The term *displacement* is useful in defining strain. The displacement of a given particle from its initial position must be such that it cannot correspond to the movement of the whole body. In other words, the body must be deformed or distorted in shape rather than just moved as a rigid body.

Strain is a fundamental concept in continuum mechanics where the body being deformed is continuous, or a continuum. A strained or deformed body may be examined with respect to the kind of strain that has occurred, whether the strain is reversible (or recoverable) or permanent or a combination; if reversible, the body is examined with respect to the moduli that define a variety of shape changes. Each of these aspects of strain will now be addressed. For simplicity, strain will be examined two-dimensionally as in the discussion on stress, although deformation commonly involves all three principal axes.

Kinds of Strain

Strain may occur axially and result in contraction or elongation, depending on whether axial compressive or tensile stresses have been the cause of deformation. A body also may be strained by shear stresses. The sign convention used in our discussion of stress is continued here. Again, it should be noted that sign conventions differ among rock mechanics, soil mechanics, and geological literature. Fortunately, there are no resulting major conceptual differences.

In Figure 1.28*a*, the contraction or shortening in length that is shown results from the application of stress normal to the upper and lower planar surfaces of the body. The deformation is commonly referred to as *normal strain* because it is due to normal stress on the surface. Other terms are *natural strain* and *elongation.* By the convention used here, the strain shown is positive and has been the result of compressive stress application. The strain in Figure 1.28a is a unitless value defined by

$$\epsilon = \Delta l / l \tag{1.21}$$

where:

$$l = \text{length}$$

The lateral strain shown does not enter into the calculation of normal strain but may be expressed numerically by Poisson's ratio which is an elastic constant (see p. 37).

The shearing strain shown in Figure 1.28*c* is a product of the shear stresses shown in Figure 1.28*b* that follow the convention used here. Shear strain has resulted in an angle $>90°$ at the intersection of the X and Y axes. This is positive shearing strain by our convention. A reversal of shear stress directions would result in negative shear strain and an angle $<90°$. The angle formed between either the X or Y axis and the displaced side of the shear strained body is $\frac{1}{2}\gamma$, and the new angle at the intersection of the X and Y axes is $90° + \gamma$. The shear deformation shown in Figure 1.28c is often referred to as pure shear strain, which results from the pure shear stress state illustrated by Figure 1.24*c*. A pure shear stress state requires a normal stress of zero on the shear plane at maximum shear stress along the plane. A two-dimensional view of shear strain, referred to as simple shear strain, is shown in Figure 1.28*d*.

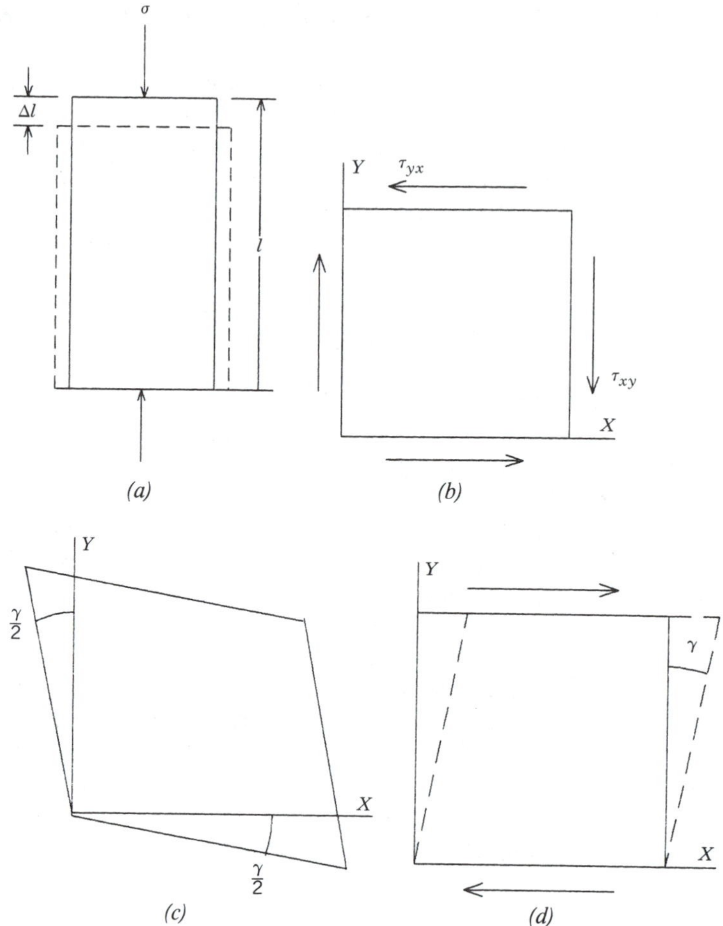

Figure 1.28 Types of strain; (*a*) Axial. (*b*) Shear stresses acting to cause pure shear. (c) Pure shear. (*d*) Simple shear, showing shear stresses.

Shear strain is quantified graphically by use of the Mohr circle of strain. The use of the Mohr circle is permissible because in the stress case, shear stress is one of the components of the circle, as are the principal stress components that cause shear stress. Rather than using normal stress and shear stress as the axes of a Mohr circle construction, strain, ϵ, and the distortion angle, $\frac{1}{2}\gamma$, are used as in Figure 1.29. In the figure, ϵ_1 and ϵ_2 are principal strains corresponding to σ_1 and σ_2. Strains ϵ_x and ϵ_y correspond to stresses σ_x and σ_y (see pp. 22–24) and $\frac{1}{2}\gamma_{xy}$ corresponds to τ_{xy}. Substitution of appropriate ϵ and $\frac{1}{2}\gamma$ values in stress equations permits calculation of state of strain at any point in a body. The state of strain at any point can be shown graphically as in the figure. For a state of pure shear, in which the principal stresses, σ_1 and σ_3, are equal and opposite and in which the maximum shear stress occurs where the normal stress $\sigma_n = 0$, the origin of the Mohr strain circle would be at the origin of the axes, as in the stress state. For such a case, $2\,\theta = 90°$ and the angle the shear plane makes with

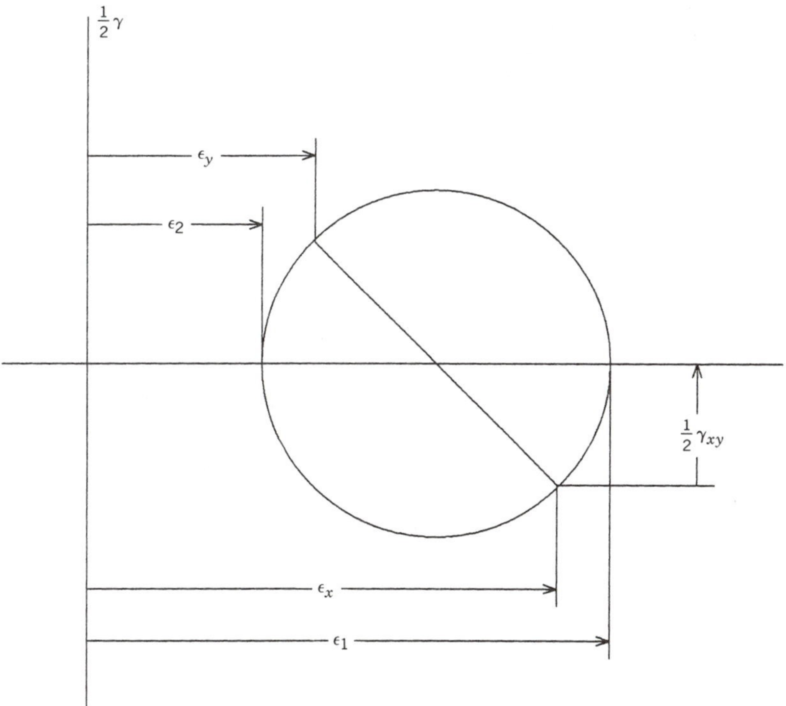

Figure 1.29 Mohr circle for state of strain showing strain components ϵ_x, ϵ_y and ½ γ_{xy} and magnitude and direction of principal strains ϵ_1 and ϵ_2.

the σ_1 axis is 45°. Symmetry requires two intersecting shear planes 90° apart.

Permanence of Strain

Strain of particles in a body may be dependent on the duration of applied stresses. If, with the removal of deforming stresses, the body returns to its original dimensions, the strain has been elastic in nature. In such cases, the strain obeys Hooke's law, which states that normal stress is proportional to extensional or axial strain. If the strain is completely recovered, the body is a purely elastic body. The purely elastic response of a material may be shown graphically by Figure 1.30*a*.

Permanent, nonrecoverable strain results when the applied normal stress exceeds the yield point of a material. The yield point is a stress level greater than that which can be tolerated elastically. If the material strains at this point without rupture or failure, the deformation is plastic or ductile strain. Figure 1.30*b* illustrates a perfectly plastic material. Most earth materials exhibit both elastic and plastic behavior (described more fully in chapter 4).

Creep, or viscous deformation, occurs when strain is time dependent. Plastic deformation occurs over a span of time as stress levels that normally would cause only elastic deformation over short periods of time. In such cases, a stress-strain rate curve illustrates the deformation (Figure 1.30*c*).

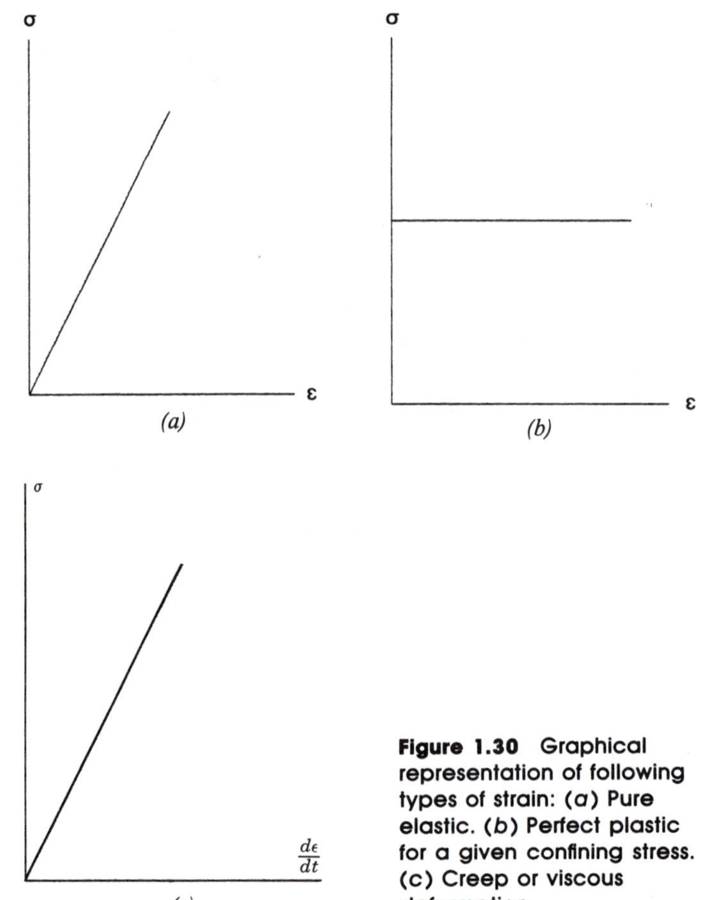

Figure 1.30 Graphical representation of following types of strain: (*a*) Pure elastic. (*b*) Perfect plastic for a given confining stress. (*c*) Creep or viscous deformation.

Elastic Constants

As indicated earlier, some strains and the stresses causing them are variously combined to form constants, or moduli, that characterize the elastic deformation responses of the materials to applied stresses. The term *constant*, though commonly used in this context, is misleading, as the value of a given elastic constant of a material is not constant for all stress fields that produce the elastic deformation. Thus, the term *modulus* is preferred by some.

The most commonly determined elastic constant is Young's modulus, or the modulus of elasticity, which is a form of Hooke's law. It defines elastic normal strain in a body by the following relationship:

$$\mathrm{E} = \sigma_n/\epsilon \tag{1.22}$$

where:

E = Young's modulus in psi or Pa

σ_n = normal stress

ϵ = axial strain

Rate of stress application will result in different values of E for the same material. If one were to solve for normal stress, it would be seen that Young's modulus, E, is a proportionality constant relating stress and strain. Modulus of deformation is used for the stress-strain ratio for non-elastic axial deformation.

Lateral or transverse strain, which occurs along with axial contraction and elongation, is represented by the unitless Poisson's ratio. It is obtained from the following equation:

$$v = \Delta l / \Delta d \qquad (1.23)$$

where:

v = Poisson's ratio

Δl = change in length

Δd = change in diameter

Poisson's ratio cannot exceed 0.5, a value obtained from an ideal, incompressible material. Earth materials, thus, have values less than 0.5.

The shear or rigidity modulus, G, is a measure of the shearing strain resulting from shear stress on a plane. Measurement units are psi or Pa. Shear modulus G (or μ) is obtained from the following:

$$G \text{ or } \mu = \tau / \gamma \qquad (1.24)$$

where:

τ = shearing stress

γ = shearing strain

Volumetric strain or dilation, which occurs when a body is subjected to hydrostatic stresses, is called the bulk or incompressibility modulus, K. The value of K is obtained by the following:

$$K = \sigma_0 / \epsilon_v \qquad (1.25)$$

where:

σ_0 = hydrostatic pressure

ϵ_v = volumetric strain

Hydrostatic pressure is the stress state in which $\sigma_1 = \sigma_2 = \sigma_3$.

The various elastic constants have an important role in many aspects of engineering geology. The response of earth materials to natural, construction, and in-service pressures resulting from structures such as tunnels, dams, and buildings are examples of applications. Seismic exploration and one of the techniques for obtaining elastic constants dynamically are dependent on the elastic properties of materials expressed as elastic constants. Chapters 4, 6, and 7 contain more detailed information on the involvement of selected elastic constants in the applications just given.

SUMMARY

Engineering geology is dependent on a thorough working knowledge of geologic principles. It is a specialty that requires great breadth of geologic knowledge and experience. In addition, knowledge and understanding of civil engineering practice and the needs of the civil engineer must accompany geologic expertise. Often one's classical training in geology must be adapted to the special and practical needs of the civil engineer. The practical engineering definitions of rock and soil and the universal application of the term *joint* to a variety of discontinuities (as noted later in chapter 4) are cases in point.

Another factor to be kept in mind is the tendency for conditions on and within the earth to attain an equilibrium state. An example is the chemical weathering of certain high-temperature, high-pressure silicate minerals such as the feldspars to form clay minerals when subjected to surface and near-surface environments. Jointing in rock may be a response to removal of overburden loads by erosion or by construction. Mass wasting is a response to a need for a soil or rock slope to reach equilibrium with gravity for a given set of conditions such as moisture content, composition, and nature of discontinuities.

The impact of geologic history on engineering geology is often not sufficiently recognized. Earlier episodes of mountain building have resulted in folding and faulting and associated igneous and metamorphic rocks. However, we may overlook the possibility that different climates in the past created conditions not expected in the present climatic setting. Occurrence of karst topography in areas presently having arid or semiarid climates has created problems for civil engineering projects. Abandoned drainageways and associated stream deposits from Pleistocene glaciation and related drainage systems are also products of past geologic events that should be recognized.

The concepts of stress and strain are held jointly by geologists and civil engineers and form the basis for understanding many responses of soils and rocks to natural and manmade conditions. The Mohr circle provides a graphical means of illustrating the interaction of the shear strength mobilized along a failure plane and normal stresses acting on that plane as failure occurs at different axial and confining stresses. Understanding of these relationships is basic to a working knowledge of why soils and rocks fail under certain conditions.

By contrast, an understanding of elastic and plastic strain provides the basis for instrumenting and predicting soil and rock deformation characteristics under natural and construction-caused loads. Material strain or deformation is often overshadowed by the more spectacular failure of materials. However, it is the universal and often problem-causing component of changes in stresses on a soil or rock mass where the strength of the material is never a factor. The elastic constants typified by the modulus of elasticity play an important role in the deformation properties of soils and rocks as well as in other areas such as seismic exploration methods.

In the chapters that follow, the elements of physical and historical geology, of mechanics, and of civil engineering are combined to provide a comprehensive treatment of engineering geology.

REFERENCES

ASTM, 1983, Standard definitions of terms and symbols relating to soil and rock mechanics: Am. Soc. for Testing and Mater., Annual book of ASTM standards, Vol. 04.08, Spec. D653–82, pp. 170–198.

Gary, M., McAfee, R., Jr., and Wolf, C. L. (eds.), 1972, Glossary of geology: Am. Geol. Inst., Washington, D.C., 805 pp.

Geological Society, 1972, The preparation of maps and plans in terms of engineering geology: Geol. Soc. (London) Eng. Group Working Party, Q. J. Eng. Geol., Vol. 4, pp. 295–381.

———, 1977, The description of rock masses for engineering purposes: Geol. Soc. (London) Eng. Group Working Party, Q. J. Eng. Geol., Vol. 10, pp. 355–388.

IAEG, 1981, Rock and soil description and classification for engineering geological mapping report: Intl. Assoc. Eng. Geol. Comm. on Eng. Geol. Mapping, Intl. Assoc. Eng. Geol. Bull., No. 24, pp. 235–274.

Krynine, P. D., and Judd, W. R., 1957, Principles of engineering geology and geotechnics: McGraw-Hill, New York, 730 pp.

Terzaghi, K., and Peck, R. B., 1967, Soil mechanics in engineering practice, 2nd ed.: John Wiley & Sons, New York, 729 pp.

Wahlstrom, E. E., 1981, Summary of the engineering geology of the Harold D. Roberts Tunnel: U.S. Geol. Surv. Prof. Paper 831-E, 15 pp.

2

Investigation Fundamentals

THE ROLE OF AN ENGINEERING GEOLOGIST

Conducting investigations is at the core of engineering geologic practice. The engineering geologist applies his or her geologic skills to the practical solution of engineering problems. However, geologic data gathered in an engineering geologic investigation are of more than academic interest. Their interpretation results in recommendations that affect the lives and well-being of the general public.

Certain characteristics are a prerequisite for a geologist who conducts engineering geologic investigations. Burwell and Roberts (1950) described these characteristics in their classic paper on the role of the geologist in an engineering organization. They note that an engineering geologist needs to be a competent geologist. Second, there has to be an ability to translate geologic findings into forms that can be applied to engineering works. This requires a third characteristic, an ability to provide sound judgments and make decisions. The fourth characteristic bears on the temperament of the engineering geologist. Effectiveness often depends on possessing personal traits such as tact, levelheadedness, and practicality.

The geologic skills required of an engineering geologist are no different than for any geologic researcher (Burwell and Roberts, 1950). Structural geology, geomorphology, hydrogeology, stratigraphy, petrography, and other major branches of geology may have a bearing on a particular engineering problem. An engineering geologist must seek to maintain a general competency in these areas while developing specialized skills important in solving engineering geologic problems. The knowledge needed to recognize a potentially active fault is the same regardless of whether the geologist is conducting a dam-feasibility study or trying to explain the tectonic history of a mountain range. Determining whether the characteristics of a fractured rock mass are likely to prove troublesome during tunneling requires knowledge different from that usually possessed by a structural geologist.

The results of an engineering geologic investigation have no value unless they are transformed into terms applicable to the engineering project. The investigator is better able to accomplish this with some knowledge of engineering principles. Boyce (1982) illustrates this point by noting that the engineering properties of earth materials are often the basis for economical and safe designs. This means that descriptions and interpretations of rock and soil conditions at a site need to be stated in terms understandable to engineers. Rock descriptions that focus on age and mineral content are generally less useful than descriptions of the degree of weathering and fracture spacing. Familiarity with design requirements, construction techniques, and other factors critical to a particular project make the investigation findings usable. Lessing and Smosna (1975) describe several instances where geologic information bore little resemblance to project needs. Their examination of environmental impact statements for nuclear power plants, highway corridors, and similar major construction projects often found irrelevant geologic information such as the depositional environments of the bedrock or types of fossils found. Geologic information that might affect the choice among alternatives or suitability such as seismicity and character of surficial materials were ignored. Avoiding these failings in describing and interpreting geologic conditions is facilitated by familarity with principles of engineering design.

Successful engineering geologic investigation calls for sound judgment and the ability to make decisions on the part of the investigator. Engineering projects are developed on the basis of economic, political, engineering, and scientific considerations. Judgments based on geologic investigation that influence changes in cost or suitability of a project will be questioned. Because judgment is opinion, the investigator should employ a logical thought process in reaching a conclusion. This should clearly identify assumptions made and distinguish between facts and inferences derived from the investigation. This is especially critical in forecasting impacts of a project. It is unlikely that an investigation will generate sufficient information to eliminate all uncertainty in geologic interpretations or predictions. This is why the ability to make decisions is important. Few investigations are allowed the time or money to permit collection of every pertinent fact. The geologist may need to assess critical factors based on limited data. Reluctance to render these interpretations forces others to make them. It is unlikely that these interpretations and recommendations will be as sound as those the geologist can make.

Effective use of information generated in an engineering geologic investigation often depends on the temperament or personal qualities of the investigator. The investigator is usually a member of a team of specialists involved with a particular project. The ability to be tactful in the give and take of developing the project is a valuable asset (Burwell and Roberts, 1950). Being helpful and practical are also useful characteristics. The geologic information needs may not always be obvious to the project engineer or general manager. The geologist should be prepared to point out these needs and to offer practical advice on solving problems that arise relating to geologic conditions.

ELEMENTS OF AN INVESTIGATION

Investigation is often viewed as synonomous with drilling, sampling, mapping, and other geologic fieldwork. This tendency to equate investigation with data collection slights other less glamorous but equally important elements of investigation. Broadly viewed, investigation incorporates all of the aspects of the scientific method. This perspective is discussed in detail by Romesburg (1984). The first step of an investigation is analogous to formulating a hypothesis for testing. An investigation starts with formulating or framing the question or questions to be answered. Choices are made concerning both the scope of the study and approaches to gaining needed information. It is only after these steps are completed that data collection should begin. Developing the experimental design and data necessary to test a hypothesis in a scientific study parallels this process. Collected data, no matter how voluminous, will not complete the investigation process. The data must be interpreted in light of the questions posed for the investigation. The scientific researcher must relate experimental data to the hypothesis to determine whether to reject, accept, or reformulate it. The resulting answers based on analysis of the data must be communicated to complete the process. The state of scientific knowledge on a subject will not progress unless the researcher shares the results of his investigation, too. Failure to adequately execute any of these steps is a misapplication of the scientific method for either scientific study or engineering geologic investigation.

Formulating an Investigation

Formulating the investigation is often a forgotten or overlooked element. This oversight may cause the investigation to take more time or cost more money than necessary to produce the required information. Even worse, it may mean the right information is not obtained.

Formulating the investigation requires identification of the question or questions that the investigation needs to answer. A common mistake is to assume that restating the goal or objective of the project as a question will suffice. This is usually too general for properly basing the work to be done in the investigation. For example, a countywide planning effort may seek to identify areas suitable for housing developments. In Fairfax County, Virginia, achieving this general goal involved basing the investigation on answering the more specific questions of what conditions influence slope stability, flood frequency, and severity as well as availability of earth-construction materials (Van Driel, 1978).

Once the questions are framed, other aspects of the investigation can be defined; for instance, the detail or the amount of information needed. More detailed information is needed to enable the engineer to design a facility than is required for determining preliminary costs in a feasibility study. Another aspect is the scope of the investigation. The size of the area to be investigated is dependent on the scope of the investigation. Identifying a well site for a specific facility limits investigation to a smaller area than an investigation of groundwater aquifers present within an entire county. The scope of the investigation may place limits on the time available to complete the investigation or on the time during which fieldwork can be carried out. The time available for an investigation is often limited by some deadline. This may simply be the terms of a contract or some legal requirement. The time available for fieldwork may also be affected by the location of the area to be investigated. Investigating foundation suitability for a series of lift towers at a ski resort is necessarily limited to the summer months except under emergency conditions.

A properly formulated investigation begins with the engineering geologist knowing the questions to be answered, how detailed the information must be, the area to be studied, and the time constraints involved. Other elements of the investigation logically build on this base. It is difficult to see how an investigation lacking proper formulation can ensure that the right information is obtained at a reasonable cost.

Data Collection

Data collection usually involves both office and field studies (Kent, 1981). Office study consists of gathering all the existing information useful to your investigation. Field study involves generating new data through exploration and testing.

Published literature is a prime source for obtaining information for an office study. The number of journals and the volume of published papers has increased dramatically in recent years. It is no longer sufficient to merely review whatever is available in a good university library. Using a computer-based reference service permits more effective location of pertinent material. Frequently, these bibliographic services or university libraries offer a document-delivery service for obtaining needed references. The most difficult aspect of office studies is finding the right source for obtaining this existing information. Trautmann and Kulhawy (1983) compiled an extensive and detailed listing of sources that serve as a good starting point. Once the published material is secured, it must be read and information specific to the investigation extracted for later use.

Another part of the office study is collecting information from aerial photographs or remotely sensed imagery. Specific types of information available from this data source are detailed in chapter 7. Studies that involve changes over time or that cover a large area can make especially good use of aerial photography. Not all investigations profit from this data source. An important limiting factor is imagery scale and resolution. Some features important to a particular investigation may be difficult to discern at certain scales. At the other end of this continuum, other scales may result in only parts of the feature being visible on individual images. This makes recognition equally difficult. The best scale is one that permits easy recognition of pertinent features by using the fewest number of images to survey the area under investigation.

7.1336 (Combining 7.1336 and 7.1338)
(8.70) Bureau of Reclamation

LOG OF TEST PIT OR AUGER HOLE
FOR BORROW AND FOUNDATION INVESTIGATIONS

Feature Example Project ______ Area Designation Borrow Area L-4
Hole No. AH–455 Coordinates N 16,500 E 6,910 Ground Elevations 669.8 Approx. Dimensions 24-inch dia.
Depth to Water Level *21.9 feet Method of Excavation Auger hole Date April 5–12, 19__ Logged by ______

CLASSIFICATION SYMBOL		DEPTH (FEET)	SIZE AND TYPE OF SAMPLE TAKEN	CLASSIFICATION AND DESCRIPTION OF MATERIAL (SEE CHART—"UNIFIED SOIL CLASSIFICATION"; GIVE GEOLOGIC AND IN-PLACE DESCRIPTION FOR FOUNDATION INVESTIGATIONS)	PERCENTAGE OF COBBLES AND BOULDERS ##				
LETTER	GRAPHIC				VOLUME OF HOLE SAMPLED (CUBIC FEET)	WEIGHT OF 3 TO 5-INCH SAMPLED (LBS)	PERCENTAGE BY VOLUME OF 3 TO 5 INCH ###	WEIGHT OF PLUS 5-INCH SAMPLED (LBS)	PERCENTAGE BY VOLUME OF PLUS 5-INCH ###
ML			75-lb. sack	0′–2′ SILT, slightly organic with some alfalfa and weed roots; nonplastic; small amount of fine sand; dark brown; dry.	40	0	0	0	0
CL			175-lb. sack	2′–8.5′ Lean CLAY, medium plasticity, high dry strength approx. 25% sand and gravel to ¾ inch size; most of gravel is shale; brown; dry.	130	0	0	0	0
				8.5′–16′ Micaceous SILT, moderate amount of very fine sand, noticeable mica flakes; very slight plasticity; tan; dry.	150	0	0	0	0

ML-MH	Not required for		200-lb. sack	16′–18′ SILT, similar to material 8.5 to 16 feet, but contains about 20% shaly gravel to 1-inch size; red; dry.	40	0	0	0	0
ML			90-lb. sack	18′–25′ GRAVEL-SAND MIXTURE WITH COBBLES, well-graded; approx. 50% gravel and 50% sand, mostly hard, subrounded; approx. 9% cobbles, by volume to 8-inch maximum size; very small amount of nonplastic fines; black; dry above water table; river terrace gravel, STOPPED BY HARD MATERIAL.	80	2430	19.1	1150	9
GW-SW		G.W.L. 4/12/7	21.9						

REMARKS: In-place density test at 8 feet; dry density 89.4 p.c.f., water content 8.9 percent.
Bulk specific gravity of cobbles and boulders: 2.55 by displacement.

NOTES: Record water test and density test data, if applicable, under remarks.
Record after water has reached its natural level; give date of reading adjacent to graphic symbol or in remarks.
Applicable only to borrow pits and to foundtions that are potential sources of construction materials.

$\frac{\text{(Lbs of rock sampled) 100}}{\text{(Bulk specific gravity of rock) 62.4 (Cubic feet of hole sampled)}}$

Record bulk specific gravity in Remarks, stating how obtained (measured or estimated)

Figure 2.1 Test pit log representative of data that may be available from governmental agencies. (From USBR, 1974)

Map review is another important aspect of office data collection. This may be limited to sensing the lay of the land for planning the field stage of data collection. Often, maps will contribute specific data needed by the investigation. Information on slope, drainage networks, and existing structures can be derived from topographic maps. Topographic maps permit plotting of surface profiles across the study area. Similarly, inferences can be made for subsurface conditions from cross-sections based on geologic maps.

Various government agencies may hold data acquired under provisions of different laws and regulations. Logs of water wells or oil wells, maps of buried utilities, plans of past excavations for buildings, and records of levels in wells are typical of this type of data (Figure 2.1). In some instances, there may be in-house records or unpublished reports from public agencies containing information usable in an investigation. A recent guide provides information on obtaining such information from the U.S. Geological Survey (USGS) (Dodd et al., 1985). Addresses for state agencies and geological organizations appear annually in the American Geological Institute publication, *Geotimes.* Falk and Miller (1975) compiled a list of national earth-science agencies around the world. These sources and those obtained from knowledgeable local individuals will assist one in finding government-agency information useful to an investigation.

The office stage of data collection is complete once the available sources have been tapped, the information extracted, and the data organized in a usable form. The investigator must compare this existing information to the information needs of the study. This not only reveals gaps in the information that must be remedied by collecting data in the field, but also guides selection of appropriate methods for acquiring it.

Data are collected in the field through either exploration or sampling and testing. Depending on the needs of an investigation, both surface and subsurface exploration methods may be employed. Chapter 7 describes some exploration methods in detail. Compton (1962) and similar references explain geologic field techniques often useful in an engineering geologic investigation.

Surface exploration is almost always part of field-data collection. Locating springs or seeps, mapping landslides, and measuring the strike and dip at rock outcrops are possible surface exploration activities. The data collected are usually transferred to a map base. The map may be the end product of the investigation or serve as a means for recording data to establish the spatial distribution of certain information (Figure 2.2).

Subsurface exploration is frequently part of an engineering geologic investigation. This exploration may take the form of direct examination by means of digging pits or trenches and by drilling boreholes. A number of indirect methods are also available. Geophysical methods such as refraction or reflection seismology, electrical resistivity, and magnetic surveys are the best known of the indirect exploration methods. The subsurface data can be compiled in the form of geologic profiles, graphic logs, isopach maps, or other two or three-dimensional representations of subsurface conditions (Ragan, 1973) (Figure 2.3).

Sampling and testing may occur consecutively with the surface and

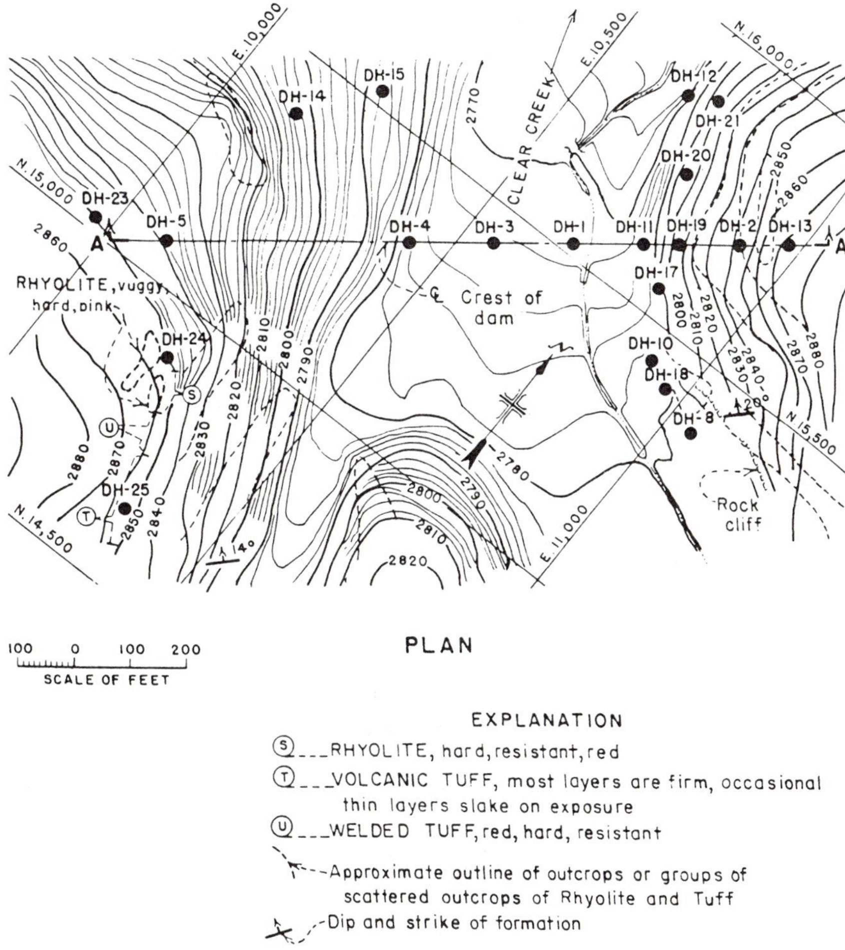

Figure 2.2 Use of topographic map to record drill hole (DH) locations in an investigation. (From USBR, 1974)

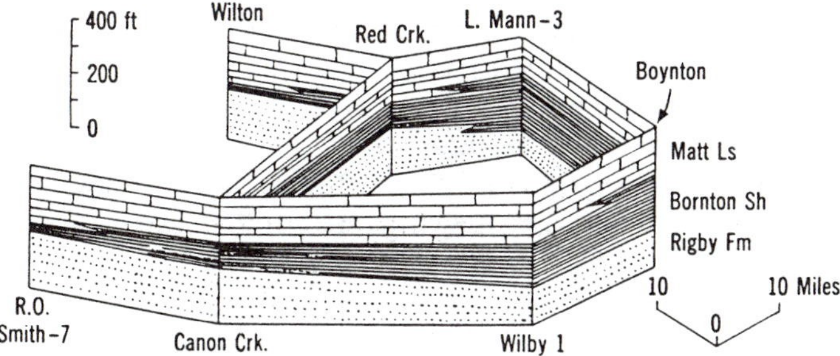

Figure 2.3 A fence diagram based on drill hole data showing the variation in thickness of bedrock units underlying an area. (Reprinted with permission from Manual of Field Geology by R. R. Compton, 1962, John Wiley & Sons, Inc.)

subsurface exploration (Boyce, 1982). One of the principal purposes of this work is to characterize the materials and conditions present. Sampling produces either qualitative or quantitative results. Identifying the different lithologies encountered while drilling an exploratory well is an example of qualitative sampling. Measuring the spacing of fractures in a core recovered from a drillhole is sampling that produces quantitative results. Testing may occur in-place (*in situ*) or under laboratory conditions. Pumping a well and measuring the rate of change in the water level is in-place sampling of the groundwater conditions. Testing of rock strength is commonly done in a laboratory setting.

Deciding whether in-place or laboratory testing will yield the best values for characterizing the condition or material property for an investigation requires careful consideration by the geologist. Some choices are simple. Changes in groundwater levels or strain rates in rock masses must be measured in-place. Sampling consists of installing instrumentation to measure and record changes in these conditions (Pratt and Voegele, 1984). Obvious choices exist in characterizing materials, too. A laboratory test of a rock sample for strength may inadequately characterize the strength of a rock mass being tunneled. The laboratory test is unable to reflect the influence on strength exerted by the number, spacing, and character of discontinuities in the rock mass. The laboratory test can only provide the maximum strength. The factor limiting suitability for tunneling or dictating the best tunneling technique is more likely to be the minimum strength value.

In other circumstances, the choice between in-place or laboratory testing is less clear. The shear strength of a soil may be as adequately characterized by a borehole shear test as by a triaxial shear test apparatus in the laboratory. Use of laboratory testing will be tempered by the ability of tests to adequately represent field conditions. Stability analysis for an earth embankment made from material carefully graded and controlled is adequately computed from values yielded by laboratory testing. However, analyzing slope stability for a natural slope involves more variability in the materials. It may require in-place testing to produce data more closely duplicating field conditions.

Documentation is important to data collection in both the office and field. It is especially important for data collection in the field. Greater expense and effort is necessary to reacquire field data lost owing to inadequate documentation than is required to look up a pertinent fact in a published report. Documentation takes a number of forms. Daily project sheets that summarize work accomplished, materials used, and other factors useful in contract control and billing are one necessary form (Kent, 1981). Field notes and drilling logs are another. Note taking in the field is a highly individual activity. Regardless of style, there are common features to good note taking. Field notes should document pertinent observations (Figure 2.4). They should indicate where and when the observation was made. The type of instrument used and any settings on it should be recorded. A reference to either a map or landmark should be used in locating the observation. A sketch or cross-section to show the setting is helpful. Always record the units for any measurement, employ some stan-

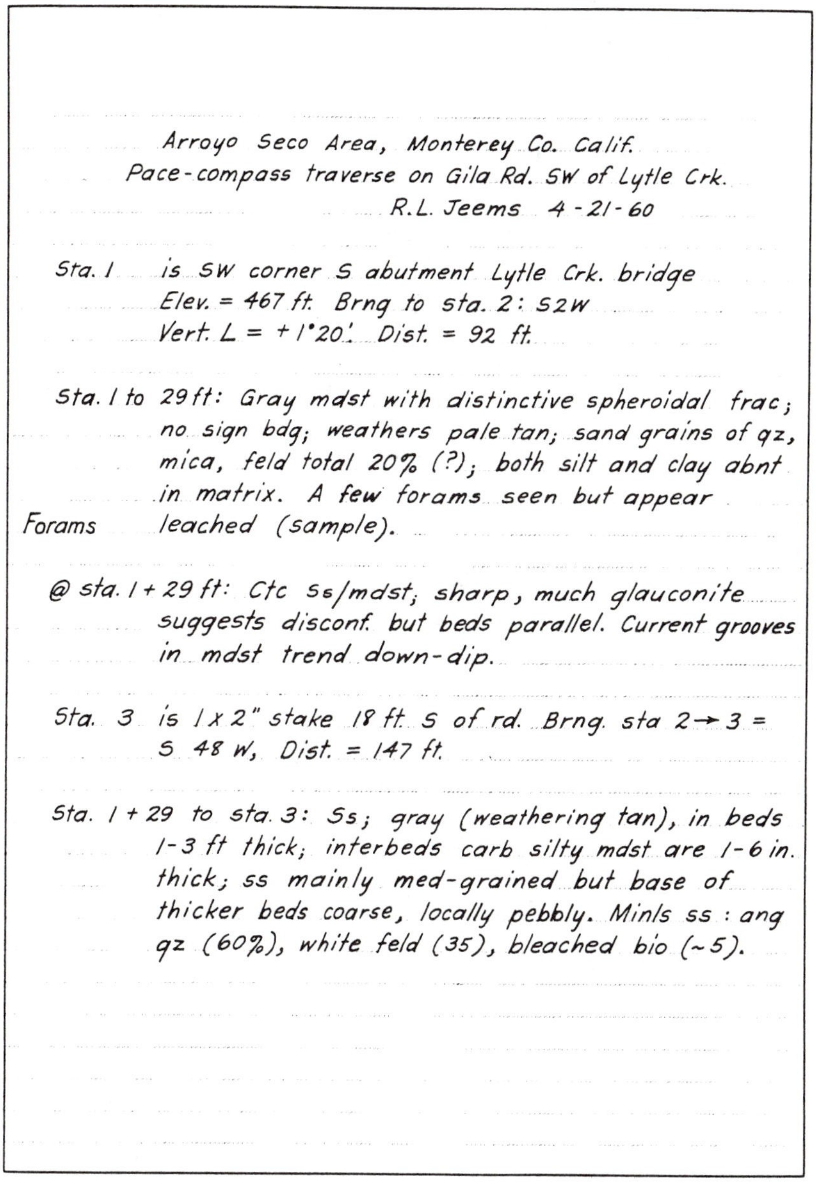

Arroyo Seco Area, Monterey Co. Calif.
Pace-compass traverse on Gila Rd. SW of Lytle Crk.
R.L. Jeems 4-21-60

Sta. 1 is SW corner S abutment Lytle Crk. bridge
Elev. = 467 ft. Brng to sta. 2: S2W
Vert. ∠ = +1°20'. Dist. = 92 ft.

Sta. 1 to 29 ft: Gray mdst with distinctive spheroidal frac; no sign bdg; weathers pale tan; sand grains of qz, mica, feld total 20% (?); both silt and clay abnt in matrix. A few forams seen but appear
Forams leached (sample).

@ sta. 1 + 29 ft: Ctc Ss/mdst; sharp, much glauconite suggests disconf. but beds parallel. Current grooves in mdst trend down-dip.

Sta. 3 is 1 x 2" stake 18 ft. S of rd. Brng. sta 2→3 = S 48 W, Dist. = 147 ft.

Sta. 1 + 29 to sta. 3: Ss; gray (weathering tan), in beds 1-3 ft thick; interbeds carb silty mdst are 1-6 in. thick; ss mainly med-grained but base of thicker beds coarse, locally pebbly. Minls ss : ang qz (60%), white feld (35), bleached bio (~5).

Figure 2.4 A representative page from a notebook recording field data. (Reprinted with permission from Manual of Field Geology by R. R. Compton, 1962, John Wiley & Sons, Inc.)

dard set of map or graphic symbols and use a consistent approach to numbering the samples taken. Electronic notebooks are among the innovations in documentation, the result of the widespread use of computers. Watts and West (1985) show through a rock slope stability study in Virginia the value of electronic notebooks in documenting fieldwork. Another important form of documentation in an investigation is borehole logging. This involves the recording of the materials and conditions found throughout each hole (Boyce, 1982). This is taken down with reference to specified depths or elevations. The information is often recorded on a form to graphically represent the borehole (Figure 2.5).

Interpretation

An investigation is incomplete without an interpretation of the data collected in both office and field studies. The interpretation starts with analysis of the data in light of the purpose and product called for by a particular investigation. The analysis results serve as the basis for developing conclusions and recommendations. It should be recognized that this sequence of events is idealized. In an actual investigation, data analysis usually occurs throughout data collection. Preliminary analysis permits changes in the amount and type of data collected as the need (or lack of need) for these data becomes evident from analysis. This may avoid collecting extraneous data or failure to collect data not initially recognized as pertinent to the investigation.

Understanding the Role of Measurement and Scales

It is important to understand the relationship of data to scales of measurement. This relationship influences the way in which an analysis is carried out and the conclusions that may be drawn. Measurement is not limited to using numbers. It can mean attaching a label such as a name or descriptive term to an observation or sample. The different scales of measurement available are: (1) nominal scale, (2) ordinal scale, (3) interval scale, and (4) ratio scale.

The nominal scale is the application of unordered descriptors or values to observations or samples. A familiar example is the identification of the rock type for each collected sample. Limestone, andesite, and schist are nominal scale descriptors. A special case of nominal scale measurement is the binary scale. Noting that a soil is saturated or unsaturated is an example of this form of nominal scale data. This is an unrefined scale that only permits the investigator to distinguish some factor between data points. Cheeney (1983) notes the only arithmetic operation valid on nominal scale values is whether values are equal or unequal to each other.

The ordinal scale applies ordered descriptors or values to data. Describing the landslide susceptibility of slopes as either low, moderate, or high not only distinguishes among the slopes, but ranks their degree of susceptibility. Although the ordinal scale ranks the data in order, it does not indicate the relative difference between points in that ranking. In our example, there is no assurance the difference in landslide susceptibility between low and moderate is the same amount as between moderate and high. Equality and greater-than or less-than operations are the only valid arithmetic operations for ordinal scale data (Cheeney, 1983).

The interval scale is a more familiar scale of measurement. It shows the order among data and the relative difference within that order. The zero point in an interval scale is arbitrarily set, permitting negative values to be obtained. Temperature is a familiar case with a zero point set in relation to the freezing point of water.

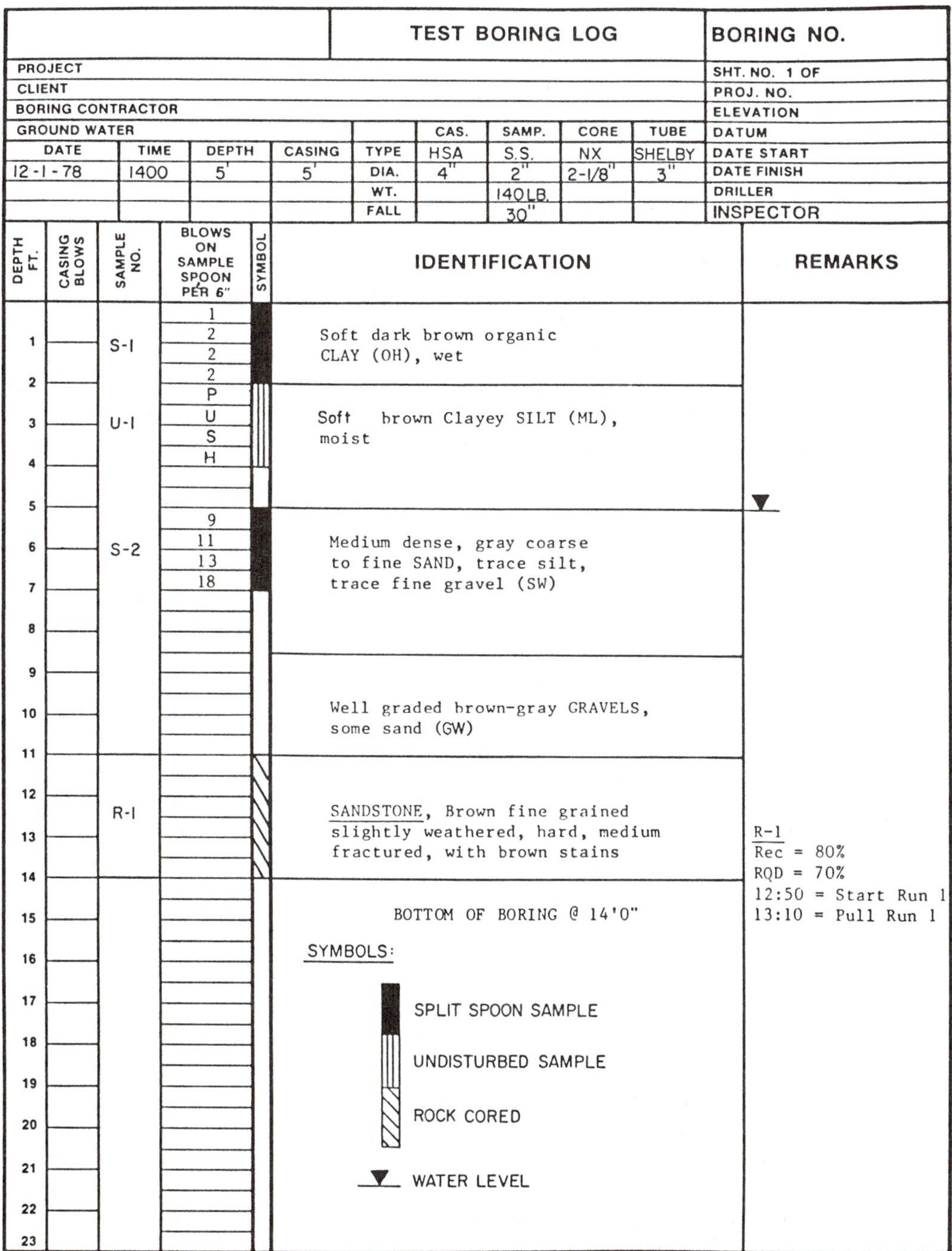

	TEST BORING LOG	BORING NO.
PROJECT		SHT. NO. 1 OF
CLIENT		PROJ. NO.
BORING CONTRACTOR		ELEVATION

GROUND WATER					CAS.	SAMP.	CORE	TUBE	DATUM
DATE	TIME	DEPTH	CASING	TYPE	HSA	S.S.	NX	SHELBY	DATE START
12-1-78	1400	5'	5'	DIA.	4"	2"	2-1/8"	3"	DATE FINISH
				WT.		140LB.			DRILLER
				FALL		30"			INSPECTOR

DEPTH FT.	CASING BLOWS	SAMPLE NO.	BLOWS ON SAMPLE SPOON PER 6"	SYMBOL	IDENTIFICATION	REMARKS
1		S-1	1 2 2 2		Soft dark brown organic CLAY (OH), wet	
2–4		U-1	P U S H		Soft brown Clayey SILT (ML), moist	
5						▼
6–7		S-2	9 11 13 18		Medium dense, gray coarse to fine SAND, trace silt, trace fine gravel (SW)	
9–11					Well graded brown-gray GRAVELS, some sand (GW)	
11–14		R-1			SANDSTONE, Brown fine grained slightly weathered, hard, medium fractured, with brown stains	R-1 Rec = 80% RQD = 70% 12:50 = Start Run 1 13:10 = Pull Run 1
15					BOTTOM OF BORING @ 14'0"	
16–23					SYMBOLS: SPLIT SPOON SAMPLE UNDISTURBED SAMPLE ROCK CORED ▼ WATER LEVEL	

Figure 2.5 Log sheet recording depths and testing results for a drill hole. (From NAVFAC, 1982)

The ratio scale differs from the interval scale by having a fixed zero point. Most of the traditional measures such as length, volume, and weight are on the ratio scale. For example, 16 yd^3 of soil are four times more material to excavate than 4 yd^3. The zero point is clear; 0 yd^3 of soil means no excavation has occurred.

Some data-analysis techniques require data sets to be in the same scale. This may mean changing more-refined scales to less-refined scales. Seismic velocity is normally measured on a ratio scale in feet per second (ft/sec). A range of values could be converted to an ordinal scale by dividing the range into three equal parts and assigning the terms high, medium, and low velocity to values falling in each third of the range. The same range of seismic velocities could be altered to a nominal scale by choosing a certain value and labeling all values below that point as slow and those above it as fast. Although there is a loss of information in changing the scale, it may permit a more valid comparison with other data collected at the less-refined scale owing to limitations imposed by the investigation or by the nature of the characteristic being measured. The important point is to understand the nature of scales to avoid rendering an invalid interpretation.

Data are often categorized as qualitative or quantitative. It should be recognized that this difference is a function of the scale of measurement. Data measured on the nominal scale is qualitative. Quantitative data may be on the ordinal, interval, or ratio scales. Qualitative data are sometimes viewed as inferior to quantitative data. This is an unfortunate misperception that can impair interpretation of collected data. Lienhart and Stransky (1981) provide a good example of the equal usefulness of qualitative and quantitative data in assessing rock for use as riprap. The suitability of a source of this kind of rock can depend on how it weathers. In their example, laboratory testing provided a quantitative measure by relating the loss of weight over time for a rock sample. A small percentage loss found for one sample suggested that this was a suitable source. However, that sample displayed extensive fracturing or severe spalling despite remaining in one piece. This qualitative data indicated the source was actually unsuitable. The test would have to employ a longer-than-normal time period to yield quantitative data leading to this same conclusion.

Making Data Manageable for Analysis

Data analysis should begin with reducing raw data to a manageable form. Once this reorganization is complete, appropriate analysis techniques can be applied to the data. The results obtained from chosen techniques are evaluated with each other and with other data or criteria to answer the specific question being addressed by the investigation. To illustrate these steps assume that a railroad alignment is being widened through a mountainous region. One section will require excavation into a rock mass. The presence of well-defined joints in the rock raises the question of whether the rock face will remain stable after excavation. Measurement of joint orientations in the rock mass are collected as the raw data. The observations collected by several geologists are gathered from field notebooks and reorganized into a table. These observations are plotted on a stereo net to identify the orientation of the major joint sets. The planes of these joint sets are compared to the orientation of the railroad alignment and slope angles of the planned excavation. This kinematic analysis indicates whether the planned cut might produce conditions conducive to a planar, wedge, or topple failure. Chapter 4 provides a detailed example of kinematic analysis of a rock mass.

Reorganizing data involves preparing tables, graphs, measured profiles, cross-sections, and other representations that permit comparison and examination. In some investigations, spatial variation may be important. The data would then be reorganized into a series of maps that would permit comparisons among these factors. Transforming seismic or drill-hole data into geologic profiles creates an easier form for examining variation in subsurface conditions. If variation in some factor over time is of interest, arranging the data sequentially in a table or on a graph with one axis representing time may be the best way to reorganize the collected data.

Applying Analysis Techniques

Once the data are in a manageable form, analysis may prove quite simple. Maps may reveal the presence or absence of some important factor. The critical depth to some subsurface layer will be evident from cross-sections. In other instances, analysis may require a more complex approach that can be characterized as either mathematical or statistical in nature.

Mathematical analysis techniques are usually based on some theoretical concepts applicable to the condition being analyzed. Certain techniques for analyzing slope stability are representative of mathematical techniques. Infinite slope-analysis techniques are based on mathematical relationships that apply principles of physics to a soil or rock mass. These mathematical analysis techniques often involve simplifying assumptions to make the formula manageable (Morgenstern and Sangrey, 1978).

Statistical techniques used in analysis are also based on mathematical theory (Davis, 1973; Lewis, 1977). But this theory is not specific to a particular process. Many statistical techniques useful in engineering geology help to identify the degree of correlation between factors of interest. This is often useful in predicting certain outcomes that are of concern to the investigation. A certain outcome may be difficult to predict because of lack of full knowledge of the process involved or inadequate records. If another dependent factor that is better known or recorded can be related reliably to the independent factor being predicted, it can serve as a substitute in prediction. The occurrence of debris flows is an outcome that is difficult to predict because of a lack of long-term records in an area. However, a correlation can sometimes be made between rainfall intensities and debris-flow occurrence. This permits use of longer-term rainfall records to express the likelihood of future debris-flow occurrence in the area. A pitfall of this type of analysis is whether the relationship defined by the correlation is a true correlation of independent and dependent factors or an association of two factors both dependent on some third factor not represented in the analysis. There should be some physical basis for assuming independence and dependence between the factors used in this type of statistical analysis.

Other statistical techniques are useful to objectively understand the data collected. The volume of information may make detection of some significant trend difficult to perceive. Other techniques can verify whether a perceived trend or relationship is significant or inconsequential.

Whether you are choosing mathematical or statistical techniques to use in analyzing your data, one important fact must be remembered. These techniques all involve some assumptions concerning the data. They will yield answers whether those assumptions are true for the data they are applied to or not. Indiscriminate use of an analysis technique without regard to these assumptions will result in answers that are incorrect or meaningless. This often results from a cookbook approach to mathematical or statistical analysis. The individual responsible for choosing analysis techniques must be sufficiently familiar with the underlying assumptions of each technique to ensure that a valid result is obtained.

The last step in interpretation is developing the conclusions and recommendations for the investigation. This is the point where the analysis results are transformed into answers sought by the investigation. For example, the answer sought may be the number of suitable sites for disposing of solid waste in a county. Data analysis would involve comparing appropriate factors to some criteria to indicate suitability or unsuitability. For example, overburden might need to exceed a certain thickness to be suitable. The number of suitable sites will be the number of areas where all the criteria defining suitability are met. These sites would be recommended for use in future waste disposal. Where some level of acceptable uncertainty exists in identifying these sites, recommendations may include the need for more detailed site investigation prior to designing the final disposal facility.

Communication

The final step in an investigation is communication. It should be obvious that without transferring the answers developed by the investigation to the user, it will have served no useful purpose. The communication process transfers the investigation results to the user for action. The user may be an individual who contracted for the information, a company manager, or a governmental board. This is a varied audience, which means that no single form of communication will be appropriate in all cases. The form of communication may be dictated by the situation. For example, an investigation for an environmental-impact statement will be written into the format of that document. Table 2.1 gives an example from California.

Some investigations support applications to zoning boards, county commissions, or similar official bodies. Sometimes, this may require presentation of investigation findings in a public meeting. Oral communication in these situations would need to conform to the rules governing communication under such circumstances. A particular case in which presentation of geologic information must conform to formal rules is court proceedings. As an expert witness or participant in a legal case, it will be necessary to understand these rules to communicate effectively. Kiersch (1969) provides a good summary of points to consider as a technical witness in a court proceeding.

Table 2.1 Suggested Guidelines for Geologic/Seismic Considerations in Environmental Impact Reports

The following guidelines were prepared by the Division of Mines and Geology with the cooperation of the State Water Resources Control Board to assist those who prepare and review environmental impact reports.

These guidelines will expedite the environmental review process by identifying the potential geologic problems and by providing a recognition of data needed for design analysis and mitigating measures. All statements should be documented by reference to material (including specific page and chart numbers) available to the public. Other statements should be considered as opinions and so stated.

I. CHECKLIST OF GEOLOGIC PROBLEMS FOR ENVIRONMENTAL IMPACT REPORTS

Geologic Problems		Could the project or a geologic event cause environmental problems?			Is this conclusion documented in attached reports?	
Problem	Activity Causing Problem	No	Yes	Environmental Problems	No	Yes
Earthquake Damage	Fault Movement					
	Liquefaction					
	Landslides					
	Differential Compaction/Seismic Settlement					
	Ground Rupture					
	Ground Shaking					
	Tsunami					
	Seiches					
	Flooding					
	(Failure of Dams and Levees)					
Loss of Mineral Resources	Loss of Access					
	Deposits Covered by Changed Land-Use Conditions					
	Zoning Restrictions					
Waste Disposal Problems	Change in Groundwater Level					
	Disposal of Excavated Material					
	Percolation of Waste Material					
Slope and/or Foundation Instability	Landslides and Mudflows					
	Unstable Cut and Fill Slopes					
	Collapsible and Expansive Soil					
	Trench-Wall Stability					
Erosion, Sedimentation, Flooding	Erosion of Graded Areas					
	Alteration of Runoff					
	Unprotected Drainage Ways					
	Increased Impervious Surfaces					
Land Subsidence	Extraction of Groundwater, Gas, Oil, Geothermal Energy					
	Hydrocompaction, Peat Oxidation					
Volcanic Hazards	Lava Flow					
	Ash Fall					

(continued)

Table 2.1 *(continued)*

II. CHECKLIST OF GEOLOGIC REPORT ELEMENTS

Report Elements	Yes	No
A. General Elements Present		
1. Description and map of project.	☐	☐
2. Description and map of site.	☐	☐
3. Description and map of pertinent off-site areas.	☐	☐
B. Geologic Element (refer to checklist)		
1. Are all the geologic problems mentioned?	☐	☐
2. Are all the geologic problems adequately described?	☐	☐
C. Mitigating Measures		
1. Are mitigating measures necessary?	☐	☐
2. Is sufficient geologic information provided for the proper design of mitigating measures?	☐	☐
3. Will the failure of mitigating measures cause an irreversible environmental impact?	☐	☐
D. Alternatives		
1. Are alternatives necessary to reduce or prevent the irreversible environmental impact mentioned?	☐	☐
2. Is sufficient geologic information provided for the proper consideration of alternatives?	☐	☐
3. Are all the possible alternatives adequately described?	☐	☐
E. Implementation of the Project		
1. Is the geologic report signed by a registered geologist?*	☐	☐
2. Does the report provide the necessary regulations and performance criteria to implement the project?	☐	☐

*Required for interpretive geologic information.
Source: CDMG, 1975.

Regardless of the form of communication or the audience, every effort should be made to employ appropriate techniques. Many good guides and references are available as aids (USGS, 1978; Cochrane et al., 1979; Heron, 1986). Also, attention to the choice of words should not be minimized. Waggoner et al. (1969) note that some important construction disputes have hinged on the interpretation of one or two words.

Effective communication is not totally a function of technique or audience. Williams (1984) provides several good recommendations for ensuring effective communication in most situations. First, whatever communication technique is used, the conclusions must be supported by the data and analysis. Second, it is best to keep data separate from interpretations. To have confidence in the conclusions, the user must be able to see how the data supports the conclusions. Third, assist the reader or listener in understanding your findings by stating the logical consequences of your conclusions. They will likely be more obvious to you than to your audience. This is especially true for audiences with a less-technical background. Be prepared to make use of pictures, diagrams, and other nonverbal means of communicating your findings to supplement written or spoken words.

TYPES OF INVESTIGATION

There are two basic types of engineering geologic investigation: regional studies and site investigation. Regional studies address questions of land and resource allocation. Heightened concern for the environment and a desire for efficient land and resource management has provided the impetus for many regional studies in recent years.

Regional studies are usually undertaken to provide information for land-use planning. Regional studies often focus on distinguishing, through use of geologic information, among locations for a particular use. Governmental entities require this type of information to avoid or to reduce losses from geologic hazards and to enhance use of geologic resources. This need arises from legal mandates, land-use zonation, and problems created by past land use. Regional studies also identify flood-plain areas for federally subsidized insurance, a planning need arising from a legal mandate. Orderly growth of a community spurs study of undeveloped areas to zone sites for future water-supply storage and treatment facilities. The evaluation of the extent of groundwater contamination resulting from past liquid-waste disposal is representative of the kind of unforeseen problem that arises from past land use. In most instances, regional studies enable one to consider the greatest range of solutions to such problems.

Site investigation is the type of investigation commonly associated with engineering geology. These studies are as varied as the kind of structures or facilities that exist in our society. Site investigations concentrate on the geologic information that affects the design and construction of a particular project at a specific location. The designers of a particular project need this information in order to ensure that their expected results can be achieved at a reasonable cost. In other cases, site investigation is a response to governmental regulation. McCalpin (1985) illustrates this point with several cases at the local governmental level. The emphasis in site investigation is on determining how to make the project work within existing geologic conditions.

Regional Studies

Regional studies are often an appraisal of possible land uses. Although there are many purposes, there are some common features to this type of investigation. Regional studies generally involve large areas of several square miles to several thousands of square miles. The information developed lacks the detail needed to base engineering design or predict geologic occurrences. It is useful for broad decision-making purposes required in regional planning. Maps or inventories are the typical products of regional studies.

Young et al. (1982) conducted a regional study that yielded an inventory. Monroe County in New York State needed to assess the extent of possible groundwater contamination arising from inactive waste-disposal sites. Too many sites and little knowledge of contents made comprehensive drilling and testing inappropriate because of the expense and time involved. Priorities had to be set to concentrate detailed investigation on the most important sites. The study initially classified waste-disposal sites according to their former use. Geologic data were gathered to assess geologic characteristics influencing leachate production, containment, attenuation, or migration. Figure 2.6 is the geologic ranking sheet used for recording and evaluating specific sites. Table 2.2 describes the individual

GEOLOGIC RANKING SHEET
FOR GENERAL COMPARISON OF ABANDONED LANDFILL/DUMP SITES

SYMBOLS USED IN COLUMNS

X PROBABLE EFFECT
U UNCERTAIN: LIKELY EFFECT
⊗ EFFECT OF OVERRIDING SIGNIFICANCE
Superscripts refer to footnotes.

FACTORS TO BE EVALUATED	PRESUMED EFFECT[1]		
	A	B	C
	HIGHER HAZARD	INTERMEDIATE (UNCERTAIN)	LOWER HAZARD
OVERBURDEN GEOLOGY [2]			
ESTIMATED PERMEABILITY [3]			
RELIEF, GEOMORPHOLOGY [4]			
SEPARATION OF WASTE FROM GROUNDWATER [5]			
GROUNDWATER GRADIENT [6]			
BEDROCK CHARACTER [7]			
SOIL MINERALOGY; TEXTURES [8]			
NUMBER OF ENTRIES			
MULTIPLIER	3	2	1
ENTRIES X MULITPLIER			

SUBTOTAL ____
ADDITIONAL FACTORS ____
TOTAL POINTS: SITE RANK ____

SITE NAME/NO. __________

SITE RANK
(CHECK ONE)

.01 HIGHEST PRIORITY (17-21 PTS)	
.02 INTERMEDIATE PRIORITY (12-16 PTS)	
.03 LOWEST PRIORITY (7-11 PTS)	

___ TENTATIVE ___ FINAL

NOTE: IN CASES WHERE MORE THAN HALF THE CRITICAL FACTORS MUST BE RATED AS UNCERTAIN (U), THE RANK SHOULD BE TENTATIVE.

ADDITIONAL FACTORS
(CIRCLE AND ADD TO CHART)

THESE POINTS MAY INCREASE (+1),
DECREASE (−1),
OR NOT AFFECT (0) SCORE

VERY LARGE SITES (20+ ACRES)	+1
ENGINEERING/GEOLOGIC DATA ON OR NEAR SITE	0, −1, +1
GEOLOGY EXTRAPOLATED CONFIDENTLY FROM NEARBY	0, −1, +1

DESCRIBE IMPORTANT OR OVERRIDING FACTORS BELOW IF APPROPRIATE (DESCRIBE SPECIAL CONDITIONS): ______

Figure 2.6 Geologic ranking sheet used to inventory abandoned landfill dump sites in Monroe County, New York. (Reprinted with permission from Am. Chem. Soc. Symp. Ser. No. 204, R. A. Young, A. B. Nelson, and L. A. Hartshorn, Methodology for Assessing Uncontrolled Site Problems at the County Level *in* Risk Assessment at a Hazardous Waste Site. Copyright © 1982 American Chemical Society.)

Table 2.2 Description of Factors on Geologic Ranking Sheet[a]

1. PRESUMED EFFECT: A decision is required as to whether each inferred or documented FACTOR would increase or decrease the hazard relative to leachate production, migration, or attenuation. No simple, uniform guidelines can be set forth that cover all situations or geohydrologic complexities.
2. OVERBURDEN GEOLOGY: From inferred nature of unconsolidated sediments would leachate occurrence be likely to increase or decrease human exposure to pollutants?
3. ESTIMATED PERMEABILITY: Is estimated permeability of unconsolidated materials likely to increase or decrease exposure risks. Include estimated effect of either aquifers or aquicludes or inferred combinations.
4. RELIEF, GEOMORPHOLOGY: Does relief on or adjacent to the site influence the occurrence or migration of leachate so as to increase or decrease the exposure hazard?
5. SEPARATION OF WASTE FROM GROUNDWATER: Does the estimated depth to the water table imply a high or low risk for contamination or leachate production. Relate to permeability and gradient factors.
6. GROUNDWATER GRADIENT: Gradient is dependent on local relief, estimated permeability, aquifer characteristics and rainfall patterns. Steep or flat gradients by themselves cannot be presumed to have similar effects in each case. Judgment is required on local conditions.
7. BEDROCK CHARACTER: Is local bedrock an important factor in local hydrologic system? If so, do textures or structures in bedrock produce asymmetry or enhanced flow of potential leachate plume (flow along bedding, joints, faults, or solution channels).
8. SOIL PROPERTIES, TEXTURE AND BEHAVIOR: Are there known textural or mineralogical factors that could enhance or diminish leachate migration such as strong cation exchange or swelling/shrinking clays (cracking)? Are seasonal effects such as rainfall duration, infiltration capacity, freeze-thaw conditions, vegetation cover, etc., of significance?

[a] Factors used to rank abandoned landfill dump sites.

Source: Reprinted with permission from Am. Chem. Soc. Symp. Ser. No. 204, R. A. Young, A. B. Nelson, and L. A. Hartshorn, Methodology for Assessing Uncontrolled Site Problems at the County Level *in* Risk Assessment at Hazardous Waste Site. Copyright © 1982 American Chemical Society.

factors on the geologic ranking sheet. Matrices combining this geologic ranking with current land use and proximity to drinking-water sources prioritized sites for future detailed study.

Regional studies define capability for various land-use activities. Engineering geology identifies geologic factors in the affected area that are either a constraint or a resource (Laird et al., 1979). Constraints are factors that make one location less suitable for a use than another. A resource is a geologic material necessary to land-use activities. Van Driel (1978) describes some constraints and resources that applied to planned-development housing. Table 2.3 shows representative constraints and resources found in a Fairfax County, Virginia, study area.

Geologic constraints and resources are found in both the natural and man-altered environment. Land-use capability is generally thought of in terms of the natural environment. Proximity to an active fault is a constraint on land use attributable to the natural environment. Exploitable

Table 2.3 Examples of Geologic Constraints and Resources

Condition	Problem	Solution
Unstable slopes	Landsliding; damage to structure	Avoid unstable areas or use special construction methods
Sand and gravel resources	Potential resource may be preempted	Extract resource before development
Flood-prone areas	Periodic flooding; damage to structures	Prohibit construction on or unauthorized alteration of 100-yr flood plain
Vegetation	Valuable open space may be destroyed	Retain stands of mature trees for storm-water retention and environmental corridors
Artificial fill	Decomposition of organic materials	Avoid filled areas; remove organic material and compact fill

[a] Based on geologic conditions related to planned-development housing in Fairfax County, Virginia.

Source: Reprinted with permission from Geology, Vol. 6, J. N. Van Driel, Practical Use of Geologic Information by Planners, pp. 592–596, 1978. Published by Geological Society of America.

gravel deposits from late Pleistocene glaciation are a resource resulting from the natural environment. Alteration of the natural environment results from past land use. For example, past coal mining creates subsidence-prone areas that pose a constraint on surface uses. Alteration of the natural environment can also be a consequence of reservoir construction. In some cases, regional groundwater levels have been raised, making shallower, less-expensive wells possible.

Use of computers to prepare engineering geology and land-capability maps as part of a regional study is increasingly common. Comparison studies demonstrate that computer-generated maps are less expensive and take no longer to prepare than manually compiled maps (Tilmann et al., 1975; Newman et al., 1978). Additional benefits of computerization include (1) enabling geologists to spend more time investigating geologic conditions instead of preparing maps, (2) establishing data files that are easily revised as new information is discovered, (3) reducing errors that result from human miscalculation, and (4) producing displays of land capability in both traditional and nontraditional formats. Figure 2.7 is part of a computer composite map of inferred relative stability during earthquakes for the Salt Lake City, Utah, area (Van Horn and Van Driel, 1977). To produce this map, stability-evaluation numbers were assigned to units found on (1) a map of relative-slope stability, (2) a map of thickness of loosely packed sediments and depth to bedrock, (3) a map of the ratio of gravel-bearing intervals to total depth of wells, and (4) a map of the depth to the top of the principal aquifer. Computer manipulation totaled the

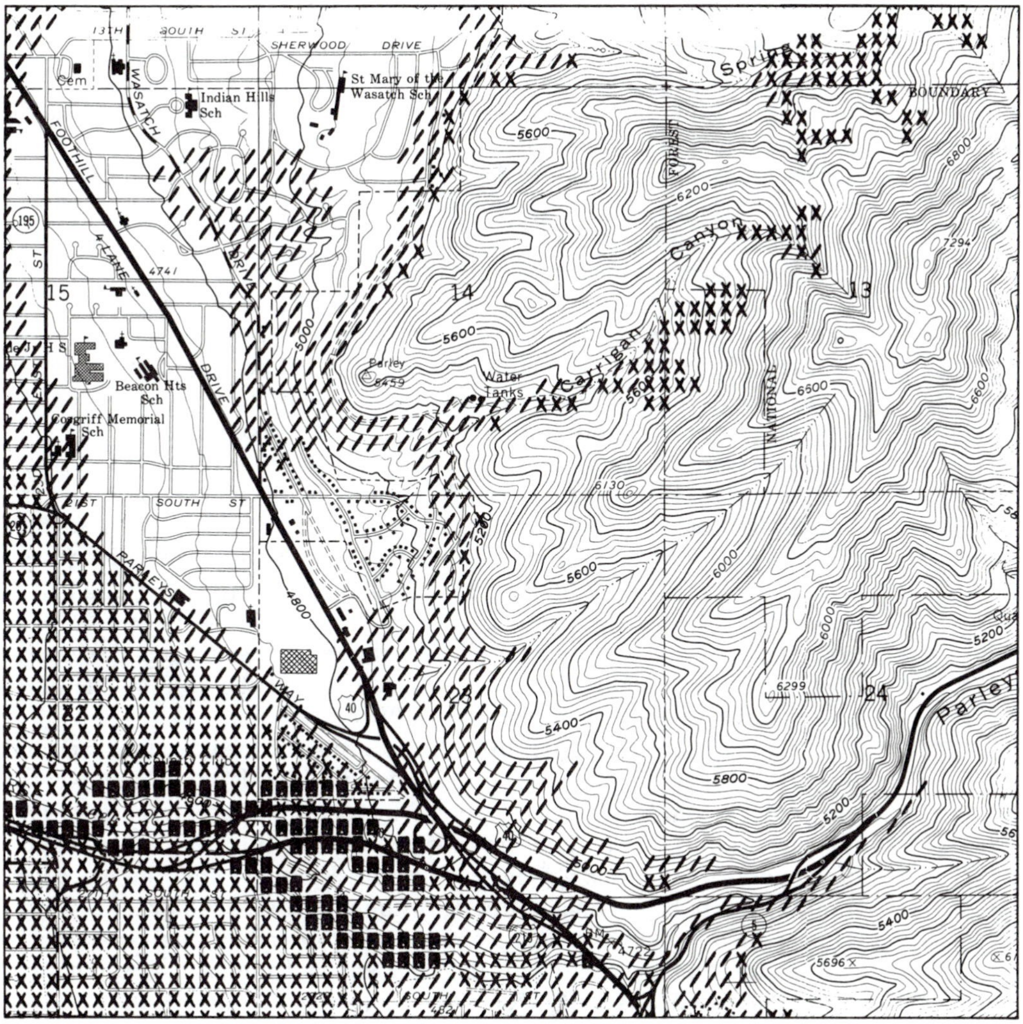

DESCRIPTION OF MAP UNITS

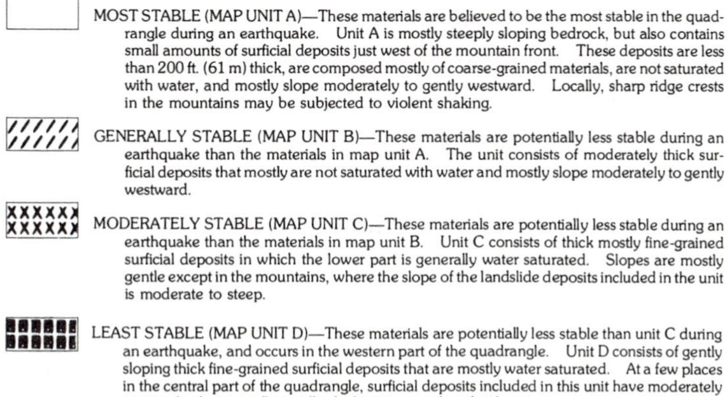

MOST STABLE (MAP UNIT A)—These materials are believed to be the most stable in the quadrangle during an earthquake. Unit A is mostly steeply sloping bedrock, but also contains small amounts of surficial deposits just west of the mountain front. These deposits are less than 200 ft. (61 m) thick, are composed mostly of coarse-grained materials, are not saturated with water, and mostly slope moderately to gently westward. Locally, sharp ridge crests in the mountains may be subjected to violent shaking.

GENERALLY STABLE (MAP UNIT B)—These materials are potentially less stable during an earthquake than the materials in map unit A. The unit consists of moderately thick surficial deposits that mostly are not saturated with water and mostly slope moderately to gently westward.

MODERATELY STABLE (MAP UNIT C)—These materials are potentially less stable during an earthquake than the materials in map unit B. Unit C consists of thick mostly fine-grained surficial deposits in which the lower part is generally water saturated. Slopes are mostly gentle except in the mountains, where the slope of the landslide deposits included in the unit is moderate to steep.

LEAST STABLE (MAP UNIT D)—These materials are potentially less stable than unit C during an earthquake, and occurs in the western part of the quadrangle. Unit D consists of gently sloping thick fine-grained surficial deposits that are mostly water saturated. At a few places in the central part of the quadrangle, surficial deposits included in this unit have moderately to steeply sloping valley walls, fault scarps, and artificial cuts.

Figure 2.7 Part of a slope stability map for the Sugar House Quadrangle, Utah, prepared by using computer mapping methods. (From Van Horn and Van Driel, 1977)

stability numbers from each map for every location in the study area. A range of totaled values defined the composite stability rating. Considerable judgment by the geologist is required to assign stability values and subdivide the range of totals in a rational and meaningful manner. Regional studies rarely define land capability solely on engineering geology factors or for a single land-use consideration. The large areas, multiple uses, and changing economic and sociopolitical forces in regional planning make computerization of data especially useful.

The San Francisco Bay Region Environment and Resources Planning Study was designed to develop and test methods to facilitate the use of engineering geologic information in regional planning (Laird et al., 1979). In the study, a series of maps for different engineering geologic information was generated; and a novel approach was used to communicate the land capability derived from this information. Maps were generated to show different levels of cost for various land uses. These costs represented the cumulative social costs for geologic hazards, constraints, and resources applicable to each land-use category. Included in the calculations were costs for expected damage, mitigation (including investigation), and loss of opportunity (Table 2.4). Figure 2.8 compares maps that show cost for rural or agricultural use and for industrial use. Such representation of engineering geologic information makes it easier to answer the bottom-line question in land-use issues: How much will it cost?

In the case of geologic hazards, the cost may be measured in lives and property damage. With geologic resources, the cost is the added expense incurred to community development in securing these resources. This may, for example, take the form of increased construction cost as a result of the importing of rock aggregate. Costs are especially important in determining land policy. The amount of insurance required to offset expected damage or the relative saving from investing in construction of specially engineered structures can only be decided by comparing the potential extent and likelihood of a geologic hazardous event. Likewise, the potential loss to the tax base resulting from zoning an area for a waste-disposal site, for example, must be considered in the light of costs for waste-treatment alternatives. The expenditures of substantial sums of money and/or the commitment of significant resources are policy decisions with far-reaching consequences. Flawed or shortsighted policy can waste this effort and bring unfortunate consequences on an unsuspecting public. In contrast, land-use policy based on appropriate engineering geologic data can foster a harmonious balance between humankind and the natural environment.

Obviously, the quality and quantity of engineering geologic data available on a regional basis is not the sole factor in saving lives and property from a disasterous earthquake or ensuring that needed rock aggregate is available for future construction. The response of governmental bodies and the affected citizenry is often the ultimate factor. It is possible to have excellent engineering geologic data available and still have land-use policies that foster development in hazardous areas or limit the availability of geologic materials. With poor or minimal engineering geologic information, these consequences are almost a certainty.

Table 2.4 Summarized Costs of Geologic Hazards, Constraints, and Resources for Both Agricultural and Industrial Uses for Part of the Santa Clara Valley, California

Hazard, constraint, or resource	*Summary of costs for rural or agricultural use, in dollars per acre* Severe ←				→ Slight	*Summary of costs for industrial use, in dollars per acre* Severe ←				→ Slight
Surface rupture	$ 20	$ 2	$ 0	—	—	$ 80	$ 80	$ 0	—	—
Ground shaking—San Andreas, Hayward	40	30	10	$ 5	$1	40,000	30,000	10,000	$ 3,000	$700
Ground shaking—Southern Hayward	4	3	1	1	0	4,000	3,000	1,000	300	70
Ground shaking—Calaveras	30	10	5	1	—	30,000	10,000	3,000	700	—
Stream flooding	200	0	—	—	—	40,000	0	—	—	—
Dam failure	0	—	—	—	—	0	—	—	—	—
Dike failure	500	0	—	—	—	70,000	0	—	—	—
Shrink/swell soils	50	20	0	0	—	10,000	4,000	0	0	—
Settlement	20	20	10	5	—	10,000	10,000	6,000	700	—
Liquefaction	10	9	1	0	0	3,000	2,000	200	20	0
Subsidence	0	—	—	—	—	0	—	—	—	—
Landslides	40	30	20	10	0	100,000	70,000	30,000	10,000	0
Soil creep	100	30	0	0	—	20,000	8,000	0	0	—
Erosion and sedimentation	4,000	500	200	0	—	200	30	10	0	—
Septic tanks	0	—	—	—	—	0	—	—	—	—
Sand and gravel	0	—	—	—	—	20,000	0	—	—	—
Mercury	0	—	—	—	—	0	—	—	—	—
Agricultural land	0	—	—	—	—	5,000	0	—	—	—

EXPLANATION (rural or agricultural use)

Level	Symbol	Total cost range (in dollars per acre)	
1		0.01	10.00
2		10.01	31.60
3		31.61	100.00
4		100.01	316.00
5		316.01	1,000.00
6		1,000.01	10,000.00

EXPLANATION (industrial use)

Level	Symbol	Total cost range (in dollars per acre)	
1		0.01	10,000.00
2		10,000.01	17,783.00
3		17,783.01	31,623.00
4		31,623.01	56,234.00
5		56,234.01	100,000.00
6		100,000.01	215,278.00

Source: Laird et al., 1979.

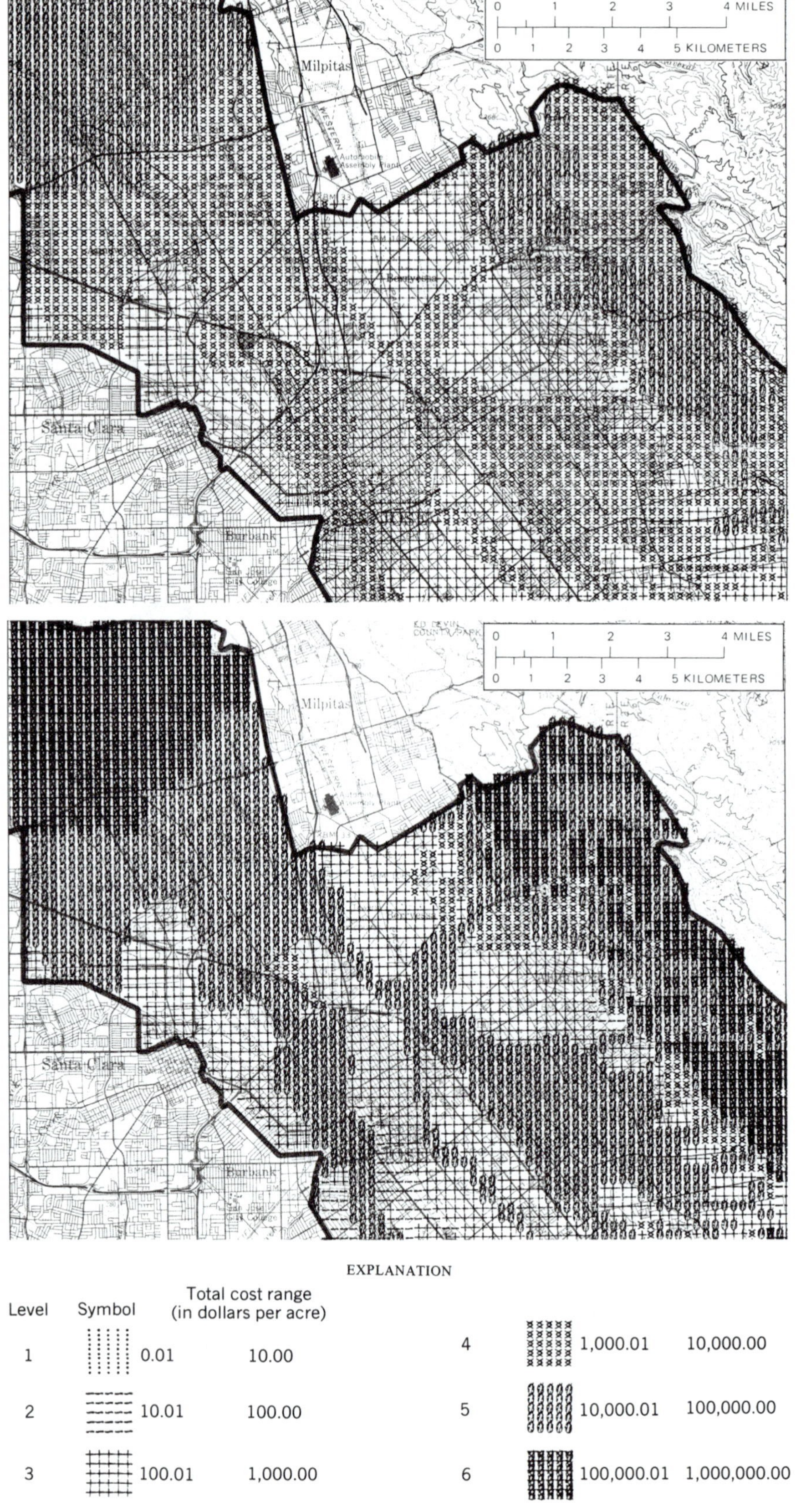

Figure 2.8 Two land-capability maps for the area in the Santa Clara Valley, California—summarized in Table 2.4. The upper map shows costs for rural or agricultural use; the lower map shows costs for industrial use. See Table 2.4 for explanation and data used to compile each map. (From Laird et al., 1979)

Site Investigations

Traditionally, engineering geologists have conducted site investigations. Kent (1981, p. 71) states:

> The purpose of a site investigation is to evaluate the impact of construction on existing site conditions, and of existing site conditions on proposed construction, to anticipate what can be expected during construction, and to develop criteria for design and construction based upon determined site specific physical parameters.

There are three distinct phases of site investigation paralleling typical development of a project: (1) the initial investigation is a preliminary study that supports the feasibility stage of a project, (2) site exploration involves detailed investigation and provides information necessary for project design and construction planning, and (3) implementation studies permit changes in expected conditions that are incorporated in design modifications and evaluation of completed work, thus supporting the actual construction of the project.

Preliminary Studies

Studies in support of project feasibility usually address geologic factors that are important in determining: (1) the relative suitability of alternative sites or project designs, (2) the extent of detailed investigation required for construction, and (3) geologic information for basing reasonable cost estimates (Boyce, 1982). Preliminary studies need to concentrate on key criteria for a particular project (Kent, 1981), generating information needed to complete the project and to prepare environmental documents or applications for project approval by governmental bodies. Table 2.5 outlines the required geologic and soils information for a hydroelectric project license application to the Federal Energy Regulatory Commission. A preliminary study that includes too much detail will waste effort not useful at this flexible stage of project development.

Kreig and Reger (1976) provide an example of investigation applied to determining the geologic feasibility among alternative locations. These authors used terrain evaluation for investigation along the corridor for the Trans-Alaska Pipeline project. A 789-mile-long corridor might initially appear to be a regional study rather than a site investigation. However, pipelines, roads, and similar projects are a special case of site investigation that address narrow linear study areas. To ensure sufficient data for basing modification of the route and documenting potential problem areas, detailed soils data collected at selected sites were used to characterize terrain units that were identifiable by less-expensive aerial photography. When combined with a computerized retrieval system, the data could be examined to ensure reasonable consistency within each terrain unit. Thoughtful application of statistical methods supported this examination. In this way, sufficient detail was gained for basing decisions on rerouting sections without collecting data inappropriate to a feasibility study.

Preliminary studies should specify how detailed subsequent site exploration will be. It is not always possible to choose sites free of potentially

Table 2.5 Excerpt from Regulations Specifying Geologic Information Required in an Application to the Federal Energy Regulatory Commission for Licensing a Hydroelectric Project

(6) Report on Geological and Soil Resources. The applicant must provide a report on the geological and soil resources in the proposed project area and other lands that would be directly or indirectly affected by the proposed project on those resources. The information required may be supplemented with maps showing the location and description of conditions. The report must contain:

(i) A detailed description of geological features, including bedrock lithology, stratigraphy, structural features, glacial features, unconsolidated deposits, and mineral resources.

(ii) A detailed description of the soils, including the types, occurrence, physical and chemical characteristics, erodability and potential for mass movement.

(iii) A description showing the location of existing and potential geological and soil hazards, and problems, including earthquakes, faults, seepage, subsidence, solution cavities, active and abandoned mines, erosion, and mass soil movement, and an identification of any large landslides or potentially unstable soil masses which could be aggravated by reservoir fluctuation.

(iv) A description of the anticipated erosion, mass soil movement and other impacts on the geological and soil resources due to construction and operation of the proposed project; and

(v) A description of any proposed measures of facilities for the mitigation of impacts on soils.

Source: Federal Register, 1981.

difficult geologic conditions. Wahlstrom and Hornback (1968) note that siting of the Dillon Dam in Colorado was based mainly on geographic and topographic factors. Preliminary studies revealed that the dam and support facilities were sited in a geologically complex region. Preliminary investigation noted the need for detailed subsurface investigation of the dam foundation. One of the findings supporting that need was the presence of severely fractured rock likely to promote leakage at the dam unless properly mitigated. Later, detailed study guided grouting that placed 175,957 sacks of concrete using 57,877 feet of drillhole.

It is difficult to estimate the cost of a project in a feasibility study without a good idea of the types of construction and design problems likely to arise from geologic conditions. For example, preliminary investigation of San Francisco's Bay Area Rapid Transit (BART) identified several geologic factors that greatly influenced construction and cost of the project (Taylor and Conwell, 1981). The selected routes would encounter areas of saturated artificial fill and bay mud, saturated sands, high groundwater table, unconsolidated sediments, and faulted and sheared rock. This required adapting construction to deal with potential problems of soft running ground, excess water, settlement of adjacent structures, and difficult excavation and support in rock tunnels. The cost per mile for different tunnel sections reflected the differing design and construction necessitated by these geologic conditions (Figure 2.9).

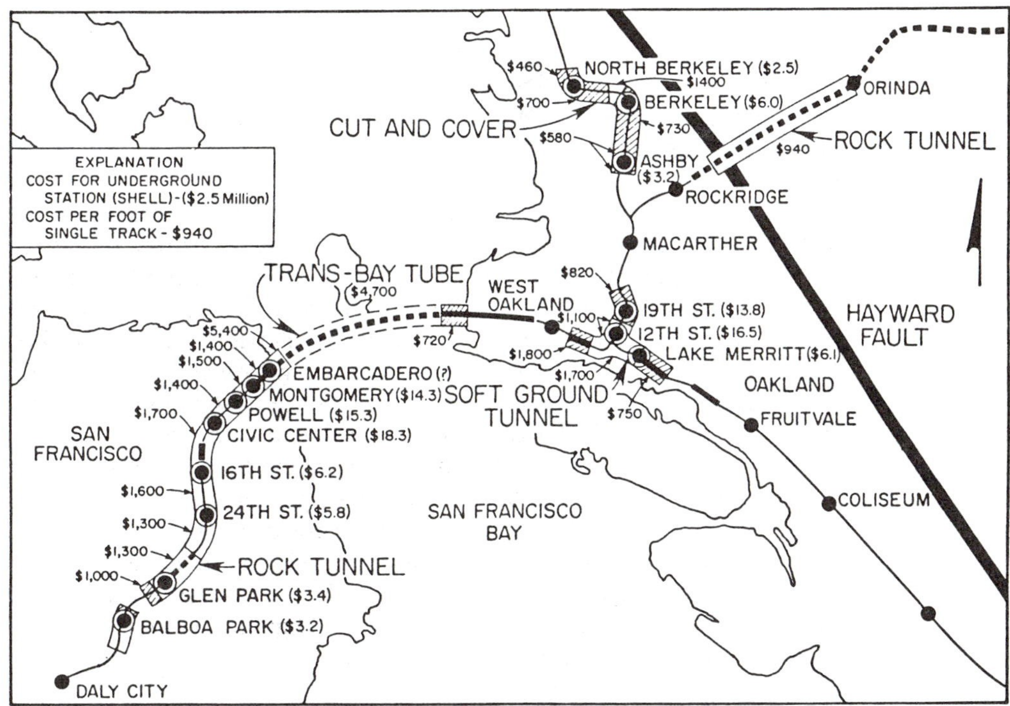

Figure 2.9 Map of the Bay Area Rapid Transit (BART) routes with the cost per mile and construction methods dictated by the geologic conditions present. (From C. L. Taylor and F. R. Conwell, BART—Influence of Geology on Construction Conditions and Cost, *Bulletin of the Association of Engineering Geologists*, Vol. 28, No. 2, May 1981, pp. 195–205. (*Used by Permission.*)

Site Exploration

Site exploration is the intensive investigation of a site. Its results are incorporated into the final design and construction of a project. Often some of the information is used in preparing contracts. This influences the bidding and final cost of constructing the project. Boyce (1982) notes that site exploration includes two distinct goals. The first is determining and interpreting surface and subsurface conditions that influence design and construction. The second is evaluating the behavior characteristics and engineering significance of earth materials present or those intended for use in construction. Building a road illustrates this. Exploration along the proposed route would include areas that require special designs. Wet areas needing drainage, stream channels needing bridges or large fills, and rock outcrops requiring blasting might be among the factors identified during exploration. The evaluation of earth materials would include both the materials along the route and the sources of construction materials. Along the route, the behavior characteristics of the materials may influence the height and steepness of cut slopes and the amount of excavation needed. Sources of earth materials for constructing the road could include rock aggregate used to surface the road. The ability of the aggregate to withstand repeated wetting and drying would be one of the characteristics evaluated to determine a suitable aggregate source. Site exploration requires that the geologist do more than merely supply information on geologic conditions. The information should identify ways to mitigate problems discovered or opportunities to improve the design or decrease the costs of the project. Obviously, this requires a close working relationship with the project engineers.

Fookes et al. (1985) illustrate many of these aspects of site exploration in an extensive discussion on building low-cost roads in mountainous terrain. For example, the Kumlintar hairpins in the Dharan–Dhankuta road in eastern Nepal were necessitated by steep terrain that restricted the road corridor to a strip often no more than 500-meters wide. Figure 2.10 shows the results of both surface and subsurface investigation along that corridor. Note that special conditions such as the existence of shallow slides as well as the types of material encountered are included in the figure. Spoil-disposal sites and sources of construction material are also part of site investigation. Figure 2.11 shows details for the upper part of the hairpins. The site investigation discovered remnants of an old rockslide, the scar of a rockslide, and scars of fresh debris slides. Working with the engineers, a preliminary design to mitigate these potential problem areas was developed. The final as-built road included most of these mitigating features and some additional ones added during construction.

Implementation Studies

Once construction begins, it is sometimes assumed that the engineering geologist has no further role in the project. This is a misconception. As Russell (1981) notes, Karl Terzaghi advocated continued study during construction. It was referred to as his "learn as you go method." There are two good reasons for following this prudent practice (Kent, 1981). First, geologic conditions encountered during project construction may differ from what was expected. This is sometimes called *changed condition.* Most contracts provide for billing increased costs owing to changed condition in addition to the agreed on price for the work. The geologist can assist in determining whether this increased cost is truly a consequence of a changed condition and an appropriate item for payment. Waggoner et al. (1969) and Waggoner (1981) describe geologic conditions and considerations associated with some specific construction claims.

Continued investigation during construction will serve as the basis for changing the project design to avoid major problems in project performance. Sherman (1963) explained how adverse geologic conditions encountered during construction of three highway tunnels near Yellowstone National Park in Wyoming resulted in a design change. In two tunnels, extensive fault zones and joint sets creating blocky rock conditions were encountered. This caused unsafe roof conditions and seepage necessitating a design modification that included a reinforced concrete lining and drains.

Investigation during construction will aid in ensuring compliance with specifications for the project. This information is important for the contract inspectors to effectively discharge their responsibilities. Lund and Euge (1984) describe their excavation mapping and inspection during construction at the Palo Verde nuclear generating station in Arizona. Some

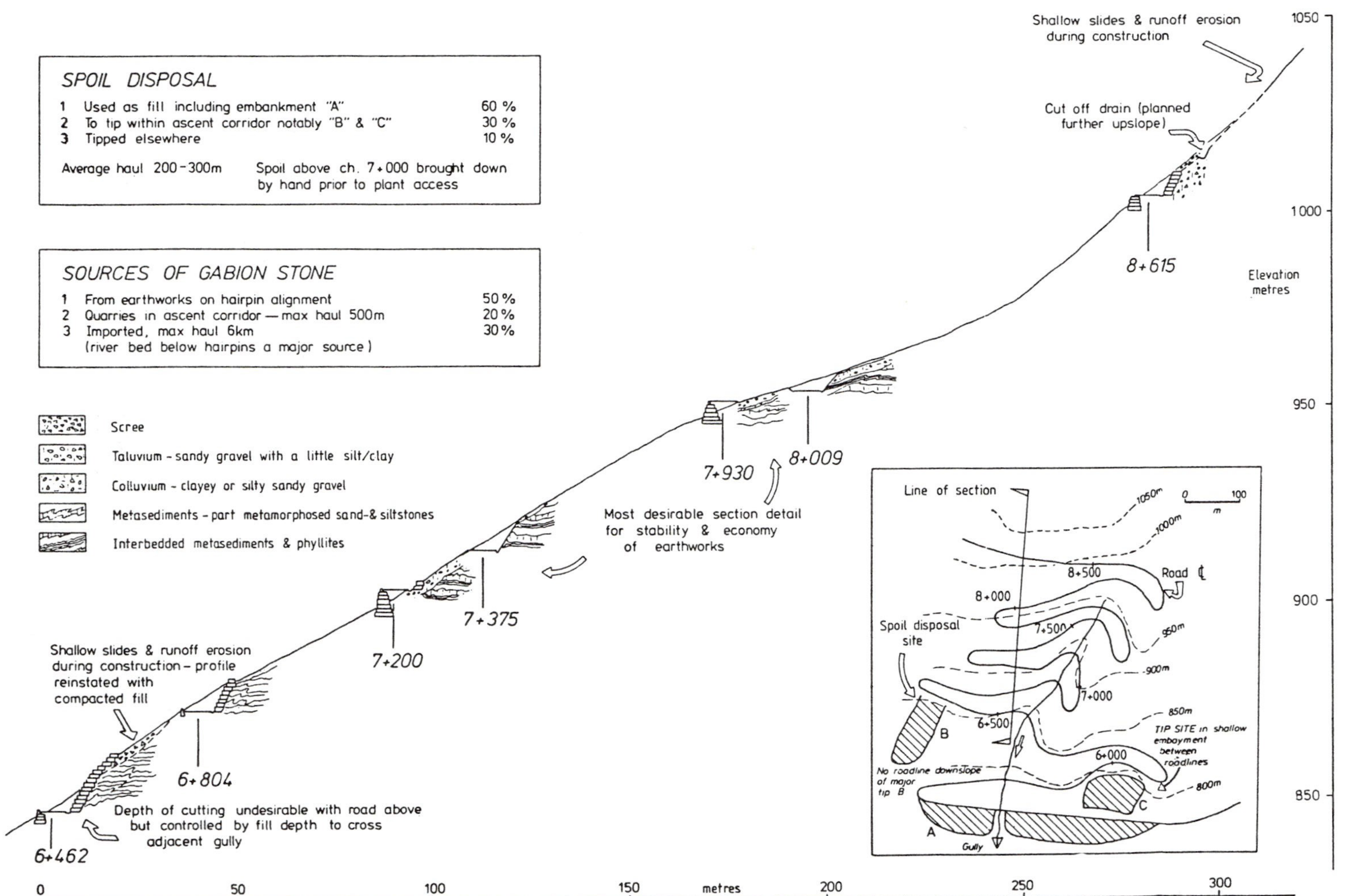

Figure 2.10 Cross-section for the Kumlintar hairpins, eastern Nepal. Both surface and subsurface data developed in site investigation are represented. (Reprinted with permission from Eng. Geol., Vol. 21, P. G. Fookes et al., Geological and Geotechnical Engineering Aspects of Low-cost Roads in Mountainous Terrain, 1985, Elsevier Science Publishers B.V.)

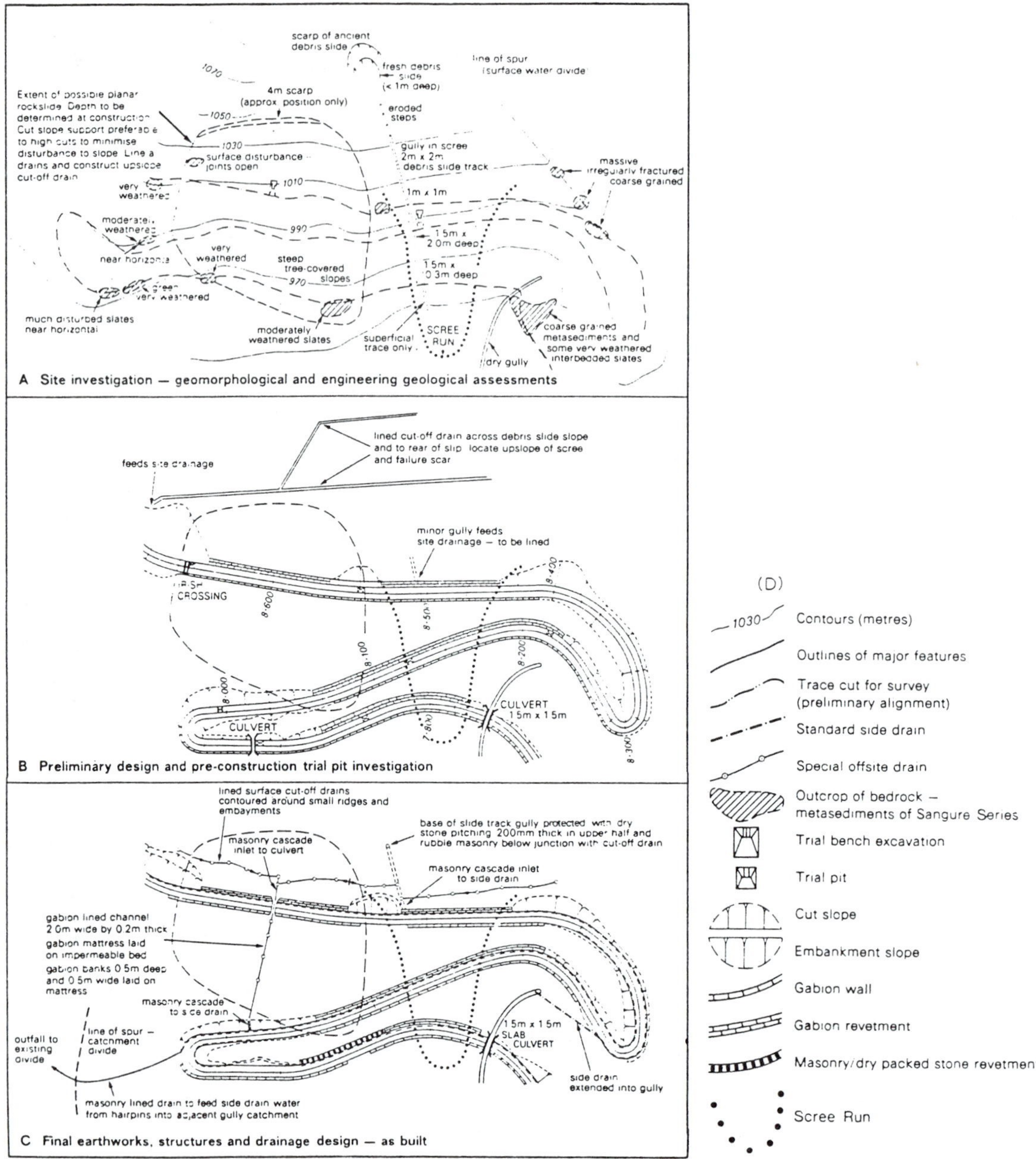

Figure 2.11 Map showing upper portion of the Kumlintar hairpins, eastern Nepal. The top map shows the geologic conditions important to designing the road. The middle one displays a preliminary design encompassing measures to mitigate geologic problems identified in the site investigation. The bottom map shows how the road was actually built. (Reprinted with permission from Eng. Geol., Vol. 21, P. G. Fookes et al., Geological and Geotechnical Engineering Aspects of Low-cost Roads in Mountainous Terrain, 1985, Elsevier Science Publishers B.V.)

facilities are critical to plant operation, including the safe shutting down of the reactors. These are designated as Category I facilities under seismic-design criteria. At Palo Verde, investigation during construction was necessary to ensure that evidence of faulting or other geologic processes likely to compromise the integrity of the facility did not go undetected. An additional benefit of implementation studies is the preparation of an as-constructed geologic map. Project additions at a later date or other projects nearby will benefit from mapping that shows what was actually encountered during construction.

In some projects, implementation studies may continue into the postconstruction period. These studies may evaluate the overall project performance (Kent, 1981). This is especially useful when new techniques or unusual conditions were part of a project. Some projects may include postconstruction studies to meet regulatory requirements (Paris and Berry, 1984). For example, a new sewage treatment facility may be required to maintain a series of monitoring wells to ensure that effluent is not entering the local groundwater. A number of the monitoring and testing techniques described by Pratt and Voegele (1984) may be applicable to implementation studies during and after construction.

SUMMARY

Engineering geologic practice centers on conducting investigations. It is the means by which the engineering geologist applies geologic skills to solve an engineering problem. It requires that an engineering geologist be: (1) a competent geologist, (2) able to translate geologic findings into forms applicable to engineering needs, (3) capable of providing sound judgments and make decisions, and (4) tempermentally inclined to work in a team situation.

An investigation is more than collecting data through mapping, drilling, and sampling. In general terms, a good investigation incorporates elements comparable to those that make a good scientific study. The initial step in an investigation is formulation. It identifies the questions to be answered and indicates how detailed the information must be, the area to be studied, and the time constraints involved. This step is followed by data collection. Data collection is the familiar gathering of information through both office and field studies. The next step is interpretation. The information acquired is interpreted in the light of the purpose for the investigation. Analysis serves as the basis for drawing conclusions and developing recommendations. Effective analysis requires the ability to transform the data into manageable form and make proper use of numerical and statistical methods. The final step is communication. The answers generated by the investigation must be transferred to the user. Without this step, the investigation has served no useful purpose.

Engineering geologic investigations consist of two types: regional studies and site investigation. Regional studies are typically undertaken to provide information for land-use decisions. These decisions may relate to zoning requirements, environmental mandates, or general planning concerns. Regional studies often involve large areas. The information developed lacks the detail necessary to guide engineering design or predict geologic occurrences except in the broadest sense. Maps and inventories are common products of regional studies.

Site investigation is traditionally associated with engineering geology. Although site investigations are as varied as the engineering needs in a modern society, they all involve gathering information that affects the design and construction of a specific project at a particular location. Site investigations support all phases of a project from feasibility studies to design and construction planning to actual construction and postproject monitoring.

REFERENCES

Boyce, R. C., 1982, An overview of site investigations: Bull. Assoc. Eng. Geol., Vol. 19, pp. 167–171.

Burwell, E. B., Jr., and Roberts, G. D., 1950, The geologist in the engineering organization: Engineering Geology (Berkey) Vol., Geol. Soc. Am., Boulder, Colo., pp. 1–9.

CDMG, 1975, Guidelines for geologic/seismic considerations in environmental impact reports: Calif. Div. Mines Geol. Note No. 46, Sacramento.

Cheeney, R. F., 1983, Statistical methods in geology: George Allen & Unwin, Boston, Mass., 169 pp.

Cochrane, W., Fenner, P., and Hill, M., 1979, Geowriting: Am. Geol. Inst., Falls Church, Va., 80 pp.

Compton, R. R., 1962, Manual of field geology: John Wiley & Sons, New York, 378 pp.

Davis, J. C., 1973, Statistics and data analysis in geology: John Wiley & Sons, New York, 550 pp.

Dodd, K., Fuller, H. K., and Clarke, P. F., 1985, Guide to obtaining USGS information: U.S. Geol. Surv. Circ. 900, 35 pp.

Falk, A. L., and Miller, R. L., 1975, Worldwide directory of national earth-science agencies: U.S. Geol. Surv. Circ. 716, 32 pp.

Federal Register, Vol. 46, No. 219, Friday, Nov. 13, 1981.

Fookes, P. G., Sweeney, M., Manby, C. N. D., and Martin, R. P., 1985, Geological and geotechnical engineering aspects of low-cost roads in mountainous terrain: Eng. Geol., Vol. 21, pp. 1–152.

Heron, D. (ed.), 1986, Figuratively speaking: American Association of Petroleum Geologists, Tulsa, Okla., 110 pp.

Kent, M. D., 1981, Site investigation for construction excavation: Bull. Assoc. Eng. Geol., Vol. 18, pp. 71–76.

Kiersch, G. A., 1969, The geologist and legal cases—responsibility, preparation, and the expert witness, *in* Legal aspects of geology in engineering practice, Geol. Soc. Am., Boulder, Colo., pp. 1–6.

Kreig, R. A., and Reger, R. D., 1976, Preconstruction terrain evaluation for the trans-Alaska pipeline project, *in* Geomorphology and engineering: Dowden, Hutchinson & Ross, New York, pp. 55–76.

Laird, R. T., Perkins, J. B., Bainbridge, D. A., Baker, J. B., Boyd, R. T., Huntman,

D., Staub, P. E., and Zucker, Melvin B., 1979, Quantitative land-capability analysis: U.S. Geol. Surv. Prof. Paper 945, 115 pp.

Lessing, P., and Smosna, R. A., 1975, Environmental impact statements—worthwhile or worthless?: Geology, Vol. 3, pp. 241–242.

Lewis, P., 1977, Maps and statistics: John Wiley & Sons, New York, 318 pp.

Lienhart, D. A., and Stransky, T. E., 1981, Evaluation of potential sources of riprap and armor stone—methods and considerations: Bull. Assoc. Eng. Geol., Vol. 18, pp. 323–332.

Lund, W. R., and Euge, K. M., 1984, Detailed inspection and geologic mapping of construction excavations at Palo Verde nuclear generating station, Arizona: Bull. Assoc. Eng. Geol., Vol. 21, pp. 179–189.

McCalpin, J., 1985, Engineering geology at the local government level: planning, review, and enforcement: Bull. Assoc. Eng. Geol., Vol. 22, pp. 315–327.

Morgenstern, N. R., and Sangrey, D. A., 1978, Methods of stability analysis, *in* Landslides—analysis and control, Transp. Res. Bd., Natl. Acad. Sci., Washington, D.C., pp. 315–327.

NAVFAC, 1982, Design manual: soil mechanics: U.S. Dept. of Defense, NAVFAC DM–7.1, Dept. of the Navy, Washington, D.C., 360 pp.

Newman, E. B., Paradis, A. R., and Brabb, E. E., 1978, Feasibility and cost of using a computer to prepare landslide susceptibility maps of the San Francisco Bay region, California: U.S. Geol. Surv. Bull. 1443, 27 pp.

Paris, W. C., Jr., and Berry, R. M., 1984, Considerations for bench mark subsidence monitoring: Bull. Assoc. Eng. Geol., Vol. 21, pp. 207–213.

Pratt, H. R., and Voegele, M. D., 1984, In situ tests for site characterization, evaluation and design: Bull. Assoc. Eng. Geol., Vol. 21, pp. 3–22.

Ragan, D. M., 1973, Structural geology: John Wiley & Sons, New York, 208 pp.

Romesburg, H. C., 1984, Cluster analysis for researchers: Lifetime Learning Publications, Belmont, Calif., 334 pp.

Russell, H. A., 1981, Instrumentation and monitoring of excavations: Bull. Assoc. Eng. Geol., Vol. 18, pp. 91–99.

Sherman, W. F., 1963, Engineering geology of Cody Highway tunnels, Park County, Wyoming, *in* Engineering Geology Case Histories No. 4, Geol. Soc. Am., Boulder, Colo., pp. 27–32.

Taylor, C. L., and Conwell, F. R., 1981, BART—influence of geology on construction conditions and costs: Bull. Assoc. Eng. Geol. Vol. 18, pp. 195–205.

Tilmann, S. E., Upchurch, S. B., and Ryder, G., 1975, Land use site reconnaissance by computer—assisted derivative mapping: Bull. Geol. Soc. Am., Vol. 86, pp. 23–34.

Trautmann, C. H., and Kulhawy, F. H., 1983, Data sources for engineering geologic studies: Bull. Assoc. Eng. Geol., Vol. 20, pp. 439–454.

USBR, 1974, Earth manual: U.S. Bur. of Reclamation, Denver, Colo., 810 pp.

USGS, 1978, Suggestions to authors of the reports of the United States Geological Survey: U.S. GPO, Washington, D.C., 273 pp.

Van Driel, J. N., 1978, Practical use of geologic information by planners: Geology, Vol. 6, pp. 592–596.

Van Horn, R., and Van Driel, J. N., 1977, Computer composite map showing inferred relative stability of the land surface during earthquakes, Sugar House Quadrangle, Salt Lake County, Utah: U.S. Geol. Surv. Map I–766–0.

Waggoner, E. B., Sherard, J. L., and Clevenger, W. A., 1969, Geologic conditions and construction claims on earth- and rock-fill dams and related structures,

Engineering Geology Case Histories No. 7, Geol. Soc. Am., Boulder, Colo., pp. 33–43.

Waggoner, E. B., 1981, Construction claims—spurious and justified: Bull. Assoc. Eng. Geol., Vol. 18, pp. 147–150.

Wahlstrom, E. E., and Hornback, V. Q., 1968, Engineering geology of Dillon Dam, spillway shaft, and diversion tunnel, Summit County, Colorado, Engineering Geology Case Histories No. 6, Geol. Soc. Am., Boulder, Colo., pp. 13–21.

Watts, C. F., and West, T. R., 1985, Electronic notebook analysis of rock slope stability at Cedar Bluff, Virginia: Bull. Assoc. Eng. Geol., Vol. 22, pp. 67–85.

Williams, J. W., 1984, Engineering geology information for varied audiences: the professional–the planner–the general public: Bull. Assoc. Eng. Geol., Vol. 21, pp. 365–369.

Young, R. A., Nelson, A. B., and Hartshorn, L. A., 1982, Methodology for assessing uncontrolled site problems at the County level, *in* Risk assessment at hazardous waste site, Am. Chem. Soc. Symp. Ser. No. 204, pp. 55–71.

3

Engineering Soil

Engineers, geologists, and soil scientists have different definitions for the term *soil*. Figure 3.1 provides a visual comparison of these definitions. An engineer clearly views any mineral material that lacks high strength as being a soil. Engineering soil is roughly equivalent to *regolith*, a term used by geologists to describe all unconsolidated material mantling the surface of the earth. Regolith may include saprolite, in-place bedrock that is chemically altered and coherent and that retains its original texture. To a soil scientist, soil is the part of engineering soil or regolith that contains living matter and supports or is capable of supporting plants. Generally, pedogenic soil shows a differentiation into distinct horizons. The lowermost or C-horizon is commonly equivalent to either rock or saprolite as recognized by geologists.

Chapter 1 points out the different weathering processes. Soils, as a product of weathering, reflect these differences in their character and properties. Soil formed under primarily physical weathering processes will often be distinguishable from soil formed by chemical weathering. In many instances, soil is a weathering product that has been transported and deposited some distance away from its source. Transportation and deposition of the material also affects the resulting soil.

Parent material exerts an influence on soil independent of weathering processes. The nature of the rock subjected to weathering will place limits on the resulting soil. For example, the secondary mineral composition of a soil is limited by the primary mineral composition of the parent material.

The rate of soil development is dependent on both weathering processes and parent material and varies with: (1) grain size of the parent material, (2) mineral composition of the parent material, (3) temperature during weathering, and (4) the presence of water. Larger mineral grains weather faster than smaller grains. Weathering is more rapid for parent material composed of less-stable minerals; and it is faster at higher temperatures. The presence of water enhances certain forms of chemical weathering and

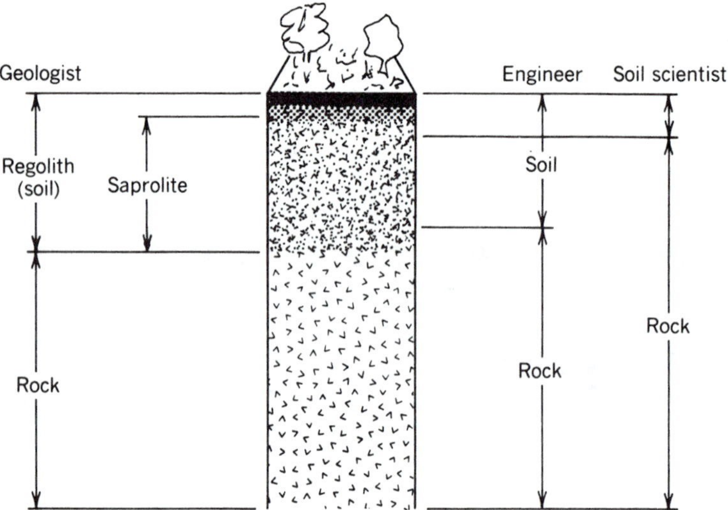

Figure 3.1 A comparison of definitions used by geologists, engineers, and soil scientists for soil and its components.

increases weathering by removing soluble weathering products. Obviously, these generalities represent idealized cases without regard to modifying conditions. A basic understanding of soil-forming relationships will aid the engineering geologist in evaluating soils and their uses.

The importance of soil to the work of the engineering geologist stems from its role in construction. First, soil is one of the most common building materials. It is present, usually in abundant quantities, almost any place something is to be built. Levees and earthen dams are good examples of structures made largely from soil. Second, structures are commonly founded on soil. This influences the design of a structure. It may require that a building have a particular type of foundation such as a spread footing. The soil may dictate the need for a structural design more tolerant of deformation. Third, it is important to know whether soil on natural slopes adjacent to construction will remain in place. Changes that cause soil masses to move downslope generally pose problems for nearby engineered works.

The engineering geologist brings geologic principles to bear in addressing engineering concerns associated with the use of soil in construction. This geologic perspective facilitates exploration for soil with desirable qualities and in sufficient quantities. It also ensures recognition of special soil characteristics important to design and operation.

DESCRIBING SOIL FOR ENGINEERING PURPOSES

Volume and Weight Relationships

A soil mass consists of material that represents three states of matter. Most of the soil mass is composed of solids. The solids are usually mineral material, although organic material may be present. The space between the individual solid grains is occupied by either air, a gas, or water, a liquid. A unit mass of soil can be divided into these component parts on the basis of either volume or weight relationships among these states of matter. Figure 3.2 is a representation of volume and weight relationships. This conceptual view, or phase diagram, enables certain physical characteristics to be determined. These descriptive characteristics enable the geologist to determine the engineering properties of a particular soil: The basic ones are unit weight, specific gravity, porosity and void ratio, water content, and degree of saturation. These descriptive characteristics, viewed in the context of the phase diagram, enable most weight and volume relationships to be determined.

Unit Weight

Unit weight, or soil density, is a basic measurement. Using the phase diagram, a unit of soil would have a specific weight and volume (Figure 3.3*a*). The weight would include the air and water present as well as the solids. The unit weight is obtained by the formula:

$$\gamma = \frac{\text{weight of material}}{\text{volume of material}} \tag{3.1}$$

This unit weight is sometimes referred to as the moist unit weight of soil (γ_m).

If the soil had been completely dried in an oven, the unit weight would be the volume and weight of the air and solids (Figure 3.3*b*). This yields a dry unit weight (γ_d). In contrast, complete saturation of the soil would replace the volume occupied by air with water (Figure 3.3*c*). The resulting unit weight is the saturated unit weight (γ_{sat}).

The unit weight of water (γ_w) is a value useful in many engineering geology computations. This value is a constant based on the mass of water at 4°C: 1 ft^3 of water weighs 62.4 lb, yielding a unit weight of 62.4 lb/ft^3.

Whether the unit weight represents natural or artificial conditions is an

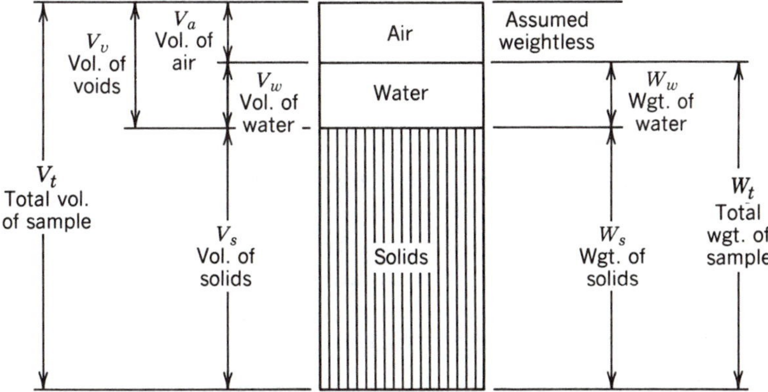

Figure 3.2 Phase diagram of a soil mass representing the volume and weight components for air, water, and solids.

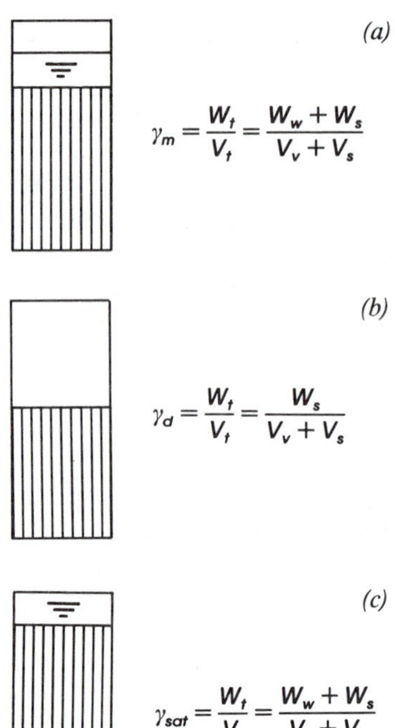

Figure 3.3 (*a*) Phase diagram representing moist unit weight of soil. (*b*) Phase diagram representing dry unit weight of soil. (*c*) Phase diagram representing saturated unit weight of soil.

important factor in some calculations. For embankments and other engineering works, the degree of moisture in the soil mass can be controlled during construction. The unit weight often used in related computation is the maximum dry unit weight or density (γ_{max}). This is readily obtained by using standard laboratory methods. For some calculations, the unit weight needed is the natural density or unit weight of the in-place soil. This is usually a moist unit weight that requires measurement in the field.

Specific Gravity

Specific gravity is a fundamental relationship based on physics. It is defined as:

$$G_s = \frac{\text{weight of unit volume}}{\text{weight of unit volume of water at } 4\,^\circ\text{C}} \tag{3.2}$$

Although both volume and weight relationships are represented by the phase diagram, weight is the more common measurement. Specific gravity embodies a relationship that involves both weight and volume. It provides a means for using weight measurements to determine soil properties normally defined by volume relationships.

Porosity and Void Ratio

Both air and water within a soil occupy the spaces between the solid particles. Two measures exist to represent this space, porosity and void ratio. Porosity *(n)* represents the proportion of the total volume of the mass occupied by space:

$$n = \frac{V_v}{V_T} \times 100 \qquad (3.3)$$

where:

V_v = volume of voids
V_T = total volume

Void ratio *(e)* expresses the relationship between the volume of space and the volume of solids within the volume of the total mass:

$$e = \frac{V_v}{V_s} \qquad (3.4)$$

where:

V_s = volume of solids

Porosity will range in value from $0 \leq n \leq 1$. Void ratio has a possible range of $0 < e < \infty$. Each property can be expressed in terms of the other property:

$$n = \frac{e}{1 + e} \qquad (3.5)$$

$$e = \frac{n}{1 - n} \qquad (3.6)$$

Using specific gravity, weight measurements can be used to determine these volume relationships. Porosity is computed by:

$$n = 1 - \frac{\gamma_d}{62.4\ G_s} \qquad (3.7)$$

Void ratio is defined by:

$$e = \frac{62.4\ G_s}{\gamma_d} - 1 \qquad (3.8)$$

where:

n = porosity
e = void ratio
G_s = specific gravity of the soil
γ_d = dry density

Table 3.1 provides representative values for porosity and void ratio in different materials.

Water Content

The presence of water is an important consideration in many engineering geology problems. Water or moisture content is a useful expression that enables an investigator to estimate soil mass responses. Water content *(w)* is expressed as:

Table 3.1 Some Porosity and Void Ratio Values for Different Soils

	Voids[1]				
	Void Ratio			Porosity (%)	
	e_{max}	e_{cr}	e_{min}	n_{max}	n_{min}
	loose		dense	loose	dense
Granular Materials					
Uniform Materials					
a. Equal spheres (theoretical values)	0.92	—	0.35	47.6	26
b. Standard Ottawa SAND	0.80	0.75	0.50	44	33
c. Clean, uniform SAND (fine or medium)	1.0	0.80	0.40	50	29
d. Uniform, inorganic SILT	1.1	—	0.40	52	29
Well-graded Materials					
a. Silty SAND	0.90	—	0.30	47	23
b. Clean, fine to coarse SAND	0.95	0.70	0.20	49	17
c. Micaceous SAND	1.2	—	0.40	55	29
d. Silty SAND & GRAVEL	0.85	—	0.14	46	12
Mixed Soils					
Sandy or Silty CLAY	1.8	—	0.25	64	20
Skip-graded Silty CLAY with stones or rk fgmts	1.0	—	0.20	50	17
Well-graded GRAVEL, SAND, SILT & CLAY mixture	0.70	—	0.13	41	11
Clay Soils					
CLAY (30%–50% clay sizes)	2.4	—	0.50	71	33
Colloidal CLAY (−0.002 mm: 50%)	12.0	—	0.60	92	37
Organic Soils					
Organic SILT	3.0	—	0.55	75	35
Organic CLAY (30%–50% clay sizes)	4.4	—	0.70	81	41

Source: Reprinted with permission from Basic Soils Engineering, 2nd ed., B. K. Hough, 1969, John Wiley & Sons, Inc.

$$W = \frac{W_w}{W_s} \times 100 \tag{3.9}$$

The range of values possible for this variable are $0 < w < \infty$. Water content is expressed as a percentage. Standard testing can determine the optimum water content (w_o). This value is the water content that will produce a maximum dry density for a given soil under a certain compactive effort.

Degree of Saturation

Degree of saturation *(S)* defines the proportion of total space in the soil mass containing water. It is computed by:

$$S = \frac{V_w}{V_v} \tag{3.10}$$

A saturated soil contains only a volume of voids and a volume of water. This is 100% saturation. At the other end of the range of possible values, saturation can be 0%. The soil is completely dry at this value, with the volume of voids occupied only by air. This volume relationship can be determined using weight measurements by:

$$S = \frac{W G_s}{e} \tag{3.11}$$

where:

W = water content

S = degree of saturation

G_s = specific gravity of the soil

e = void ratio

The preceding discussion can only be considered a brief description of the volume and weight relationships useful in describing engineering soil. Table 3.2 is a handy reference to the computation and interrelationship of these characteristics. Module 3.1 illustrates how these relationships permit known values to be used to determine unknown values. For more detailed treatment of these physical characteristics, consult references such as Terzaghi and Peck (1967), Hough (1969), USBR (1974), and Bowles (1979).

Gradation

The simplest means for describing soil is by the size distribution of its particles. This is termed *gradation.* Typically, the range of possible particle sizes is subdivided with a descriptive name applied to each subdivision. Table 3.3 shows the size ranges that define different particle sizes as used by engineers, geologists, and soil scientists. Each profession has a slightly different assignment of size range for defining the same particle. This evolved from the different purposes of each profession in describing the particle-size distribution of a soil.

Table 3.2 A Summary of Formulas Used to Determine Key Physical Properties of Soils

		Property	Saturated Sample (W_s, W_w, G Are Known)	Unsaturated Sample (W_s, W_w, G, V Are Known)	Supplementary Formulas Relating Measured and Computed Factors			
Components	V_s	Volume of solids	$\frac{W_s}{G\gamma_w}$		$V-(V_a+V_w)$	$V(1-n)$	$\frac{V}{(1+e)}$	$\frac{V_v}{e}$
	V_w	Volume of water	$\frac{W_w}{\gamma_w}$		V_v-V_a	SV_v	$\frac{SV_e}{(1+e)}$	SV_se
	V_a	Volume of air or gas	zero	$V-(V_s+V_w)$	V_v-V_w	$(1-S)V_v$	$\frac{(1-S)V_e}{(1+e)}$	$(1-S)V_se$
	V_v	Volume of voids	$\frac{W_w}{\gamma_w}$	$V-\frac{W_s}{G\gamma_w}$	$V-V_s$	$\frac{V_sn}{1-n}$	$\frac{V_e}{(1+e)}$	V_se
Volume	V	Total volume of sample	V_s+V_w	Measured	$V_s+V_a+V_w$	$\frac{V_s}{1-n}$	$V_s(1+e)$	$\frac{V_v(1+e)}{e}$
	n	Porosity	$\frac{V_v}{V}$		$1-\frac{V_s}{V}$	$1-\frac{W_s}{GV\gamma_w}$	$\frac{e}{1+e}$	
	e	Void ratio	$\frac{V_v}{V_s}$		$\frac{V}{V_s}-1$	$\frac{GV\gamma_w}{W_s}-1$	$\frac{W_wG}{W_sS}$	$\frac{n}{1-n}$ $\frac{wG}{S}$
Weights for Specific Sample	W_s	Weight of solids	Measured		$\frac{W_t}{(1+W)}$	$GV\gamma_w(1-n)$	$\frac{W_wG}{eS}$	
	W_w	Weight of water	Measured		wW_s	$S\gamma_wV_v$	$\frac{eW_sS}{G}$	
	W_t	Total weight of sample	W_s+W_w		$W_s(1+w)$			

Table 3.2 (*continued*)

	Property	Saturated Sample (W_s, W_w, G Are Known)	Unsaturated Sample (W_s, W_w, G, V Are Known)	Supplementary Formulas Relating Measured and Computed Factors			
Weights for Sample of Unit Volume	γ_d Dry unit weight	$\frac{W_s}{V_s + V_w}$	$\frac{W_s}{V}$	$\frac{W_t}{V(I + w)}$	$\frac{G\gamma_w}{(I + e)}$	$\frac{G\gamma_w}{I + wG/S}$	
	γ_t Wet unit weight	$\frac{W_s + W_w}{V_s + V_w}$	$\frac{W_s + W_w}{V}$	$\frac{W_t}{V}$	$\frac{(G + Se)\gamma_w}{(I + e)}$	$\frac{(I + w)\gamma_w}{w/S + I/G}$	
	γ_{sat} Saturated unit weight	$\frac{W_s + W_w}{V_s + V_w}$	$\frac{W_s + V_v\gamma_w}{V}$	$\frac{W_s}{V} + \left(\frac{e}{I + e}\right)\gamma_w$	$\frac{(G + e)\gamma w}{(I + e)}$	$\frac{(I + w)\gamma_w}{w + I/G}$	
	γ_{sub} Submerged (buoyant) unit weight	$\gamma_{sat} - \gamma_w$		$\frac{W_s}{V} - \left(\frac{I}{I + e}\right)\gamma_w$	$\left(\frac{G + e}{I + e} - 1\right)\gamma_w$	$\left(\frac{I - I/G}{w + I/G}\right)\gamma_w$	
Combined Relations	w Moisture content	$\frac{W_w}{W_s}$		$\frac{W_t}{W_s} - I$	$\frac{S_e}{G}$	$S\left[\frac{\gamma_w}{\gamma_d} - \frac{I}{G}\right]$	
	S Degree of saturation	1.00	$\frac{V_w}{V_v}$	$\frac{W_w}{V_v\gamma_w}$	$\frac{wG}{e}$	$\frac{w}{\left[\frac{\gamma_w}{\gamma_d} - \frac{I}{G}\right]}$	
	G Specific gravity	$\frac{W_s}{V_s\gamma_w}$		$\frac{S_e}{w}$			

Source: NAVFAC, 1982.

MODULE 3.1

A soil that weighed 62.1 gm wet was found to weigh 49.8 gm after being dried in an oven. It is determined the soil has a dry unit weight of 86.5 pcf and a specific gravity of 2.68. Find the void ratio and degree of saturation for this soil.

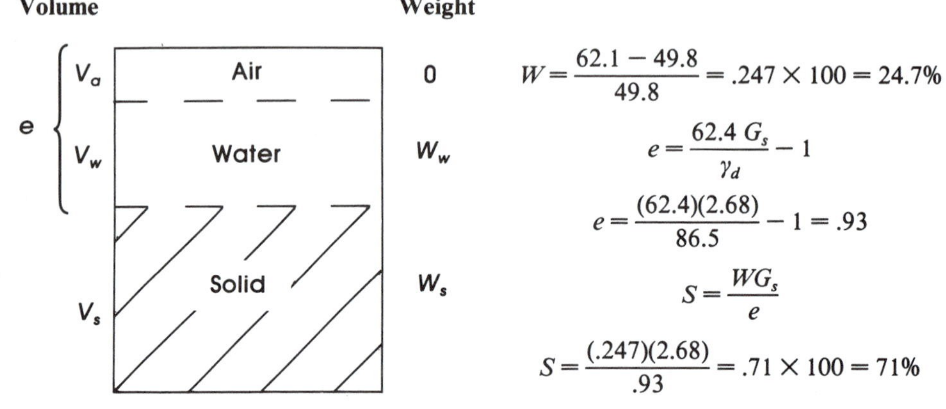

$$W = \frac{62.1 - 49.8}{49.8} = .247 \times 100 = 24.7\%$$

$$e = \frac{62.4\, G_s}{\gamma_d} - 1$$

$$e = \frac{(62.4)(2.68)}{86.5} - 1 = .93$$

$$S = \frac{WG_s}{e}$$

$$S = \frac{(.247)(2.68)}{.93} = .71 \times 100 = 71\%$$

Gradation of a soil is based on sieve analysis. The percentage of a soil passing through sieves that define different particle-size classes is often plotted to show the particle-size distribution of the soil. Figure 3.4 illustrates the cumulative distribution curves for several different soils.

Just as professional differences exist for the size ranges that define different particles, a difference also exists in the description of the distribution of particles in a soil. Engineers describe a soil in which all particle sizes from the smallest to the largest are well represented as being well-graded. A poorly graded soil is composed predominantly of a particular size class. Where particle-size distribution has a good range but lacks some intermediate-size classes, it is referred to as a gap or skip-graded soil. In contrast, geologists are trained to use gradation to characterize the sorting of an unconsolidated material. Sorting is described in the opposite sense than that an engineer uses to describe grading. This can lead to confusion in communication that involves a geologist and an engineer. Geologists describe a soil composed predominantly by one size class as well-sorted. A poorly sorted soil has particles that represent the full range of particle sizes. Engineering geologists need either to adopt engineering terminology in their work with engineers or in their communications to take great care in their use of terms that describe gradation.

Cohesive and Noncohesive Soils

An important distinction can be made between soils that is based on their response to being wet or dry. Soils consisting of particles that stick together

Table 3.3 A Comparison of Particle-Size Classes Used by Geologists, Engineers, and Soil Scientists

Geology	Engineering	Soil Science
	Boulders	
Boulders		
	-305 mm-	
-256 mm—		
	Cobbles	
Cobbles		
	-76.2 mm-	-76.2 mm-
-64 mm-		
	Gravel	
Pebbles		Gravel
	-4.75 mm-	
-2 mm-		-2 mm-
	Sand	
Sand		Sand
	-0.074 mm-	
-0.062 mm-		
		-0.050 mm-
	Silt	
Silt		
		Silt
	-0.005 mm-	
-0.004 mm-		
	Clay	
Clay		-0.002 mm-
		Clay

in either a dry or wet state display cohesion. Cohesion is dependent on the particle-size distribution of the soil. Cohesive soils have a significant proportion of fine-grained particles. The fine-grained particles in a cohesive soil attract each other even in the absence of water. This is especially true when the fine-grained particles include a significant amount of clay. Soils lacking a significant proportion of fine-grained particles will stick together when wet, but they disaggregate into individual grains or pieces after drying. These soils are cohesionless or noncohesive.

Atterberg Limits

The distinction between cohesive and noncohesive soils is important to an understanding of how a soil may or may not deform under stress. The amount of water in the soil controls deformation behavior for that soil. This behavior is called *soil consistency.* The soil may respond to stress in a nonplastic, plastic, or viscous manner. A cohesive soil will change from a nonplastic to a plastic to a viscous state as water content increases. Noncohesive soils are nonplastic for almost all ranges of water content. High water content can cause a noncohesive soil to deform as a viscous fluid.

Consistency for a particular soil is defined by the water content present when it changes its response to stress. These changes are called the *Atterberg limits.* Figure 3.5 shows the Atterberg limits in relation to increasing water content. Between 0% water content and the plastic limit (PL), the soil is nonplastic. Between the plastic limit and the liquid limit (LL), the soil behaves as a plastic material. This range of water content is termed the *plasticity*, or *plastic index* (PI), and is calculated by:

$$PI = LL - PL \tag{3.12}$$

Above the liquid limit, the soil behaves as a liquid or viscous material.

Shrinkage limit (SL) is essentially the fourth of the Atterberg limits. It is a measure of the possible volume change in a wetted soil. The shrinkage limit is defined as the water content expressed as a degree of saturation. The smaller the shrinkage limit, the greater the likelihood that the soil will change in volume. For example, a soil with a shrinkage limit of 25% will

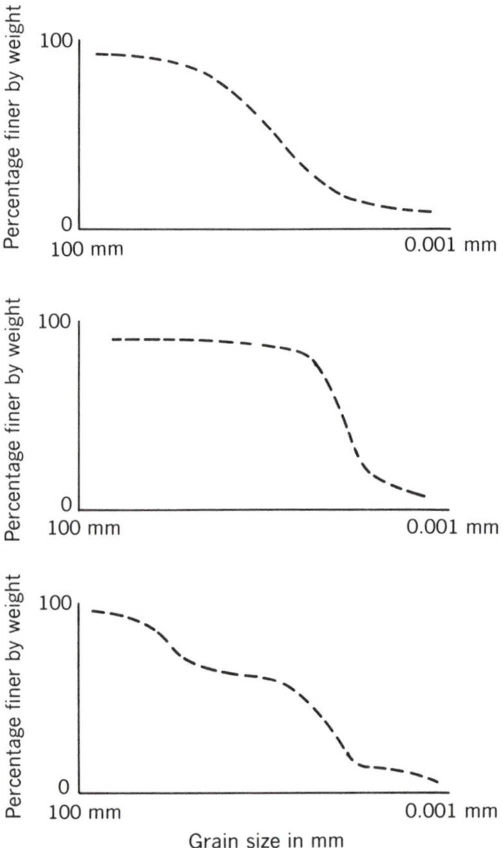

Figure 3.4 Cumulative particle-size distribution curves for three soils. Uppermost curve is uniform or well-graded, middle curve shows a poorly graded soil, and the lower curve is a gap, or skip-graded, soil.

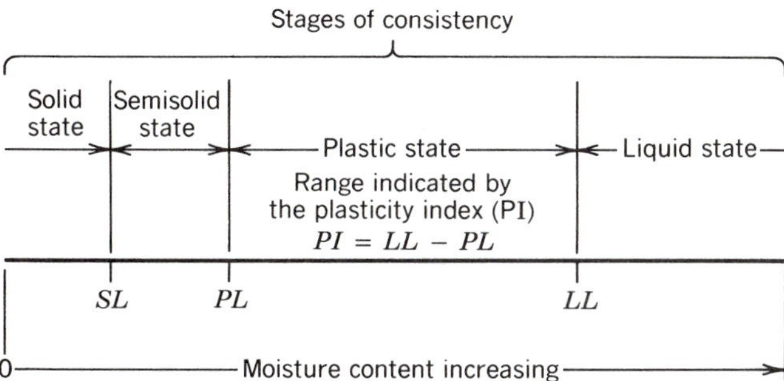

Figure 3.5 Diagram showing the Atterberg limits and the states of soil consistency defined by them. (From USBR, 1968)

increase in volume when the water content exceeds that degree of saturation. This soil is less susceptible to a volume change than a soil with a shrinkage limit of 10%.

The Importance of Clay Mineralogy

Clay mineralogy can cause a soil to increase in volume or expand in the presence of water. Clay minerals are platy, with a negative surface charge (Fig. 3.6*a*). Water in contact with particles becomes organized in a manner dictated by the charges around the plate. This effectively makes the water part of the clay particle and is termed *adsorption.* Deposition of clay normally occurs in quiet water. Sufficient ions are present in natural water to permit altering of the surface charge on some clay particles. The presence of opposing charges causes electrical attraction to form clusters, or flocs, of clay particles that then settle (Figure 3.6*b* and 3.6*c*). Different clay minerals exhibit stronger or weaker negative charges that influence this behavior. This is called the *activity* of a clay and is computed by the following formula:

$$\text{Activity} = \frac{PI}{\text{Percent clay}} \tag{3.13}$$

The activity state of a soil can be estimated by using the chart in Figure 3.7.

Clay mineralogy is also the dominant factor in controlling the plasticity of a clay soil. The nonexpansive clay minerals, kaolinite and illite, have lower PI values than expansive, water-attracting montmorillonite. In turn, montmorillonite has a wide range of PI values that depend on the adsorbed cations available for satisfaction of the charge deficiencies that characterize montmorillonite. Calcium (Ca^{++}) and sodium (Na^{+}) are common pore-water constitutents. Calcium-rich pore water is more effective in satisfying the charge deficiencies than sodium-rich water because of the calcium valence. The resultant improvement in charge satisfaction and interparticle bonding reduces the amount of pore water needed for complete charge neutralization. As a result, a calcium-rich montmorillonite has a lower PI than does a sodium-rich montmorillonite, which requires more water with the lower-valence sodium for charge neutralization. The part played by

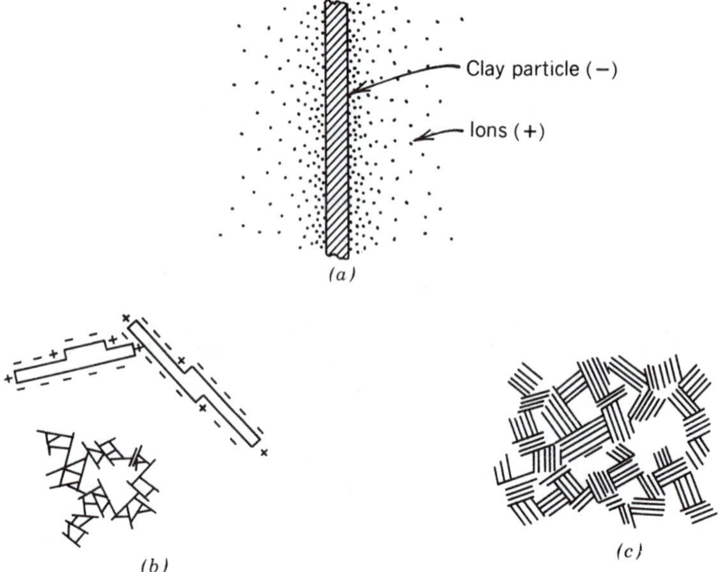

Figure 3.6 (*a*) Clay particle showing the distribution of electrical charges. (*b*) Flocculated structure and (*c*) parallel structure of clay particles. (Reprinted with permission from Soil Mechanics in Engineering Practice, 2nd ed., K. Terzaghi and R. B. Peck, 1967. Copyright © by John Wiley & Sons, Inc.)

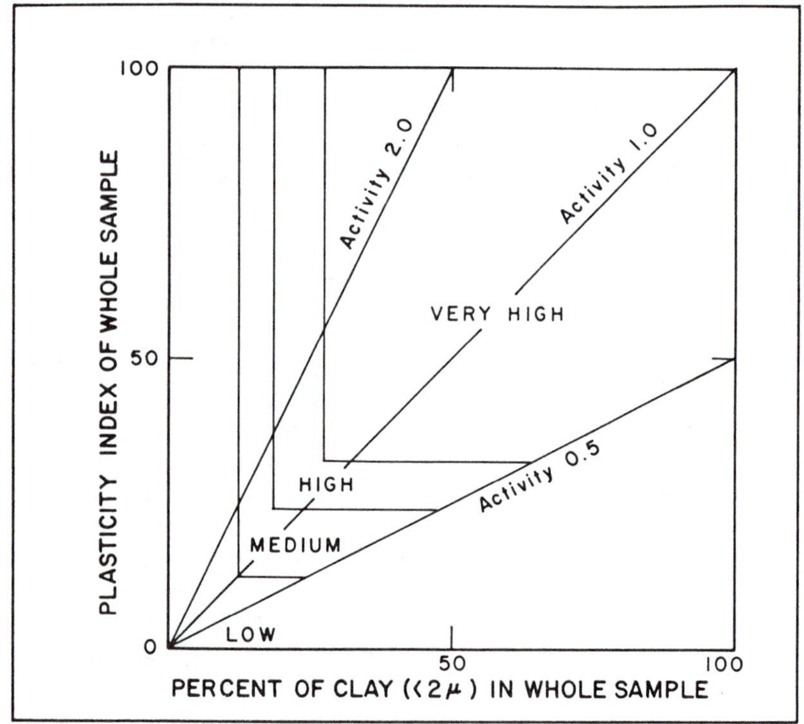

Figure 3.7 Chart for determining the activity level of a soil mass. (From NAVFAC, 1982)

clay minerals in the properties and uses of soils for engineering purposes is more complex than indicated here. Gillott's (1968) *Clay in Engineering Geology* should be referred to for more comprehensive information.

Classification of Engineering Soils

As noted earlier, engineering soils include the "soil" and unconsolidated materials of the geologist and the soil horizons of the soil scientist. Some rocks by geological definition are classed as engineering soils where there is little or no cementing material or where chemical weathering has destroyed the "strong and permanent cohesive forces" required for rock under the engineering definition.

Although many geologists find a soil classification based purely on particle-size classification satisfactory for their professional needs, the civil engineer requires a classification that has engineering applications. The result has been the development of a number of engineering soil classifications. All widely used engineering soil classifications involve a combination of particle size and measures of plasticity and are textural-plasticity soil classifications.

In addition to providing an orderly system for classification, the use of particle size and plasticity permits the engineer to estimate the engineering properties of soils such as compaction, settlement, drainage, frost susceptibility, placement, excavation, and embankment characteristics. As grain size decreases, engineering problems associated with soils tend to increase. Also the difficulty with which particle-size distribution in a soil sample is determined also increases. As a result, the proportions and properties of the so-called *fines* (silt and clay sizes) present in a soil are evaluated by their plasticity rather than by more time-consuming sedimentological procedures. The measures of plasticity, the Atterberg limits, are directly applicable to design and construction uses of a soil, whereas strict size ranges and amounts are not.

Engineering soils are subdivided into two main groups as a function of their predominant sizes and associated plasticity. The coarse-grained soils are composed of sand size and larger particles. They are separated into size ranges by sieving of materials up to cobble size. Except for minor fractions of plastic fines, they characteristically are nonplastic. The fine-grained soils consist predominantly of silt- and clay-sized particles with differing degrees of plasticity measured by their Atterberg limits rather than by sieving and settling velocity methods.

The "standard" size ranges used by geologists have not been used in most engineering soil classifications. The most obvious difference is in the size range of sand, as will be observed in the discussion that follows. Differences in silt and clay boundaries within the fine-grained soils are of less importance as the boundaries are not defined rigorously by size for engineering purposes, as noted previously.

The most widely used classification schemes are those that divide soils into an orderly, easily remembered system of groups, or classes, that have similar physical and engineering properties and that can be identified by simple and inexpensive tests. These groups ideally provide estimates of both the engineering characteristics and performance of soils for design and construction engineers. The descriptions of soils within the groups of a given classification typically are represented by alphabetical or alphanumeric symbols for rapid identification in written material, graphic boring logs, and on engineering drawings. The continued use of a few engineering soil classifications is the result of the provision in each for the needs of the civil engineer as well as the adaptability of the classification to the variety of soils encountered in engineering practice.

Unified Soil Classification

The Unified Soil Classification system (USC) was developed cooperatively by the U.S. Army Corps of Engineers (USAE) and the U.S. Bureau of Reclamation (USBR) assisted by Dr. A. Casagrande and based on his earlier Airfield Classification (Casagrande, 1948). The USC classification was published in 1953 by both agencies (USAE, 1953; USBR, 1953). It has since been adopted by the American Society for Testing and Materials (ASTM) as the standard classification of soils for engineering purposes (ASTM, 1983). The success of the USC is indicated by its routine use worldwide and its acceptance for international geotechnical communication (Mirza, 1982).

The USC system is a textural-plasticity classification scheme. Soils are divided into two major groups (Table 3.4), coarse-grained and fine-grained soils, using the No. 200 sieve (Table 3.5) as the size criterion. When more than half of the soil sample is larger than the No. 200 sieve, it is classified as coarse-grained and is further subdivided by sieving and gradation. When more than half of the soil sample is smaller than the No. 200 sieve, it is classified as fine-grained and is subdivided primarily by liquid limit values and degree of plasticity. The presence of organic material is an additional classification factor for fine-grained soils (see Table 3.4).

Paired letter symbols (Table 3.4) are used for each soil group in the USC system. The first symbol refers to the predominant particle size (with the exception of organics). The second symbol for coarse-grained soils refers to gradation for clean (little or no fines) soils and the presence of silt- and clay-size particles for soils with appreciable amounts of fines. The second symbol for fine-grained soils subdivides on the basis of low (L) or high (H) plasticity. Table 3.6 summarizes these symbols.

Laboratory determination of liquid limit and plasticity indexes for a soil sample permits assignment of fine-grained soils (including the fine fraction of coarse-grained soils) to the proper group by use of the plasticity chart, or A-line diagram (Casagrande, 1948), as illustrated by Figure 3.8. Field-test procedures may be used to estimate the group to which a fine-grained soil should be assigned prior to more definitive laboratory testing. The tests are measures of crushing strength, dilatancy, and toughness, all measures of relative proportions of silt and clay sizes and plasticity. The three tests are summarized and related to USC groups in Table 3.7.

The USC system includes typical soil names with the classification system. Soils that are intermediate between two groups may be identified symbolically by combined notation such as SM–ML and SC–CL. The chart in Table 3.4 and the tests required for soil-group assignment have

Table 3.4 USC System Chart

Major Divisions			Group Symbols	Typical Names
Coarse-Grained Soils More than 50% retained on No. 200 sieve*	Gravels 50% or more of coarse fraction retained on No. 4 sieve	Clean Gravels	GW	Well-graded gravels and gravel-sand mixtures, little or no fines
			GP	Poorly graded gravels and gravel-sand mixtures, little or no fines
		Gravels With Fines	GM	Silty gravels, gravel-sand-silt mixtures
			GC	Clayey gravels, gravel-sand-clay mixtures
	Sands More than 50% of coarse fraction passes No. 4 sieve	Clean Sands	SW	Well-graded sands and gravelly sands, little or no fines
			SP	Poorly graded sands and gravelly sands, little or no fines
		Sands With Fines	SM	Silty sands, sand-silt mixtures
			SC	Clayey sands, sand-clay mixtures
Fine-Grained Soils 50% or more passes No. 200 sieve*	Silts and Clays Liquid limit 50% or less		ML	Inorganic silts, very fine sands, rock flour, silty or clayey fine sands
			CL	Inorganic clays of low to medium plasticity, gravelly clays, sandy clays, silty clays, lean clays
			OL	Organic silts and organic silty clays of low plasticity
	Silts and Clays Liquid limit greater than 50%		MH	Inorganic silts, micaceous or diatomaceous fine sands or silts, elastic silts
			CH	Inorganic clays of high plasticity, fat clays
			OH	Organic clays of medium to high plasticity
Highly Organic Soils			PT	Peat, muck and other highly organic soils

*Based on the material passing the 3-in. (75-mm) sieve.

Source: Reprinted with permission from the Annual Book of ASTM Standards, 1983. Copyright © ASTM, 1916 Race Street, Philadelphia, PA 19103.

Table 3.5 Selected Sieve Sizes and Corresponding Particle Sizes Used in Engineering Soil Classifications

U.S. Standard Sieve Size	Particle Size (diameter in mm)
4	4.75
10	2.0
40	0.42
200	0.075

Table 3.6 Key to Symbols Used by USC System

Primary Group Symbols	Modifying Group Symbols
G—Gravel size	W—Well-graded coarse materials
S—Sand size	P—Poorly graded coarse materials
M—Silt size	M—Silt fines
C—Clay size	C—Clay fines
O—Organic material	L—Low plasticity
	H—High plasticity

been refined recently to formally include combinations such as these (ASTM, 1985). The USC soil groups are associated with specific engineering properties, applications, and performance (see Table 3.8). Thus illustrates the practicality of the system for civil engineers.

The USC system is not without problems. One such problem is reliance by some field personnel on field examination and description of a soil without performing the necessary sieving and Atterberg limits tests as defined by the ASTM D 2487 (ASTM, 1983) standard test method. Another related problem is the result of complete reliance on the standard USC tests and the resulting group assignments while ignoring a complete description of a soil with possible geological association at a site. Prokopovich (1977) reported a case where an SM soil from an auger hole in decomposed granite was misinterpreted as remnants of river-terrace deposits. The difference is of some importance when dealing with foundation design or proposing excavation procedures. This example is not a fault of the classification scheme but of the personnel using it.

Origin of a soil may have engineering significance, but it is not included or implied in the USC system. Mirza (1982) has suggested the addition of lower-case suffixes (Table 3.9) to the USC symbols to indicate the geological origin of soils. This is a refinement of an expansion of the classification originally suggested by Casagrande (1948). Thus, as in the case cited by

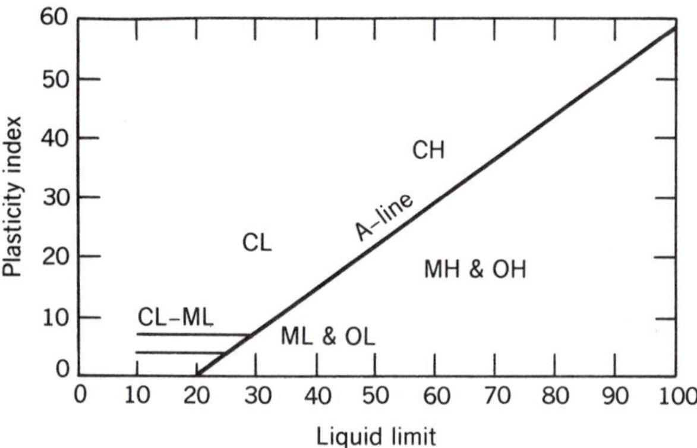

Figure 3.8 Plasticity chart for classification of fine-grained soils from Atterberg limits using USC system notation.

Table 3.7 Summary of Field Tests for Assignment of Fine-grained Soils to USC Groups

		Identification Procedures on Fraction Smaller Than No. 40 Sieve Size			
		Dry Strength (Crushing Characteristics)	Dilatancy (Reaction to Shaking)	Toughness (Consistency Near Plastic Limit)	
Fine-Grained Soils More than half of material is smaller than No. 200 sieve size	Silts and Clays Liquid limit less than 50	None to slight	Quick to slow	None	ML
		Medium to high	None to very slow	Medium	CL
		Slight to medium	Slow	Slight	OL
	Silts and Clays Liquid limit greater than 50	Slight to medium	Slow to none	Slight to medium	MH
		High to very high	None	High	CH
		Medium to high	None to very slow	Slight to medium	OH
Highly Organic Soils		Readily identified by color, odor, spongy feel and frequently by fibrous texture.			PT

DRY STRENGTH (Crushing characteristics)

After removing particles larger than No. 40 sieve size, mold a pat of soil to the consistency of putty, adding water if necessary. Allow the pat to dry completely by oven, sun, or air drying, and then test its strength by breaking and crumbling between the fingers. This strength is a measure of the character and quantity of the colloidal fraction contained in the soil. The dry strength increases with increasing plasticity.

High dry strength is characteristic for clays of the CH group. A typical inorganic silt possesses only very slight dry strength. Silty fine sands and silts have about the same slight dry strength, but can be distinguished by the feel when powdering the dried specimen. Fine sand feels gritty whereas a typical silt has the smooth feel of flour.

DILATANCY (Reaction to shaking)

After removing particles larger than No. 40 sieve size, prepare a pat of moist soil with a volume of about one-half cubic inch. Add enough water if necessary to make the soil soft but not sticky.

Place the pat in the open palm of one hand and shake horizontally, striking vigorously against the other hand several times. A positive reaction consists of the appearance of water on the surface of the pat with changes to a livery consistency and becomes glossy. When the sample is squeezed between the fingers, the water and gloss disappear from the surface, the pat stiffens, and finally it cracks or crumbles. The rapidity of appearance of water during shaking and of its disappearance during squeezing assist in identifying the character of the fines in a soil.

Very fine clean sands give the quickest and most distinct reaction whereas a plastic clay has no reaction. Inorganic silts, such as a typical rock flour, show a moderately quick reaction.

TOUGHNESS (Consistency near plastic limit)

After removing particles larger than the No. 40 sieve size, a specimen of soil about one-half-inch cube in size is molded to the consistency of putty. If too dry, water must be added and if sticky, the specimen should be spread out in a thin layer and allowed to lose some moisture by evaporation. Then the specimen is rolled out by hand on a smooth surface or between the palms into a thread about one-eighth inch in diameter. The thread is then folded and rerolled repeatedly. During this manipulation the moisture content is gradually reduced and the specimen stiffens, finally loses its plasticity, and crumbles when the plastic limit is reached.

After the thread crumbles, the pieces should be lumped together and a slight kneading action continued until the lump crumbles.

The tougher the thread near the plastic limit and the stiffer the lump when it finally crumbles, the more potent is the colloidal

(continued)

Table 3.7 *(continued)*

clay fraction in the soil. Weakness of the thread at the plastic limit and quick loss of coherence of the lump below the plastic limit indicate either inorganic clay of low plasticity, or materials such as koalin-type clays and organic clays which occur below the A-line.

Highly organic clays have a very weak and spongy feel at the plastic limit.

Source: USBR, 1974.

Prokopovich, the soil properly classified and geologically interpreted would have been an SM–r soil to show the soil's residual origin.

Other Classifications

In addition to the USC system, an engineering soil classification was developed by the U.S. Bureau of Public Roads in 1928. It is the American Association of State Highway (and Transportation) Officials [AASHO (AASHTO)] classification of highway subgrade materials (AASHO, 1961). It also is a textural-plasticity classification that uses sieved fractions and Atterberg limits for assignment of soils to seven main groups and several subgroups (Table 3.10). The classification is more specific than the USC system in the limits placed on size ranges and amounts and ranges of liquid limits and plasticity indexes for fines. As with the USC system, these limits are placed on groups within both the granular (coarse-grained) and silt-clay (fine-grained) soils as required by soil gradations.

Rather than using the No. 4 sieve (4.75 mm) of the USC system as the upper limit of the sand-size range, the AASHTO (old AASHO) classification uses the No. 10 sieve (2.0 mm) as the upper size limit of sand as does the Wentworth scale. However, the No. 200 sieve (0.075 mm) used in the USC system is retained to separate the finer fractions from sand (Figure 3.9).

The increased number of soil groups in the AASHTO classification compared with the USC system as well as the different upper size limit of sand make comparisons of the two systems difficult. Table 3.11 provides comparisons of the two systems, as do the source data in Table 3.12. The soil sample given would be classed as A-2-7 in the AASHTO system and SC in the USC system. A plasticity chart similar to the A-line chart for the USC system is shown in Figure 3.10.

Extensive modifications of the USC system in size limits, symbols, and soil names have been proposed by the International Association of Engineering Geology (IAEG) Commission on Engineering Geological Mapping (Matula, 1981). The IAEG classification is shown in Table 3.13 for comparison purposes. Assessment of the usefulness of such major revisions of the USC system cannot be made until practicing geotechnical engineers and engineering geologists find them to be of value.

Other engineering soils classifications have been developed such as the Federal Aviation Administration (FAA) classification and British Standard Code of Practice CP2001. Summaries of these and other classifications are in papers by Liu (1967) and Matula (1981).

Table 3.8 Selected Engineering Performance Characteristics of USC Groups

Major Divisions		Letter		Value as Subgrade When Not Subject to Frost Action	Value as Subbase When Not Subject to Frost Action	Value as Base When Not Subject to Frost Action	Potential Frost Action
Coarse-Grained Soils	Gravel and Gravelly Soils	GW		Excellent	Excellent	Good	None to very slight
		GP		Good to excellent	Good	Fair to good	None to very slight
		GM	d	Good to excellent	Good	Fair to good	Slight to medium
			u	Good	Fair	Poor to not suitable	Slight to medium
		GC		Good	Fair	Poor to not suitable	Slight to medium
	Sand and Sandy Soils	SW		Good	Fair to good	Poor	None to very slight
		SP		Fair to good	Fair	Poor to not suitable	None to very slight
		SM[a]	d	Fair to good	Fair to good	Poor	Slight to high
			u	Fair	Poor to fair	Not suitable	Slight to high
		SC		Poor to fair	Poor	Not suitable	Slight to high
Fine-Grained Soils	Silts and Clays LL is Less Than 50	ML		Poor to fair	Not suitable	Not suitable	Medium to very high
		CL		Poor to fair	Not suitable	Not suitable	Medium to high
		OL		Poor	Not suitable	Not suitable	Medium to high
	Silts and Clays LL is Greater Than 50	MH		Poor	Not suitable	Not suitable	Medium to very high
		CH		Poor to fair	Not suitable	Not suitable	Medium
		OH		Poor to very poor	Not suitable	Not suitable	Medium
Highly Organic Soils		PT		Not suitable	Not suitable	Not suitable	Slight

[a] SM_d are soils with LL $\leq$ 28 and PI $\leq$ 6; SM_u are soils LL $>$ 28.

Source: USAE, 1953.

(continued)

Table 3.8 *(continued)*

Compressibility and Expansion	Drainage Characteristics	Value for Embankments	Compaction Characteristics	Value for Foundations	Requirements for Seepage Control
Almost none	Excellent	Very stable, pervious shells of dikes and dams	Good, tractor, rubber-tired, steel-wheeled roller	Good bearing value	Positive cutoff
Almost none	Excellent	Reasonably stable, pervious shells of dikes and dams	Good, tractor, rubber-tired, steel-wheeled roller	Good bearing value	Positive cutoff
Very slight	Fair to poor	Reasonably stable, not particularly suited to shells, but may be used for impervious cores or blankets	Good, with close control, rubber-tired, sheepsfoot roller	Good bearing value	Toe trench to none
Slight	Poor to practically impervious				
Slight	Poor to practically impervious	Fairly stable, may be used for impervious core	Fair, rubber-tired, sheepsfoot roller	Good bearing value	None
Almost none	Excellent	Very stable, pervious sections, slope protection required	Good, tractor	Good bearing value	Upstream blanket and toe drainage or wells
Almost none	Excellent	Reasonably stable, may be used in dike section with flat slopes	Good, tractor	Good to poor bearing value depending on density	Upstream blanket and toe drainage or wells
Very slight	Fair to poor	Fairly stable, not particularly suited to shells, but may be used for impervious cores or dikes	Good, with close control, rubber-tired, sheepsfoot roller	Good to poor bearing value depending on density	Upstream blanket and toe drainage or wells
Slight to medium	Poor to practically impervious				
Slight to medium	Poor to practically impervious	Fairly stable, use for impervious core for flood control structures	Fair, sheepsfoot roller, rubber tired	Good to poor bearing value	None
Slight to medium	Fair to poor	Poor stability, may be used for embankments with proper control	Good to poor, close control essential, rubber-tired roller, sheepsfoot roller	Very poor, susceptible to liquefaction	Toe trench to none
Medium	Practically impervious	Stable, impervious cores and blankets	Fair to good, sheepsfoot roller, rubber tired	Good to poor bearing	None
Medium to high	Poor	Not suitable for embankments	Fair to poor, sheepsfoot roller	Fair to poor bearing, may have excessive settlements	None
High	Fair to poor	Poor stability, core of hydraulic fill dam, not desirable in rolled fill construction	Poor to very poor, sheepsfoot roller	Poor bearing	None
High	Practically impervious	Fair stability with flat slopes, thin cores, blankets and dike sections	Fair to poor, sheepsfoot roller	Fair to poor bearing	None
High	Practically impervious	Not suitable for embankments	Poor to very poor, sheepsfoot roller	Very poor bearing	None
Very high	Fair to poor	Not used for construction		Remove from foundations	

Table 3.9 Proposed Lower-case Suffixes to Identify Geological Origin of Engineering Soils of USC System

Suffix	Geological Origin
a	aeolian
f	fluvial
g	glacial
l	lacustrine
m	marine
r	residual
t	tectonic
v	volcanic

Source: Reprinted with permission from Can. Geotech. J., Vol. 19, C. Mirza, A Case for the Extension of the Unified Soil Classification System, 1982.

THE ENGINEERING PROPERTIES OF SOIL

The suitability of a soil for a particular use depends on its response to that use. Suitability usually depends on one or more engineering properties of a soil. These properties are determined through the use of the physical characteristics (described earlier) and their interrelationships. The performance of engineering works will depend on the correct assessment of engineering properties to determine suitability and to predict performance of a soil for its intended use.

Obermeier and Langer (1986) examined the relationship of geology and engineering properties for soil and rock in Fairfax County, Virginia. Their mapping and testing served to establish the range of physical characteristics for different soils and rock found in this area. These attributes were then used to define engineering properties such as suitability as compacted material, road-performance characteristics, excavation conditions, and nature of surface and internal drainage. Although their mapping and testing were intended to examine how geology affects engineering properties, they also illustrate how exploration and characterization of engineering properties of soils should take place for engineering projects.

Two engineering properties are especially important to many types of engineering works and situations involving soil. The first property is the degree to which soil will change volume under a load. This is termed *compressibility.* A structure placed on a highly compressible soil is likely to suffer settlement damage as the soil volume decreases under the application of this static load. The second engineering property is shear strength. This property is the resistance of soil to sliding of one mass against another. A familiar example is the ability of a road fill to remain in place when placed on a soil-covered hillslope.

Table 3.10 AASHTO (AASHO) Soil Classification Chart

General Classification	Granular Materials (35% or less passing No. 200)							Silt-Clay Materials (More than 35% passing No. 200)			
Group Classification	A-1		A-3	A-2				A-4	A-5	A-6	A-7
	A-1-a	A-1-b		A-2-4	A-2-5	A-2-6	A-2-7				A-7-5; A-7-6
Sieve Analysis: Percent passing:											
No. 10	50 Max.	—	—	—	—	—	—	—	—	—	—
No. 40	30 Max.	50 Max.	51 Min.	—	—	—	—	—	—	—	—
No. 200	15 Max.	25 Max.	10 Max.	35 Max.	35 Max.	35 Max.	35 Max.	36 Max.	36 Min.	36 Min.	36 Min.
Characteristics of fraction passing No. 40:											
Liquid limit	—		—	40 Max.	41 Min.	40 Max.	41 Min.	40 Max.	41 Min.	40 Max.	41 Min.
Plasticity Index	6 Max.		N.P.[a]	10 Max.	10 Max.	11 Min.	11 Min.	10 Max.	10 Max.	11 Min.	11 Min.[b]
Group Index	0		0	0		4 Max.		8 Max.	12 Max.	16 Max.	20 Max.
Usual Types of Significant Constituent Materials	Stone Fragments Gravel and Sand		Fine Sand	Silty or Clayey Gravel and Sand				Silty Soils		Clayey Soils	
General Rating as Subgrade	Excellent to Good					Fair to Poor					

[a] N.P. = not plastic.

[b] Plasticity index of A-7-5 subgroup is equal to or less than LL minus 30. Plasticity index of A-7-6 subgroup is greater than LL minus 30.

Source: AASHO, 1961.

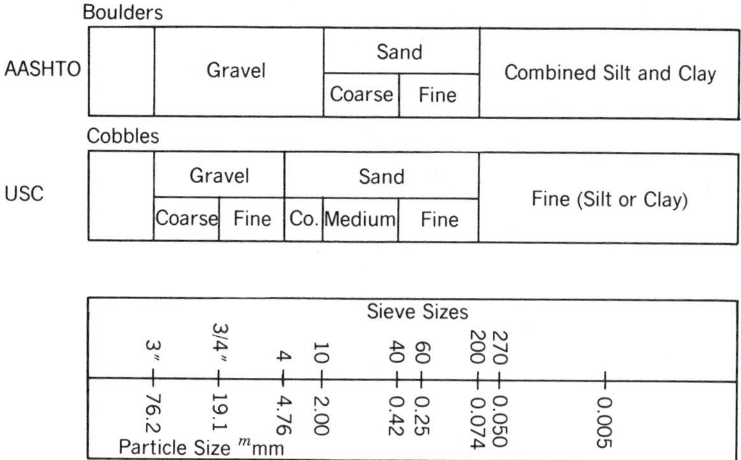

Figure 3.9 Comparison of soil-size limits of AASHTO (AASHO) and USC systems. (From Liu, 1967)

Table 3.11 Comparison of Soil Groups in USC and AASHTO Systems

Soil Group in USC System	Comparable Soil Groups in AASHTO System: Most Probable	Possible	Possible but Improbable
GW	A-1-a	—	A-2-4, A-2-5 A-2-6, A-2-7
GP	A-1-a	A-1-b	A-3, A-2-4, A-2-5, A-2-6, A-2-7
GM	A-1-b, A-2-4 A-2-5, A-2-7	A-2-6	A-4, A-5, A-6, A-7-5, A-7-6, A-1-a
GC	A-2-6, A-2-7	A-2-4, A-6	A-4, A-7-6, A-7-5
SW	A-1-b	A-1-a	A-3, A-2-4, A-2-5, A-2-6, A-2-7
SP	A-3, A-1-b	A-1-a	A-2-4, A-2-5, A-2-6, A-2-7
SM	A-1-b, A-2-4, A-2-5, A-2-7	A-2-6, A-4, A-5	A-6, A-7-5, A-7-6, A-1-a
SC	A-2-6, A-2-7	A-2-4, A-6, A-4, A-7-6	A-7-5
ML	A-4, A-5	A-6, A-7-5	—
CL	A-6, A-7-6	A-4	—
OL	A-4, A-5	A-6, A-7-5, A-7-6	—
MH	A-7-5, A-5	—	A-7-6
CH	A-7-6	A-7-5	—
OH	A-7-5, A-5	—	A-7-6
Pt	—	—	—

Source: Liu, 1967.

Table 3.12 Source Data for Classification of Soils by USC and AASHTO Systems[a]

Sieve No.	Particle Size, mm	Percent Finer
4	4.76	100
10	2.00	45
40	0.42	37
200	0.074	29
270	0.050	27
—	0.002	23
	Liquid Limit 62	
	Plasticity Index 41	

[a] Classifications: USC—SC soil; AASHTO—A-2-7 soil.

Source: Liu, 1967.

Compressibility

Construction often involves the use of soil to make a structure or the placement of a structure made of other materials on a soil foundation. In either case, the compressibility of the soil used is an important consideration. Compressibility is the decrease in volume of a soil mass as a consequence of either natural or artificial means. This volume change is primarily due to a change in the volume of voids. To a lesser extent, it can result from a change in the volume of solids.

Consolidation is the form of compressibility that occurs under a static load. Consolidation is basically the process of driving water from the voids in a soil mass. If consolidation is not taken into consideration, it can lead to settlement that may seriously impair the structure being founded on the soil.

Compaction is an artificial densification of soil. It is a consideration whenever soil is used as a construction material. Reducing the volume of voids for this purpose is commonly accomplished by vibrating or loading and unloading the soil mass. Just as moisture in the soil is a factor in

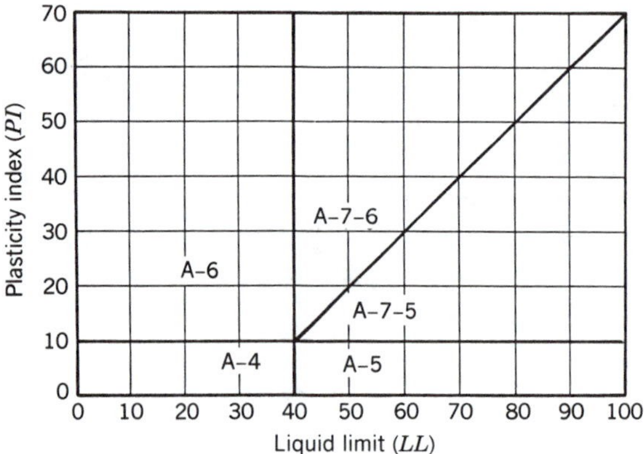

Figure 3.10 Plasticity chart for classification of fine-grained soils using the AASHTO (AASHO) system. (From Liu, 1967)

Table 3.13 Soil Classification for Engineering Geological Mapping

Main Division of Soil Groups		Fines Percentage finer than 0.06 mm	Soil Groups	Group Symbols	Sub-Group Symbols	Soil Name
Coarse Soils More than 35 per cent of the material less than 60 mm is larger than 0.06 mm	Gravel More than 50 per cent of coarse materials is larger than 2 mm	0–5	Gravel	G	GW GP GPu GPg	Gravel, well graded Gravel, poorly graded Gravel, uniformly graded Gravel, gap graded
		5–35	Gravel, silty	GM	GML etc.	Gravel, silty, of low plasticity
			Gravel with Fines	GF		
			Gravel, clayey	GC	GCL etc.	Gravel, clayey, of low plasticity
	Sand More than 50 per cent of coarse materials is smaller than 2 mm	0–5	Sand	S	SW SP SPu SPg	Sand, well graded Sand, poorly graded Sand, uniformly graded Sand, gap graded
		5–35	Sand, silty	SM	SML etc.	Sand, silty, of low plasticity
			Sand with Fines	SF		
			Sand, clayey	SC	SCL etc.	Sand, clayey, of low plasticity
Fine Soils More than 35 per cent of the material less than 60 mm is larger than 0.06 mm	Silts and Clays, Gravelly or Sandy	35–65	Silt, gravelly	MG	MLG etc.	Silt, gravelly, of low plasticity
			Fine Soil, gravelly	FG		
			Clay, gravely	CG	CLG etc.	Clay, gravelly, of low plasticity
			Silt, sandy	MS	MSL etc.	Silt, sandy, of low plasticity
			Fine Soil, sandy	FS		
			Clay, sandy	CS	CLS etc.	Clay, sandy, of low plasticity
	Silts and Clays	65–100	Silt (M-Soil)	M	ML etc.	Silt of low plasticity
			Fine Soil	F	CL	Clay of low plasticity
			Clay	C	CI CH CV CE	Clay of intermediate plasticity Clay of high plasticity Clay of very high plasticity Clay of extremely high plasticity
Organic Soils	Organic Sand, silt or clay			Descriptive letter O Suffixed to any symbol		
Peat	Predominantly plant remains which may be fibrous or amorphous			PT		

Note: Material coarser than 60 mm is removed and recorded as cobbles (60–200 mm) or boulders (over 200 mm).

Source: Reprinted with permission from Bull. Int. Assoc. Eng. Geol., No. 24, M. Matula, Recommended Symbols for Engineering Geological Mapping, 1981.

consolidation, it plays a role in compaction. Altering the moisture content of the soil during the densification process can facilitate obtaining maximum compaction.

Consolidation

Consolidation occurs when a soil is placed under a static load. It is a consideration whenever a structure is founded on soil. Consolidation will produce some degree of settlement that may vary over the site. This differential settlement may or may not pose a problem to the structure. A metal tank for storage of water or petroleum products will be more tolerant of differential settlement than a multistory, rigid concrete-framed building. Recognition of this fact influenced construction of Interstate – 15 north of Ogden, Utah. Differential settlement could be expected along this stretch of highway because of the fine-grained sediments present and the shallow groundwater influenced by the nearby Great Salt Lake. By early placement of the interchange embankments and subgrade, most of the final load could be placed on underlying soil. Settlement was allowed to take place, with addition of material to maintain the final grade. Once settlement ceased, the rigid paved surface and other features less tolerant of settlement were added to complete construction.

Consolidation is dependent on the time necessary for water to leave the soil mass. For coarse-grained materials, drainage is sufficiently rapid, so that consolidation is essentially instantaneous. It is more important in dealing with fine-grained, saturated soils. Any load is borne initially by the water in the pore spaces as well as the solid part of the layer. Little or no compression takes place because water is incompressible. As water drains from the compressible layer, the load is transferred to the solid part of the layer.

Two factors are of prime concern in addressing consolidation. First, the total consolidation expected should be computed. Total consolidation is the amount of vertical displacement or settlement that occurs between the start and finish of consolidation. The nature of the soil rather than the size of the load is critical to the amount of consolidation. An overconsolidated clay will likely display less consolidation than an equal thickness of normally consolidated clay. Second, the amount of time required for this to occur will be important to know. Increased load will not accelerate the time of consolidation. Like total consolidation, time of consolidation depends on the nature of the soil. More time will be required for consolidation of a thick soil mass compared to a thin one. It will take less time for a soil with high permeability than one with low permeability.

Laboratory tests yield plots of void ratio to load for soils of interest. The change in void ratio can be determined directly from these tests or a compression index (C_c) can be interpreted from a series of semilogarithmic curves of void ratio to load (Figure 3.11). In either case, these values can be used to compute the expected settlement (ΔH):

$$\Delta H = \frac{C_c H_t}{1 + e_o}\left[\frac{(\text{Log } P_o + \Delta P)}{P_o}\right] \tag{3.14}$$

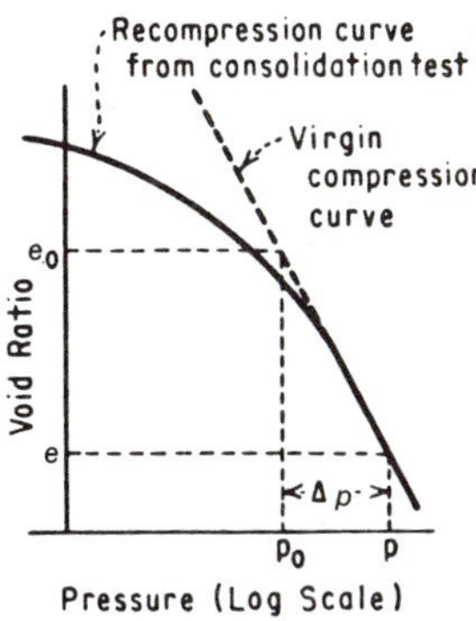

The virgin compression curve or the field consolidation curve, for clayey soils, appears on a semi-logarithmic diagram as a straight line as shown at left. This line can be represented by the equation

$$e = e_0 - C_c \log_{10} \frac{p_0 + \Delta p}{p_0}$$

in which C_c (dimensionless) is the Compression Index.
The virgin compression curve is established by extending the straight-line part of the recompression curve. By selecting two points (e_0, p_0) and (e, p) and substituting in the above equation, C_c can be determined.

$$C_c = \frac{e_0 - e}{\log_{10} \frac{p_0 + \Delta p}{p_0}}$$

(A) METHOD OF DETERMINING THE COMPRESSION INDEX, C_c

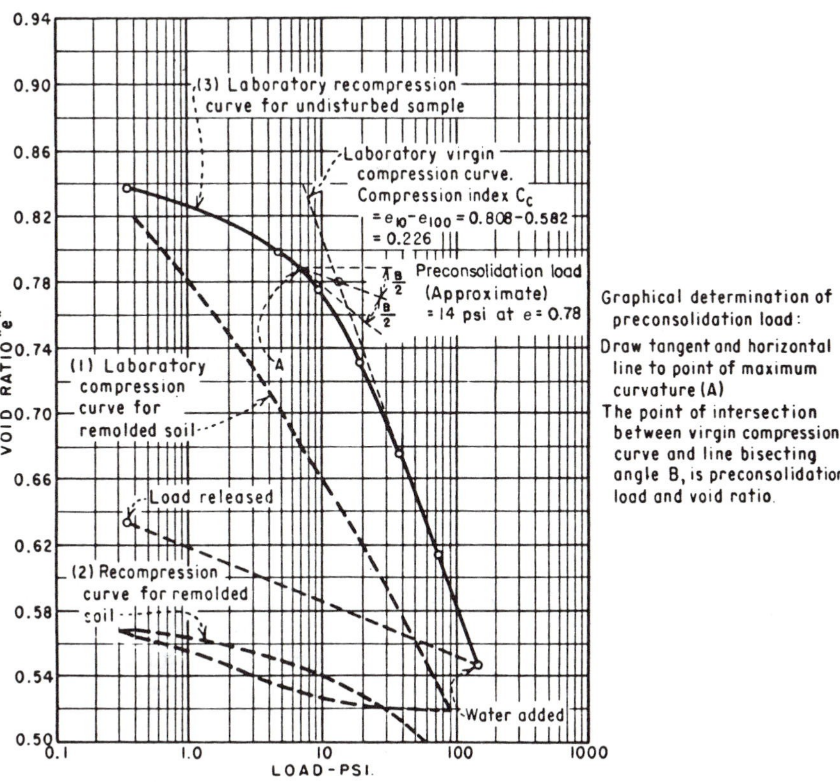

(B) VOID RATIO-LOAD CURVES AND PRECONSOLIDATION LOAD

Figure 3.11 Diagram showing plot of void ratio to load curves for determining the compression index of a soil. (From USBR, 1968)

where:

ΔH = total settlement or consolidation
C_c = compression index
H_t = height of the layer
P_o = original load
ΔP = change in load
e_o = original void ratio

The first step in computing the time for settlement or consolidation to occur is to determine the coefficient of consolidation (C_v). This is achieved by substituting the permeability of the soil and compression index in the following formula:

$$C_v = \frac{k}{C_c \gamma_w} \tag{3.15}$$

where:

C_v = coefficient of consolidation
k = permeability of the soil layer
C_c = compression index
γ_w = unit weight of water

The time for settlement can be computed for any portion of the expected total settlement. Of interest, in most cases, is the time until 90% of the settlement has occurred. The formula is:

$$t = \frac{T_v H^2}{C_v} \tag{3.16}$$

where:

t = time required for the proportion of total settlement or consolidation to occur
T_v = time factor
H = maximum length of the drainage path
C_v = coefficient of consolidation

The coefficient of consolidation is obtained by the formula in equation 3.15. The time factor depends on the percentage of consolidation of interest and drainage conditions. Figure 3.12 provides time-factor values. Module 3.2 provides examples of calculations for amount and time of consolidation using formulas in equations 3.14 and 3.16.

As noted earlier, consolidation is a time-dependent process independent of the size of the load imposed. However, means for reducing the amount of settlement or accelerating its occurrence do exist. The amount of settlement can be reduced by removing the compressible foundation soil. This is only practical where the soil layer is fairly thin. When sufficient time permits, a surcharge could be placed prior to actual construction. This surcharge might be stockpiled material for use in construction. A surcharge is applicable with a relatively thin compressible layer. Consolidation can be accelerated by combining the surcharge with different types of vertical drains. Each of these methods has advantages and disadvantages related to both the nature of the compressible layer and the constraints associated with the proposed construction that need to be considered (NAVFAC, 1982).

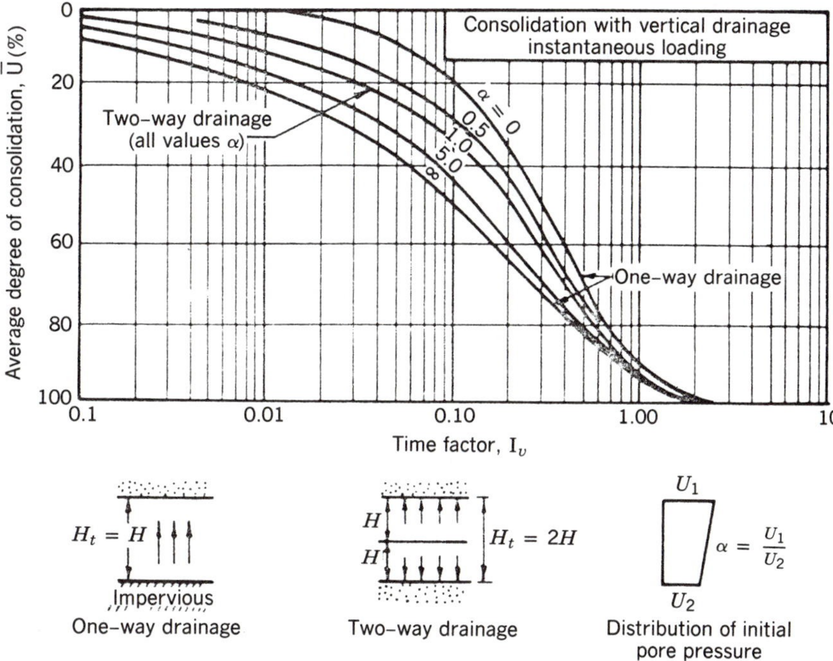

Figure 3.12 Chart for determining time rate of consolidation for vertical drainage owing to instantaneous loading. (Modified from NAVFAC, 1982)

Compaction

Compaction is achieved as soil particles become reoriented to a configuration that contains fewer voids. This reduction in voids may also include fracturing of grains and bending or distortion of individual particles. The principal purposes for compaction are to increase the strength of the soil and to reduce permeability.

The techniques and equipment used to compact soil vary with the type of project and its purpose. In general, soil is applied to the construction site in layers of a specified thickness. These layers are called lifts. This material is then repeatedly loaded or vibrated by equipment to achieve the desired compaction. Equipment used to compact the soil may simply apply pressure by rolling over the material a number of times. A tamper may be used to vibrate the material, sometimes in conjunction with a roller. Figure 3.13 shows a sheepsfoot roller being pulled over 18-in. lifts of soil to achieve compaction in building a reinforced fill on a road.

Achieving desired compaction requires controlling the material, its water content, and the energy applied to compact the soil. It is necessary to overcome frictional resistance during reorientation of solid grains and reduction of voids. The particle-size distribution of the soil used will place limits on the degree of compaction that can be reasonably achieved. Selection of material with a size distribution suitable for obtaining the compaction desired is an important step.

Water present in the soil increases resistance to reorientation of particles by the force of surface tension of water between grains. Increasing water

MODULE 3.2

A compressible soil layer 12-ft (H_t) thick is subjected to a static load of 2,000 lb (Δ_p).

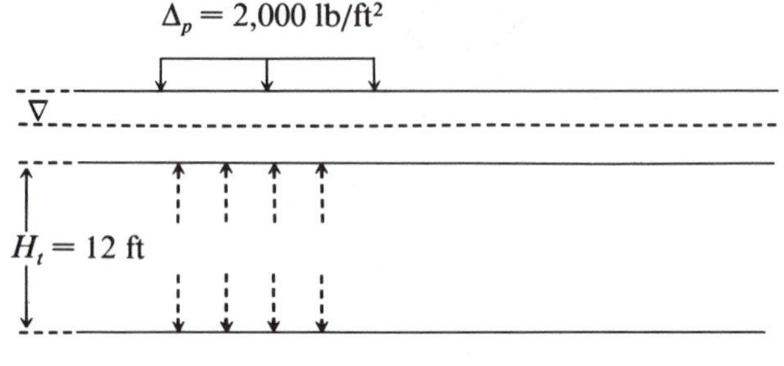

Consolidation testing of the material yields the following values:

$$e_o = 0.78$$
$$C_c = 0.23$$
$$P_o = 2{,}016 \text{ lb/ft}$$
$$C_v = 0.02 \text{ ft/day}$$
$$d = H_t/2$$

Determine the amount of consolidation that will occur:

$$\Delta H = \frac{H_t C_c}{1 + e_o} \text{Log} \frac{(P_o + \Delta_p)}{P_o}$$

$$= \frac{(12)(0.23)}{1.78} \text{Log} \left(\frac{2{,}016 + 2{,}000}{2{,}016} \right)$$

$$= \frac{(12)(0.23)}{1.78} (0.30) = 0.47 \text{ ft or } 5.58 \text{ in.}$$

Assuming two-way drainage, compute the time for 90% (0.9) consolidation to occur:

$$t = \frac{T_v d^2}{C_v} = \frac{(0.88)(6)^2}{0.02} = 1{,}584 \text{ days}$$

where:

$$T_v = 0.88 \text{ (90\% consolidation)}$$

content in the soil can reduce this frictional resistance and facilitate compaction. At the same time, the soil should be as dry as possible. Testing using the Proctor or modified Proctor compaction tests can determine the densities obtained for different water contents. Optimum water content will be identified by this testing. This is the water content of the soil when it is compacted to the maximum dry density.

Compaction testing involves placing a prepared soil sample in an apparatus and subjecting it to repeated blows. The hammer weight used and the height from which it is dropped from differ for various versions of the Proctor compaction test. For exact descriptions of this testing, consult AASHTO (1974a, 1974b). Density is determined from the sample by using a penetration-resistance test. The moisture content is computed for this

Figure 3.13 A bulldozer pulling a sheepsfoot roller to compact the initial soil layers on a reinforced fill for a mountain road.

density. A plot showing the variation in dry density compared to moisture content produces a curve that identifies the maximum dry density and associated optimum moisture content. Figure 3.14 shows a representative test result for soil upon which a single-story building is to be founded.

Laboratory results serve as a standard for field measurements. Inspection during construction will enable one to evaluate whether compaction is meeting design specifications. The specification will depend on the nature of the structure. Two common field-testing techniques are the sand-cone procedure and neutron-probe measurement. The sane-cone procedure is destructive sampling that requires excavation of a small hole in the compacted fill. The material removed from this hole is weighed and its moisture content determined in the laboratory. Dry sand of known density is placed in the hole. By relating the densities of the soil and sand, the volumes of material removed and replaced, the measured moisture content of the soil, and the maximum dry density and optimum moisture content determined by the Proctor method, the percentage of compaction can be calculated. Module 3.3 shows the necessary computations for such a test.

Field-compaction testing with a neutron probe is nondestructive sampling. It involves inserting a neutron radiation source into the compacted surface to a prescribed depth. The neutrons reaching a sensor are affected by the density and moisture of the material in the fill. The measured value is compared to established charts to determine the percentage of compaction for comparison to the maximum dry density determined in the Proctor testing of the soil used.

Shear Strength

Compared to other material used to build structures, soil is a very low-strength material. The widespread use of soil as a building material can be attributed to its abundance, availability, and ease of handling. The com-

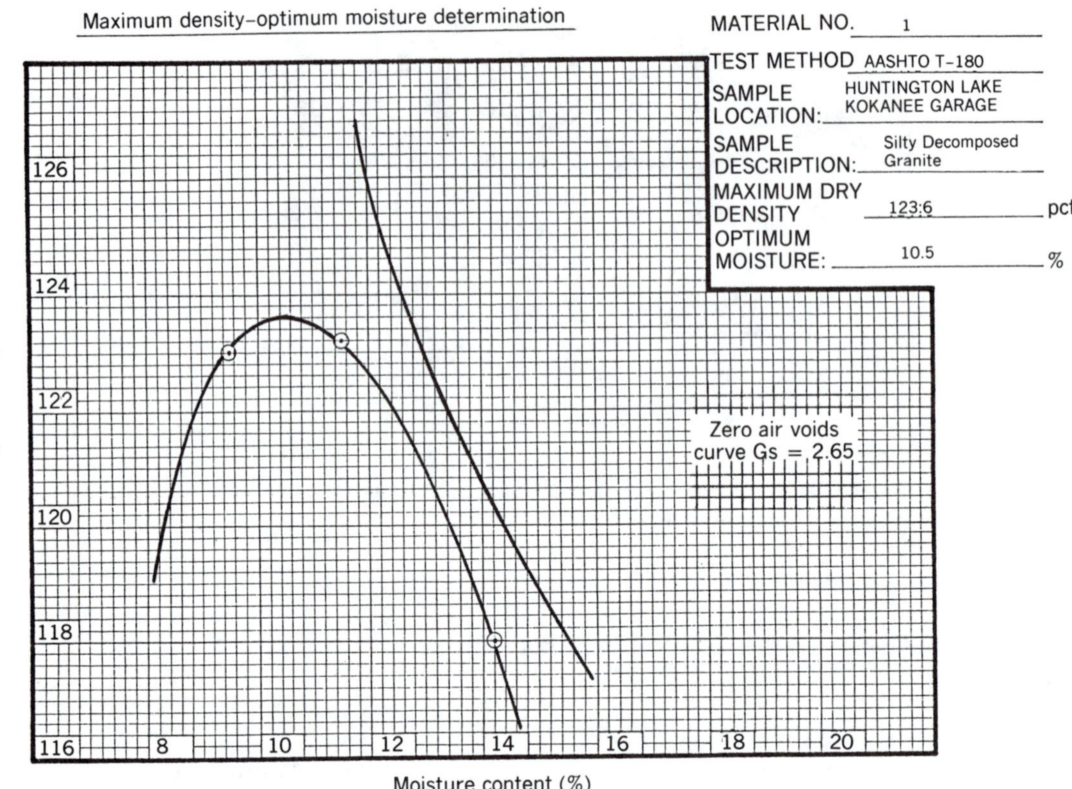

Figure 3.14 A plot of maximum density and optimum moisture determination based on Modified Proctor compaction test results for a coarse-grained soil.

mon use of soil in building fostered our understanding of strength in soils and how to calculate strength parameters. Soil strength varies greatly between different soils and even within a particular soil. Wu and Sangrey (1978) provide a concise overview of shear strength characteristics of some common soils.

The shear strengths of engineering soils have an important role in design, construction, and long-term stability of structures on, in, and with soil materials. In addition, the stability of natural slopes in soil material is dependent on soil strength.

The strengths of rocks described in chapter 4 may be given as compressive, tensile, and shear strengths, depending on the test procedures and engineering need. By comparison, the shear strength of a given engineering soil is the objective of testing procedures because it is the shear strength that is of greatest value to the engineer. Shear strengths are measured directly by direct shear testing and indirectly by uniaxial and triaxial compressive testing. These tests will be described later. The Mohr shear strength envelope described in chapter 1 is used to graphically define the shear strength characteristics of a soil for a given set of test conditions. These conditions include the use of disturbed versus undisturbed samples, soil moisture content, and moisture drainage conditions imposed either by

MODULE 3.3

A road fill is being placed and compacted. To determine whether contract specifications are being met, compaction is tested in the field by the sand-cone method. A hole is dug in the fill and 5.85 lb of soil is removed. It takes 5.32 lb of loose, dry sand with a known density of 97 pcf to refill the hole. Previous laboratory testing of the soil found its maximum dry density to be 101 pcf, optimum moisture content to be 5%, and specific gravity to be 2.7. A 58.1-gm sample of wet soil weighed 52.4 gm after oven drying.

Determine the percentage of compaction for the fill at the test site.

(1) $V_{\text{hole}} = \dfrac{5.32 \text{ lb}}{97 \text{ pcf}} = 0.05 \text{ ft}^3$

(2) $\gamma_m = \dfrac{5.85 \text{ lb}}{V_{\text{hole}}} = \dfrac{5.85 \text{ lb}}{0.05 \text{ ft}^3} = 106.7 \text{ pcf}$

(3) $W = \dfrac{58.1 \text{ gm} - 52.4 \text{ gm}}{52.4 \text{ gm}}$
$= 0.109 \times 100 = 10.9\%$

(4) $\gamma_d = \dfrac{\gamma_m}{(1+W)} = \dfrac{106.7 \text{ pcf}}{(1+0.109)}$
$= \dfrac{106.7 \text{ pcf}}{1.109} = 96.2 \text{ pcf}$

(5) $\% \text{ comp} = \dfrac{\gamma_d}{\gamma_{\max}} = \dfrac{96.2 \text{ pcf}}{101 \text{ pcf}}$
$= 0.95 \times 100 = 95\%$

testing procedures or by soil physical characteristics. The intent of testing is to duplicate the conditions to which the soil will be subjected during and following construction.

The Mohr-Coulomb failure criterion,

$$\tau_f = c + \sigma_n \tan \phi, \tag{3.17}$$

is a linear equation that defines sufficiently the shear stress for soils on a failure plane (τ_f) and the angle of internal friction (ϕ) for a given normal stress (σ_n) acting on the failure plane as shown by Figure 1.27. Shear strength at failure, τ_f, is often reported as "s" because it is a unique failure shear stress. As noted in chapter 1, (Figures 1.26 and 1.27), the angle of the failure plane, β, may be obtained graphically from the Mohr circle diagram. Another notation used for this angle is θ, as in the following equation:

$$\theta = 45° + \phi/2 \tag{3.18}$$

The additional factor, c, in equation 1.19 is the value for cohesion, which when added to $\sigma_n \tan \phi$ results in the shear strength, τ_f, at failure for a given test.

Engineering soils exhibit differences in physical characteristics that have an effect on the graphical representation of test results by the Mohr circle diagram as well as on the components of equation 1.19. Earlier, we were introduced to the fact that soils can be cohesive and noncohesive. Cohesive soils are silts and clays that display cohesion between the soil particles and have a value of $c > 0$. Noncohesive or cohesionless soils are sands and gravels, or granular soils, for which $c = 0$. The Mohr failure envelope constructed from triaxial testing of noncohesive soils is characterized by Figure 3.15, where $c = 0$. Equation 1.19 becomes

$$\tau_f = \sigma \tan \phi \tag{3.19}$$

The practical aspect of this is that granular soils exhibit shear strengths only when confined and in amounts proportional to the confining stress and controlled by the angle of internal friction.

Shear Strength in Noncohesive Soils

The shear strength of noncohesive soils is determined by a variety of factors that can be separated into two groups. One group includes factors related to sample preparation and test procedures. The other set of factors is dependent on the physical properties of the soil. In the first group, the void ratio to which a soil has been compacted and the confining stresses during testing are the principal factors influencing shear strength of a given soil. Interacting with these are a combination of factors resulting from the physical characteristics of a given soil. These include particle size, particle shape, particle surface roughness, and gradation, or grain-size distribution —all of which control the interlocking of the particles. The denser the packing (lower void ratio), the greater the influence of interlock on the angle of internal friction and the corresponding shear strength (Figure 3.16). Soil physical characteristics and packing density, therefore, influence the frictional resistance to relative particle movement as shown by variations in the angle of internal friction. For a given set of test conditions, that is, void ratio and confining stresses, it will be the physical state of the soil that will set one soil apart from another in terms of angle of internal friction and corresponding shear strength.

If a granular, or noncohesive, soil contains water in the pore spaces, or voids, there is little or no significant influence on the angle of internal friction provided pore-water pressures can dissipate as loading takes place. This is commonly referred to as a drained state. If pore water pressures

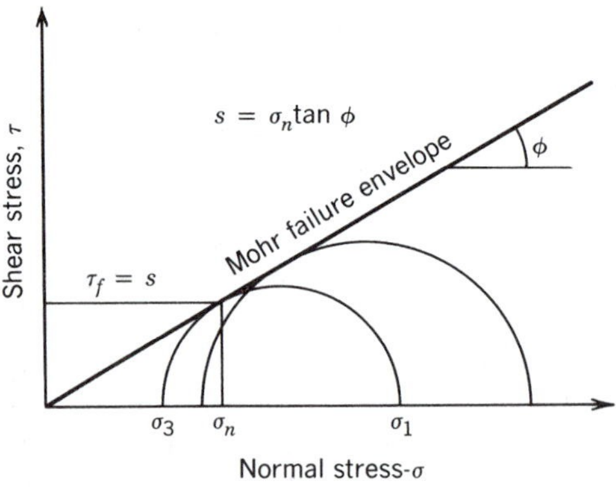

Figure 3.15 Typical Mohr diagram for a noncohesive soil.

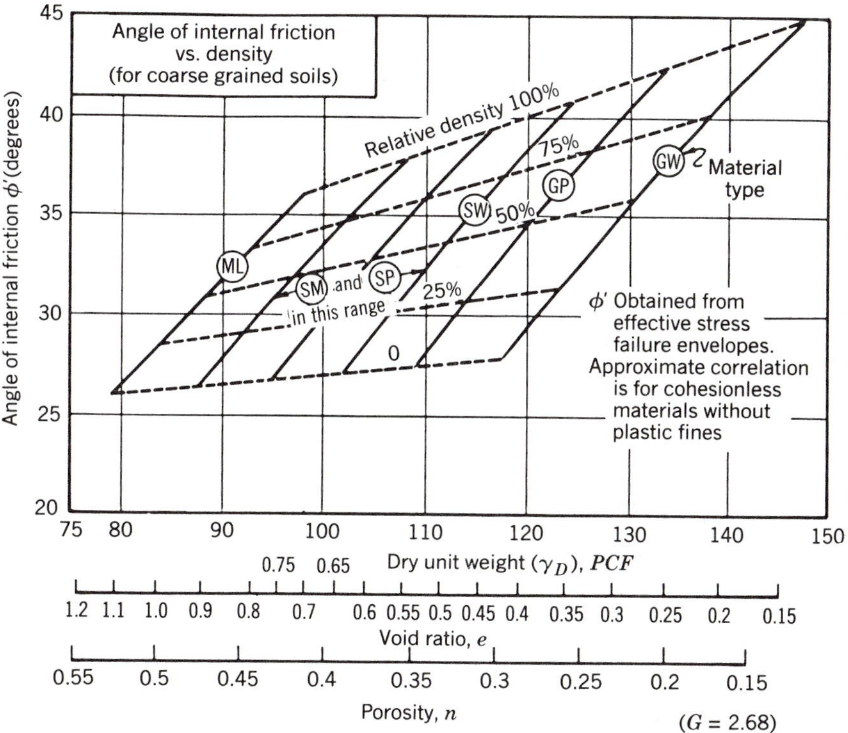

Figure 3.16 Dependence of angle of internal friction on density of cohesionless soils. See Table 3.6 for definition of soil symbols used. (From NAVFAC, 1982)

increase with loading, significant changes occur in the angle of internal friction. This is an undrained state. The resulting hydrostatic stresses developed in the pore water act against the loading and confinement imposed on the soil particles. The result is effective stresses on the sample that are less than those imposed by loading and confinement by an amount equal to the pore-water pressure. The Mohr-Coulomb equation for this condition becomes:

$$\tau_f = (\sigma_n - \mu) \tan \phi \tag{3.20}$$

where μ is the pore pressure.

Figure 3.17 illustrates the changes occurring in the stress states, that is, σ_1 versus σ_1', as a result of pore pressure. This figure graphically shows why proper drainage and monitoring of pore pressure by instruments are important factors in the construction and maintenance of soil structures such as embankments.

Soil-testing procedures provide for examining both the drained and undrained states, so that design and construction that involve granular soils utilize data compatible with site conditions. Granular materials subject to pore-pressure buildup, especially under dynamic loading, are silty sands and fine sands that have lower permeabilities than coarser materials.

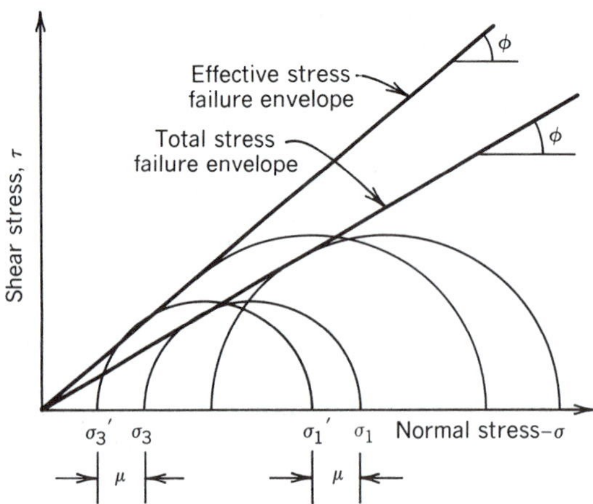

Figure 3.17 Mohr diagram showing influence of pore pressure, μ, on axial and confining stresses, σ_1 and σ_3, and resultant effective stress failure envelope.

Shear Strength in Cohesive Soils

Cohesive soils exhibit plasticity in varying degrees, depending on particle size, moisture content, and for the clays, their mineralogy. Plasticity index PI (defined earlier) is a measure of the moisture-content range over which a cohesive soil is plastic. Silts are less plastic than clays and have lower PI values. Clays characteristically are more plastic and have greater ranges of PI values owing to their mineralogy (as explained earlier).

In cohesive soils, the plasticity index is inversely proportional to the shear strength. Given the general statement that soils with high liquid limits have the least strength, the remaining controlling factor for a given soil is the amount of moisture present in the soil. When sampling a soil for testing, it is imperative to retain the original moisture content. This is achieved by sealing the sample or the ends of a sampling tube with wax or a similar material. It is logical and correct to assume that a reduction in moisture content at a site is a major factor in improving cohesive soil strength even in montmorillonite-rich soils.

Factors other than particle size and mineralogy influence shear strengths of cohesive soils. As in the case of noncohesive soils, decreases in void ratio result in increases in shear strength. This is of greater importance when one is confronted with clay soils. The platy clay mineral aggregates typically are deposited randomly that result in high void ratios (Figure 3.6). This makes clay soils highly susceptible to reductions in volume, or consolidation, from a variety of causes.

A normally consolidated soil is one that has not been subjected to effective stresses greater than those occurring at the time of sampling. If the effective stresses have been higher in the past, the soil is overconsolidated, resulting in a reduction in void ratio and an increase in strength. Overconsolidation can occur for many reasons. A common geologic cause is removal of overburden pressure by erosion of the overlying material or, in the case of glaciation, by melting of the overlying ice sheet. Other causes include variations in moisture content over time and construction loads.

Most clay soils retain approximately the same strength characteristics in an undisturbed state as when they have been mechanically disturbed or remolded. Some unique clays, however, undergo significant reductions in strength after remolding. An extreme case occurs when the sample liquifies, having contained water in excess of the liquid limit while in a stable state. Clays that exhibit reduced remolded strengths are classed as sensitive clays.

Sensitivity is attributable to the structure or arrangement of particles in a clay soil. As stated earlier, the deposition of clay typically results in deposition of clusters, or flocs, of clay particles. Clay sensitivity is determined quantitatively in terms of shear strength as follows:

$$\text{Sensitivity} = \text{undisturbed/remolded} \tag{3.21}$$

A highly sensitive or quick clay that liquifies on remolding typically has a sensitivity value of 20. An insensitive clay by comparison ideally has a value of 1. Remolding of highly sensitive clays by testing or construction or by natural occurrences such as earthquakes can result in liquifaction of the clay mass and catastrophic failure. Notable examples are described by Mitchell and Klugman (1979) and Gregersen and Loken (1979).

A number of test conditions are employed when determining the engineering properties of clay-rich soils. These include various combinations of soil consolidation and drainage conditions during testing. Additional information about these test conditions and many other aspects of cohesive soil-engineering properties can be found in texts on soil mechanics and geotechnical engineering such as Lambe and Whitman (1969), Dunn et al. (1980), and Das (1985).

Measuring Shear Strengths

In the preceding section, direct shear and triaxial testing were given as a means of obtaining shear strengths of soil samples. Selection of the testing method depends on the type of material (cohesive or noncohesive) and conditions of drainage and confinement required for a given soil use. Both are laboratory tests with provisions for control of test conditions and gathering of test data.

As implied by its name, direct shear testing involves the application of shear stress directly to a sample rather than indirectly as in the case of compressive testing. The shear stress required for failure, s or τ_f, is obtained for various values of stress that are normal to the shear plane, σ_n, providing coordinates for construction of a Mohr shear strength envelope and measurement of the internal angle of friction, ϕ.

Direct shear equipment is the same as the indirect shear box illustrated in Figure 4.36. Instead of a discontinuous rock sample, a soil sample is placed in the box. Only shear stresses until failure for given normal stresses are measured during soil testing.

The direct shear test has a number of limitations. It has no provision for confinement of a sample or for testing in an undrained state. Changing normal stresses may simulate various states of axial stress at failure for corresponding confinement, but lack of control of pore pressures is a serious handicap. However, the direct shear test is commonly used with granular soils where the influence of pore pressure is more predictable than for cohesive soils. Granular, or noncohesive, soils typically are disturbed during sampling, so that loading of the shear box and sample preparation at a desired density and moisture content are not as critical as with undisturbed cohesive soils.

The triaxial test is performed in a cell that permits control of compressive and confining stresses (Figure 3.18). The confining stresses, σ_2 and σ_3, are equal because confinement is achieved by air or hydraulic pressure in a chamber surrounding the sample. The unconfined compressive strength of a sample is a limiting case in the triaxial testing where $\sigma_3 = 0$. The compressive strength for each confining stress is used to construct a Mohr diagram from which the angle of internal friction is obtained as well as the shear strength, s or τ_f, for each test. Shear strength or the shear stress along a plane at failure is measured indirectly (described in chapter 1) as a result of application of axial or compressive stresses.

The triaxial test is used for both noncohesive and cohesive soils because of its versatile test conditions. It is used almost exclusively for testing cohesive soils. Undisturbed samples are used whenever possible because of the effects remolding has on cohesive soils. Sample void ratio can be measured before and after testing to determine the degree of consolidation resulting from testing. Moisture content normally is that of the natural soil, but it may be adjusted prior to testing. A sample may be prepared to duplicate desired construction or postconstruction conditions prior to testing.

The results of triaxial testing of a cohesive soil are illustrated by Figure 3.19. The tests were conducted on undisturbed samples with original moisture contents of approximately 19%. The soil samples' liquid limits averaged 38%, with a plasticity index of 17%. These values conform to the

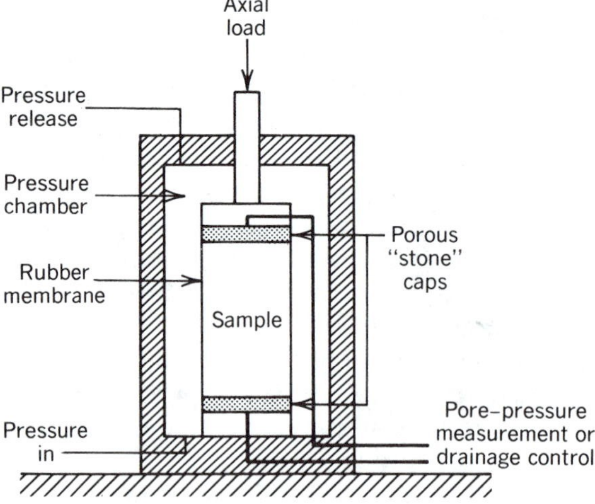

Figure 3.18 Schematic diagram of a triaxial cell.

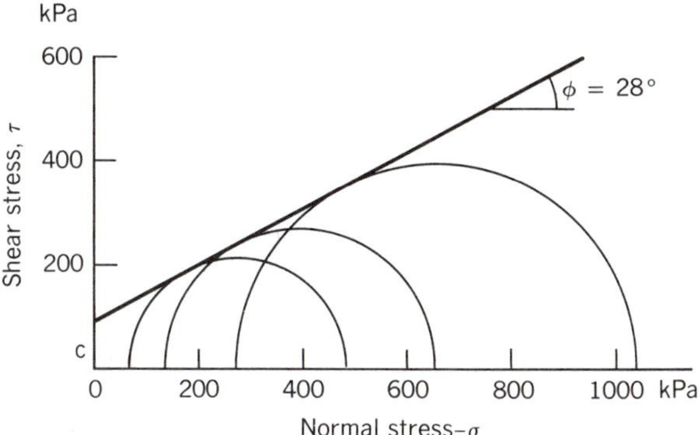

Figure 3.19 Mohr diagram for the CL soil described in text.

definition of a CL soil as defined in the discussion of the USC system. The compressive strengths for the three confining stresses, the shear strengths with increasing confinement, and the 28° angle of internal friction are all characteristic of low-plasticity cohesive soil.

The response of a cohesive soil to differing degrees of consolidation and drainage during testing is of paramount interest to the engineer. The triaxial cell provides for a variety of combinations of consolidation and drainage tests unavailable by any other test procedure. These test procedures are described in the texts referenced on pp. 124–125.

USES OF SOIL SCIENCE CLASSIFICATION

Soil science emphasizes soil as a component of the biological system (Soil Survey Staff, 1975). Classification in soil science focuses on understanding how to use soil to produce food and fiber. As a consequence, factors considered important to distinguishing different soils are those that play a role in observing and correcting problems of agricultural soil uses. These factors are largely confined to the upper layer of soil most affected by climate and biologic activity. Discriminating between different soil types under this classification system rests mainly on the characteristics observable in the upper 1 m or 2 m of soil.

Soil Development and Classification

Jenny (1941) defined the five basic factors responsible for soil formation: (1) climate, (2) biotic activity, (3) parent material, (4) topography, and (5) time. Climate exerts an obvious control on weathering through the temperature and amount of water involved with the process. Biotic activity will influence soil through the alteration of the effectiveness of weathering and the addition of organic material to the chemical activity occurring. Parent material, as the medium being altered, places constraints on the range of end products that may result from weathering. For example, the resulting soil will depend on the stability of minerals that make up the parent material and its texture (grain size) (Birkeland, 1984). Topography will influence soil formation through slope orientation and steepness. Orientation affects microclimate and vegetation, varying the way climate and biotic activity may affect the same parent material. Slope steepness influences the degree to which soil particles stay or move between points on the landscape. Time is often the factor of greatest interest to the engineering geologist. A soil will only form if some amount of time passes that permits the other four factors to interact.

Soil formation is clearly a dynamic process resulting from the interaction of the five soil-forming factors. An engineering geologist must recognize that the soil found at a particular location is the product of additions and removals at the ground surface as transformations and transfers within the soil (Simonson, 1978). Figure 3.20 represents this activity. It should be noted that removals must occur at a rate slower than soil formation in order for differentiation to occur.

The taxonomy of soil classification as currently practiced in the United States employs the distinctive layers or horizons created by soil formation.

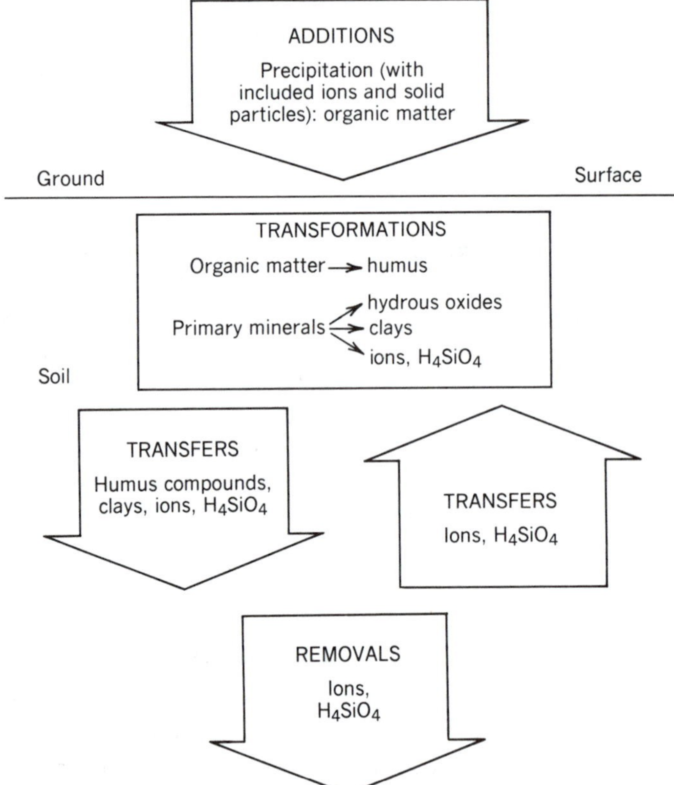

Figure 3.20 Diagram illustrating the additions and removals occurring in soil formation. (Reprinted with permission from Soils and Geomorphology, P. W. Birkeland, 1984, Oxford University Press. As adapted from R. W. Simonson, a multiple-process model of soil genesis, 1978, in Quarternary Soils: Geo Abstracts, University of East Anglia, by permission from © W. C. Mahaney, ed.)

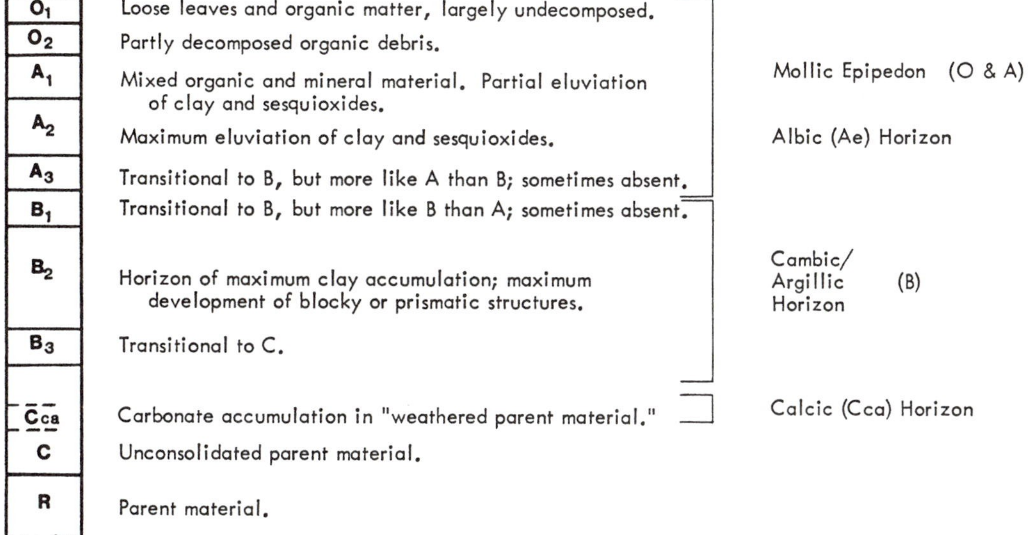

Figure 3.21 Soil profile with some of the diagnostic horizons used in soil science classification represented. (From R. J. Shlemon, Application of Soil-Stratigraphic Techniques to Engineering Geology, *Bulletin of the Association of Engineering Geologists*, Vol. 22, No. 2, May 1985, pp. 129–142. (*Used by Permission*).)

Presence or absence of diagnostic horizons and their physical characteristics define the different soils in this taxonomy. The vertical stacking of soil horizons forms the soil profile (Figure 3.21). A soil is classified according to the horizons and properties in that soil profile. Soils extend over an area in addition to their presence in a vertical section. This is analogous to recognizing that a particular rock unit described at an outcrop extends over some area around that location. Lateral variation in a soil and among adjacent soils across the landscape is called a *soil catena* or *toposequence*.

It is inappropriate to fully describe the basics of soil taxonomy here. The Soil Survey Staff (1975) provides a detailed and readable manual for this purpose. Initially, the taxonomic names may seem quite strange. Soil names reflect the hierarchical nature of this classification. Table 3.14 gives several examples. Like stratigraphic nomenclature for rock units, there are specific rules for naming individual soils. These rules enable engineering geologists to interpret soils classified under this system, providing them with much useful information.

Applications to Engineering Geology

Soil science classification can be applied to engineering geology in two ways. The first is through the use of soil survey mapping. This involves gleaning information usable in an engineering geologic investigation from data that are compiled for a very different purpose. The second way is through soil stratigraphy. In this latter application, soil classification techniques are used to produce engineering geologic data.

Table 3.14 Some Soil Science Classification Names Illustrating Different Taxonomic Levels

Order	Suborder	Great group	Subgroup	Family	Series
Alfisols					
Aridisols					
Entisols	Aquents				
	Arents				
	Fluvents	Cryofluvents	Typic Cryofluvents	Coarse-loamy, mixed, acid	Susitna.
		Torrifluvents	Typic Torrifluvents	Fine-loamy, mixed (calcareous), mesic.	Jocity and Youngston.
			Vertic Torrifluvents	Clayey over loamy, mixed, (calcareous) hyperthemic.	Glamis.
	Orthents	Cryorthents	Typic Cryorthents	Loamy-skeletal, carbonatic	Swift Creek.
			Pergelic Cryorthents	Loamy-skeletal, mixed (calcareous)	Durelle.
Histosols					

Source: Soil Survey Staff, 1975.

Soil Survey Mapping

Soil classification is used to produce soil survey mapping. These surveys are generally on a countywide basis and are carried out by soil scientists of the Soil Conservation Service or other governmental agencies such as the Forest Service.

One use of these maps is to produce a surficial geology map suitable for geologic purposes. Traditional geologic mapping emphasizes bedrock relationships that often provide little information on the nature of overlying unconsolidated material. But the nature of the surficial materials may play a more important role in road construction, building foundations, waste-disposal siting, and canal construction than the bedrock underneath. Prokopovich (1984) provides a good illustration of this use of soil survey mapping. By grouping mapped soil units in a soil survey for eastern Fresno County, California, he was able to generate a map of surficial materials with units that reflected: (1) composition of parent material, (2) age, and (3) topographic-geomorphic location. These surficial geology maps were used in reconnaissance and feasibility studies for several canals. Then, for a minimal amount of time and expense, alternative locations over large areas can be compared by taking into account considerations such as the likelihood of encountering perched water tables or potential salt crystallization problems. Thus, areas where more detailed study might be necessary can be identified early in project planning.

Another use of soil survey maps is to identify potential problems or favorable conditions in siting specific projects. Most soil surveys include some engineering test data for each type of soil mapped: Typical data are particle-size analysis, liquid limit, optimum moisture content, and so on (see Table 3.15). In addition, the relationship of a classified soil to the AASHO or USC systems may be provided. Soil surveys also commonly include interpretations of the suitability or potential problems that might be expected for different uses of mapped soil units. Although there are insufficient data to permit decisions on design recommendations, such surveys provide the geologist with an indication of what to expect (Table 3.16). Thus, the kind of factors to be included in more detailed investigation can be identified by use of these interpretations.

Soil Stratigraphy

Soil stratigraphy can be applied in engineering geology to problems that require dating of some critical process or reconstructing the geomorphic history for a particular locality (Shlemon, 1985). Both modern soils and paleosols can be used. Paleosols are soils formed in the past and subsequently buried, effectively stopping further soil formation. Three types of paleosols are found: (1) the buried paleosol under deposits of a younger age, (2) exhumed paleosols that were buried but that are now exposed by erosion, and (3) relict paleosols that are preserved on old stable surfaces which represent strong development under different conditions than present (Shlemon, 1985). Use of soils as stratigraphic units employs the basic principles of stratigraphy taught to all geologists (Birkeland et al., 1971; Finkl, 1980).

Several approaches are typically used in determining the relative age of processes or sediments. Age may be based on certain surface or subsurface horizons or on the relative degree of development present among soils in an area. The organic horizon of a modern or paleosol soil can be used for radiocarbon dating. These bulk samples yield a mean residence time for that soil (Shlemon, 1985). This represents the minimum age of the soil. Soils will be younger than the surfaces or deposits on which they are formed and older than any overlying deposits. Therefore, the mean residence time for soil makes it a useful marker for distinguishing among the deposits or surfaces where it is present.

Under certain climatic conditions, a diagnostic subsurface horizon may form within the soil profile. The thickness of this profile and its appearance may coincide in a few localities with datable material. Coexistence with datable material enables time to be related to these horizon characteristics. Once this is done, the horizon can be used as a time marker in areas where these datable materials are absent. Machette (1978) provides an example of this technique. He was able to calibrate the age of calcareous paleosols; this permitted dating fault movement near Albuquerque, New Mexico.

In some areas, sufficient study can establish a soil chronosequence, which relates time to the degree of soil development among a number of soils present in an area. Muhs (1982) illustrates the type of soil-forming factors and considerations required to successfully establish a soil chronosequence. His work with soil developed on Quaternary-age marine terraces on San Clemente Island, California, included the use of radiocarbon,

Table 3.15 Engineering Data for the Academy Soil from the Eastern Fresno Area Soil Survey

Soil Series and Map Symbols	Depth to—		Shrink-swell Potential	Depth from Surface (typical profile)	Classification	
	Bedrock or Hardpan	Seasonal High Water Table			Dominant USDA Texture	Unified
	Feet	*Feet*		*Inches*		
Academy: AaA, AaB,	1½–4½	>10	Moderate	0–12	Loam	ML or CL
			Moderate	12–30	Clay loam	CL
				30	Mixed sandy loam material (consolidated)	

Classification—Continued	Percentage Passing Sieve—			Permeability	Available Water Capacity	Reaction	Salinity
AASHO	No. 4 (4.7 mm)	No. 10 (2.0 mm)	No. 200 (0.074 mm)				
				Inches per hour	*Inches per inch of soil*	*pH value*	
A-4	95–100	90–100	50–60	0.8–2.5	0.18–0.20	5.6–6.5	<1
A-6	95–100	90–100	60–70	0.05–0.2	0.24–0.26	6.1–7.3	<1

Source: SCS, 1971.

Table 3.16 Recommendations on Suitability and Uses for the Academy Soil from the Eastern Fresno Area Soil Survey

Soil Series and Map Symbols	Suitability as Source of—				Hydrologic Soil Group
	Topsoil	Sand	Gravel	Road Fill	
Academy: AaA, AaB	Fair: Clay loam less than 40 inches thick.	Unsuitable: More than 50 percent material passing No. 200 sieve.	Unsuitable: Less than 5 percent gravel throughout profile.	Fair and poor: A-4 and A-6.	C

Soil Features Affecting—				
	Water Retention			
Road Location	Embankments	Reservoir Area	Agricultural Drainage	Irrigation
In many places a temporary perched water table develops above dense subsoil.	Fair to good stability; medium to high compressibility.	Low seepage; underlying material broken by vertical and horizontal cracks in places.	Slow permeability	Moderate to high water-holding capacity; slow intake rate.

Source: SCS, 1971.

uranium-series, and amino acid racemization techniques to provide time values for differing degrees of soil development.

Fault displacement is one of the important processes to which soil stratigraphy is applied. Adequate design for a structure requires knowing whether any fault present at a site should be considered active and the likelihood of future movement on an active fault. Spellman et al. (1984) employed relative soil development to determine that one of four faults present at the site of a proposed water-filtration plant was active. Shlemon (1985) was able to date the last displacement of slip surfaces at a nuclear reactor site on the basis of a buried paleosol, a stone line truncating that paleosol, and a modern soil with an organic horizon (Figure 3.22).

The geomorphic history of a locality may be important to siting a facility. This is especially true for hazardous waste sites and nuclear-waste repositories that are expected to remain intact for extended periods of time. Janda and Croft (1967) provide an illustration of how the geomorphic history of a locality can be understood by use of a soil chronosequence. They established a chronosequence in the northeastern San Joaquin Valley of California that permitted alluvial deposits in the valley to be correlated to glacial stages in the nearby Sierra Nevada. Figure 3.23 summarizes the chronosequence. This history could be used as a basis for identifying relative site stability where long-term geomorphic stability is a consideration.

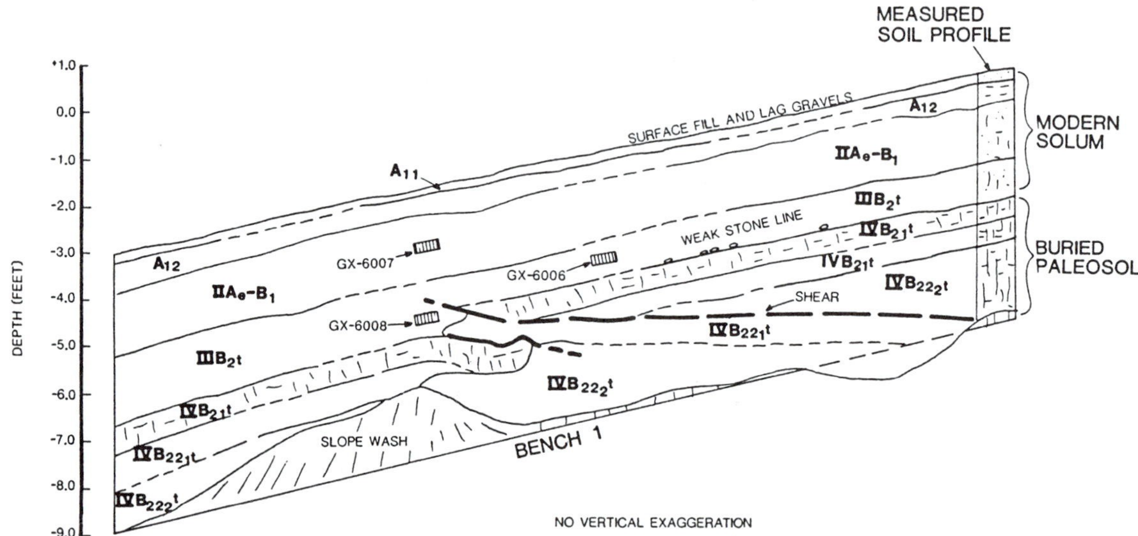

Figure 3.22 Cross section showing soil relationships at the GETR site. (From R. J. Shlemon, Application of Soil-Stratigraphic Techniques to Engineering Geology, *Bulletin of the Association of Engineering Geologists*, Vol. 22, No. 2, May 1985, pp. 129–142. (*Used by Permission*).)

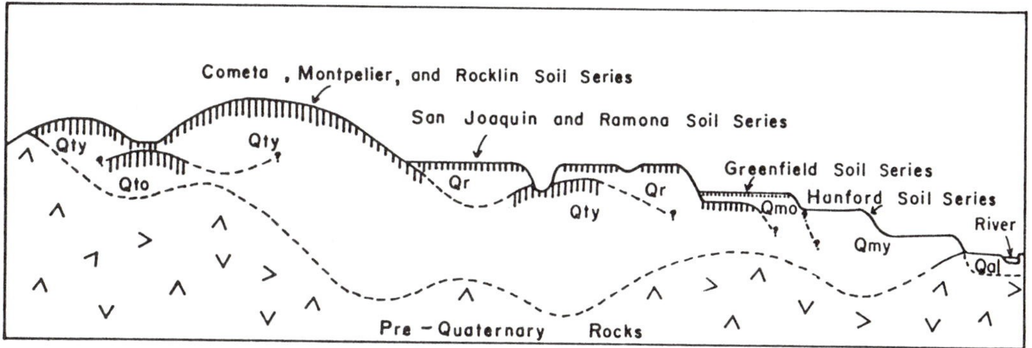

Figure 3.23 Diagrammatic representation of the soil chronosequence for the northeastern San Joaquin Valley, California. Names are the soil series most diagnostic of the Quaternary formations. Qto, older portion of Turlock Lake Formation. Qty, younger portion of Turlock Lake Formation. Qr, Riverbank Formation. Qmo, older portion of Modesto Formation. Qmy, younger portion of Modesto Formation. Qal, Recent alluvium. (From Janda and Croft, 1967)

SUMMARY

The abundance and widespread presence of soil makes it an important construction material. Unless placed on rock, most engineering works must be founded on soil. Soil displays considerable variability in its characteristics and properties owing to a variety of geologic factors.

Volume and weight relationships are the basis for describing soils. Individual characteristics such as unit weight, void ratio, and water content help define relationships that form the basis for engineering properties. Other soil characteristics permit classification for rapid identification of soils that are suitable or unsuitable for various uses.

Engineering properties of soil define the suitability of a soil for a particular use. They facilitate estimating the response of that soil to different conditions that may be applied to it as a construction material or foundation. Engineering properties of special note are compressibility and shear strength. Compressibility refers to the reduction in volume a soil may undergo. One form of compressibility is consolidation because of a static load (e.g., a building) being placed upon the soil. Another form is compaction: The reduction of void spaces by mechanical means in constructing with soil, which is achieved by repeated loading.

The engineering geologist applies geologic principles to the use of soils in engineering works. The physical characteristics and engineering properties of a soil are derived from the parent material and the geologic processes that produced it. Weathering produces soil development usable in dating and mapping soils of interest. A geologic perspective facilitates effective exploration for soils both with desirable qualities and in sufficient quantities needed for various projects. It also permits identification of special characteristics critical to the design and operation of engineering works.

REFERENCES

AASHO, 1961, The classification of soils and soil-aggregate mixtures for highway construction purposes, *in* Standard specifications for highway materials and methods of sampling and testing, 8th ed., Part 1, Specifications, Am. Assoc. State Highway Officials, pp. 45–51.

AASHTO, 1974a, The moisture-density relations of soils using a 5.5-lb (2.5-kg) hammer and a 12-in. (305-mm) drop: Am. Assoc. State Highway and Transp. Officials Designation T 99–74, *in* Standard specifications for transportation materials and methods of sampling and testing, 11th ed., Part II, Methods of sampling and testing, Am. Assoc. State Highway and Transp. Officials, pp. 301–308.

———, 1974b, Moisture-density relations of soils using a 10-lb (4.54-kg) hammer and an 18-in. (457-mm) drop: Am. Assoc. State Highway and Transp. Officials Designation T 180–74, *in* Standard specifications for transportation materials and methods of sampling and testing, 11th ed., Part II, Methods of sampling and testing, Am. Assoc. State Highway and Transp. Officials, pp. 601–608.

ASTM, 1983, Standard test method for classification of soils for engineering purposes: Am. Soc. for Testing and Mater., ASTM Designation D 2487–69, Annual book of ASTM standards, Sec. 4, Vol. 04.08, pp. 392–396.

———, 1985, Standard test method for classification of soils for engineering purposes: Am. Soc. for Testing and Mater., ASTM Designation D 2487–83, Annual book of ASTM standards, Sec. 4, Vol. 04.08, pp. 395–408.

Birkeland, P. W., Crandell, D. R., and Richmond, G. M., 1971, Status of correlation of Quaternary stratigraphic units in the western conterminous United States: Quat. Res., Vol. 1, pp. 208–227.

Birkeland, P. W., 1984, Soils and geomorphology: Oxford Univ. Press, New York, 372 pp.

Bowles, J. E., 1979, Physical and geotechnical properties of soils: McGraw-Hill, New York, 478 pp.

Casagrande, A., 1948, Classification and identification of soils: Am. Soc. Civ. Eng., Trans., Vol. 113, pp. 901–930.

Das, B. M., 1985, Principles of geotechnical engineering: PWS Engineering, Boston, Mass., 571 pp.

Dunn, I. F., Anderson, L. R., and Kiefer, F. W., 1980, Fundamentals of geotechnical analysis: John Wiley & Sons, New York, 414 pp.

Finkl, W. W., 1980, Stratigraphic principles and practices as related to soil mantles: catena, Vol. 7, pp. 169–194.

Gillott, J. E., 1968, Clay in engineering geology: Elsevier, 296 pp.

Gregersen, O., and Loken, T., 1979, The quick-clay slide at Baastad, Norway, 1974: Eng. Geol., Vol. 14, pp. 183–196.

Hough, B. K., 1969, Basic soils engineering, 2nd ed.: Ronald Press, New York, 634 pp.

Janda, R. J., and Croft, M. C., 1967, The stratigraphic significance of a sequence of noncalcic brown soils formed on the Quaternary alluvium of the northeastern San Joaquin Valley, California: *in* Quaternary soils, Desert Res. Inst., Univ. of Nevada, Reno, pp. 157–190.

Jenny, H. 1941, Factors of soil formation: McGraw-Hill, New York, 281 pp.

Lambe, T. W., and Whitman, R. V., 1969, Soil mechanics: John Wiley & Sons, New York, 553 pp.

Liu, T. K., 1967, A review of engineering soil classification systems: Highway Res. Rec., No. 156, pp. 1-22.

Machette, M. N., 1978, Dating Quaternary faults in the southwestern United States by using buried calcic paleosols: J. Res. U.S. Geol. Surv., Vol. 6, pp. 369-381.

Matula, M., 1981, Recommended symbols for engineering geological mapping: IAEG Comm. on Eng. Geol. Mapping, Intl. Assoc. Eng. Geol. Bull., No. 24, pp. 227-234.

Mirza, C., 1982, A case for the extension of the unified soil classification system: Can. Geotech. J., Vol. 19, pp. 388-391.

Mitchell, R. J., and Klugman, M. A., 1979, Mass instabilities in sensitive Canadian soils: Eng. Geol., Vol. 14, pp. 109-134.

Muhs, D. R., 1982, A soil chronosequence on Quaternary marine terraces, San Clemente Island, California: geoderma, Vol. 28, pp. 257-283.

NAVFAC, 1982, Design manual: soil mechanics: U.S. Dept. of Defense, NAVFAC DM-7.1, Dept. of the Navy, Washington, D.C., 360 pp.

Obermeier, S. F., and Langer, W. H., 1986, Relationships between geology and engineering characteristics of soils and weathered rocks of Fairfax County and vicinity, Virginia: U.S. Geol. Surv. Prof. Paper 1344, 30 pp.

Prokopovich, N. P., 1977, Discussion: The unified soil classification system: Bull. Intl. Assoc. Eng. Geol., Vol. 14, No. 3, pp. 183-185.

———, 1984, Use of agricultural soil survey maps for engineering geology mapping: Bull. Assoc. Eng. Geol., Vol. 21, pp. 437-447.

SCS, 1971, Soil survey of eastern Fresno area, California: U.S. Dept. of Agriculture, Soil Conservation Service, Washington, D.C., 323 pp.

Shlemon, R. J., 1985, Application of soil-stratigraphic techniques to engineering geology: Bull. Assoc. Eng. Geol., Vol. 22, pp. 129-142.

Simonson, R. W., 1978, A multiple-process model of soil genesis: *in* Quaternary soils, Geo Abs., Univ. of East Anglica, Norwich, England, pp. 1-25.

Soil Survey Staff, 1975, Soil taxonomy: U.S. Dept. of Agriculture, Soil Conservation Service, Agricultural Handbook 436, U.S. GPO., Washington, D.C., 754 pp.

Spellman, H. A., Stellar, J. R., Shlemon, R. J., Sheahan, N. T., and Mayeda, S. H., 1984, Trenching and soil dating of Holocene faulting for a water filtration plant site, Sylmar, California: Bull. Assoc. Eng. Geol., Vol. 21, pp. 89-100.

Terzaghi, K., and Peck, R. B., 1967, Soil mechanics in engineering practice, 2nd ed.: John Wiley & Sons, New York, 729 pp.

USAE, 1953, The unified soil classification system: U.S. Army Eng. Waterw. Exp. Stn., Tech. Memo. No. 3-357, Vol. 1, 30 pp.

USBR, 1953, Unified soil classification system: U.S. Bur. Reclamation, Supplement to the earth manual, Denver, Colo., 26 pp.

———, 1968, Earth manual: U.S. Bur. Reclamation, Denver, Colo. 783 pp.

———, 1974, Earth manual, 2nd ed.: U.S. Bur. Reclamation, Denver, Colo., 810 pp.

Wu, T. H., and Sangrey, D. A., 1978, Strength properties and their measurement, *in* Landslides, analysis and control, Spec. Rep. 176, Transp. Res. Bd. Natl. Acad. Sci., Washington, D.C., pp. 139-154.

4

Engineering Properties of Rocks

Rock is involved in many civil engineering projects. Rock properties inherently are a part of the exploration, design, construction, and in-service (postconstruction) phases of such projects. The rock classifications in chapter 1 (Tables 1.1, 1.2, and 1.3), suggested by the International Association of Engineering Geology (IAEG), provide rock names and geologic characteristics for most engineering applications. In addition to the rock properties included in the classifications, engineering uses of rock require a more generic subdividing of rocks into two groups. These are intact rock and rock mass. *Intact rock* is the term applied to rock containing no discontinuities such as joints and bedding. It is synonymous with the term *rock material.* As the name implies, a *rock mass* is a mass of rock interrupted by discontinuities, with each constituent discrete block having intact rock properties. Significant engineering-related differences exist between these two groups.

INTACT ROCK

An intact specimen may be described by standard geologic terms such as rock name, mineralogy, texture, degree and kind of cementation, and weathering. Each term implicitly carries factors of value in engineering use of rock.

The typical intrusive igneous rock will have larger crystals than the typical extrusive igneous rock while sharing similar crystalline interlocking textures and silicate mineral compositions. The metamorphic rock name provides information about mineralogy and the degree of foliation, if present. A sedimentary rock name may imply certain physical features while leaving others undefined. Sandstones and shales, for instance, are clastic rocks defined only by the size of their predominant discrete grains. Textural differences in rock types are illustrated by Figures 4.1*a* and 4.1*b*.

Descriptive adjectives must be used to further define this type of rock's physical properties such as sandy shale and calcareous sandstone. A limestone, by comparison, is defined by its composition rather than grain size. It may be composed of cemented individual detrital grains of calcite

Figure 4.1 Photomicrographs of rock textures. (*a*) Igneous rock showing crystalline interlock. (*b*) Sedimentary rock showing clastic texture.

derived from wave action on a reef or it may be a dense crystalline limestone from the precipitation of a limy mud on a sea or lake bottom. Differing amounts of clay minerals, silt, and/or sand-sized grains of quartz and fossils may be present in a given specimen, none of which alters the rock's name—all affecting its engineering properties. Descriptive adjectives such as finely crystalline, clastic, clayey (argillaceous), or sandy (arenaceous) are applied. In each case, the engineer would know only that variability in the physical properties exists.

Geologic terminology is informative, but it does not provide the engineer with the quantitative data that are needed. Although quantitative estimates of physical properties may be made for a typical rock type, the properties for a typical rock are of little value on a given engineering project when one considers the many variables that influence the physical properties of a specific rock. The problem for the engineer is further complicated by the lateral and vertical variations that occur within a rock mass as a result of differing environmental conditions either at the time of rock formation or at some later time. Anisotropy in rock, therefore, is the rule rather than the exception. This anisotropy usually is recognized and described by geologists at a scale that may not be sufficiently detailed for engineering purposes.

Rock Strength

The engineering use of rock—whether as foundation material, in excavations and tunnels, or in maintaining stable slopes—involves rock masses in which the presence of discontinuities often determines the engineering character to a greater degree than do the physical properties of the intact rock. However, there remain many cases where intact rock properties are of great engineering value. Prediction of deformation amounts and rates in openings made in highly stressed, highly elastic rocks is a typical example. The performance of tunnel-boring machines is, in part, dependent on such intact rock factors as mineralogy, texture, grain size, and foliation (Tarkoy, 1975; Cording and Mahar, 1978). The strength and elasticity of intact rock are used in the design of dams and pressure tunnels where deformation and failure characteristics are useful in the design and predicted performance of a structure. The need, and the time available, for covering a freshly exposed rock surface to prevent slaking or physical deterioration also are functions of intact rock properties.

In the material that follows, physical properties of value to the engineer (e.g., strength and elastic modulus) and tests devised to determine them will be described. It should be remembered, however, that all are a function of the geologic properties of the rocks such as mineral composition, grain size, grain interlock, kind and amount of cement, and the many other geologic parameters that apply to a given rock. These parameters will be related to the engineering properties where appropriate. The Geological Society Engineering Group Working Party has summarized (Table 4.1) the geologic and engineering properties or indexes that define intact rocks for engineering purposes (Geological Society, 1977). The International Society for Rock Mechanics has published suggested methods for petrographically

Table 4.1 Properties and Indexes That Define Intact Rock Properties

Rock type	Alteration	Strength
Color	Hardness	Sonic velocity
Grain size	Durability	Young's modulus
Texture and fabric	Porosity	Poisson's ratio
Weathering	Density	Primary permeability

Source: Modified and reprinted with permission from Q. J. Eng. Geol., Vol. 10, Geol. Soc. (London) Eng. Group Working Party, The Description of Rock Masses for Engineering Purposes, 1977.

examining those rock features that have a bearing on mechanical behavior that can only be observed with a microscope (ISRM, 1978e).

Strength is a fundamental quantitative engineering property of a rock specimen. By definition, it is the amount of applied stress at rock failure, or rupture. The applied stress may be compressive, shear, or tensile in application, giving rise to compressive, shear, and tensile strengths (see chapter 1). Of these, compressive strength is greatest, most easily obtained either directly or indirectly, and the most commonly used of the three strengths in engineering applications.

Uniaxial Compressive Strength

Compressive strength is not a single value for a given rock specimen, but rather it is directly proportional to the confining stress and stress application rate. The influence of confinement on compressive strength introduced in chapter 1 will be discussed later. The compressive strength without confinement is the uniaxial or unconfined compressive strength (UCS). It is the most commonly measured and used compressive strength. Strengths at varying amounts of confining stress are triaxial strengths and must always be related to the confining stress value.

Standardized testing eliminates the problem introduced by strength variations caused by stress application rate. The suggested stress application rate for uniaxial compressive strength tests is 0.5–1.0 MPa/sec (ISRM, 1979a). Because most test specimens are obtained by the coring of rock, the cylinder has been accepted as the standard test shape. Core size and length–diameter proportions have been found to introduce variations in test results. A length-to-diameter ratio range of 2.5–3.0 and a core diameter no less than NX size (approximately 54 mm) generally are accepted standards.

Although average uniaxial strength values for given rock types are of little direct value in engineering applications, such values and their ranges provide insight into the need for detailed site investigations and subsequent rock testing. A number of uniaxial strengths for several representative igneous, metamorphic, and sedimentary rock types have been compiled from the literature and are shown in Table 4.2. Given the crystalline texture of both basalt and granite, the smaller crystal size in basalt compared with granite is a primary reason for the higher mean strength and maximum strength for basalt. The presence of vesicles, or voids from

Table 4.2 Compilation of Uniaxial Strength Data for Nine Common Rock Types[a]

	Granite	Basalt	Gneiss	Schist	Quartzite	Marble	Lime-stone	Sand-stone	Shale
Av. Strength	181.7	214.1	174.4	57.8	288.8	120.5	120.9	90.1	103.0
Max. Strength	324.0	358.6	251.0	165.6	359.0	227.6	373.0	235.2	231.0
Min. Strength	48.8	104.8	84.5	8.0	214.9	62.0	35.3	10.0	34.3
Strength Range	275.2	253.8	166.5	157.6	144.1	165.6	337.7	225.2	196.7
No. of Samples	26	16	24	17	7	9	51	46	14

[a] Values in megapascals (MPa). Data from (1) Wuerker, 1956; D'Andrea et al., 1965; (2) Birch, 1966; (3) S. P. Clark, Jr., 1966; (4) Handin, 1966; (5) Kulhawy, 1975.

Source: Data used by permission (1) Am. Inst. Min. Metall. & Pet. Eng., Paper 663-G, R. G. Wuerker, Annotated Tables of Strength and Elastic Properties, 1956; (2) Handbook of Physical Constants, F. Birch, Compressibility; Elastic Constants, Mem. 97, Geol. Soc. Am., pp. 97–173, 1966; (3) Handbook of Physical Constants, rev. ed., S. P. Clark, Jr. (ed.), Mem. 97, Geol. Soc. Am., 587 pp., 1966; (4) Handbook of Physical Constants, J. Handin, Strength and Ductility, Mem. 97, Geol. Soc. Am., pp. 273–289, 1966; (5) Eng. Geol., Vol. 9, F. H. Kulhawy, Stress Deformation Properties of Rock and Rock Discontinuities, 1975, Elsevier Science Publishers B.V. (Also source for Tables 4.5, 4.6.)

expanding gas, in some basalts results in significant reduction in strength and is a major factor in the range of strengths shown. In granite, crystal size is the primary strength factor. The corresponding reduction in crystal interlock and the influence of crystal cleavage with increased crystal size result in a wide strength range as one progresses from fine-grained granite to granite pegmatite.

The influence of degree of foliation on strength may be noted by comparing both the average strengths and the strength ranges for gneiss and schist data. Although the greater crystal interlock present in gneiss is a primary factor in higher mean strength and maximum value, considerable overlap in ranges is to be expected. This is due to variations in mineralogy and orientation of foliation with the applied stress direction in tests conducted on gneiss and schist specimens. An additional factor is the subjectiveness of assigning names to rock units and the common gradation of schist into gneiss even at test-specimen scale. The influence of foliation orientation during testing is examined when triaxial testing is discussed.

The higher mean strength and the maximum and minimum values observed in quartzites are to be expected. The absence of foliation and the quartz bonding of quartz grains that typically have sutured grain-to-grain contacts create a rock of exceptionally high intact strength. Marble is another nonfoliated metamorphic rock that in a pure state is composed of calcite, or dolomite, with a crystalline texture similar to an igneous rock. Its strength values will be similar to those limestones that have crystalline textures resulting from chemical precipitation and that were relatively pure calcite or dolomite prior to metamorphism. The range of marble strengths is not as great as those encountered in limestones. Limestones may range from strong, highly siliceous rocks to very clayey (argillaceous) or shaly limestones with low strengths.

The strengths of two detrital sedimentary rocks, namely sandstone and shale, are tabulated on Table 4.2. The extreme strength range in sandstones is a product of variations in degree and type of cementation of the sand-sized particles and of the proportion of other-than sand-sized particles such as clay and silt. The cohesiveness of clay minerals in shales is the cause of higher strengths compared with those of poorly cemented sandstones. At the other end of the strength range, shales may have calcareous and siliceous cementing agents that impart relatively high strengths when testing is done normal to the bedding.

The point to be made by an examination of the sampling of test data reproduced in Table 4.2 is that we can predict the strengths of rocks only in generalities if only given general names such as granite, limestone, and sandstone. Descriptive adjectives assigned to specimens improve the investigator's prediction capabilities, but only laboratory testing of representative samples can provide definitive quantitative information.

Although the engineering classification of rocks is described later, it is pertinent to consider the classification of intact rocks on the basis of uniaxial compressive strength. Table 4.3 presents selected intact rock strength classifications that span the interval between the early work of Coates (1964) and Deere and Miller (1966) to the published standards of the International Society for Rock Mechanics (ISRM, 1978f). It is evident that there is little uniformity either in the strength range or the number of strength categories utilized to subdivide the individual ranges. Given such lack of uniformity, it is important that strength descriptors used in reports or publications be identified with strength limits to avoid misunderstanding.

The lower limits of strength that appear in Table 4.3 present an additional problem. Several are 1.0 MPa or less, and they are representative of materials that may have the engineering properties of soil. It is for this reason that the Geological Society Group Working Party (Geological Society, 1977) suggested a lower limit of 1.25 MPa for the uniaxial compressive strength for rock.

Shear Strength

Compressive failure of an intact sample is the result of exceeding the shear strength of the material along an inclined failure surface shown diagrammatically in Figure 4.2. The shear strength is indirectly measured in such tests. As shown in chapter 1, the angle of the failure surface relative to the applied compressive stress plane, β, can be obtained graphically from confined and/or triaxial tests on rock specimens. The angle $\propto$ is the angle between the applied compressive stress and the failure surface. It has a role in the strength differences noted in rocks with planar anisotropy such as foliation. One would expect that an orientation of such planes parallel to the axis of applied compressive stress would result in the weakest strength value. However, the lowest strength is obtained when the planes are oriented at the angle $\propto$ along which the greatest shear stress is generated as shown in Figure 4.3.

Donath (1961), Attewell and Farmer (1976), and Hoek and Brown (1980) are among those who have illustrated this relationship for a variety of rock types having planar foliation under a number of confining stress conditions. The angle of the failure surface is virtually unaffected by changes in confining stress. The engineering implication of the angular relationship between the direction of stress application and the orientation

Table 4.3 Selected Uniaxial Compressive Strength Ranges Devised to Classify Intact Rock from Maximum to Minimum Strengths (strengths in Megapascals (MPa))

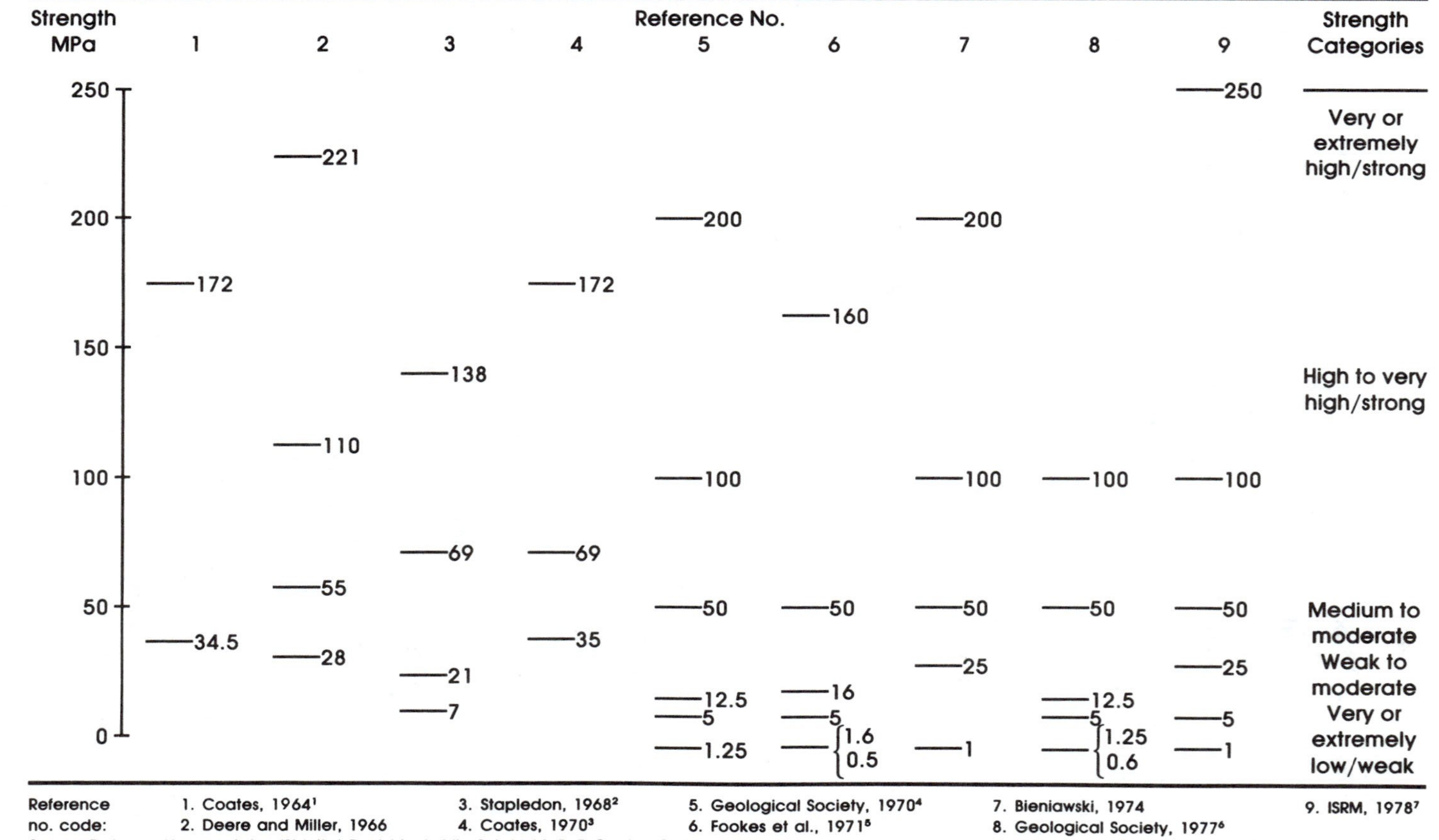

Reference No.	Strength boundaries (MPa), maximum to minimum
1	172, 34.5
2	221, 110, 55, 28
3	138, 69, 21, 7
4	172, 69, 35
5	200, 100, 50, 12.5, 5, 1.25
6	160, 50, 16, 5, {1.6, 0.5}
7	200, 100, 50, 25, 1
8	100, 50, 12.5, 5, {1.25, 0.6}
9	250, 100, 50, 25, 5, 1

Strength MPa	Strength Categories
250	Very or extremely high/strong
200	
150	High to very high/strong
100	
50	Medium to moderate
	Weak to moderate
0	Very or extremely low/weak

Reference no. code: 1. Coates, 1964[1] 2. Deere and Miller, 1966 3. Stapledon, 1968[2] 4. Coates, 1970[3] 5. Geological Society, 1970[4] 6. Fookes et al., 1971[5] 7. Bieniawski, 1974 8. Geological Society, 1977[6] 9. ISRM, 1978[7]

Source: Data used by permission (1) Intl. J. Rock Mech. Min. Sci., Vol 1, D. F. Coates, Classification of Rocks for Rock Mechanics, 1964, Pergamon Press; (2) Intl. J. Rock Mech. Min. Sci., Vol. 5, D. H. Stapledon, Discussion-Classification of Rock Substances, 1968, Pergamon Press; (3) Rock Mechanics Principles, rev. ed., D. F. Coates, Mon. 874, 1970, reproduced by permission of the Minister of Supply and Services Canada; (4) Reproduced by permission of The Geological Society of London from Geol. Soc. Eng. Group Working Party, The Logging of Rock Cores for Engineering Purposes, Q. J. Eng. Geol., Vol. 3, 1970; (5) Reproduced by permission of The Geological Society of London from P. G. Fookes, W. R. Dearman, and J. A. Franklin, Some Engineering Aspects of Rock Weathering with Field Examples from Dartmoor and Elsewhere, Q. J. Eng. Geol., Vol. 4, 1971; (6) Reproduced by permission of The Geological Society of London from Geol. Soc. Group Working Party, The Description of Rock Masses for Engineering Purposes, Q. J. Eng. Geol., Vol. 10, 1977; (7) Untl. J. Rock Mech. Min. Sci. & Geomech. Abstr., Vol. 15, Comm. on Stand. of Lab. and Field Tests, Suggested Methods for the Quantitative Description of Discontinuities in Rock Masses, 1978, Pergamon Press.

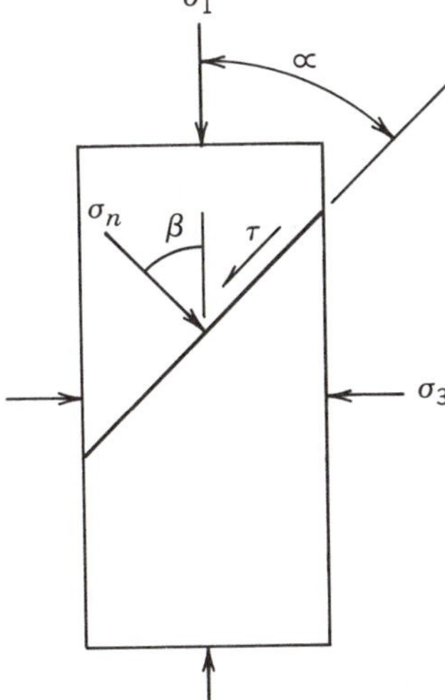

Figure 4.2 Development of shear failure surface under axial compression. ∝ is the angle between the applied axial stress and the shear failure plane. See Figure 1.26 for other notations.

of foliation may be of major consequence. An example may be found in the siting of thin-arch concrete dams in which reservoir filling causes large compressive and shear forces to be applied to the abutments, or valley walls, by the dam. It would not be desirable to have the resulting shear stress components oriented parallel to the prevailing direction of foliation, bedding, or a particular joint set.

Tensile Strength

The tensile strength of intact rock is the least commonly determined rock strength property; in part this is the result of the more common application of compressional and shear stresses—rather than tensile stresses—to rock in nature and during construction. However, tensile stress plays a part in such applications as the design of roof spans in underground excavations and the prediction of sudden failure under tension when confined, highly elastic rocks are stress relieved (as in some mines and tunnels) with resultant rock bursts.

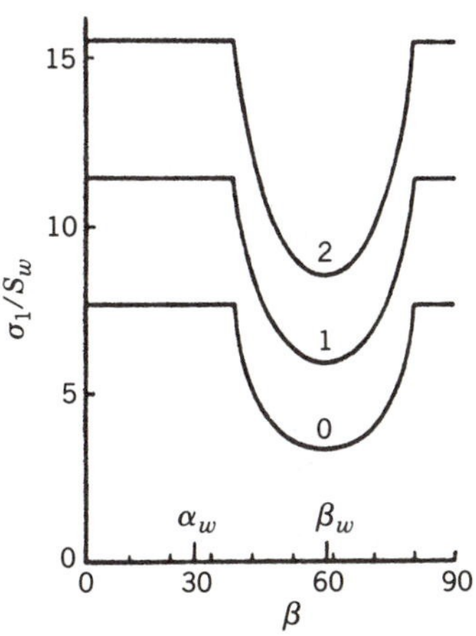

Figure 4.3 Influence of planar anisotropy on axial strength σ_1 for samples with normals to planes of weakness oriented at angle β_w from σ_1, where S_w is shear strength of planes of weakness. Numbers on curves indicate tests at different confining stresses. (Reprinted with permission from Fundamentals of Rock Mechanics, 2nd ed., J. C. Jaeger and N. G. W. cook, 1976, Chapman & Hall.)

Tensile strength of rock is controlled by the same factors that govern compressive and shear strengths, that is, composition, texture, grain size, kind and amount of cementing material, and moisture content. Invariably, tensile strength is the lowest of the three strengths for any given rock type and specimen. The presence of oriented features such as foliation and lamination within the intact specimen lowers tensile strength as it lowers compressive and shear strengths. Table 4.4 compares compressive and tensile strengths of massive rocks; it is reproduced from the definitive study by Hobbs (1964). One of the most comprehensive tabulations of tensile strengths for a wide variety of rock types has been prepared by Kulhawy (1975). The relation of tensile strength to other rock properties has been reported by D'Andrea et al. (1965).

Determination of tensile strength may be made by use of cylindrical samples similar to those used for compressive strength and indirect shear strength tests. This may be done either by applying a tensile load axially, using metal caps cemented to the cylindrical sample, or by the Brazil test. The Brazil (Brazilian) test causes a disk of rock to fail indirectly under tension upon the application of compressive stress to the opposite sides of the specimen (Figure 4.4). Standardized testing procedures for both tests are available (ISRM, 1978c).

Rock Deformation

Static Elastic Moduli

Deformation data for specimens undergoing strength tests may be obtained and used to calculate the static elastic moduli of intact rock. Static elastic modulus testing differs from dynamic elastic modulus testing, in that a much lower stress application rate is used to simulate static conditions. Two useful moduli are the modulus of elasticity *(E)*, or Young's

Table 4.4 Comparison of Mean Tensile and Compressive Strengths for Selected Sedimentary Rock Types

	Tensile Strength (MPa)	Compressive Strength (MPa)
Limestone	18.00 ± 0.62 (20)[a]	41.45 ± 3.52 (4)
Sandstone	19.17 ± 0.21 (23)	77.59 ± 1.59 (5)
Sandstone	23.10 ± 0.48 (19)	80.83 ± 2.21 (10)
Sandstone	24.21 ± 0.83 (8)	90.48 ± 3.86 (4)
Mudstone	35.17 ± 3.17 (4)	50.07 ± 3.79 (4)
Limestone	36.28 ± 1.24 (24)	142.55 ± 6.14 (5)
Limestone	38.76 ± 2.69 (23)	142.97 ± 19.10 (8)
Ironstone	44.28 ± 4.48 (5)	190.69 ± 17.93 (4)
Sandstone	65.66 ± 0.83 (11)	167.66 ± 9.86 (5)

[a] Number of samples are in parenthesis; ± values are standard errors of means.

Source: Reprinted with permission from Intl. J. Rock Mech. Min. Sci., Vol. 1, D. W. Hobbs, The Tensile Strength of Rocks. Copyright © 1964, Pergamon Press.

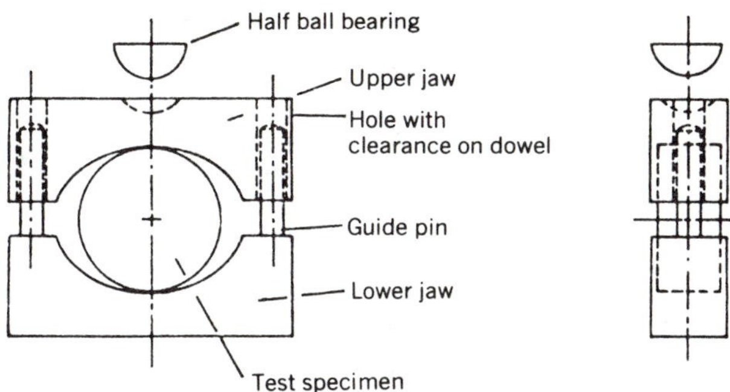

Figure 4.4 Apparatus for Brazil test. (Reprinted with permission from Intl. J. Rock Mech. Min. Sci. & Geomech. Abstr., Vol. 15, Anon., Suggested Methods for Determining Tensile Strength of Rock Materials. Copyright ©1978c, Pergamon Press.)

modulus, and Poisson's ratio (ν). The modulus of elasticity is derived from applied axial compressive stresses and resulting axial strains. Poisson's ratio is calculated from axial and diametral strains resulting from applied axial compressive stress. Both are useful in estimating elastic response of intact rock to compression from *in situ*, construction, and postconstruction stresses. Abutment stresses in a dam or those exerted against the rock by a water-pressure tunnel are examples of postconstruction stresses.

Modulus of Elasticity. The axial deformation resulting from axial compression illustrated earlier in Figure 1.26 is shown by the stress-strain diagram in Figure 4.5. The linear portion of the curve represents the elastic response of the material being tested. The modulus of elasticity *(E)* described in chapter 1 is obtained from the equation

$$E = \text{stress/strain} = \sigma/\epsilon \tag{4.1}$$

where:

σ = axial compressive stress for uniaxial tests or deviator stress ($\sigma_1 - \sigma_3$) for triaxial tests in psi, N/m^2, or Pa

ϵ = axial strain expressed as in./in., mm/mm, or as a percentage (%)

The values for E may be obtained from stress-strain diagrams (Module 4.1). The least well-defined value, the average modulus, E_{av}, is obtained from the best fit to the linear or elastic part of the curve (Figure 4.6*a*).

The secant modulus, E_s, (Figure 4.6*b*) and the tangent modulus, E_t, (Figure 4.6*c*) are more commonly used in engineering applications (Module 4.1). For standardization, the secant and tangent values are obtained at a fixed value or percentage of the ultimate strength. The 50% level is in considerable use today, resulting in the secant modulus, E_{s50}, and the tangent modulus, E_{t50}, shown in Figures 4.6*b* and 4.6*c*. The secant modulus is the more conservative of the two, that is, it predicts the maximum

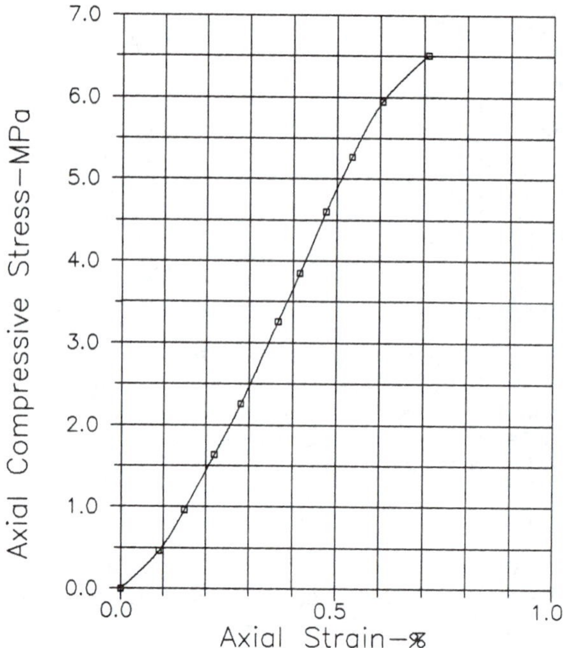

Figure 4.5 Plot of unconfined compressive test stress-strain data for a specimen of clay shale, Bearpaw Formation, Cretaceous, Montana. (Adapted from Fleming et al., 1970)

elastic deformation that would occur at the chosen percentage of ultimate strength.

The same rock property factors that control intact rock compressive strength influence the modulus of elasticity. The relationship between compressive strength and modulus of elasticity is direct and linear as documented by Judd and Huber (1962), D'Andrea et al. (1965), and Deere and Miller (1966), with a correlation coefficient of +0.72 reported by D'Andrea et al. Analysis of data from 49 uniaxial tests tabulated by Kulhawy (1975) provides a similar correlation coefficient of +0.79. Representative values of the modulus of elasticity, E, for selected rock types are shown in Table 4.5.

Poisson's Ratio. Poisson's ratio, ν, is a useful engineering property as it is a measure of change in diameter with change in length under axial compressional stress. As noted in chapter 1, it is a unitless modulus obtained from the following equation:

$$\nu = \text{unit change in diameter/unit change in length} \tag{4.2}$$

The maximum value of Poisson's ratio is 0.5 and values typically are inversely proportional to both compressive strength and modulus of elasticity (D'Andrea et al., 1965). Examination of data from Kulhawy (1975)

MODULE 4.1

Determination of elastic modulus from stress-strain diagram.

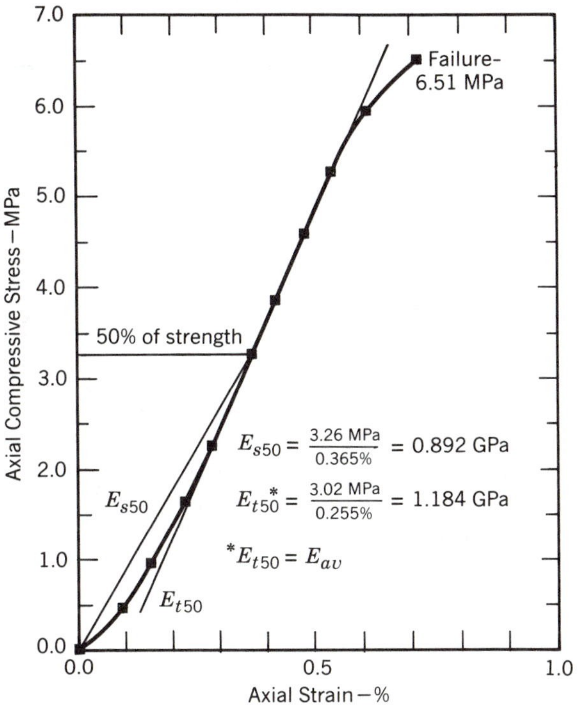

7.0
6.0
5.0
4.0
3.0
2.0
1.0
0.0
Axial Compressive Stress – MPa
0.0
0.5
1.0
Axial Strain – %
Failure- 6.51 MPa
50% of strength
E_{s50}
E_{t50}

$$E_{s50} = \frac{3.26 \text{ MPa}}{0.365\%} = 0.892 \text{ GPa}$$

$$E_{t50}^{*} = \frac{3.02 \text{ MPa}}{0.255\%} = 1.184 \text{ GPa}$$

$$^{*}E_{t50} = E_{av}$$

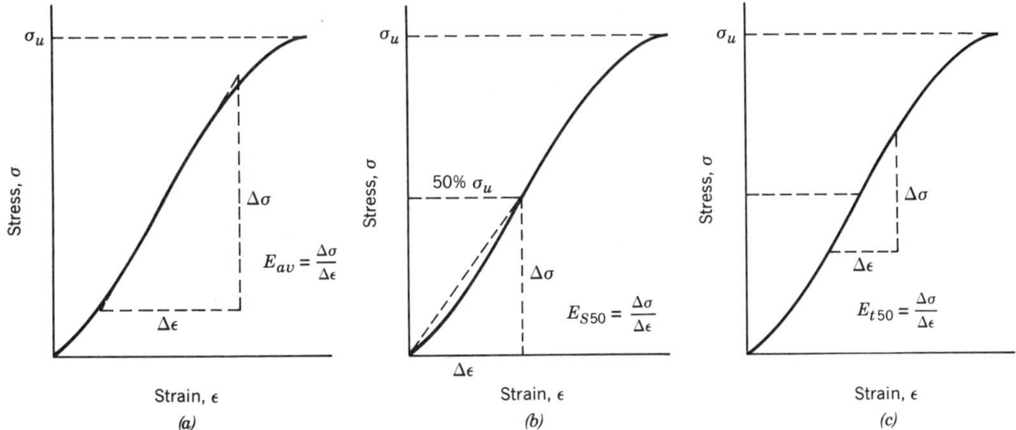

Figure 4.6 Determining of modulus of elasticity from axial stress-strain diagrams. (*a*) Average modulus, E_{av}, (*b*) Secant modulus at 50% of ultimate strength of σ_u, E_{s50}, (*c*) Tangent modulus at 50% of ultimate strength, E_{t50}.

Table 4.5 Compilation of Static Moduli of Elasticity for Nine Common Rock Types[a]

	Granite	Basalt	Gneiss	Schist	Quartzite	Marble	Lime-stone	Sand-stone	Shale
Av. *E*	59.3	62.6	58.6	42.4	70.9	46.3	50.4	15.3	13.7
Max. *E*	75.5	100.6	81.0	76.9	100.0	72.4	91.6	39.2	21.9
Min. *E*	26.2	34.9	16.8	5.9	42.4	23.2	7.7	1.9	7.5
Range	49.3	65.7	64.2	71.0	57.6	49.2	83.9	37.3	14.4
No. of Samples	24	16	17	18	10	16	29	18	9

[a] Values in gigapascals (GPa).
Source: Data used by permission, see Table 4.2

results in low correlation coefficients, namely, −0.159 with compressive strength and −0.072 with the modulus of elasticity. Figure 4.7 illustrates the typical scatter of Poisson's ratio data to strength data. Modulus of elasticity and Poisson's ratio data plot in similar manner. Representative values of Poisson's ratio for selected rock types are given in Table 4.6.

Dynamic Elastic Moduli

Modulus of elasticity (Young's modulus), shear modulus, and Poisson's ratio may be obtained by dynamic methods in addition to static laboratory tests. Dynamic elastic moduli are obtained by rapid application of stress to the sample. One method of achieving this is to subject the sample to ultrasonic compression and shear wave pulses (ISRM, 1978b). Compression and shear wave generators or transducers are attached to a prepared core specimen as in Figure 4.8. Wave velocity through the specimen is

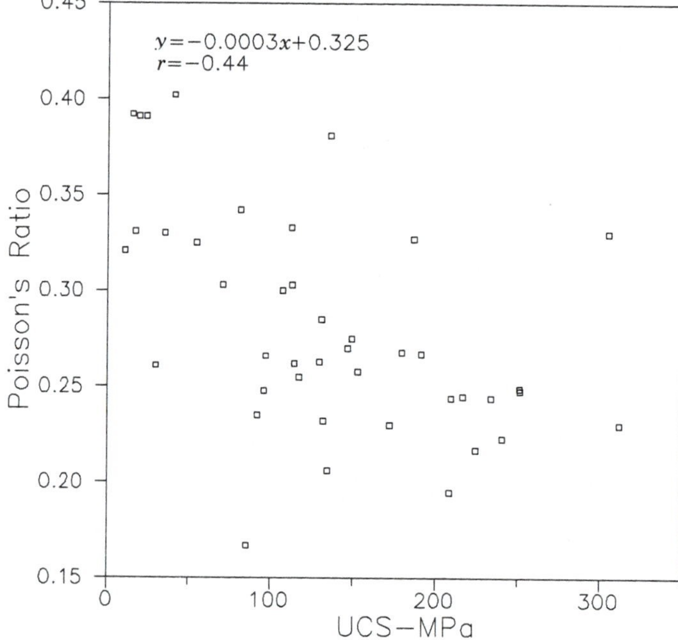

Figure 4.7 Plot of uniaxial compressive strength (UCS) versus Poisson's ratio. (Adapted from D'Andrea et al., 1965)

Table 4.6 Compilation of Poisson's Ratios for Nine Common Rock Types

	Granite	Basalt	Gneiss	Schist	Quartzite	Marble	Lime-stone	Sand-stone	Shale
Av. ν	0.23	0.25	0.21	0.12	0.15	0.23	0.25	0.24	0.08
Max. ν	0.39	0.38	0.40	0.27	0.24	0.40	0.33	0.46	0.18
Min. ν	0.10	0.16	0.08	0.01	0.07	0.10	0.12	0.06	0.03
Range	0.29	0.22	0.32	0.26	0.17	0.30	0.21	0.40	0.15
No. of samples	24	16	17	18	10	16	29	18	9

Source: Data used by permission, see Table 4.2

calculated from the travel time from the generator to a receiver at the opposite end. Shear wave velocities, V_s, are about two-thirds of the compressional wave velocities, V_p (Figure 4.9). Given the compression and shear wave velocities and specimen mass density, all of the dynamic elastic moduli listed in Figure 4.10 can be calculated as in Module 4.2 by using the equations shown in this figure. Samples may be compressionally loaded to approximate field conditions. The dynamic modulus of elasticity, E_d, increases with added axial compressional stress.

Young's modulus (modulus of elasticity) is obtained both statically and dynamically on laboratory specimens for design purposes. Typically E_d is greater than E_s (G. B. Clark, 1966) because the response of the specimen to the very short duration strain and low stress level is essentially purely elastic. The dynamic shear modulus, G_d, exhibits a similar relationship, that is, shear waves are measured at very low shear strains resulting in G_d being greater than G_s.

The velocity, V_p, at which compressional waves travel through rock has been recognized as an indicator of rock mass quality as described in the section on rock mass properties. The basis for this lies in the relationships existing between V_p and various engineering properties of intact rock samples. In most cases, the only added factor at the larger rock mass scale is the presence and number of discontinuities that lower both the velocity of wave travel and the rock mass quality. One of the most useful correlations involving V_p is the relationship with uniaxial compressive strength (Figure 4.11). Static Young's modulus, E_s, shows a similar positive linear relationship. As would be expected, V_p has a similar relationship with dynamic Young's modulus because of the use of interrelated V_p and V_s in

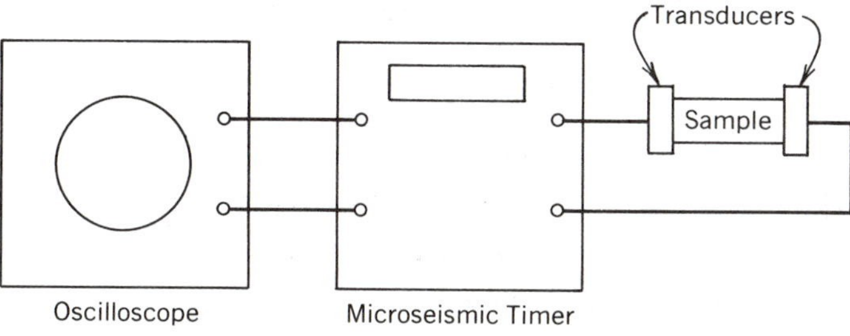

Figure 4.8 Schematic diagram of sonic velocity equipment.

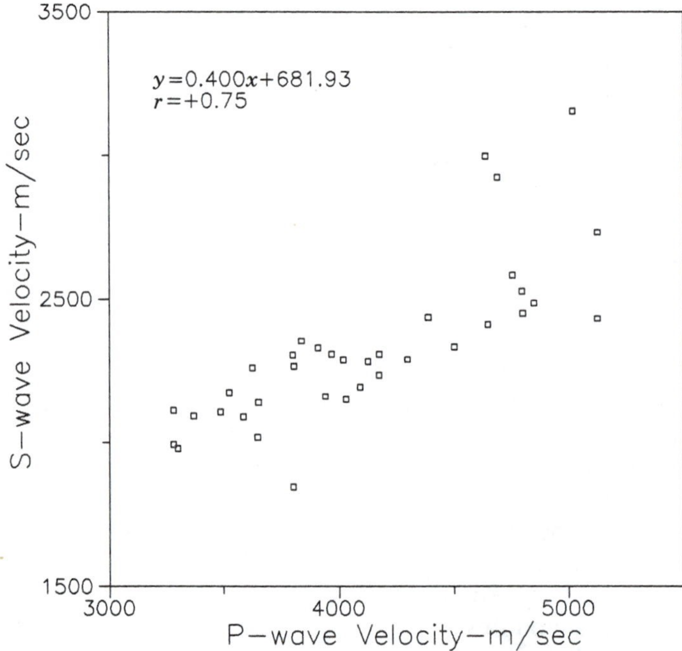

Figure 4.9 Plot of compressional (p-) wave velocity versus shear (s-) wave velocity. (From Jesch et al., 1979)

calculating E_d (Figure 4.10). By comparison, Poisson's ratio shows no well-defined linear relationships with either V_p or V_s (Figure 4.12).

Influence of Confinement on Strength and Elastic Moduli

The geotechnical properties of rocks obtained during uniaxial, or unconfined, testing are numerically unlike *in situ* rock properties. In the latter case, the rock is subjected to confining stresses along the other two orthog-

Modulus	*Equation*
Young's modulus	$= E_d = k\rho V_s^2 \dfrac{(3V_p^2 - 4V_s^2)}{(V_p^2 - V_s^2)}$
Poisson's ratio	$= \nu_d = \dfrac{V_p^2 - 2V_s^2}{2(V_p^2 - V_s^2)}$
Shear modulus	$= G_d$ or $\mu = \rho V_s^2$

Where:

ρ = mass density (rho)

V_p = P or compressional wave propagation velocity

V_s = S or shear wave propagation velocity

k = constant, depending on units used

Figure 4.10 Dynamic elastic moduli equations.

MODULE 4.2

Calculation of dynamic elastic moduli from test data.

Sample data

Pikes Peak granite, Rampart Range, Douglas County, Colorado, NX (2⅛ in. or 54 mm diameter) core

Core length = 0.123 m
Bulk density = 2.643 g/cm³
P-wave travel time through core
= 2.880×10^{-5} sec
S-wave travel time through core
= 5.426×10^{-5} sec

Velocity calculations

P-wave, $V_p = .123$ m/2.880 $\times 10^{-5}$ sec = 4,270.8 m/sec
S-wave, $V_s = .123$ m/5.426 $\times 10^{-5}$ sec = 2,266.9 m/sec

Young's Modulus, E_d, calculation (see Figure 4.10)

Value of k (conversion of bulk to mass density and Pa) = 1,000.6

$E_d = (1{,}000.6)\,(2.643)\,(2{,}266.9^2)\,[((3)(4{,}270.8^2) - (4)\,(2{,}266.9^2))/(4{,}270.8^2 - 2{,}266.9^2)] = 35.44 \times 10^9$ Pa = 35.44 GPa

Poisson's Ratio, ν_d, calculation (see Figure 4.10)

$\nu_d = ((4270.8^2) - (2)\,(2266.9^2))/((2)(4270.8^2 - 2266.9^2)) = .304$

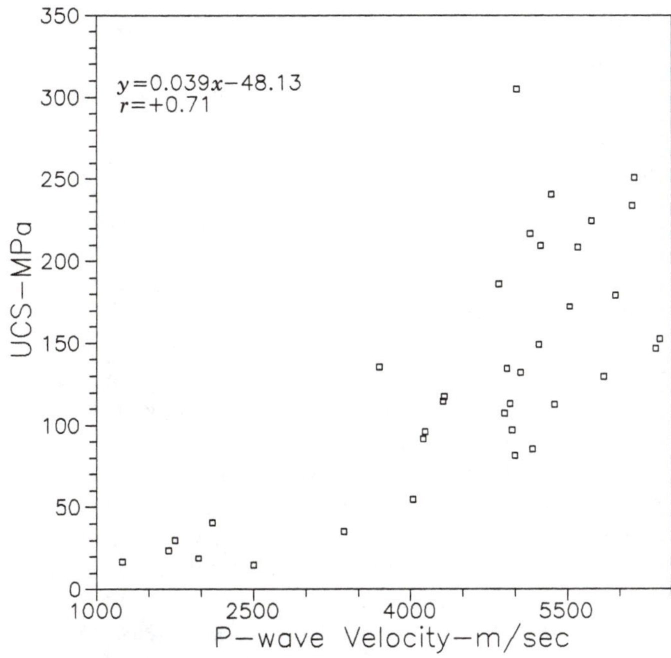

Figure 4.11 Plot of compressional (p-) wave velocity versus uniaxial compressive strength (UCS). (Adapted from D'Andrea et al., 1965).

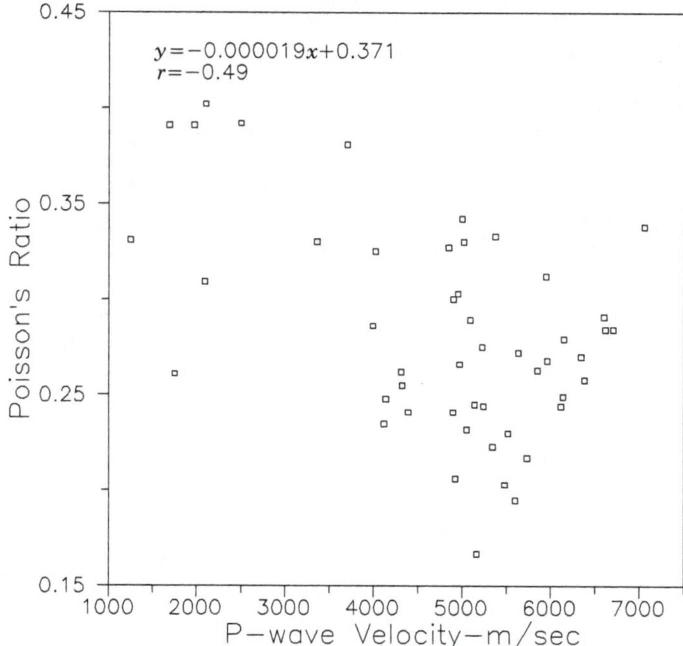

Figure 4.12 Plot of compressional (p-) wave velocity versus Poisson's ratio. (Adapted from D'Andrea et al., 1965)

onal principal axes, that is, σ_2 and σ_3 (Figure 1.26). Testing of rock specimens in the confined state is known as triaxial testing and normally refers to compressive stress application without further identification. To facilitate testing, σ_2 and σ_3 are made equal and the confining stress is applied uniformly by hydraulic pressure acting on a flexible jacket in contact with the specimen.

Triaxial testing of rock permits evaluation of rock properties under conditions that may be designed to duplicate field or *in situ* confinement. Lindner and Halpern (1978) have compiled published and unpublished *in situ* stress data for over one hundred sites in the United States and Canada. Such data are useful in estimating laboratory test requirements as well as permitting some estimation of stress-relief problems in rock-excavation projects.

Rock strength increases with increasing confining stress (Module 1.2). Although this may be illustrated by stress-strain diagrams, Figure 4.13 is an effective means of displaying the relatively linear relationship between compressive strength (axial stress) and confining stress at higher confining stresses. Axial stress may be reported and plotted as deviator stress. Deviator, or deviatoric stress, is the difference between σ_1 and σ_3, or axial and confining stresses, respectively. It is substituted for axial stress by some because only the difference between axial and confining stresses causes sample deformation. Figure 4.14 depicts the increase in deviator stress for increased confining pressure for a siltstone in the Powder River Basin, Montana.

Increased confinement of a specimen may result in increased values for

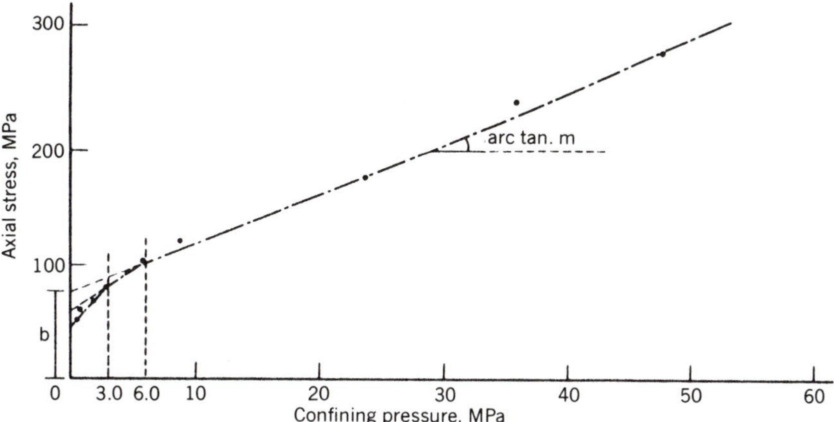

Figure 4.13 Influence of confining pressure on compressive strength where m = slope angle and b = cohesion. (Reprinted with permission from Intl. J. Rock Mech. Min. Sci. & Geomech. Abstr., Vol. 15, Anon., Suggested Methods for Determining the Strength of Rock Materials in Triaxial Compression. Copyright ©1978d, Pergamon Press.)

Young's modulus as illustrated by Figure 4.15. The influence of confinement on compressive strength and modulus of elasticity is well documented in the geologic and geotechnical literature (Griggs, 1936; Donath, 1970; Wawersik and Fairhurst, 1970; and Jaeger and Cook, 1976).

The most commonly employed graphical means of presenting the influence of confining pressure on rock strength is the Mohr diagram described in chapter 1. When the Mohr diagram is applied to rocks, all of the variables are significantly larger in value than those obtained from tests on soils. Perhaps the most striking difference, aside from stress and cohesion values, is the increase in the angle of internal friction in rocks. In addition,

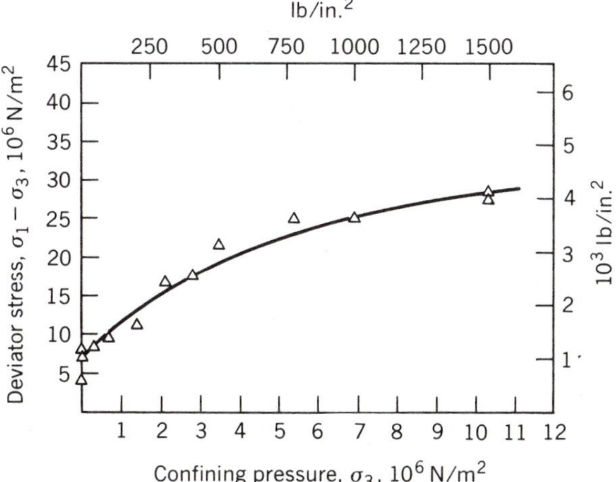

Figure 4.14 Increase in deviator stress for increasing confining stress for a siltstone unit from the Powder River Basin, Montana. (Modified from Lee et al., 1976)

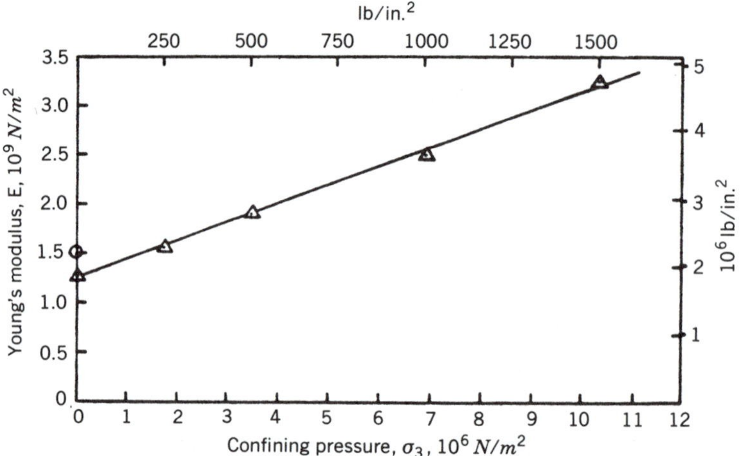

Figure 4.15 Influence of confining stress on Young's modulus, *E*, for a shale unit in the Powder River Basin, Wyoming and Montana. (Modified from Lee et al., 1976)

with an increase of only 34.5 MPa (5,000 psi) in confining pressure as shown in Figure 4.16, the strength is increased about sevenfold for the specimens of granite tested. Strength changes with increased confinement and related angle of internal friction are less for lower-strength rocks. Variations in texture, mineralogy, grain (crystal) size, and cementing material interact to cause the different responses to confinement. Data compiled by Kulhawy (1975) show an angle of internal friction of 47.0° and a standard deviation of 9.7° for eleven intrusive igneous rocks compared to 21.0° and a standard deviation of 6.8° for nine shale samples.

Triaxial testing data, as displayed in a Mohr rupture diagram, permit estimation of the shear failure plane relative to the applied compressive stress axis. The angle β (Figures 1.26 and 4.2) formed between this axis (σ_1)

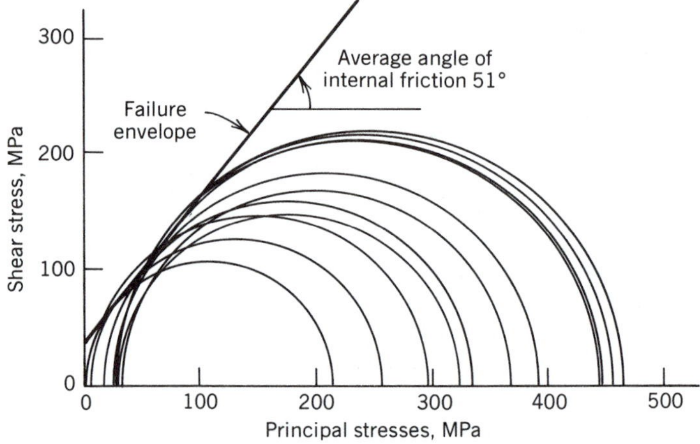

Figure 4.16 Mohr failure envelope for eleven triaxial compressive strength tests on samples of Silver Plume granite from Colorado. (Modified from Nichols and Lee, 1966)

and the stress normal to the shear failure plane (σ_n) is one-half the obtuse angle subtended between the principal stress axis and the normal to the rupture line on the Mohr diagram. Therefore, the shear failure surface and its orientation relative to the applied compressive stress, σ_1, are an intimate and predictable part of the failure of intact rock under triaxial compressive stresses.

Standardization of triaxial testing is an important consideration because of varying strain and strength responses to the application of axial loads and confining pressures on test specimens. Among the test criteria in use are those of the American Society for Testing and Materials (ASTM, 1979) and the International Society for Rock Mechanics (ISRM, 1978d). The ISRM test procedure, for example, calls for the simultaneous and essentially equal application of axial load and confining pressure until the test confining pressure is reached. From that point until sample failure, the axial load should be increased continuously at a constant stress rate within 0.5 to 1.0 MPa/sec or at a rate that will cause failure within 5 to 15 min of loading.

Engineering Aspects of Chemical Weathering

The physical properties of intact rock may be altered significantly by chemical weathering (as indicated earlier in chapter 1). Chemical weathering may be pervasive throughout the intact rock as opposed to mechanical, or physical, weathering, which causes reduction in rock-block size in rock masses without influencing the intact properties. Emphasis here will be placed on the geotechnical aspects of chemical weathering and the potential for estimating physical properties. The relation of chemical weathering to rock mass properties and intact and rock mass classification is addressed in subsequent sections.

In geotechnical practice, it is necessary to examine the long- and short-term influence of chemical weathering on rock. Long-term weathering will have had an influence on the physical state, or condition, of rock at a given site. The degree of chemical weathering will reflect time, processes, and conditions at the site, for example, rock exposure, jointing, and moisture. There are short-term weathering processes that may create problems during and following construction by exposing the rock to the elements, whether on the surface or in the subsurface, as in a tunnel. This susceptibility of rock to short-term weathering has been called weatherability (Fookes et al., 1971).

The slaking, or breakdown, of some clay-bearing rocks such as shale and siltstone when wet is an example of short-term weathering. Slaking may result from a combination of chemical and mechanical weathering processes (Franklin and Chandra, 1972).

Rock Properties

Reduction in compressive strength is the most obvious and important geotechnical factor caused by chemical weathering or alteration of intact rock. Dearman et al. (1978) have tabulated ranges of compressive strength for different weathered states of granite: fresh, >250 MPa; discolored,

100–250 MPa; weakened, 2.5–100 MPa; and soil, <2.5 MPa. There is a corresponding reduction in the modulus of elasticity with increasing degree of weathering. Hamrol (1961) has related this to an increase in porosity in granite samples as the degree of weathering increased. The changes in strength and elasticity resulting from chemical weathering or alteration are dependent on the susceptibility of rock composition to weathering—all other factors such as time and climate being equal.

A reduction in compressional wave velocity takes place with increasing chemical weathering or decomposition of intact rock. Dearman et al. (1978) have tabulated ranges of velocity for various degrees of weathering in granites and gneisses: fresh, 3,050–5,500 m/sec; slightly weathered, 2,500–4,000 m/sec; moderately weathered, 1,500–3,000 m/sec; highly weathered, 1,000–2,000 m/sec; completely weathered to residual soil, 500–1,000 m/sec.

Weathering Classifications

The degree to which chemical weathering or alteration has occurred in a given rock type is amenable to classification. The necessity for a weathering classification is apparent as one considers the influence weathering has on rock strength and elastic properties. The use of such classifications is universal in all modern rock mass strength classification. Weathering classifications of necessity remain descriptive or qualitative because of the wide range of unweathered, or fresh, rock strength and elastic properties for the many types of rocks.

There is at present some agreement on five weathering grades, ranging from fresh rock to completely weathered rock. The weathering, or alteration, grades proposed by the Geological Society Engineering Group Working Party in "The Logging of Cores for Engineering Purposes" (Geological Society, 1970) are representative of these grades and are shown in expanded form in Table 4.7. The fresh, discolored, weakened, and soil states of weathering used by Dearman et al. (1978) to describe rock strength earlier are the equivalent of grades I, II, III–V, and VI, respectively.

Granite is either explicitly or implicitly the basis for most of the weathering classification schemes, as it displays the full range of weathering grades. The influence of rock composition on weathering grade must be understood. A pure limestone, for instance, remains in grade I as the products of weathering are lost by solution. A quartzite will seldom be found at a grade lower than II. The soft clay-rich rocks such as shale occur as the products of weathering and the grade will be dependent on the amount and kind of cement present and the degree to which it has been leached or weathered from the matrix. The influence of climate on the degree and rate of chemical weathering has been the subject of research by several investigators, for example, as Peltier (1950), Saunders and Fookes (1970), and Fookes et al. (1971). The last two works have specifically addressed the impact of weathering on engineering properties of rocks under different climatic conditions.

Table 4.7 Representative Intact Rock Weathering Classification

Term	Description	Grade
Fresh	No visible sign of rock material weathering.	IA
Faintly weathered	Discoloration on major discontinuity surfaces.	IB
Slightly weathered	Discoloration indicates weathering of rock material and discontinuity surfaces. All the rock material may be discolored by weathering and may be somewhat weaker than in its fresh condition.	II
Moderately weathered	Less than half of the rock material is decomposed and/or disintegrated to a soil. Fresh or discolored rock is present either as a continuous framework or as corestones.	III
Highly weathered	More than half of the rock material is decomposed and/or disintegrated to a soil. Fresh or discolored rock is present either as a discontinuous framework or as corestones.	IV
Completely weathered	All rock material is decomposed and/or disintegrated to soil. The original mass structure is still largely intact.	V
Residual soil	All rock material is converted to soil. The mass structure and material fabric are destroyed. There is a large change in volume, but the soil has not been significantly transported.	VI

Source: Reprinted with permission from Q. J. Eng. Geol., Vol. 10, Geol. Soc. (London) Eng. Group Working Party, The Description of Rock Masses for Engineering Purposes, 1977.

Index Tests

There are a variety of field and laboratory tests that provide estimates of strength, weatherability, and weathering grade for intact rock. These are called *index tests*; they offer greater economy of preparation, testing time, and required equipment vis-à-vis standard laboratory rock tests. Although not as accurate or precise, they are sufficiently representative and quantitative for many geotechnical applications. Index tests for rock durability or response to abrasion are described in the discussion on aggregates in chapter 8.

Of the index tests available for predicting uniaxial compressive strength and modulus of elasticity, the point load strength and Schmidt rebound hammer tests are most commonly used, principally because they may be used on a greater variety of rock types with better predictability of strength and elasticity.

Point load Test

The point load test is a tensile test from which uniaxial compressive strength and modulus of elasticity are obtained empirically (Module 4.3). It is a variation of the Brazil tensile strength test (see p. 134) in which a rock

MODULE 4.3

Calculation of point load index and strength estimation.

Test data

Sample: sandstone
NX diameter (d) core (54 mm)
Force (P) at failure = 17.01×10^6 *kN*

Calculation of point load index, I_s

$I_s = P/d^2$
$I_s = 17.01 \times 10^9/2{,}916$
$I_s = 5.833 \times 10^6$ Pa or 5.833 MPa

Estimation of unconfined compressive strength (UCS)

UCS = $I_a \times 23.2$* (for NX core)
UCS = 5.833 MPa $\times$ 23.2
UCS = 135.33 MPa

* After Bieniawski, 1975 (see Figure 4.18).

cylinder is failed in tension by the application of compressive stress by two opposing curved-steel surfaces. The point load test utilizes two steel cones that apply compressive stress in opposite directions at the contact of the cone points and the specimen. Although a diametral test is preferred (Figure 4.17*a*), tests may also be conducted with less accuracy on axial and irregular rock pieces (Figures 4.17*b*, 4.17*c*). The point load index, I_s, is obtained by the following equation:

$$I_s = p/d^2 \tag{4.3}$$

Where:

P = applied load in lbf or N

d = distance between points measured by in. or mm

Tensile strength is approximately 80% of the value of I_s (ISRM, 1985).

Early applications of the point load test for determining unconfined compressive strength utilized an I_s multiplier *(k)* of either 20 (Franklin et al., 1971) or 24 (Broch and Franklin, 1972) with the value of I_s. Bieniawski (1975) has shown that the multiplier k is a linear function of the core diameter tested (Figure 4.18). Utilization of standard core size (e.g., NX) eliminates such problems in projects. For irregularly shaped test specimens, a 50-mm diameter is the standard size (Broch and Franklin, 1972). Figure 4.19 is a regression plot of published point load index and unconfined compressive strength data. The regression relation between the point load index, I_s, and the tangent modulus of elasticity, E_t, is shown by Figure 4.20. The point load test is also useful in determining maximum and minimum strengths that result from rock anisotropy, as in the case of foliated metamorphic rocks (Franklin et al., 1971).

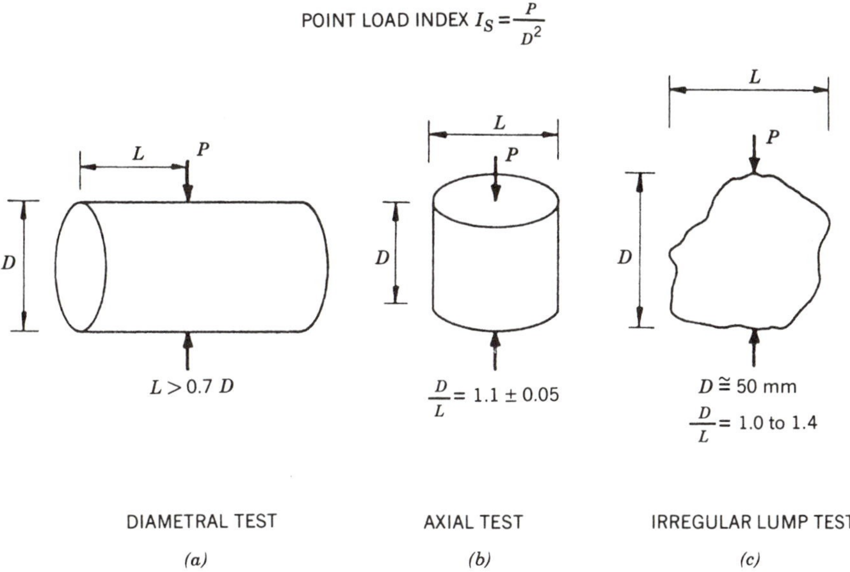

Figure 4.17 Point load test specimen shapes and related measurement parameters and limits. (Reprinted with permission from Eng. Geol., vol. 9, Z. T. Bieniawski, The Point-load Test in Geotechnical Practice, 1975, Elsevier Science Publishers B.V.)

The portability of point load test equipment, the ease with which testing may be conducted, and the correlation of test results with unconfined compressive strength combine to make the point load test a very useful one. Strength estimates may be directly applicable in the field in conjunction with rock mass parameters for excavation and tunneling work.

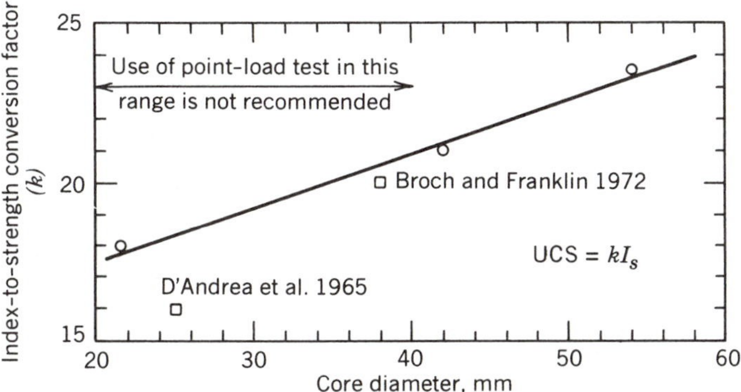

Figure 4.18 Size-correlation graph for index-strength conversion using the point load index, I_s, to obtain UCS. (Reprinted with permission from Eng. Geol., Vol. 9, Z. T. Bieniawski, The Point-load Test in Geotechnical Practice, 1975, Elsevier Science Publishers B.V.)

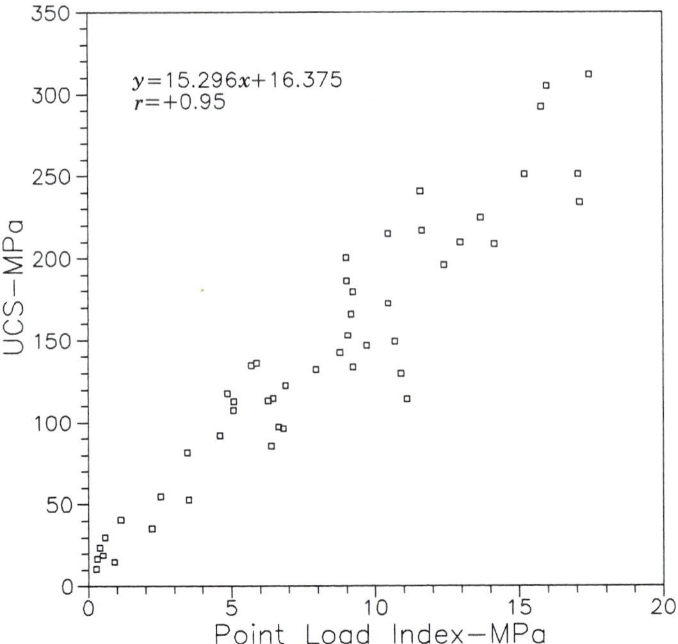

Figure 4.19 Plot of point load index, I_s, versus UCS. (Adapted from D'Andrea et al., 1965)

Schmidt Rebound Hammer Test

The Type L Schmidt rebound hammer (Figure 4.21) is another portable device that may be used to estimate unconfined compressive strength and modulus of elasticity of intact rock. The Schmidt rebound hammer (or, more commonly, the Schmidt hammer) was developed to estimate the strength of concrete. It operates on the principle that the measured rebound of a steel hammer mass when propelled with 0.075 kg-m of energy against a rock surface will be proportional to the hardness of the material,

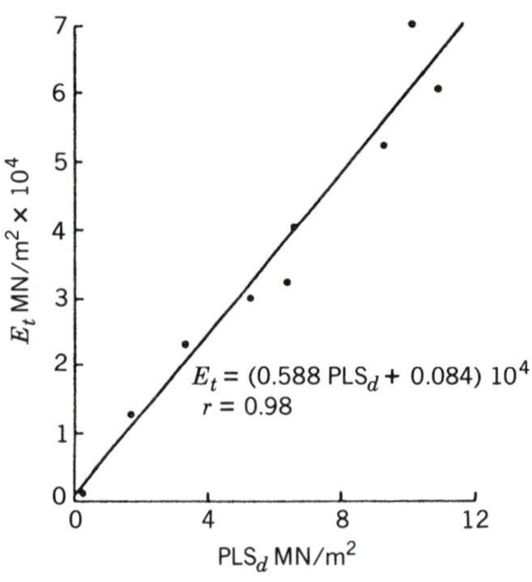

Figure 4.20 Plot of tangent Young's modulus versus point load index, I_s, where $PLS_d = I_s$. (Reprinted with permission from Bull., Intl. Assoc. Eng. Geol. No. 17, T. V. Irfan and W. R. Dearman, Engineering Classification and Index Properties of a Weathered Granite, 1978.)

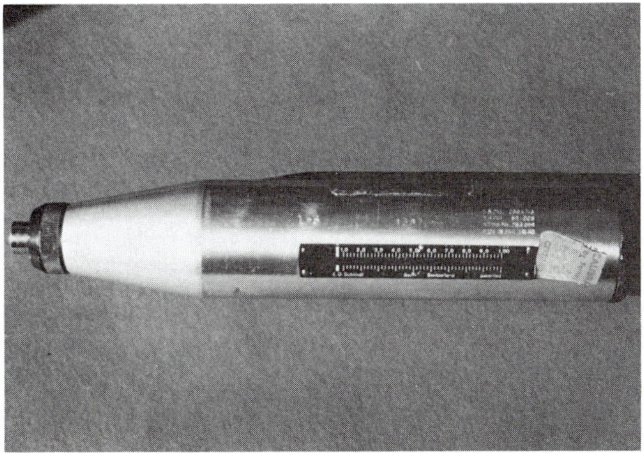

Figure 4.21 Type L Schmidt rebound hammer.

which may be correlated, in turn, with strength (Deere and Miller, 1966). Although ideally a nondestructive test, the Schmidt hammer test may cause breakage of core specimens with uniaxial strengths less than the strong category (50 MPa). Stronger specimens may also break unless tested in a special steel cradle designed for the core size being tested. The hammer may be used in the field to obtain *in situ* estimates of rock strength from rock surfaces.

Figure 4.22 illustrates the variation of Schmidt hammer values (SHV) with uniaxial compressive strength for 16 NX core specimens of Pikes Peak granite, Rampart Range, Colorado. It has been suggested that the uniaxial compressive strength can be estimated by multiplying the SHV by the dry unit weight of the specimen (Geological Society, 1977). Using the predicting equation for the Pikes Peak granite samples (Figure 4.22) having a dry unit weight of 2.643 mg/m^3, the calculated value exceeds the test strength by only 1 MPa. In general, however, it is estimated (Geological Society, 1977) that there is only a 75% probability that the laboratory determined uniaxial compressive strength will lie within 50% of the strength derived from correlation charts prepared by Deere and Miller (1966) (Figure 4.23). Thus, the Schmidt hammer test is considered to be a less desirable predictor of rock strength than the point load test. Published relationships between Schmidt hammer values and uniaxial compressive strength by Aufmuth (1974) and by Irfan and Dearman (1978) differ sufficiently to illustrate the prediction problem as follows:

$$\log \text{UCS} = 1.831 \log \text{SHV} + 1.533 \text{ (Aufmuth)}$$

$$\text{UCS} = 7.752 \text{ SHV} - 213.349 \text{ (Irfan and Dearman)}$$

Where:

UCS = uniaxial compressive strength

SHV = Schmidt hammer value

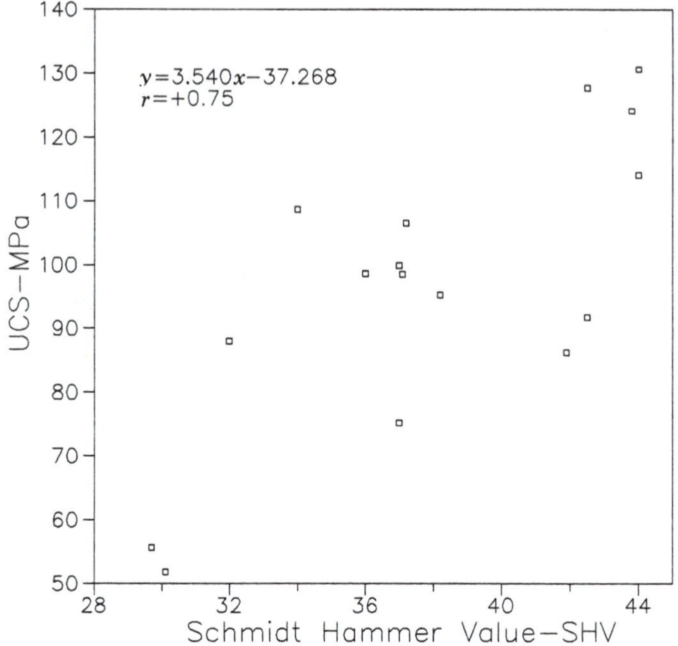

Figure 4.22 Plot of SHV versus UCS, Pikes Peak granite, Rampart Range, Colorado. (From Jesch et al., 1979)

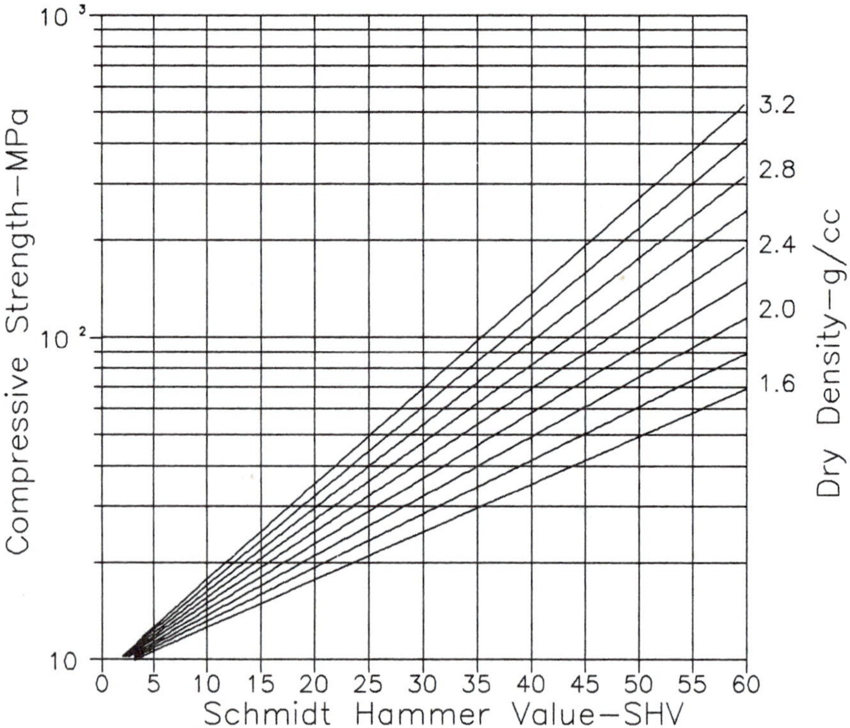

Figure 4.23 Rock strength prediction chart for Type L Schmidt hammer data. (From Deere and Miller, 1966)

Aufmuth (1974) has combined point load tensile strength and Schmidt hammer values to predict the tangent modulus of elasticity by the following equation:

$$\log E_t = 4.79 + 0.855\ \text{SHV} + 0.299 \log I_s \tag{4.4}$$

where:

E_t = tangent modulus of elasticity
SHV = Schmidt hammer value
I_s = point load tensile strength

Slake-Durability Test

The weatherability, or slaking characteristics, of the clay-rich soils and rocks is of considerable practical importance in civil engineering. Even though a minimum uniaxial compressive strength—such as the 1.25 MPa value suggested by some—may define the boundary between soil and rock, clay-rich earth materials present unique properties on both sides of this boundary. Morgenstern and Eigenbrod (1974) have summarized the weatherability problems of such materials as a result of exposure to moisture. They include the stability of excavations with time, surface durability of canal and tunnel walls, design of fills, and design and degree of compaction of mine waste dumps.

The weatherability of the clay-rich rocks probably presents the most problems because the degree of induration may cause the observer to be misled concerning their performance when exposed to the elements. As a result, the slake-durability test was devised by Franklin and Chandra (1972). The test consists of placing weighed, oven-dried samples of clay-rich rock in a drum having a 2-mm meshed exterior. Water is placed in a trough at a level below the drum axle, and the drum is rotated a specified number of times. The sample that remains in the drum is oven dried and the test is repeated for a second cycle. Following the second cycle, the slake-durability index, I_d, is obtained. It is the ratio of the weights of the dried sample before and after testing multiplied by 100. Standardized procedures for the slake-durability test have been established by the International Society for Rock Mechanics (ISRM, 1979b).

The two-cycle slake-durability test (Table 4.8) has been suggested as the standard for weatherability classification of the clay-rich rocks (Geological Society, 1977; ISRM, 1979b). However, when the second-cycle index is between 0% to 10%, the single-cycle slake-durability index is preferred, and it must be so identified.

Franklin and Chandra (1972) prefer a single-cycle classification based only on I_d, as illustrated in Table 4.9. A single-cycle classification used by Aufmuth (1974) for characterizing rock weatherability is seen in Table 4.10. Aufmuth also has shown that single-cycle slake durability is inversely related to the tangent modulus of elasticity (Figure 4.24).

A qualitative slaking test has been proposed by Wood and Deo (1975) to identify highly slake-susceptible shales prior to their use in embankment construction, which could result in embankment failure. The test involves

Table 4.8 Two-cycle Slake-Durability Classification

Slake Durability (I_d)	Classification
0-30	Very Low
30-60	Low
60-85	Medium
85-95	Medium High
95-98	High
98-100	Very High

Source: Modified and reprinted with permission from Intl. J. Rock Mech. Min. Sci. & Geomech. Abstr., Vol. 16, Anon., Suggested Methods for Determining Water Content, Porosity, Density, Absorption and Related Properties and Swelling and Slake-Durability Index Properties. Copyright © 1979, Pergamon Press.

immersing a sample in water and observing its slaking characteristics over a 24-hour period, or cycle, of wetting. This test subsequently has been identified as the jar-slake test (Lutton, 1977) for which index values, I_J, have been defined (Table 4.11). The standard slake-durability index values and jar-slake values both appear to have value for characterizing the slaking problems associated with compacted shale embankments.

Other Tests

Earlier, the relationships of compressional wave velocity in intact rock to uniaxial compressive strength and modulus of elasticity were described. The value of compressional, or P-wave, velocities as an index of rock strength and elastic moduli is well documented by D'Andrea et al. (1965), Judd and Huber (1962), and Irfan and Dearman (1978).

In addition to the index tests described, other tests are in use that are not included here either because of specialized application or less-common usage. These include the Taber abrasion test, the Shore hardness test, and the quick-absorption test. The reader is referred to papers by Deere and

Table 4.9 Single-cycle Slake-Durability Classification

Slake Durability (I_d) (%)	Classification
0-25	Very low
25-50	Low
50-75	Medium
75-90	High
90-95	Very high
95-100	Extremely high

Source: Reprinted with permission from Intl. J. Rock Mech. Min. Sci., Vol. 9, J. A. Franklin and R. Chandra, The Slake-Durability Test. Copyright © 1972, Pergamon Press.

Table 4.10 Rock Weatherability Classification Based on Single-cycle Slake Durability

Category	Word Description
A	*high* slake durability (less than 1 percent material loss)
B	*medium* slake durability (less than 3.5 percent but more than 1 percent material loss)
C	*low* slake durability (more than 3.5 percent material loss)

Source: Reprinted from Aufmuth, 1974, with permission from the American Society for Testing and Materials, 1916 Race Street, Philadelphia, PA 19103.

Miller (1966), Maidl (1972), Aufmuth (1974), Tarkoy (1975), ISRM (1978a), and Irfan and Dearman (1978) for descriptions, correlations, and applications of these and other tests.

ROCK MASSES

Large volumes of rock are involved in the design and construction of structures and excavations in rock such as dams, tunnels, underground power plants, road cuts, and open-pit mines. The stability and deformability of the rock is dependent on the strength and deformability of the

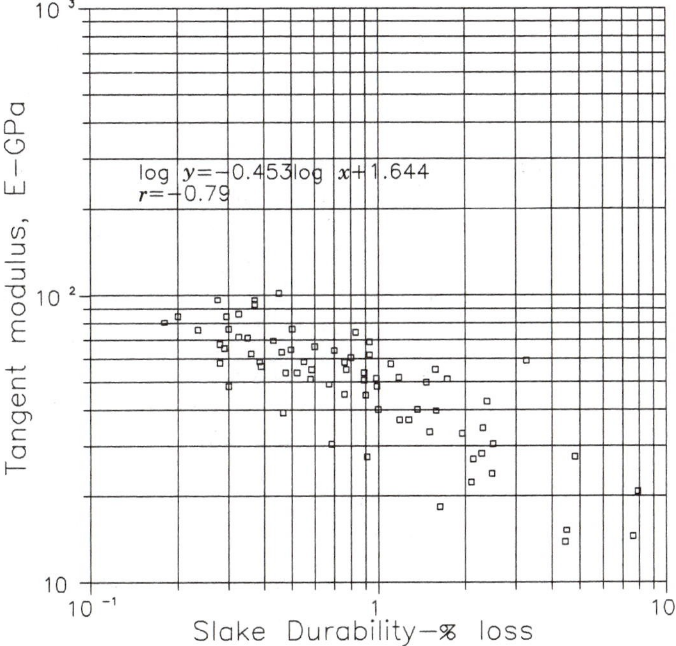

Figure 4.24 Relationship between one-cycle slake durability shown as percentage loss and tangent modulus of elasticity, *E*. (Modified and reprinted from Aufmuth, 1974, with permission from the ASTM, 1916 Race Street, Philadelphia, PA 19103.)

Table 4.11 Classification of Slaking Rock by the Jar-slake Index, I_J

I_J	Behavior
1	Degrades to a pile of flakes or mud
2	Breaks rapidly and/or forms many chips
3	Breaks slowly and/or forms few chips
4	Breaks rapidly and/or develops several fractures
5	Breaks slowly and/or develops few fractures
6	No change

Source: Lutton, 1977.

rock mass. The index properties of an intact sample are not the only criteria on which a design is based. A rock mass is typically more heterogeneous and anisotropic than intact rock.

The most universally occurring anisotropic characteristic of all rock masses is the presence of distinct breaks, or discontinuities, in the physical continuity of the rock. These include bedding surfaces, joints, faults, and well-developed metamorphic foliation. The resultant rock mass is a discontinuous aggregation of blocks, plates, or irregular geometric forms that will have significantly different physical properties compared with an intact sample from the same rock mass.

Apart from the reduction in strength from pervasive chemical weathering, the presence of discontinuities in a rock mass is the primary controlling factor of mass strength and deformability. The reduction in shear strength of rock caused by a joint is illustrated in Figure 4.25. Many investigators have noted that the strength of a rock mass is more a function of discontinuities within a rock mass than of the intact rock strength unaffected by discontinuities. By comparison, failures in soil masses are not so dependent on the presence of discontinuities because soils and soil-like rocks such as shale are intrinsically weak (Piteau, 1971).

The evaluation of the engineering properties of a rock mass includes knowledge of the intact rock properties, the occurrence and nature of discontinuities, and the degree and extent of chemical weathering. Many have noted the interaction of these factors in developing rock mass quality-classification schemes (see discussion on rock classification).

Discontinuities in Rock Masses

More information concerning discontinuities other than spacing is needed when considering the influence of discontinuities on the many types of engineering construction on and within rock masses. At this point, it is sufficient to recognize that in addition to spacing, such factors as orientation; surface roughness; surface weathering, or alteration; and presence of water influence the strength and/or deformability of a rock mass. Compressive strength of an intact sample, thus, is an inadequate measure of rock mass strength characteristics for many engineering applications that involve stability. Instead, the shear strength of a discontinuous rock mass is of primary importance. In applications where deformation of a rock mass

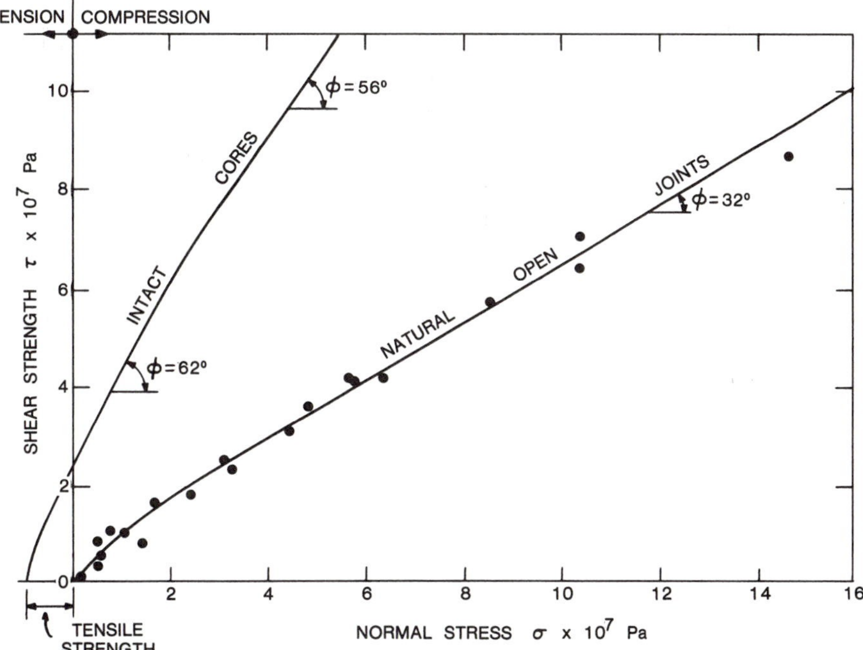

Figure 4.25 Comparison of Mohr strength envelopes of intact cores and natural open-joint shear strengths for quartz monzonite. (Reprinted with permission from Proc. 6th Symp. on Rock Mechanics, K. S. Lane and W. J. Heck, Triaxial Testing for Strength of Rock Joints, 1964, University of Missouri, Rolla.)

is critical, as in dam foundations and abutments and in pressure tunnels, the deformation moduli from intact samples become part of the total evaluation of the rock mass in addition to the deformation that may result from closure of and movement along discontinuities.

Rock masses are heterogeneous and anisotropic because of differing rock types, presence of discontinuities, and varying degrees of weathering. It is possible to set geologic limits in the field on rock masses that are similar in these respects. The boundaries commonly are lithologic contacts or major discontinuities such as faults. The advantage of this procedure is increased efficiency of sampling, which permits gathering of representative data for a given mass. In the engineering geologic evaluation of an area, rock masses with similar major lithologic and structural characteristics are known as *structural regions* (Jennings and Robertson, 1969; Piteau, 1973).

In structural regions, discontinuities have a dominant role in defining rock mass properties. As a result, a given structural region is limited to a rock mass with a statistically uniform distribution and character of discontinuities. Systematic sets of joints within a structural region are termed *design joints* (Steffen and Jennings, 1974). All joints in a given set roughly parallel one another. A joint system consists of two or more joint sets in an area.

Although joints, bedding surfaces, foliation, faults, and shear zones are all forms of discontinuities, the term *joint* is loosely, or nongenetically, used in the geotechnical literature for all or part of the various discontinuities in rock masses. Usage ranges from the geologic definition of a joint being a fracture without displacement to an all-inclusive grouping of joints, bedding surfaces, faults, and other surfaces of weakness within a rock mass (Bieniawski, 1973). Care must be taken to determine whether the terminology used by others is genetic or nongenetic. Each genetic or geologic discontinuity will have characteristic features implied by its name. The nongenetic, or generalized, use of the term *joint* is the primary source of potential misinterpretation of the physical properties of discontinuities. Problems with terminology have led to the development of a number of features that characterize a discontinuity regardless of its origin.

Here, the term *joint* is used in its broadest, nongenetic sense except when indicated otherwise. This will emphasize the need for thorough observation and measurement of discontinuity features without the bias introduced by genetic or geologic terminology.

Characteristics of Discontinuities

Orientation, spacing, continuity, surface characteristics, the separation of discontinuity surfaces, and the accompanying thickness and nature of filling material (if present) are the most consistently measured joint properties, or factors. Strengths of rock masses imparted to a large degree by shear strengths of discontinuity surfaces are typically dependent on one or more of these factors. They have been summarized by Bieniawski (1973) and Cording et al. (1975).

Orientation. The most readily apparent influence of the orientation of discontinuities on rock mass strength is evident in the failure of rock slopes along one or more discontinuities. The influence of joint orientations on slope stability (Figure 4.26) has been documented in major failures at Vaiont, Italy (Müller, 1964); Madison Canyon, Montana (Hadley, 1974);

Figure 4.26 Wedge failure caused by intersecting discontinuities, Libby Dam, Libby, Montana. Instruments monitoring rock mass stability are visible in the photo.

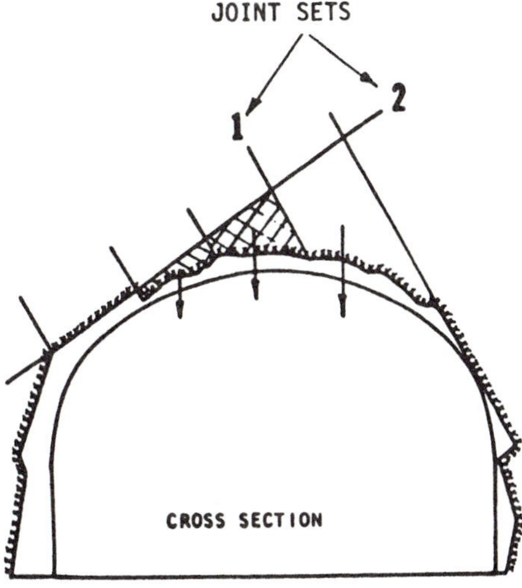

Figure 4.27 Unstable tunnel roof conditions caused by interaction of orientation and spacing of two joint sets. (From Cording et al., 1975)

Libby Dam, Montana (Hamel, 1974); and Frank, Alberta (Cruden and Krahn, 1973). Similar but less well-known rock-slope failures that occur in road cuts, open-pit mines, and in natural slopes are no less important from the engineering geologic standpoint. The spacing and orientation of multiple joint sets are of major concern in tunnel design and construction and in the stability of rock slopes (Figures 4.27 and 4.28). The influence of joint orientation on roof stability in tunnels has been summarized by Cording and Mahar (1978).

Two factors of primary importance relative to the influence of joint orientation on rock-slope stability are: (1) whether joints or joint intersections cut or daylight the slope at less than the natural or manmade slope

Figure 4.28 Rock falls caused by closely spaced multiple joint sets, with one set daylighted, Golden Gate Canyon, Colorado.

angle (Figure 4.28) and, if so, (2) whether the dip angles of the joints or the plunge angles of the joint intersections exceed the angle of friction along the joint surfaces as illustrated in Module 4.4. Thus the daylighting of certain joints in a rock mass greatly influences the stability of a rock mass.

The orientation of joints within a rock mass also influences the strength anisotropy of the mass. Rock masses with irregularly oriented joints have a greater degree of block interlock and less mechanical anisotropy than those masses with regularly oriented joints (McMahon, 1968). The anisotropy increases with increased regularity of orientation because preferred directions of weakness are generated within the rock mass. The design joint concept described earlier is based on joint regularity.

Jointing may be shown graphically on an equal-area stereographic projection of joint poles (Figure 4.29). The clustering of poles for each of several systematic planar discontinuities (sets) makes the stereographic projection a useful means of obtaining average orientation data for the sets. In Figure 4.29, the presence of three planar discontinuity sets is obvious from the clustering of the poles. In this example, a gently dipping bedding surface and two steeply dipping, orthogonal joint sets are illustrated (see Module 4.5). Increasing variation of joint orientation for a given set results in a greater scatter of poles. Of the three discontinuities, the bedding exhibits the least variability. The evaluation of joint orientation measurements has been investigated in detail by Kohlbeck and Scheidegger (1977).

The principles and techniques of plotting planes and poles on a stereographic projection are in textbooks on structural geology. Hoek and Bray (1981) have prepared similar instructional materials with the additional advantage of including engineering applications. There are abundant literature references to geotechnical applications of stereographic projection of discontinuities. Those by John (1968), Heuze (1974), and Goodman (1976) serve to introduce the reader to this literature.

Spacing. The spacing of discontinuities affects overall rock mass strength or quality. Even the strongest intact rock is reduced to one of little strength when closely spaced joints are encountered (Figure 4.28). Conversely, where the spacing is great, the behavior of the rock mass will be strongly influenced by the intact rock properties. An exception exists in the case of widely spaced daylighting joints that create a kinematically unstable state. Discontinuity spacing and intact rock strength have been used by Franklin et al. (1971) to suggest excavation methods for rock masses (Figure 4.30).

Orientation and frequency of joints combine to influence the response of a rock mass to construction. Where intersecting and daylighting joint surfaces are encountered in a cut face, closely spaced joint surfaces tend to cause numerous rock falls or raveling of the surface (Figure 4.28), whereas widely spaced joints may tend to cause massive catastrophic block failures. In tunneling, wedge-forming orientations (Figure 4.27) perform quite differently with variations in spacing. Closely spaced joints permit the wedges to fall, requiring more support than widely spaced joints, whose blocks may exceed tunnel size and may need little or no support (Cording and Mahar, 1978).

MODULE 4.4

Stereo plot illustrating kinematic relationships of two intersecting joints daylighting a cut slope.

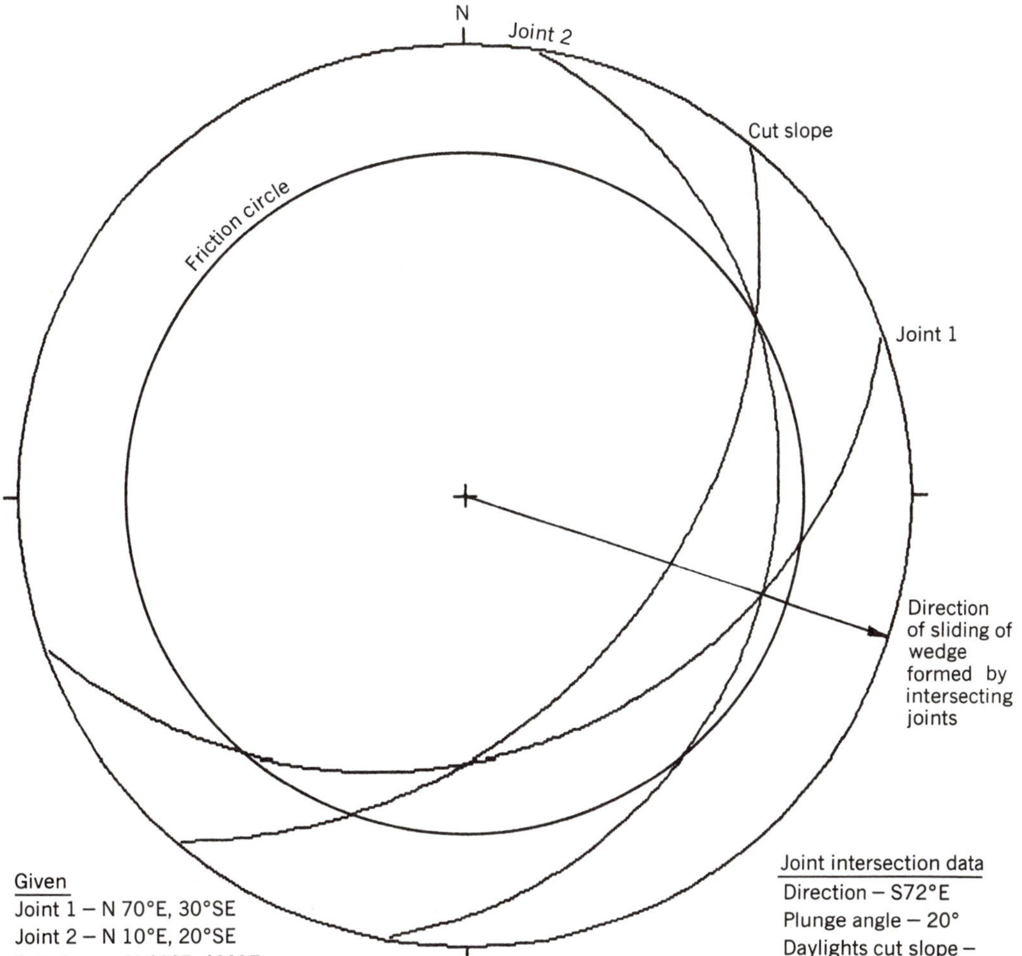

N

Joint 2

Cut slope

Friction circle

Joint 1

Direction of sliding of wedge formed by intersecting joints

Given
Joint 1 – N 70°E, 30°SE
Joint 2 – N 10°E, 20°SE
Cut slope – N 40°E, 40°SE
Friction angle (circle) – 15°

Joint intersection data
Direction – S72°E
Plunge angle – 20°
Daylights cut slope – cut slope angle > plunge angle
Unstable wedge formed – plunge angle > friction angle

Spacing distance or frequency of joints is not necessarily constant throughout a rock mass for any given joint set. Joint distributions within sets have been examined and found to follow: (1) even spacing, (2) random spacing, and (3) clustered distributions (Priest and Hudson, 1976). In some

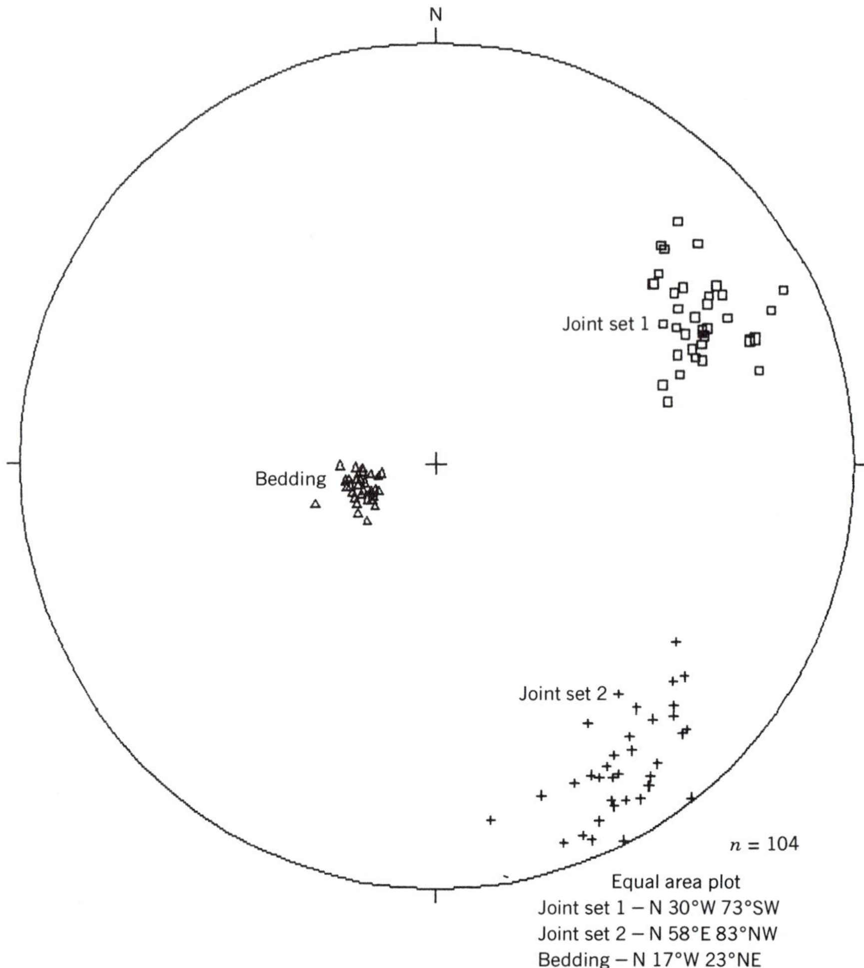

Figure 4.29 Pole plot of two orthogonal joint sets and bedding in sandstone, Larimer County, Colorado.

cases there is a tendency toward clustering or swarming of joints at regular intervals. Many factors contribute to joint spacing or distribution, including lithology, tectonic stresses, overburden stresses, and depth. Prior to excavation, joint spacing can be expected to increase with increasing depth except where the rock mass has been influenced by faulting.

Continuity. Although it may appear to be paradoxical, discontinuities are themselves discontinuous. Length or continuity is measured on an exposed rock surface. The average measured continuity, persistence, or size of joints in a given set may not be representative of actual conditions because of limited surface exposures. Except where a joint is seen to terminate in the rock mass, all lengths measured are minimum lengths.

With the exception of very continuous, throughgoing discontinuities, some of the strength of the intact rock is transferred to the rock mass

MODULE 4.5

Plot of planes and associated poles for Figure 4.29.

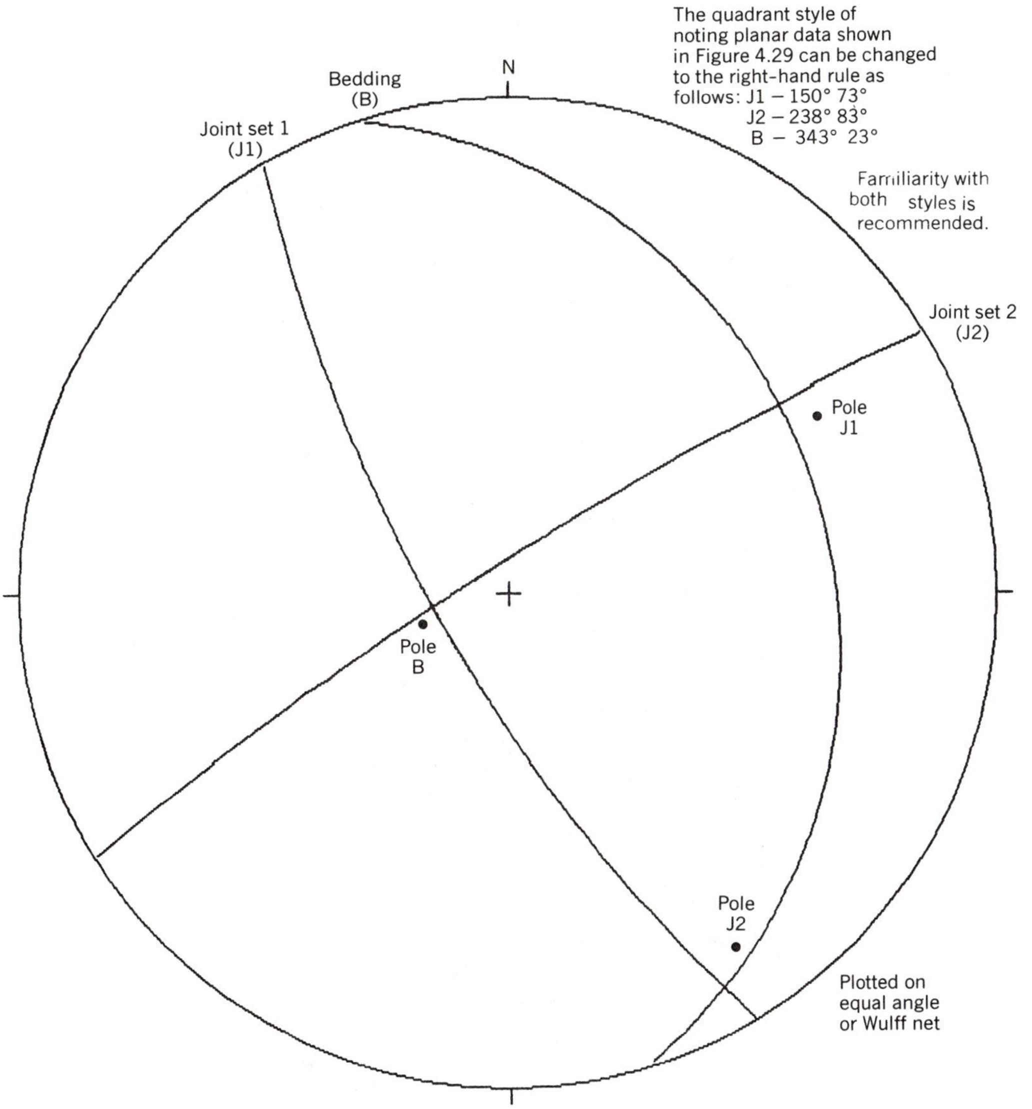

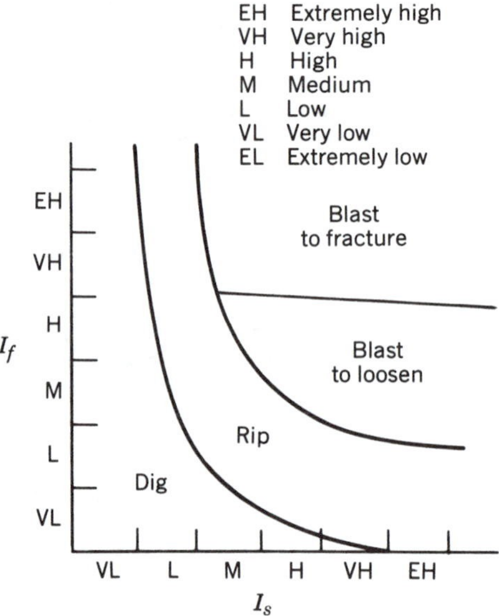

Figure 4.30 Excavation methods as a function of intact strength, estimated by point load index (I_s) and fracture spacing (I_f). (Reprinted with permission from Trans., Institution of Mining and Metallurgy, Sec. A, Vol. 80, Bull. 770, J. A. Franklin, E. Broch, and G. Walton, Logging the Mechanical Character of Rock, 1971.)

through the intervening rock or rock bridges. Depending on potential movement directions within the rock mass, either the shear or the tensile strength of the intact rock will be the factor to be considered. Maximum reduction in rock mass strength in such cases is not achieved until the intact rock or rock bridge fails, permitting propagation of failure planes or development of steplike surfaces between adjacent joints.

Where the strength contribution of the rock bridges is critical to the continued stability of a rock mass, maintenance of the bridges during construction involving blasting is imperative. Through the use of linearly arranged, closely spaced drillholes, a preferred fracture is created between the drillholes by explosive charges in the drillholes. This procedure is known as *presplitting.*

Surface Characteristics. Three factors are involved when the surface characteristics of discontinuities are considered. They are: (1) the waviness or undulation of the surface, which results in variations in orientation or attitude along a given discontinuity; (2) the smaller scale roughness of the surface, which provides friction between two adjacent blocks; and (3) the physical properties of any material that may fill the space between the two bounding surfaces of a discontinuity.

Waviness and roughness were originally defined by Patton (1966) as first- and second-order irregularities based on their relative magnitudes. Waviness and roughness are illustrated in Figure 4.31. Roughness generally is the more qualitative of the two. The projections from a rough discontinuity surface are called *asperities*, and they constitute surface roughness. Visually determined joint roughness measures, or coefficients (Figure 4.32), have been proposed by Piteau (1971), Barton and Choubey (1977), and Tse and Cruden (1979). Tracings of roughness for visually estimating roughness may be made in the field by the contour gauge used by woodworkers and other craftspersons to duplicate a pattern (Stimpson, 1982). Precise measurements of roughness may be made photogrammetrically, optically, or by mechanical surface-profiling instruments. Waviness, with its longer wavelength, may be measured in the field with a Brunton compass or similar device.

Waviness has a greater influence on rock mass strength than roughness where slope stability is involved. For two adjacent blocks to move along a wavy intervening surface, there must be displacement or dilation normal to the surface. The amount of dilation is controlled by the angle i and the wave length of the wave involved [Figures 4.31 (A–first order) and 4.33]. The waviness angle in effect decreases the slope angle β and decreases the tendency to slide. By comparison, the asperities on a rough surface must be sheared off by movement; after that has been accomplished, the resulting smoother surface has a lower friction angle, which is quite independent from any restriction in movement imposed by waviness. The reduction in shear strength from asperity shearing is illustrated in Figure 4.34.

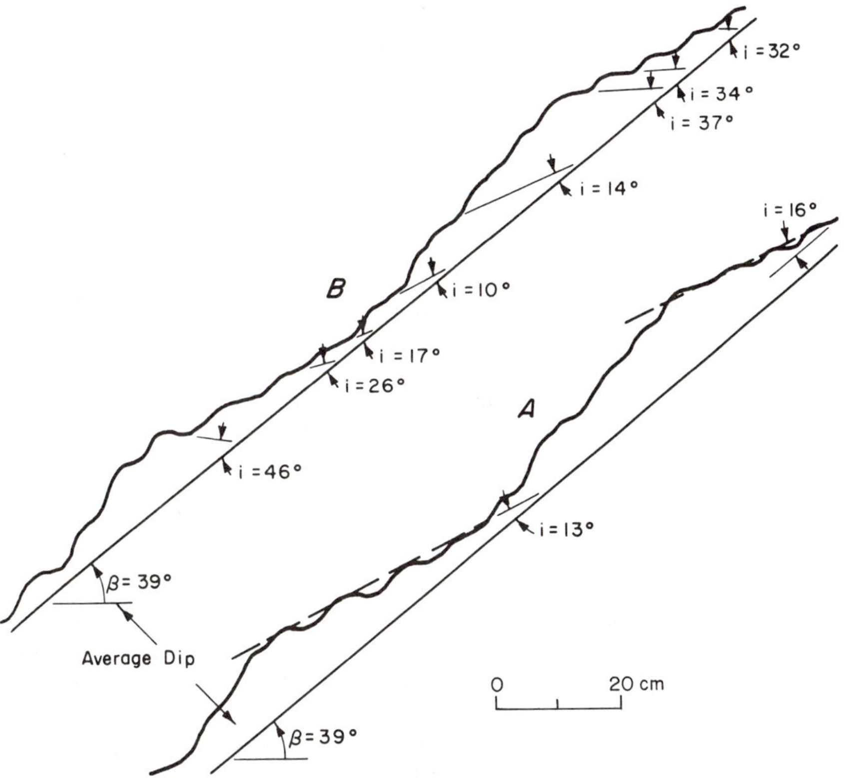

Figure 4.31 Irregularities on a discontinuity. *A*–first order and *B*–second order, where β = average slope angle; *i* = angle between surface feature and average dip of discontinuity. (Reprinted with permission from Proc. 8th Symp. on Rock Mechanics—Failure and Breakage of Rock, D. U. Deere et al., Design of Surface and Near-Surface Construction in Rock, 1967, Am. Inst. Min. Metall. & Pet. Eng.)

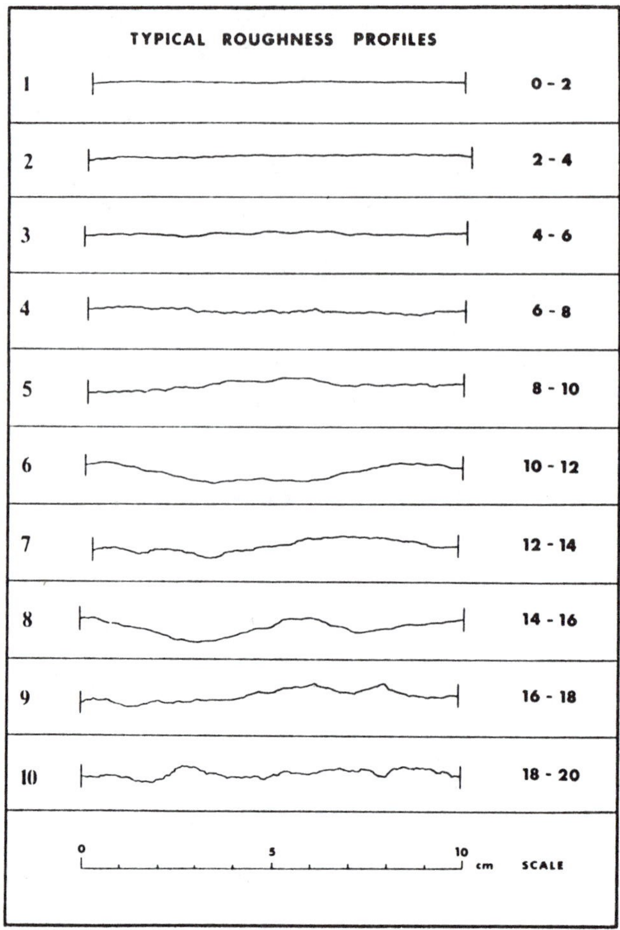

Figure 4.32 Typical roughness profiles for a range of joint roughness coefficients. (Reprinted with permission from Rock Mechanics, vol. 10, n. Barton and V. Choubey, The Shear Strength of Rock Joints in Theory and Practice, 1977.)

Separation and Filling of Joints. The amount of separation, or space, between joint surfaces and the presence of filling material may have a profound influence on the strength of a jointed rock mass. Combined with roughness, they constitute the character of the joint (Brekke and Howard, 1972). The separation of joint walls may result from the tensile stresses that created the joints, solution widening of joints, or shearing movements that can separate the surfaces by generating gouge material or by movement along wavy surfaces (dilation).

Where separations occur, the space may be empty, partially filled, or completely filled. The filling material may be clay, silt, sand, or coarse fragmental material or mixtures of them resulting from depositional filling, faulting, or wall-rock weathering. These materials typically have low shear strengths. Cording et al. (1975) have reported friction angles of 8° to 15°

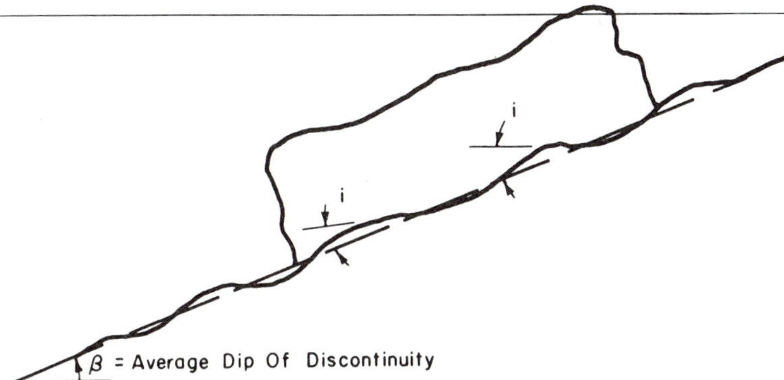

Figure 4.33 Sketch of a rock block resting on an inclined irregular discontinuity, where angles *i* and *β* are defined in Figure 4.31. (Reprinted with permission from Proc. 8th Symp. on Rock Mechanics—Failure and Breakage of Rock, D. U. Deere et al., 1967 Am. Inst. Min. Metall. & Pet. Eng.)

where shearing has occurred along clay-filled joints. This compares with angles of 22° to 35° for sheared, unfilled joints. Filling material can be mineral precipitate that heals the joint with material that may be as strong or stronger than the wall rock, a situation that may significantly improve the rock mass strength.

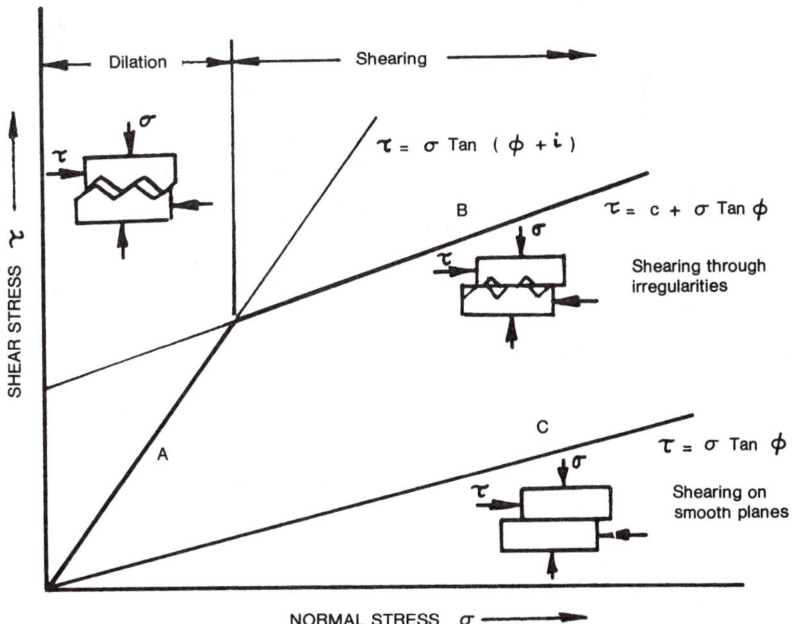

Figure 4.34 Simplified relationships between shear strength and normal stress for rough surfaces, where ϕ = friction angle and *i* = inclination of surface irregularity to plane surface. (Reprinted with permission from Rock Slope Engineering, 1st ed., E. Hoek and J. M. Bray, 1974, Institution of Mining and Metallurgy.)

The separation width of a filled joint influences the shear strength along the joint. If the filled separation width is narrow enough to permit the asperities to interlock, the peak strength of the joint will be influenced more by the shear strength of the rock and degree of roughness of the joint than the shear strength of the filling material. If the separation is so wide that there is no rock contact, the friction properties of the joint will be those of any filling material present and failure will follow soil mechanics criteria. A comparison of these relative strength conditions is shown in Figure 4.35.

Shear Strength of Discontinuities

A sketch of a direct shear box used to obtain the discontinuity shear characteristics illustrated by Figures 4.34 and 4.35 is shown in Figure 4.36. For a given normal stress, σ_n, the shear strength mobilized by a rough discontinuity surface peaks at the point of asperity shearing caused by differential horizontal movement of the blocks. This point, the peak shear strength of the discontinuity, is illustrated in Figure 4.37. It represents the same physical state illustrated in Figure 4.34 at the break in slope between the dilation and shearing portions of shear strength envelope A–B. The portion of the curve on Figure 4.37 that has constant shear strength for continuing horizontal displacement is known as residual strength and represents the friction that exists following asperity shearing. It is shown on Figure 4.34 by the shearing portions, B and C, of the shear strength envelopes, B and C. *In situ* determination of the shear strength of a discontinuity can be accomplished as illustrated in Figure 4.38. A portable shear box developed at the Imperial College of Science and Technology,

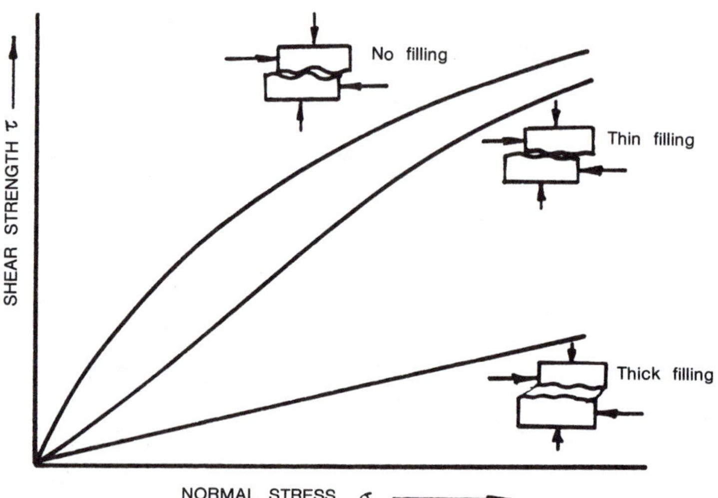

Figure 4.35 Relationships between shear strength and normal stress for discontinuities with different thicknesses of gouge infilling. (Reprinted with permission from Rock Slope Engineering, 1st ed., E. Hoek and J. M. Bray, 1974, Institution of Mining and Metallurgy.)

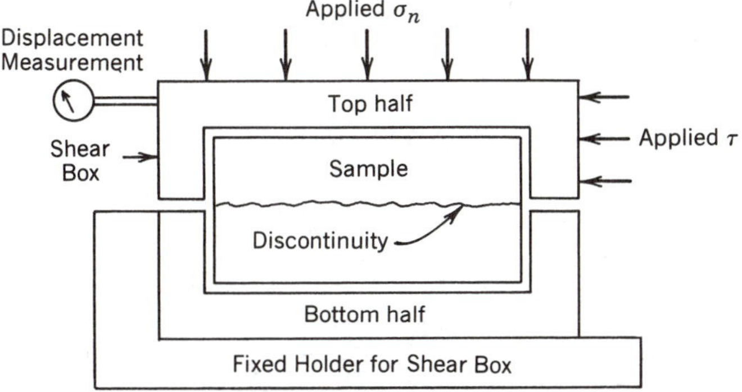

Figure 4.36 Diagram of a direct shear box for determining shear strength characteristics of discontinuities.

University of London, permits similar testing of small samples separated by discontinuities (Hoek and Bray, 1981).

The degree of roughness is related to the origin of the discontinuity and the rock type involved. In general, discontinuities result from tension on the rock mass, by shearing between adjacent rock blocks, and from irregular primary surfaces such as bedding surfaces. In forming true joints, tension imparts no smoothing to the surface and the grain size and cementation, or interlock, of constituent minerals govern roughness. Shearing associated with faulting typically creates a smoother surface (slickensides) that will be at or near residual strength when encountered during construction.

Fracturing of rock by tension creates clean surfaces. Shearing characteristically results in varying amounts and thicknesses of detritus. Joints and bedding surfaces tend to have appreciable waviness compared with the more planar surfaces caused by shearing. Thus the stability of a sheared surface at residual strength is reduced by low inherent waviness.

The foregoing material on discontinuities only briefly considers the features that significantly affect the strength of a rock mass. Johnson

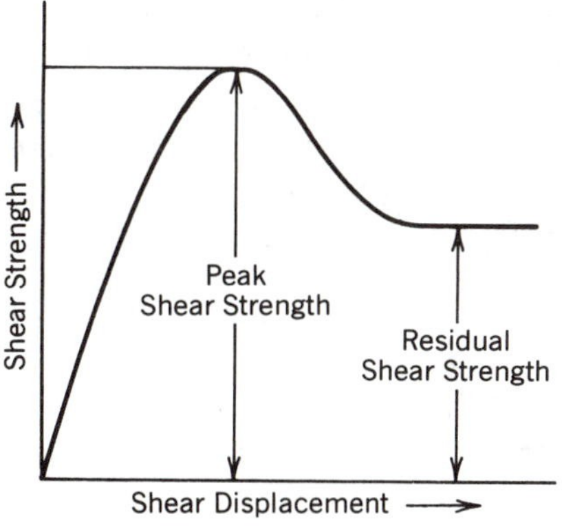

Figure 4.37 Plot of shear stress mobilized along a discontinuity for increasing shear displacement at constant normal stress.

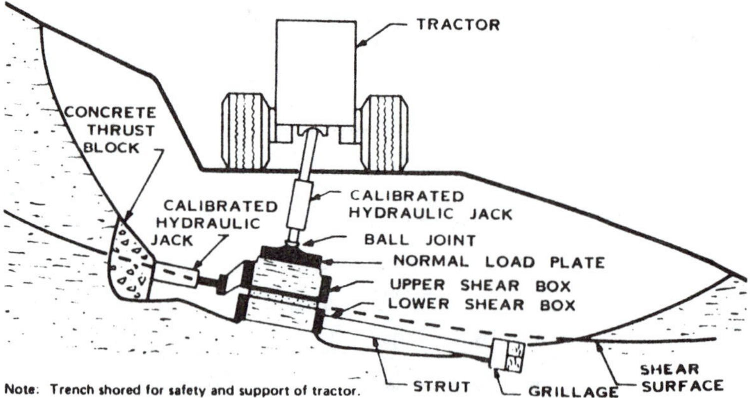

Figure 4.38 *In situ* determination of shear strength of rock along a discontinuity. (From Sowers and Royster, 1978)

(1979) has prepared a comprehensive literature review of discontinuities and other geologic factors that affect the stability of rock slopes. The influence of discontinuities on the behavior of rock in tunnels creates problems unique to, and representative of, underground construction in general. These problems have been addressed by Cording and Mahar (1974). The direct shear strength properties of a variety of discontinuity conditions and associated rock types have been summarized by Kulhawy (1975). Quantitative description and measurement of the many attributes of discontinuities have been prepared by the Geological Society Engineering Group Working Party (Geological Society, 1977) and the International Society for Rock Mechanics Commission on Standardization of Laboratory and Field Tests (ISRM, 1978f).

Weathering of Rock Masses

The weathered state of rock has a significant influence on the engineering properties of a rock mass. Physical weathering results in changes in size and number of discontinuities present in a rock mass. Chemical weathering of the rock mass is enhanced by movement of groundwater through the networks of discontinuities present in rock masses (Figure 4.39). The availability of water depends on local climate (moisture and temperature) and drainage (controlled by local topography) in conjunction with either primary or secondary rock mass permeability.

Degrees of chemical weathering, ranging from total decomposition to fresh rock, may be noted in susceptible rock masses that have been subjected to weathering over a long time under ideal climatic conditions (Figures 4.40 and 4.41). The control of water movement exerted by bedding surfaces, joints, and faults may result in localized and often deeply penetrating zones of weathering. Gradation from fresh to decomposed rock

Figure 4.39 Weathering along joints in granite, Estes Park, Colorado.

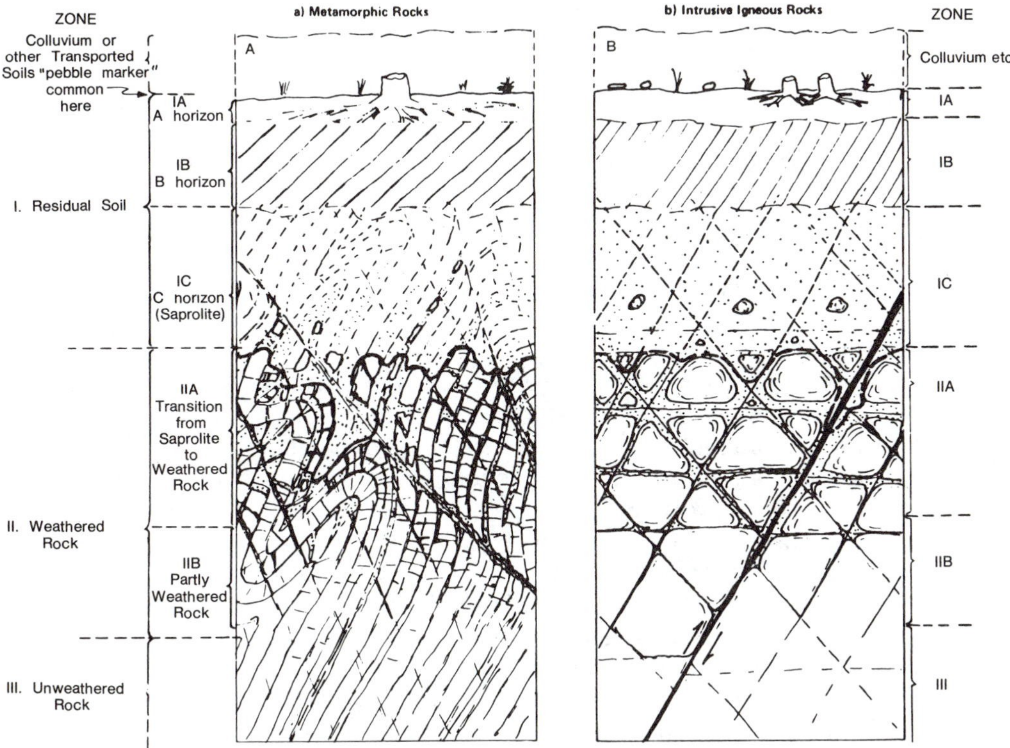

Figure 4.40 Typical weathering profiles for (*a*) metamorphic and (*b*) intrusive igneous rocks. (Reprinted with permission from Proc. 4th Pan American Conference on Soil Mechanics and Foundation Engineering, D. U. Deere and F. D. Patton, Slope Stability in Residual Soils, 1971, Fig. 1, p. 90, Am. Soc. Civil Eng.)

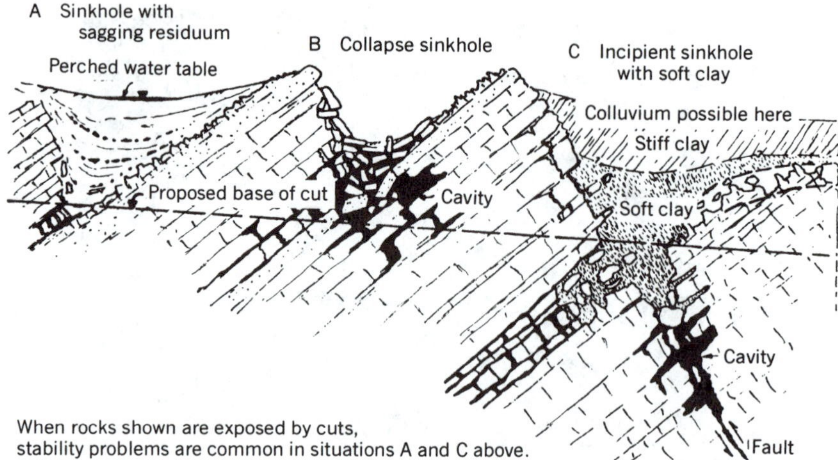

Figure 4.41 Typical weathering feature of carbonate rocks. (Reprinted with permission from Proc. 4th Pan American Conference on Soil Mechanics and Foundation Engineering, D. U. Deere and F. D. Patton, Slope Stability in Residual Soils, 1971, Fig. 10, p. 114, Am. Soc. Civil Eng.)

along discontinuities will be relatively perpendicular to the orientation of the discontinuity and may bear no relation to surface topography.

Hydrothermal chemical alteration of a rock mass may occur in different degrees without any relationship to depth relative to the surface. Logging of drill cores should include evaluation of chemical weathering and alteration changes with depth for a correct interpretation of local conditions. Papers by Saunders and Fookes (1970), Deere and Patton (1971), and Dearman (1974) are recommended for more information on the chemical weathering of rock masses.

The negative influence of physical and chemical weathering on rock mass strength or quality is apparent. The impact of physical weathering may be quantified by taking field measurements (e.g., spacing and frequency of jointing). The degree to which a rock mass has been weakened by chemical weathering and/or alteration is less adaptable to quantitative measurement. As early as 1946, Terzaghi recognized the importance of chemical decomposition on rock mass quality. Since that time, many have included chemical weathering in rock mass quality classifications, each with a descriptive system for evaluating the degree of weathering.

The weathering classification of the Geological Society Engineering Group Working Party (Geological Society, 1977) described in Table 4.7 is typical of current practice. It reflects the influence of discontinuities present in a rock mass on chemical weathering. An idealized diagram of the classification is shown in Figure 4.42. The role of discontinuities in the development of rock mass chemical weathering is shown in Figures 4.39 through 4.42.

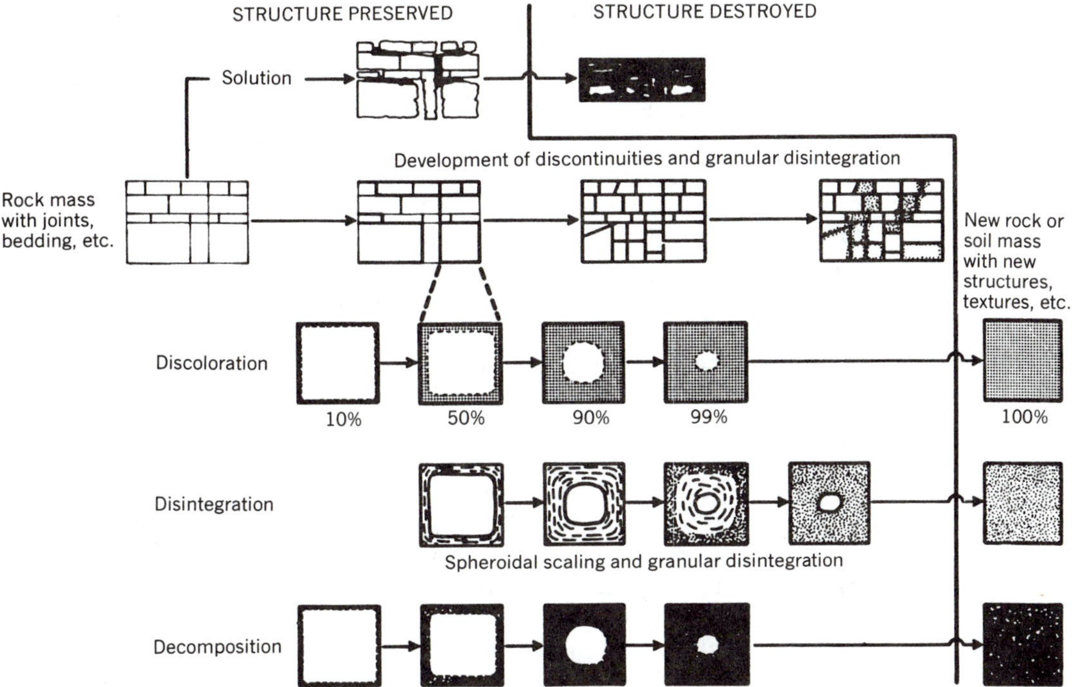

Figure 4.42 Idealized diagram of the stages of weathering of a rock mass. (Reprinted with permission from Bull. Intl. Assoc. Eng. Geol. No. 9, W. R. Dearman, Weathering Classification in the Characterization of Rocks for Engineering Purposes in British Practice, 1974.)

Rock Mass Deformation

The strengths of rock masses are of obvious importance in engineering projects that require fail-safe conditions, as in mine- and tunnel-roof spans, open-pit mine slopes, and highway and railroad rock cuts. The deformability of a rock mass, however, also is of major consequence in many types of construction because it may be structurally intolerable. Nuclear power plants, arch dams with reservoir impoundments, and tunnels carrying water under pressure are examples of structures that require minimal deformation of rock masses over the designed life of the structure.

Rock mass deformation results primarily from the closure of discontinuities and the elastic and plastic deformation of the intact rock that comprises the rock mass. The modulus of elasticity typically is used as a measure of intact rock elastic deformation. In rock masses, the modulus of deformation, E_d, is used as a measure of deformation. The modulus of deformation is defined as the sum of the deformation that occurs with closure of joints in the rock mass under compression (plastic) and the deformation that occurs with continued stress application after crack closure (elastic).

The modulus is obtained by *in situ* testing during which the rock mass is subjected to several cycles of compressive stress loading and unloading. A typical plot of stress and strain (deformation) for a single loading and unloading cycle of a rock mass in which deformation from crack closure and elastic response are shown on the deformation axis (Figure 4.43). Several cycles are utilized to simulate deformation during and following construction. Deere et al. (1967) have suggested that the measured plastic (δ_p) and elastic (δ_e) displacements obtained from *in situ* testing may define rock mass quality by the expression, rock quality $= \delta_e/(\delta_p + \delta_e)$, where unity implies a discontinuity-free rock mass.

The conservative secant deformation modulus, E_{sm} (Figure 4.43), is used in engineering design as is the secant modulus of elasticity for intact rock. It is obtained during the loading cycles of *in situ* deformation testing. Note that E_{sm} is greatly influenced by discontinuity or crack closure in the rock mass and that E_{tm} is the tangent deformation modulus during the unloading cycle. It is a measure of rock mass elastic properties. Ideally E_{tm} is the same as the laboratory-determined intact rock modulus of elasticity, E_t. The expression E_{sm}/E_{tm} defines the reduction in rock mass elasticity caused primarily by crack closure. It is the reduction factor or modulus of deformation ratio.

Although the secant modulus of deformation is conceptually a measure of the modulus of deformation, many factors complicate the calculation of a value for the modulus, E_d. These factors include the pressure applied to the rock, modulus of elasticity, Poisson's ratio of the rock, the radius of the plate applying the load, the displacement of the rock from the applied load, and the depth at which the displacement measurement is taken. The latter two are of significance because they illustrate that the modulus of deformation is not a constant for a given site. There are many three-dimensional variations in the orientation, frequency, and separation of the discontinuities in a rock mass that influence the magnitude of deformation laterally and at depth. The abutment areas for a thin-arch concrete dam are examples in which discontinuity characteristics at depth in the rock are not representative of near-surface conditions that include differing degrees and angles of exfoliation jointing. Measurement of displacement with depth is accomplished by multiple position borehole extensometers described in chapter 6.

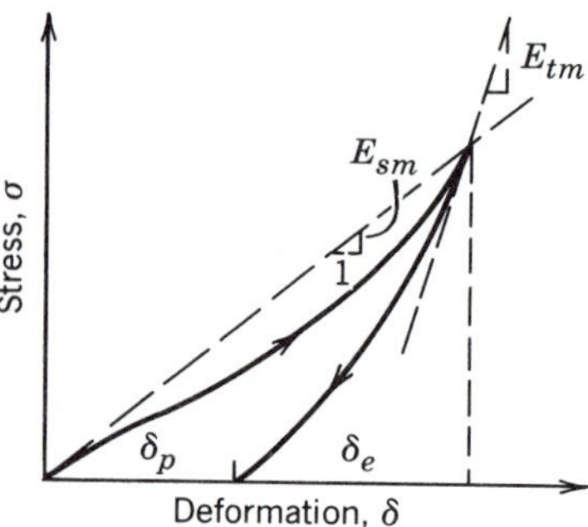

Figure 4.43 Typical stress-strain relationships for a rock mass where E_{sm} = secant modulus, E_{tm} = tangent modulus, δ_p = plastic deformation and δ_e = elastic deformation. (Reprinted with permission from Proc. 8th Symp. on Rock Mechanics—Failure and Breakage of Rock, D. U. Deere et al., Design of Surface and Near-Surface Construction in Rock, 1967, Am. Inst. Min. Metall. & Pet. Eng.)

Measurement of Deformation

Determination of the deformation modulus of a rock mass is made at the design stage of a structure. Ideally, large-scale *in situ* tests are conducted that will simulate both the magnitude and direction of structural loads at depths and volumes of rock appropriate for the structure. In lieu of the ideal, smaller-scale tests may be conducted involving smaller, less-representative volumes of rock. In either case, loading cycles are obtained by jacking, or physically compressing, the rock. Provision must be made in each case for measurement of the magnitude of deformation.

Large-Scale Tests. The deformation data most representative of a rock mass in place are obtained by large-scale uniaxial [plate (plate-bearing)] and radial jacking tests. Diagrammatic views of the jack configuration used for these tests are shown in Figures 4.44 and 4.45. In either test, loading of the rock is achieved by hydraulic flat jacks installed in the jack assembly near the rock surface. Flat jacks consist of two parallel steel plates welded together so that an intervening space may be filled with hydraulic fluid or oil. Each jack is separated from the rock by thin pads of material such as concrete or shotcrete, which eliminate rock surface irregularities. The loading of the rock mass in radial jack tests is uniformly distributed.

To be of value, the tests must involve a rock mass of sufficient volume to contain enough discontinuities to be representative of the large-scale

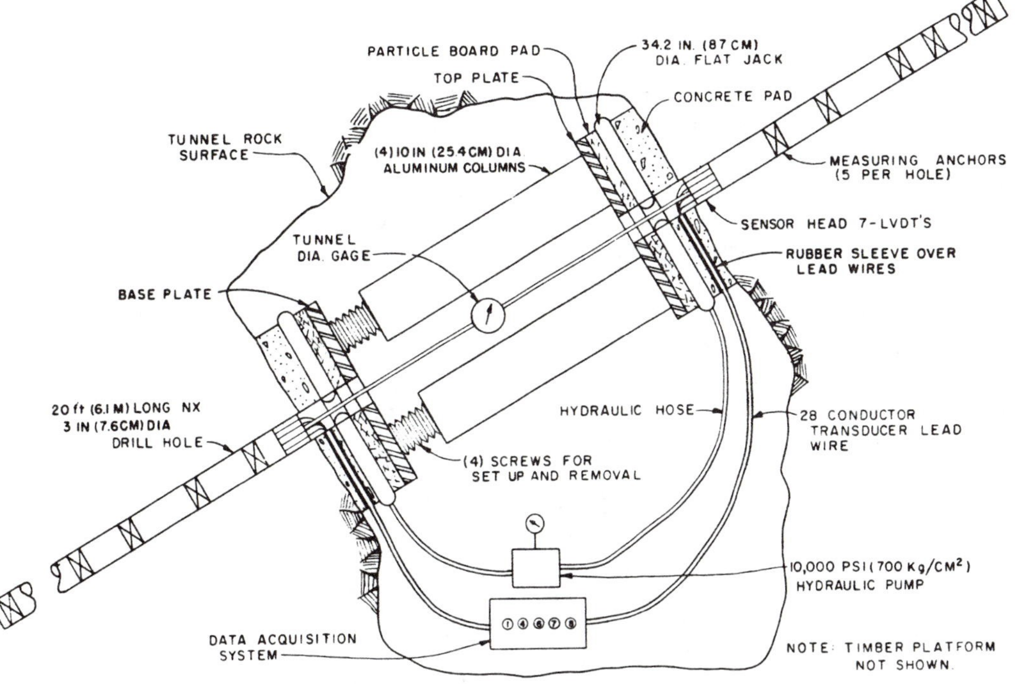

Figure 4.44 Uniaxial plate bearing jack assembly with MPBXs extending axially into rock mass. (Copyright ©, ASTM, 1916 Race Street, Philadelphia, PA 19103. Reprinted, with permission from Misterek et al., 1974.)

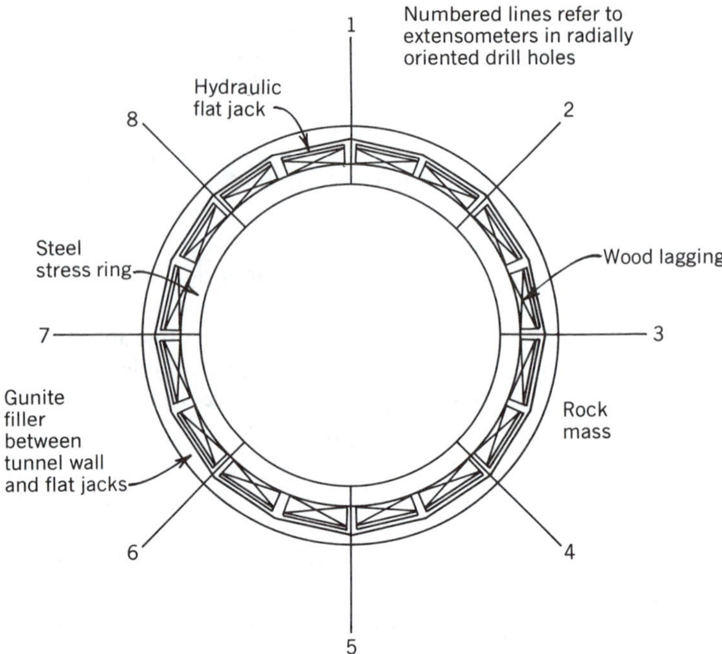

Figure 4.45 Schematic cross section of radial jack assembly in a circular tunnel.

deformation possible at large construction sites. Radial jacking tests inherently involve the greatest volume of rock of any of the tests. Although the use of both tests is limited because of cost, the radial jacking test is the more expensive, and its use is restricted to sites where simultaneous measurements of radial deformation is critical to the design. G. B. Clark (1966) has summarized these and other large-scale deformation tests. Suggested methods for conducting the uniaxial and radial jacking tests described here as well as others have been compiled by the International Society for Rock Mechanics (ISRM, 1979e).

Axial deformation from uniaxial tests and radial deformation from radial jacking tests are measured by multiple position borehole extensometers (MPBXs) installed in axially oriented and radially oriented drillholes, respectively (Figures 4.44 and 4.45). The drillholes for extensometer placement are cored to obtain rock samples for determining intact strength and elastic moduli and measurement of discontinuity characteristics. Examples of measured deformation from uniaxial and radial jack tests are shown in Figures 4.46 and 4.47. Figure 4.46 illustrates the hysteresis curves that result from differing recoverable and nonrecoverable deformation over multiple loading and unloading cycles. Figure 4.47 is the product of data from multiple cycles related to site geology.

Small-Scale Tests. Measurement of smaller-scale deformation is made by borehole dilatometers, borehole jacks, and flat jacks. The borehole devices are designed to operate in NX-size drill holes (76-mm diameter). Deforma-

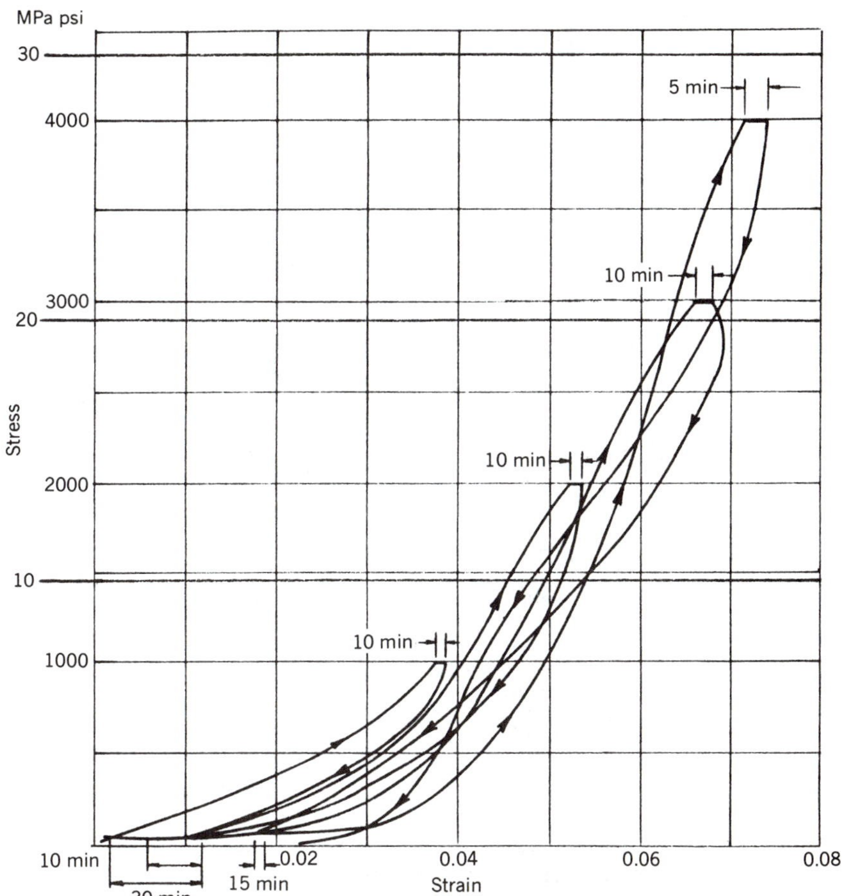

Figure 4.46 Stress-strain curves from four loading and unloading cycles using uniaxial plate loading test. (Copyright ©, ASTM, 1916 Race Street, Philadelphia, PA 19103. Reprinted, with permission from Dodds, 1974.)

tion characteristics are obtained by measuring borehole deformation resulting from compressional stress applied by the device on the borehole wall. The volume of rock affected is small and the measured deformation is not as representative of the rock mass as that obtained from plate and radial jack testing. The data, however, do provide an indication of *in situ* rock mass properties at depths unaffected by surficial crack opening and elastic unloading. The *in situ* deformation characteristics commonly are calculated and expressed as modulus of elasticity (Young's modulus), E.

Of the several kinds of dilatometers available, only the LNEC-type will be described. It is named for the Laboratorio Nacional de Engenharia Civil in Portugal where it was developed. As a typical dilatometer, it exerts uniform radial pressure around the circumference of the borehole by means of a hydraulically inflated rubber bladder or membrane. Pressures up to 14.7 MPa have been reached. Deformation along each of four diameters at 45° intervals is measured by four linear variable differential

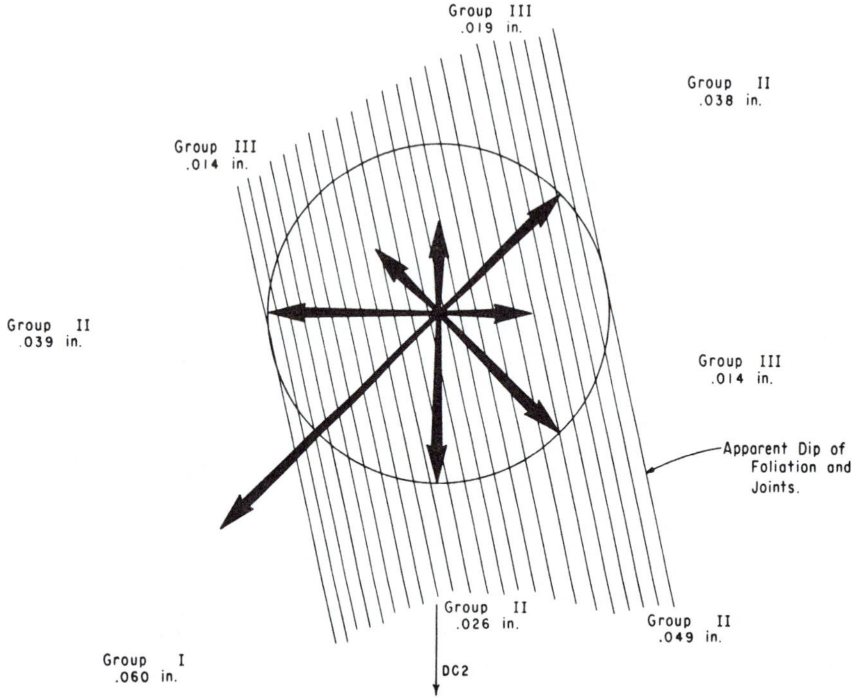

Figure 4.47 Anisotropic deformations resulting from radial jacking test. Axis of the tunnel is perpendicular to the center of the strain vectors. (Copyright ©, ASTM, 1916 Race Street, Philadelphia, PA 19103. Reprinted, with permission from Wallace et al., 1972.)

transformers (LVDTs). The operation of an LVDT is explained in chapter 6. The LNEC dilatometer membrane is 54.5 cm in length and the LVDTs have a 5-mm measurement range with an accuracy of 1 μm (Rocha, 1970).

Young's modulus—or, more correctly, the modulus of deformation—for the surrounding rock mass is calculated by using change in applied pressure, diametral deformation, borehole diameter, and Poisson's ratio of the rock. Changes in deformation magnitude and direction and deformation modulus may be determined for various depths in a borehole.

Borehole jacks have been developed to exert greater pressure on borehole walls than that obtainable with dilatometers. They differ from dilatometers in that unidirectional pressure is exerted on two diametrically opposed portions of the borehole wall by curved steel plates. The Goodman jack is considered here to be typical of the types of borehole jacks in use.

The Goodman jack is designed for use in an NX borehole. Its curved bearing plates are hydraulically operated and can apply pressures up to 69 MPa to the borehole wall. The two bearing plates are 20.3-cm long and their displacement, that is, rock deformation, is measured by two LVDTs having a measurement range of 0.5 cm and an accuracy of 2.5×10^{-4} cm. The applied hydraulic pressure and measured deformation are combined with an estimate of Poisson's ratio to calculate the modulus of deformation. Goodman et al. (1972) have estimated that the zone of deformation around the borehole owing to use of the Goodman jack does not exceed 30.5 cm.

The diametrically opposed plates permit selected orientations in a borehole dependent on known geologic conditions or construction-related stresses and their direction. Comparisons of the moduli of elasticity and deformation from unconfined compression tests on intact samples and *in situ* plate jacking and borehole jack tests are shown on Table 4.12. As noted earlier, the moduli are all calculated as Young's modulus. The expected decrease in modulus magnitude for *in situ* tests compared with intact rock is shown. Variations between the large-scale plate-jacking and small-scale borehole-jacking data are related to variations in joint spacings at the different scales.

Flat jacks, as their name implies, consist of two flat plates welded together with a space of known volume between them (see application in Figure 4.55). They have been used for some time to determine the deformation characteristics of rock near exposed faces. When confined, as in a slot cut in rock, hydraulic pressure applied to the jack will be transferred to the surrounding rock. The jacks may be grouted in place or simply placed in the slot, if tightly fitted. Deformation from small jacks, that is, approximately 0.3 m^2, may be measured by volumetric changes in hydraulic fluid during the jacking procedure. Larger jacks utilize strain gauges for measurement of jack deformation.

Flat jack tests are greatly influenced by near-surface stress relief and resultant cracks. Rocha (1970) has described large flat jacks developed by the Laboratorio Nacional de Engenharia Civil that may be inserted into deep slots cut with a 1-m diameter diamond saw in an attempt to obtain more representative *in situ* rock data. Because of their size, the jacks have

Table 4.12 Comparison of Intact Core and *in Situ* Plate-bearing and Borehole Jack Test Data at Three Engineering Project Sites

			Young's modulus, GPa		
Site	Rock Type	Poisson's Ratio	Unconfined Compression (Intact-lab)	Plate Bearing (*in situ*)*	Borehole Jack (*in situ*)*
Tehachapi Tunnel	Diorite gneiss; fractured and seamy	0.35	77.93	3.66–5.72	4.21–7.10
Dworshak Dam	Granite gneiss; massive to moderately jointed	0.20	51.72	3.45–34.48	10.62–18.62
Crestmore Mine	Marble; massive	0.25	47.59	12.00–18.76	9.31–11.72

* In same pressure range: 0–20.7 MPa

Source: Modified and reprinted with permission from Proc. 10th Symp. on Rock Mechanics, Basic and Applied Rock Mechanics, R. E. Goodman, T. K. Van, and F. E. Heuze, Measurement of Rock Deformability in Boreholes, 1972, Am. Inst. Min. Metall. & Pet. Eng.

multiple strain gauges mounted on them to measure differences in rock deformation across the large jack surface. Portable diamond saw apparatus is designed to cut slots in excess of 1.5-m deep to reach relatively undisturbed rock. Adjoining slots have been used with as many as three adjoining jacks placed in the rock. Pressures in excess of 9.8 MPa have been applied to the rock by the jacks.

As defined earlier, the ratio of the *in situ* modulus of deformation, E_d, to the laboratory-determined intact rock modulus, E, is the modulus ratio or reduction factor (Deere et al., 1967). The moduli are normalized by the values of the intact moduli to eliminate intact rock properties. The reduction factor provides a numerical measure of the influence crack closure has on the strength of a given rock mass. The need for some kind of jacking procedure to obtain the modulus of deformation limits the practical use of the factor, however. Bukovansky (1970) has compared the results of the various jacking and dilatometer procedures with the elastic rock properties obtained by uniaxial static laboratory testing. His work should be consulted if such information is needed.

Rock Mass Quality

Although the modulus of deformation and the modulus ratio, or reduction factor, are measures of rock deformability, they are also measures of the engineering performance or quality of a given rock mass. The difficulty in obtaining the required modulus of deformation has led investigators to devise other more easily determined and economical measures of rock quality. Each has been compared to more quantitative parameters such as rock strength, modulus of deformation, and reduction factor so that each has empirical, if not quantitative, applications to civil engineering practice. Rock mass quality, whether defined by moduli or by other measures, is dictated by intact rock strength, orientation and frequency (spacing) of discontinuities, and chemical weathering of the rock mass.

Rock Quality Designation (RQD)

In 1967 a modified core recovery procedure was developed to provide a Rock Quality Designation (RQD) for a given cored interval of NX core (Deere et al., 1967). The RQD value is the percentage obtained by dividing the summed lengths of all core pieces equal to or greater than 10-cm (4-in.) long by the cored interval length (see Module 4.6). Smaller pieces and/or core loss, are assumed to result from closely spaced discontinuities, shearing, faulting, or weathering, all of which decrease rock mass quality. An RQD of 100% indicates 100% core recovery with all pieces equal to or greater than 10 cm in length. Thus, it does not imply an unjointed rock mass.

The RQD percentages are directly proportional to the various measures of rock mass quality such as fracture frequency and *in situ* modulus of deformation (Figures 4.48 and 4.49). They have been found to be of value in estimating rock loads on tunnel supports where RQD provides some measure of block loosening that may occur; this, in turn, affects the loads the support system must withstand (Cording et al., 1975). An estimation

MODULE 4.6

RQD calculation from core measurements.

Data from 10-ft (drilled length) core in shale
Discontinuities-bedding surfaces and two high-angle joints
All measurements in inches (in.) and feet (ft.)

Recovered core measurements (in.)

3.4	1.3	1.3	5.7	1.3	0.5	1.1
2.8	2.9	4.3	1.2	2.3	0.3	2.0
2.6	3.0	3.2	3.2	1.2	1.2	110.2
3.5	1.2	6.8	1.0	0.5	1.1	
3.1	2.9	0.9	2.0	0.5	0.3	
4.2	6.4	0.9	2.3	1.1	0.9	
2.2	2.6	4.4	2.8	0.3	1.7	
0.5	2.1	1.5	3.6	2.0	2.1	

Core-loss calculation

Total length cored (10 ft.)	120.0 in.
Recovered core	−110.2
Core loss	9.8 in.

RQD calculation: sum of pieces = or >4 in. (underlined in recovered core measurements)

4.2
6.4
4.3
6.8
4.4
5.7
31.8 in.

$$\text{RQD} = 31.8/120 = 0.265 \times 100 = 26.5\%$$

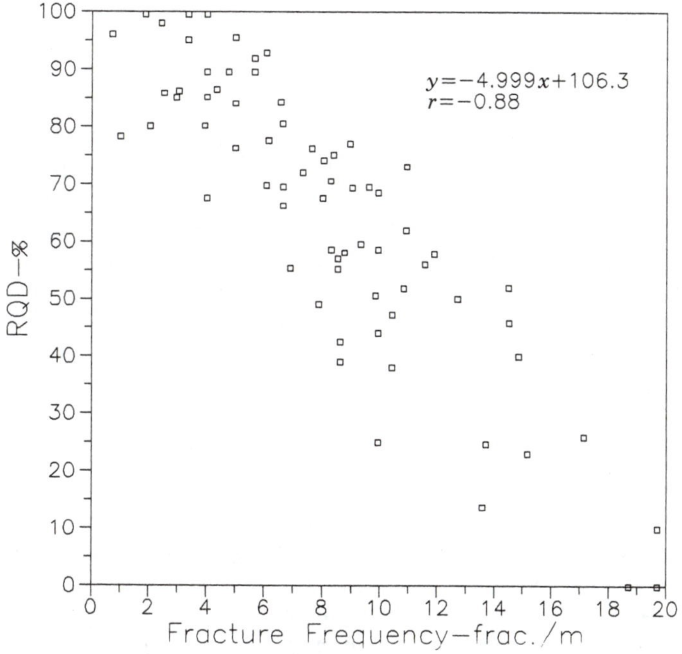

Figure 4.48 Comparison of fracture frequency and RQD from cores and tunnel exposure. (Modified and reprinted with permission from Proc. 8th Symp. on Rock Mechanics—Failure and Breakage of Rock, D. U. Deere et al., Design of Surface and Near-Surface Construction in Rock, 1967, Am. Inst. Min. Metall. & Pet. Eng.)

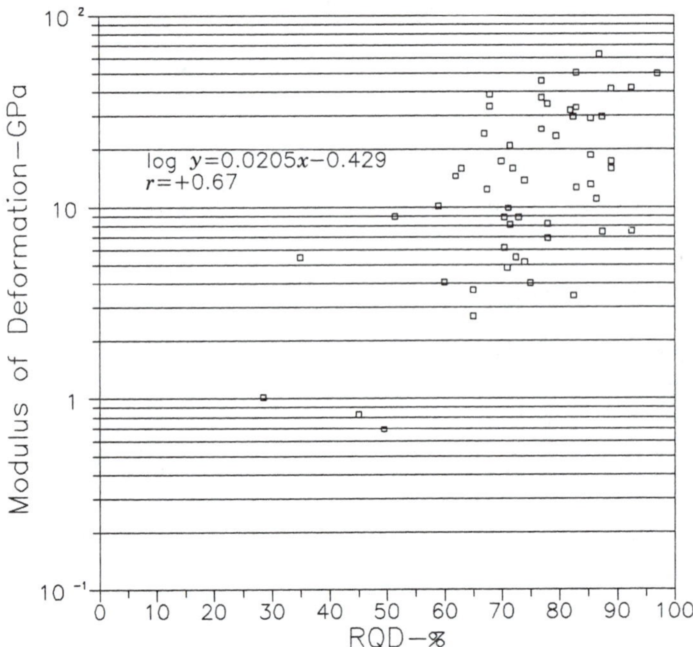

Figure 4.49 Comparison or RQD and *in situ* static modulus of deformation, E_d. (Copyright ©, ASTM, 1916 Race Street, Philadelphia, PA 19103. Modified and reprinted, with permission from Coon and Merritt, 1970.)

based on RQD is enhanced by knowledge of the orientations of the discontinuities that define the blocks. The RQD and uniaxial compressive strength have been used to determine the most appropriate method of tunneling in rock (Figure 4.50).

The RQD values may not be representative of rock mass properties for reasons other than the lack of orientation data. If smaller diameter core is obtained, more breakage will occur than with NX core. Excessive breakage and possible core loss may result from improper drilling and handling. Orientation of the drillhole parallel to discontinuities also may result in excessive breakage and loss (Cording and Mahar, 1978).

Influence of Rock Quality on Seismic Velocities

Velocities obtained by the refraction seismic method (chapter 7) or by recording shock wave travel through a geologic unit between two or more drillholes are functions of the primary and secondary properties of earth materials, for example, lithology, jointing, and weathering. As a general statement, the higher the velocity of wave travel through a rock, the better the rock quality. The gradual decrease in joint frequency with depth causes a gradual increase in velocity, all other factors remaining equal. Wantland (1964) reported a velocity of 4,300 m/sec for fresh, hard metadiabase at a dam site and 600 m/sec for the same rock in a weathered state. In highly weathered rock, the velocity may be independent of the frequency and orientation of discontinuities.

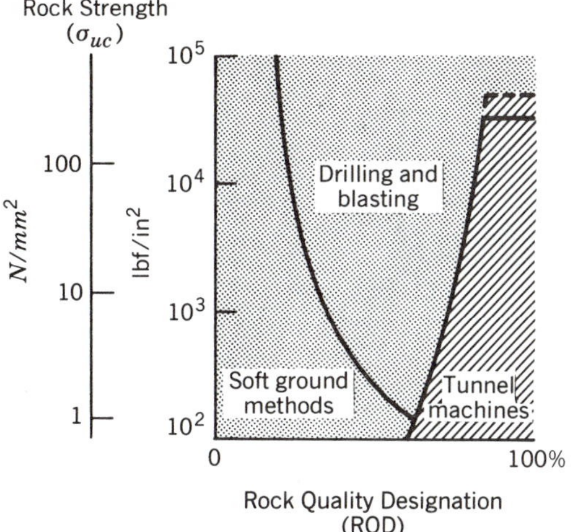

Figure 4.50 Diagram of tunneling methods most appropriate for given rock strengths and RQD conditions. (Reproduced by permission of the Geological Society of London from A. M. Muir Wood, Tunnels for Roads and Motorways, Q. J. Eng. Geol., Vol. 5, No. 1, 1972.)

A more definitive measure of the influence of jointing or fracturing on *in situ* rock strength is obtained by comparing shock wave velocity through rock in the field with that through an intact sample of the same rock unit in the laboratory. Although the *in situ* velocity may be obtained by surface refraction techniques, it more often is obtained in or between two or more drillholes. The holes drilled to obtain core samples for laboratory testing are used. The core samples are used for obtaining intact specimen velocities. Figure 4.51 depicts the uphole and crosshole methods of obtaining *in situ* data. Shock waves may be generated by use of explosives or mechanically to avoid destruction of drillholes.

The ratio of the field, or *in situ* velocity (V_F), to the laboratory velocity (V_L), V_F/V_L, is the velocity ratio, or fracture index (Deere et al., 1967; Knill and Price, 1972). As the number of discontinuities decreases, V_F will approach and ultimately equal V_L for a discontinuity-free rock mass. Conversely, Knill (1974) found that, in general, a velocity ratio of less than 0.5 indicated significantly fractured rock.

In view of the variations in velocity caused by weathering, it is not possible to arbitrarily interpret changes in fracture frequency from velocity changes alone. Rather, the value of the velocity ratio lies in its being an indicator of *in situ* conditions that require more detailed examination. Because of the control exerted by both discontinuities and chemical weathering, field velocity measurements are reliable indicators of rock mass quality.

As might be expected, the velocity ratio is related to RQD values obtained from coring that combine the influence of jointing and weathering. The squared velocity ratio (velocity index) has been found to be

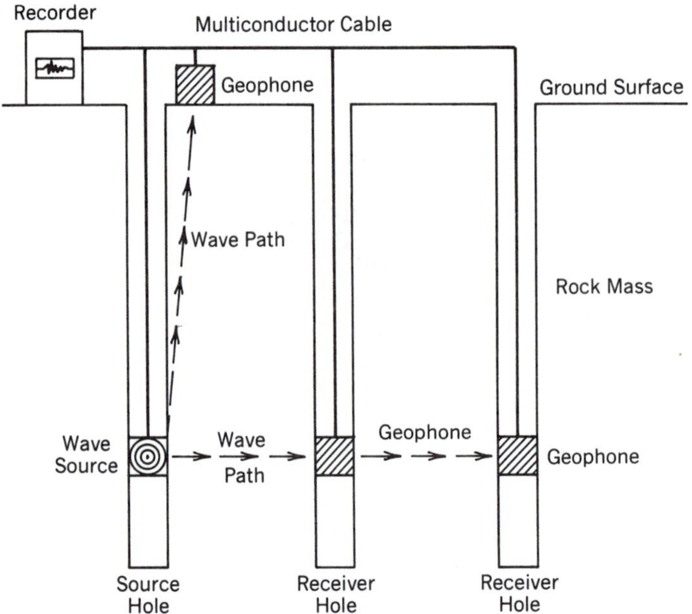

Figure 4.51 Generation and recording of crosshole and uphole seismic velocities in a rock mass.

approximately the same as RQD in some rock masses (Deere et al., 1967; Coon and Merritt, 1970). Composite data from igneous, sedimentary, and metamorphic rocks confirm this relationship (Figure 4.52). The velocity index also has been found to be linearly related to the modulus ratio, or reduction factor (Figure 4.53), although the low correlation coefficient precludes any rigorous prediction possibilities. The data used for the velocity ratio (or index) must be normalized by the laboratory velocities to reduce the influence of lithology.

Rock Mass Dynamic Elastic Moduli

The elastic moduli considered in the preceding discussion on intact rock properties are usually obtained from laboratory samples by static methods in which stresses are applied slowly, resulting in measurable strains. Dynamic elastic moduli may also be obtained by more rapid application of stress to the sample as described in the previous discussion on intact rock properties. For *in situ* conditions, dynamic moduli such as Young's modulus are usually higher (G. B. Clark, 1966). They differ for the same reasons that cause intact-specimen dynamic moduli to exceed static moduli in magnitude.

In addition, *in situ* dynamic moduli do not reflect the influence of crack closure, which affects static deformation values. They do compare favorably with the tangent modulus obtained from *in situ* testing (Deere et al., 1967; Bernaix, 1974). *In situ* dynamic moduli inherently are more representative of a larger sample than those from most jacking tests. Results also are obtained more rapidly, economically, and with fewer operational problems.

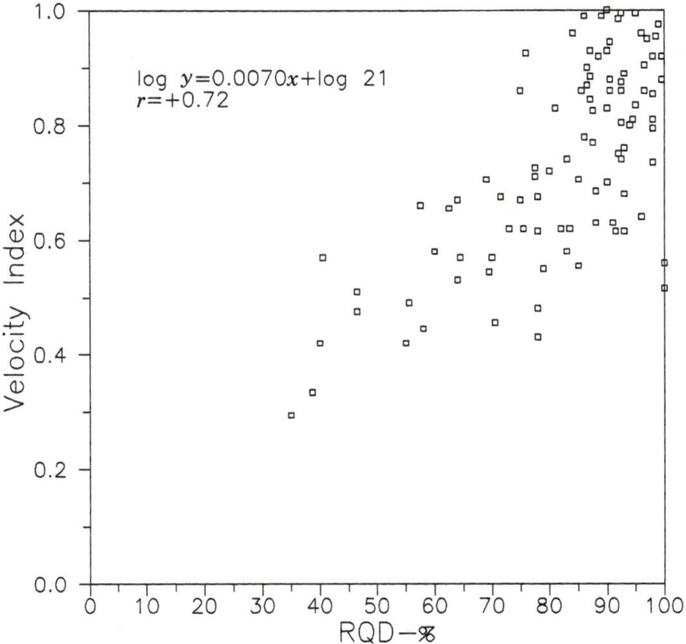

Figure 4.52 Relationship of RQD to velocity index. (Copyright ©, ASTM, 1916 Race Street, Philadelphia, PA 19103. Modified and reprinted, with permission from Coon and Merritt, 1970.)

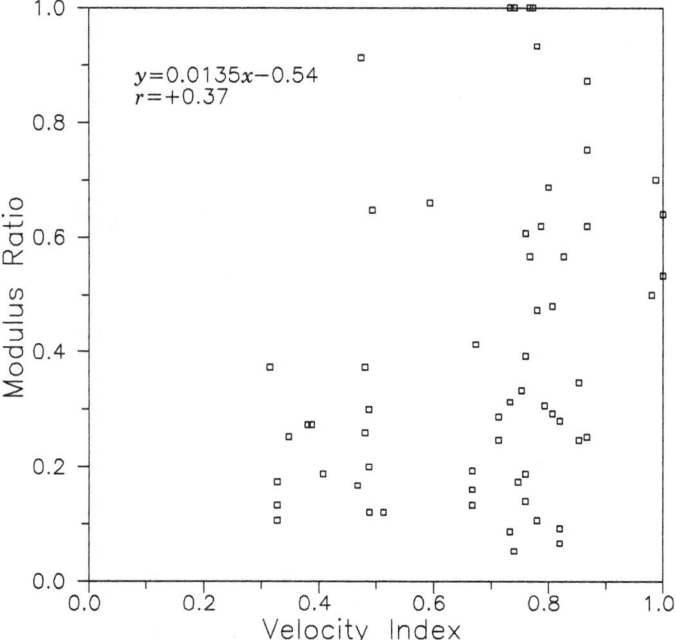

Figure 4.53 Relation of velocity index to reduction factor for a combination of igneous, sedimentary, and metamorphic rocks. (Copyright ©, ASTM, 1916 Race Street, Philadelphia, PA 19103. Modified and reprinted, with permission from Coon and Merritt, 1970.)

The generation and reception of shear waves needed for dynamic-moduli calculations have proven to be a major problem in obtaining *in situ* dynamic moduli. Improved shear wave sources and recording systems have eliminated many of the problems. Special geophones responsive to transverse (shear) particle motion must be used.

Vertically oriented (SV) shear waves are generated in boreholes for crosshole measurements of shear wave velocities. One technique involves impacting an anchoring device by moving a weight up or down in the hole. Multicomponent geophones are located in adjacent drillholes for reception of compression and shear waves.

Although the methods of generating shear waves have proved to be effective, doubt may remain about the wave forms recorded and, thus, the calculation of shear wave velocities. The current methods of shear wave generation permit reversing the direction or polarity of the first shear wave cycle. A reversal in the recorded first arrival identifies a shear wave from a compression wave because the latter will not undergo a polarity change. A comprehensive paper on *in situ* generation and measurement of shear waves by Hoar and Stokoe (1978) is recommended for a more complete treatment of the subject.

In situ dynamic elastic moduli are used in a variety of engineering applications where *in situ* static testing is not possible or feasible. These include estimation of dam-foundation and abutment deformations, sensitivity of structure foundations to dynamic compression and shear moments, response of earth dams and fill structures to dynamic shear moments, and design of pressure tunnel lining. Onodera (1963) equated rock mass soundness to the ratio of dynamic moduli, E_{field}/E_{lab}. Conceptually, the ratio is the same as the statically determined reduction factor described earlier. Although the more easily obtained velocity ratio or velocity index serve the same purpose, this relationship illustrates the consistency of the ratio concept based on mass versus intact properties of rocks.

From the foregoing it can be seen that there are several ways of approximating the data obtained from costly *in situ* static tests for use in evaluation of rock mass quality and deformability. Sufficient information has been evaluated to permit Coon and Merritt (1970) to devise an *in situ* classification relating RQD, velocity index, and rock mass deformability, represented by the reduction factor (Table 4.13). The point scatter of the

Table 4.13 *In Situ* Rock Mass Classification.[a]

RQD Classification	RQD	Velocity Index	Reduction Factor
Very poor	0–25	0–0.20	<0.20
Poor	25–50	0.20–0.40	<0.20
Fair	50–75	0.40–0.60	0.20–0.50
Good	75–90	0.60–0.80	0.50–0.80
Excellent	90–100	0.80–1.00	0.80–1.00

[a] Based on RQD, velocity index $(V_F/V_L)^2$, and reduction factor (E_d/E_{t50}).

Source: Copyright ©, ASTM, 1916 Race Street, Philadelphia, PA 19103. Reprinted with permission from Coon and Merritt, 1970.

data used indicates the need for adequate site testing to confirm local parameter relationships with published data.

In Situ Stress Measurement

The determination of the *in situ* state of stress of rock is of critical importance in the design of underground openings in rock. Assuming the rock to be elastic, stresses from overburden loads and tectonic forces, for instance, will have caused a given volume of rock to have deformed proportionally to those stresses. Upon relief of the stresses by rock excavation, strains will occur in response to the unloading. The strain may culminate in instantaneous failure (rock bursts) if the initial stresses exceed the intact rock tensile strength. Otherwise, the resultant rock strains may seriously disrupt underground structures if provision is not made for them in the design.

In situ stresses are relieved in the immediate rock mass surrounding an opening. The degree of stress relief is influenced by excavation practice such as the increase in fracturing from the use of explosives. Typically, determination of *in situ* stresses is desired outside of the zone of excavation influence and methods must be employed to sample the rock at depth. Stresses unaffected by excavation are called *absolute* in situ *stresses.* If a rock mass does not behave elastically, determination of *in situ* stresses has no engineering value.

Measurement of *in situ* stresses in rock of necessity involves the use of strain gauges and instruments and their applications. Normally, such uses are included in discussions of instrumentation. But the relevance of *in situ* stress measurement to an understanding of rock mass characteristics requires that this material be examined at this point in our discussion.

With the exception of the flat jack method of determining stress, stress is not measured directly. Strain resulting from stress relief is measured and converted to stress by elastic theory (Jaeger and Cook, 1976). The modulus of elasticity, E, and Poisson's ratio, ν, of the intact rock being tested are required for most calculations. These values are obtained in the laboratory by uniaxial compression tests of samples. Stress relief is achieved by cutting or drilling planar slots or by isolating a square or circular block or a cylinder in the rock face, or by a variety of deeper tests utilizing boreholes. Figure 4.54 illustrates an early method of relieving stress at the rock face by use of overlapping drillholes. Two-dimensional strain determinations were obtained by measurement of reference-pin combinations before and after stress relief. Successful calculations of *in situ* rock stresses by use of such methods date back to 1932 at Hoover Dam, Nevada (Olsen, 1957).

Jacking Methods

The flat jack method of determining *in situ* rock stress employs the same flat jacks described in the section on rock mass deformation. The typical flat jack procedure involves grouting a 0.3-m-square flat jack into a slot cut perpendicularly to the rock face, either by drilling overlapping holes or by cutting with a diamond saw (Figure 4.55). Prior to cutting the slot, measurement pins are embedded in the rock on either side of the slot location.

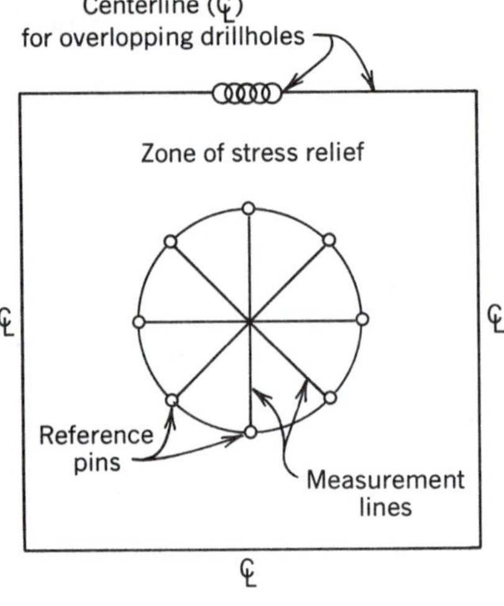

Figure 4.54 Schematic diagram for relief of *in situ* rock stresses by use of overlapping drillholes and strain measurement by use of reference pins.

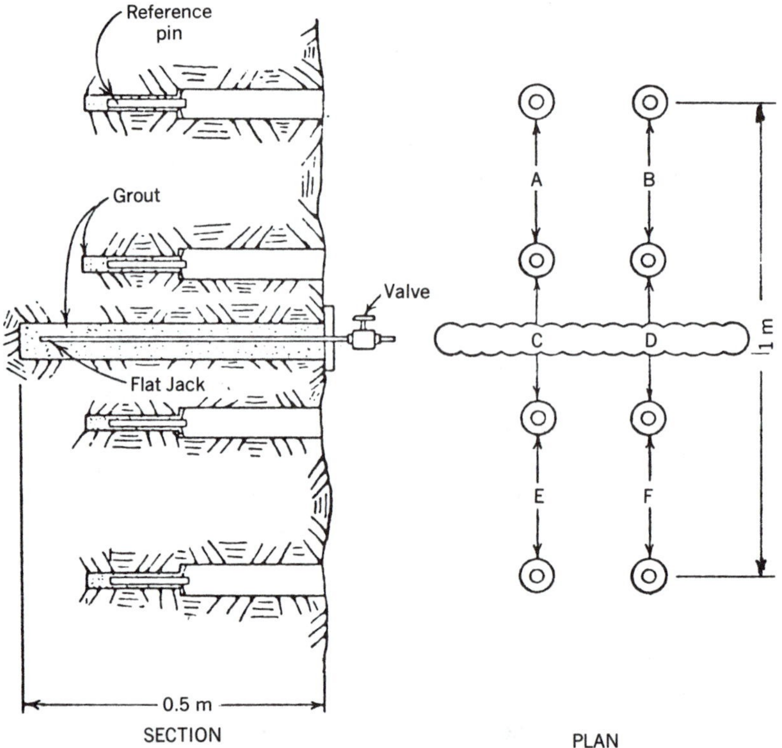

Figure 4.55 Plan and section of flat jack and reference pins. (Reprinted with permission of author, editor W. R. Judd and Rand Corp. from Proc. Intl. Conf. on State of Stress in the Earth's Crust; R. H. Merrill, In Situ Determination of Stress by Relief Techniques, 1964.)

Distances separating the pins are measured by either a bar- or tape-type surface extensometer (chapter 6).

The cutting of the slot permits stress relief into the slot. The resulting strain is carefully measured between the pins. The flat jack is inserted into the slot, grouted, and then hydraulic pressure is applied to the jack. The hydraulic pressure converted to force per unit area required to return the measured pin strains to zero is taken as the *in situ* state of stress normal to the jack and at the depth of the embedded pins. Bonded strain gauges and strain rosettes cemented to the surface and vibrating-wire strain gauges (chapter 6) attached to pins set surficially in similar fashion to the embedded pins have been used to indicate the return of the rock to a null state. The flat jack pressure required to achieve the null state is also known as the cancellation pressure. Computed stress values may be obtained for comparison with the cancellation pressure by using Young's modulus, Poisson's ratio, and measured strain from pins or strain gauges. *In situ* stress data from flat jacks are influenced by the near-surface measurement of stress (Hoskins, 1966; Jaeger and Cook, 1976).

A more complete determination of *in situ* stresses by the flat jack method may be achieved by cutting one or more slots that are mutually perpendicular to the first slot, and then following the same procedures. The variations in stress for two orthogonal flat jacks at each of several sites are shown in Table 4.14. Unless the slots are fortuitously cut normal to the orthogonal stresses in the rock, the stresses measured will be only those stress components perpendicular to the slot orientations. Rocha et al. (1966) have reported using as many as four cuts oriented 45° apart. The larger (1-m-diameter) flat jacks inserted into deep slots provided stress data less influenced by surface conditions than the more commonly used shallow installations of smaller jacks.

Table 4.14 Comparison of *in Situ* Stresses Obtained from Cancellation Pressures and by Computation

	Flat jack Method[a]			
	Cancellation Pressure (MPa)		Computed Stress (MPa)	
Site No.	Vertical	Horizontal	Vertical	Horizontal
1	4.76	3.10	4.10	4.04
2	12.70	8.90	9.04	8.62
3	4.55	3.38	5.45	3.52
4[b]	10.07	7.79	9.72	8.14

[a] At approximately 23-cm depth.
[b] Results from flat jacks in roof and wall.

Source: Reprinted by permission from author, editor, W. R. Judd and Rand Corp. from Proc. Intl. Conf. on State of Stress in Earth's Crust; R. H. Merrill, In Situ Determination of Stress by Relief Techniques, 1964.

Overcoring Methods

The use of a borehole in determining *in situ* stresses in a rock mass permits measurements to be taken at a distance from the zone influenced by excavation. Although a variety of approaches can be used to measure strain, the common practice employed to achieve relief of rock stresses is to overcore a previously drilled and instrumented borehole. Overcoring, or trepanning, may take several forms. Two commonly used techniques are shown in Figures 4.56 and 4.57. In each case, the original borehole is cored to permit collection of intact samples for laboratory determination of Young's modulus and Poisson's ratio that are required for calculation of stress from measured strain.

Bonded Strain Gauge Applications. The bottom of a borehole may be used as in Figure 4.56 for the placement of a strain gauge rosette. Overcoring of the rosette is done by the same-size coring bit used to drill the borehole. It results in stress relief from the core into the annulus. Typical strain-relief curves for each rosette axis, measured in strain gauge voltage changes, are shown in Figure 4.58.

The measurement of strain along the three diameters of the rosette at the bottom of a borehole is two-dimensional and does not provide sufficient data for calculation of the principal stresses at that point in the rock mass. Thus, only components of two of the three principal stresses may be measured unless, as in the case of the flat jack method, the borehole is fortuitously oriented along one principal axis. In which case, two principal stresses can be calculated. Assuming this is *not* the case, measurements must be made in more than one borehole to obtain the desired principal stress data. Typically three boreholes are drilled in a horizontal plane at angles other than 90° to one another to overlap (intersect) at the point of interest in the rock mass. Calculation of *in situ* stresses from bottomhole strains in boreholes is described by Leeman (1964) and Jaeger and Cook (1976).

A modification of the bottomhole deformation measurement has been developed by Leeman (1968) to obtain the complete orthogonal state of stress from a single borehole. This has been achieved by cementing three specially designed strain-gauge rosettes on the EX borehole wall, as shown in Figure 4.57, prior to overcoring. Placement of rosettes on the borehole wall solves the problem of obtaining strain measurements parallel to the

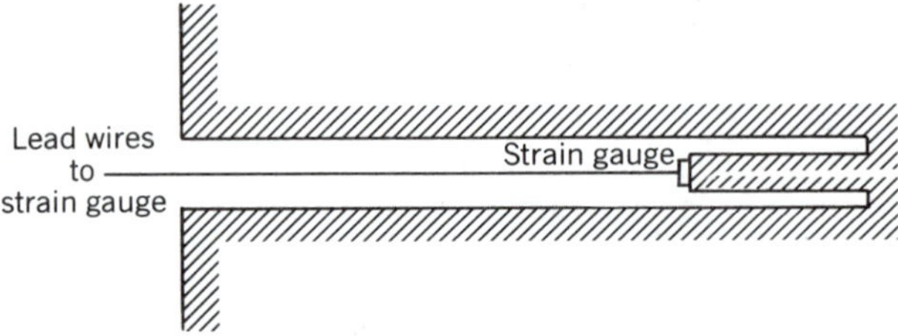

Figure 4.56 Cross section of borehole with mounted strain gauge and subsequent overcoring.

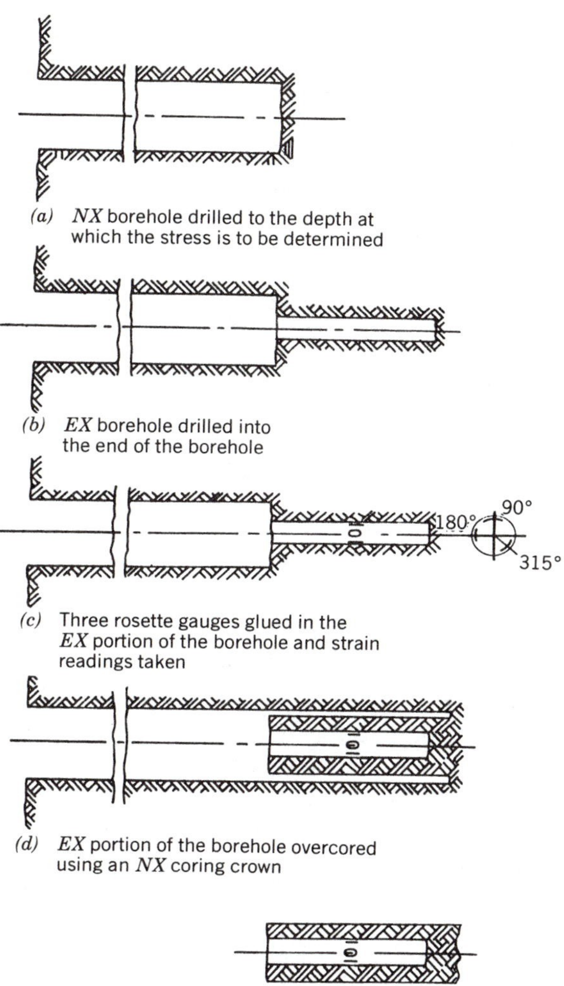

Figure 4.57 Overcoring of instrumented small-diameter borehole with larger diameter hole, where NX = 7.6 cm and EX = 3.8 cm. (Reprinted with permission from Intl. J. Rock Mech. Min. Sci., Vol. 5, E. R. Leeman, The Determination of the Complete State of Stress in Rock in a Single Borehole-Laboratory and Underground Measurements. Copyright © 1968, Pergamon Press.)

borehole axis. Plane-strain measurements made at the bottom of a borehole are perpendicular to the borehole axis.

Strain gauge bonding is a problem even though some ingenious devices have been developed for placement and cementing of strain gauge rosettes to rock surfaces in boreholes. Every method makes provision for gauge orientation prior to cementing and threading of strain gauge lead wires through the drill pipe during the overcoring operation. The latter permits monitoring of strains during overcoring as shown in Figure 4.58.

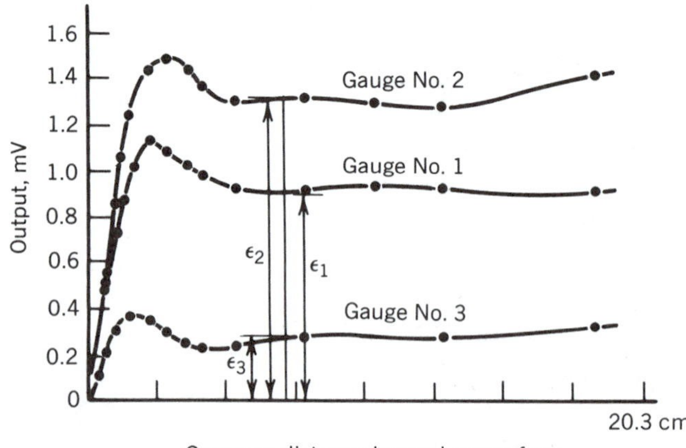

Figure 4.58 Typical strain-relief curves from bottomhole-mounted strain-gauge rosette. Strains are recorded as output voltage from readout instrument. (Reprinted with permission from Intl. J. Rock Mech. Min. Sci., Vol. 7, B. R. Stephenson and K. J. Murray, Application of the Strain Rosette Relief Method to Measure Principal Stresses Throughout a Mine. Copyright © 1970, Pergamon Press.)

Borehole Deformation Gauge. Strains in a plane normal to a borehole that result from overcoring also may be measured by recording changes in borehole diameter. The instrument most commonly used to do this is the three-component borehole deformation gauge developed by the U.S. Bureau of Mines (Merrill, 1967). Strains along three equally spaced diameters are measured by the gauge shown in Figure 4.59. Calibrated transducer arms measure the movement of the six buttons that are in contact with the borehole wall. The gauge is inserted into an EX hole (38 mm) and overcored with a 15.25-cm diameter hole. To obtain three-dimensional data, readings must be taken in three boreholes, as in the case of strains measured in the bottoms of boreholes. The mechanical contact between the transducer buttons and the borehole wall is seen by some to be a problem because the instrument may not remain stationary during the overcoring operation.

Photoelastic Methods. It has been known for many years that certain materials are photoelastic. They become doubly refracting under stress-caused strain. The distribution of stresses in the material may be seen by transmitting polarized light through the material (polarizer) and viewing the transmitted light through polarizing material oriented normal to the polarizer (analyzer). The principle of light polarization is basic to optical mineralogy, and viewing of stress fields by polarized light is covered adequately in basic books on physics.

The application of photoelasticity in determining plane stresses normal to a borehole is a logical consequence. Both uniaxial and biaxial stresses may be obtained by the use of photoelastic materials in a borehole. This is achieved by either mounting a birefringent disk, patch, or gauge on the prepared end of a borehole or by inserting a high elastic modulus glass cylinder into the borehole. The latter instruments are classed as solid inclusions and are known as stress meters.

All photoelastic methods depend on the generation of colored fringes, isochrons, or interference patterns being set up in the photoelastic material by strains induced by stress relief in the rock. The number of fringes is proportional to the strain. Each disk or stress meter is calibrated for the stress-caused strain that will produce one fringe, for example, 1.2 MPa per fringe. As in other methods of calculating *in situ* stress amounts, Young's modulus and Poisson's ratio obtained from core samples are used. Both disks and stress meters are overcored to relieve *in situ* stresses.

A typical photoelastic disk, or patch, setup is shown in Figure 4.60. In this case, only the periphery of the disk is cemented to the borehole face. The disk has a center hole that concentrates the strain and, in turn, the fringe patterns. The disk is composed of a birefringent plastic with a reflective backing. A polariscope transmits polarized light to the disk. The reflected light is examined by the analyzer portion as shown in the figure.

The direction of the two principal stresses in the plane normal to the borehole axis is exhibited by the fringe pattern of a biaxial gauge illustrated in Figure 4.61. The isotropic points lie on the major principal axis. Details of the use of such photoelastic gauges, including fringe counting and principles of operation, may be found in papers by Hawkes and Moxon (1965) and Hawkes (1971).

The solid inclusion-type of photoelastic stress meter, or gauge, is typi-

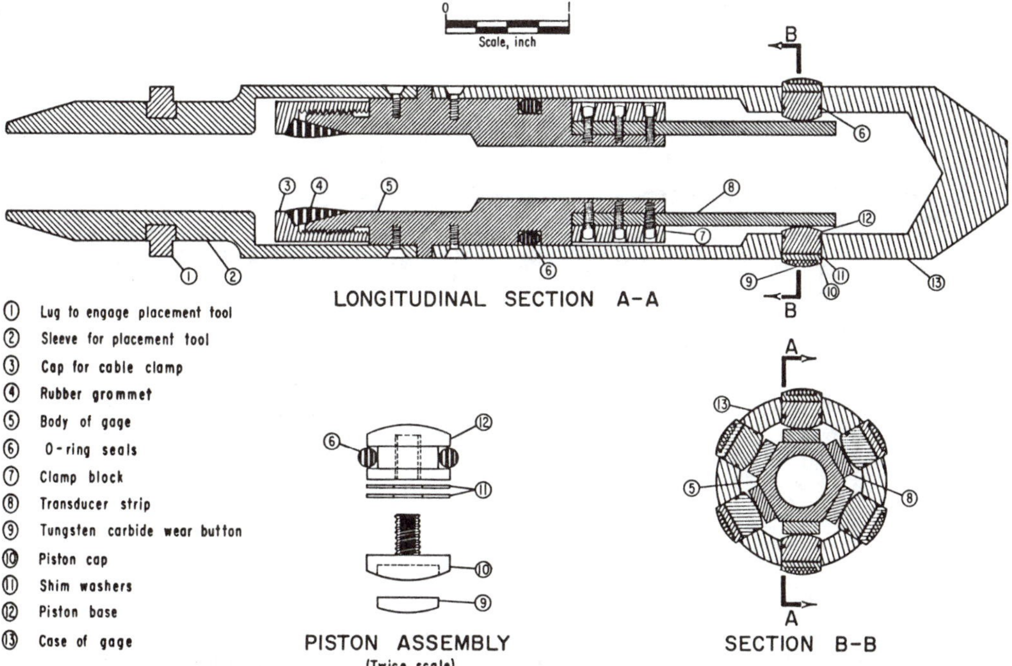

Figure 4.59 Diagrams of the three-component borehole deformation gauge. (From Merrill, 1967)

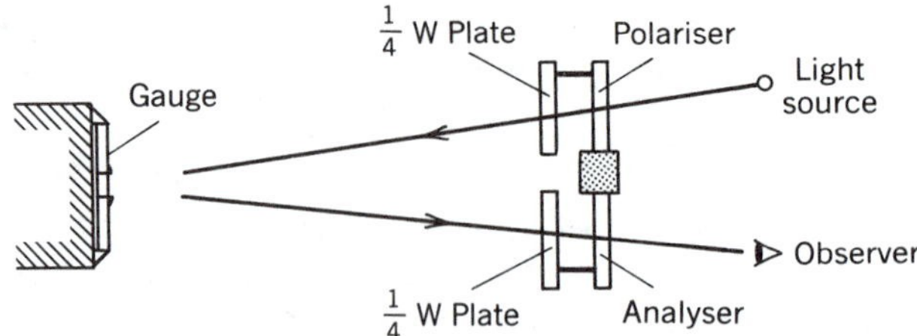

Figure 4.60 Diagram of photoelastic patch (gauge) on overcored rock cylinder and polariscope for fringe viewing. (Reprinted with permission from Intl. J. Rock Mech. Min. Sci., Vol. 2, I. Hawkes and S. Moxon, The Measurement of In Situ Rock Stress Using the Photoelastic Biaxial Gauge with the Core-Relief Method. Copyright © 1965, Pergamon Press.)

Major
Minor
Isotropic points

(a) Strain ratio +1 : $-\frac{1}{4}$

Major
Minor
Isotropic points

(c) Strain ratio +1 : $+\frac{1}{2}$

Major
Minor
Isotropic points

(b) Strain ratio +1 : $+\frac{1}{4}$

(d) Strain +1 : +1.

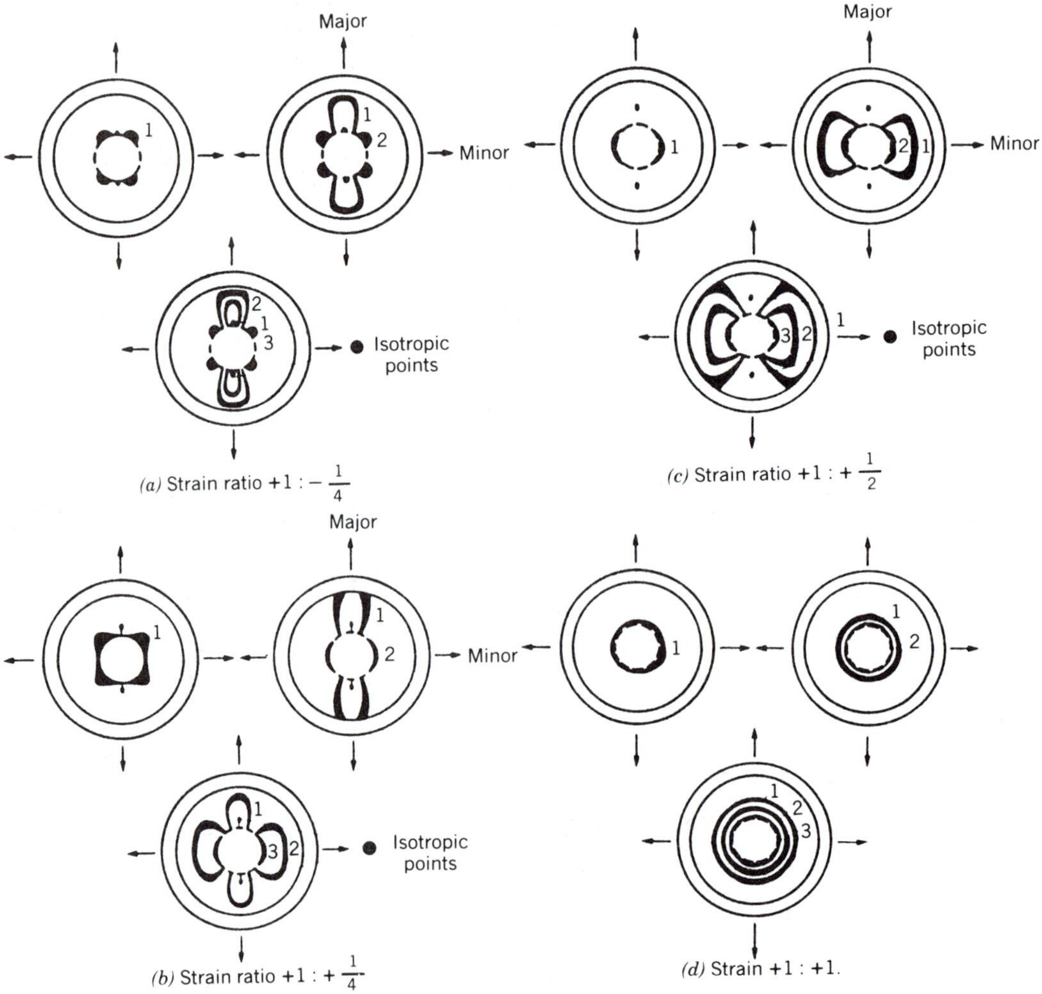

Figure 4.61 Biaxial fringe patterns of peripherally cemented photoelastic gauges for four strain ratios between major and minor principal stress directions. Numbers refer to fringe numbers. (Reprinted with permission from Intl. J. Rock Mech. Min. Sci., Vol. 2, I. Hawkes and S. Moxon, The Measurement of In Situ Rock Stress Using the Photoelastic Biaxial Gauge with the Core-Relief Method. Copyright © 1965, Pergamon Press)

cally a biaxial one in which a hollow glass or plastic cylinder is cemented into the borehole. The cylinder may be backed by a light source and a polarizing filter for transmission of polarized light through the cylinder to enable observation of fringe patterns. Plane strain normal to the borehole axis is displayed by the fringe pattern developed in the cylinder. Analysis of the fringe order, that is, number of fringes and direction of the two principal stresses, is made with an analyzer unit. The sensitivity of the meter is a function of the cylinder length and diameter, borehole diameter, and stress optical properties of the glass or plastic. Roberts et al. (1964) and Hawkes (1971) are recommended for pertinent details.

Other Methods

The use of flat jacks, strain gauge rosettes, borehole deformation gauges, and various photoelastic applications described here does not exhaust the methods available for calculation of *in situ* stress in rock. Others include the borehole deepening method (de la Cruz and Goodman, 1969), borehole fracturing method (de la Cruz, 1977), hydraulic fracturing method (Haimson, 1976), and the U.S. Geological Survey (USGS) solid inclusion, three-dimensional borehole probe (Nichols et al., 1968). Their use has not become sufficiently widespread to warrant treatment here.

In conclusion, it is important to note that highly fractured rock and highly weathered rock are not conducive to stress measurements. Thus, all such measurements should be restricted to essentially discontinuity-free, unweathered zones within a rock mass.

ENGINEERING CLASSIFICATION OF ROCKS

The inherent need for classification of rocks in the geological sciences is apparent even to the beginning geologist. Introductory and advanced textbooks in geology abound in differing kinds of classifications and associated terminology—with various degrees of complexity. As detailed as these schemes have become over the years, there is little of quantitative geotechnical value to be gained from them. Granted, one implicitly knows that a granite is stronger than a shale. However, strength is not a basis for geologic rock classification. The strength differences between an intact rock sample and that of a rock mass of the same rock type is an example of the problem. In addition, the influence of weathering or hydrothermal alteration on the strength of a particular rock type may not change the geologic classification of the rock. A granite, for instance, remains a granite regardless of variations in degree of weathering. The presence of intersecting joints reduces the strength of the rock mass (as noted earlier). Jointing has no influence on how a rock is geologically classified.

Geotechnical classifications for intact rock and rock masses have been devised and these will now be examined. Many of the classifications have been, and will continue to be, designed for specific engineering applications such as tunneling, open-pit mining, and foundations. Although some feel there is a need for an acceptable universal classification of rock for geotechnical purposes, the difficulty in arriving at such a classification

—indeed, even the feasibility of trying to do so—has been noted by specialists in rock mechanics (Deere and Miller, 1966; Voight, 1970; Müller, 1974; Goodman, 1976; Bieniawski, 1980). The presently available specialized classifications provide for the unique requirements for which each was devised. Bieniawski (1980) has summarized the classifications, their applications, and the prospects for standardization.

The more universal the intended use of a classification, the greater the number of intact rock and rock mass properties that must be measured for the classification to meet the intended purposes. Economics, measured both in time and money, becomes a critical factor in the usefulness of multipurpose classifications. Some classifications are kept rather basic and economically feasible with provisions for selection of added tests for particular engineering applications. Simple field tests of rock properties are substituted for more costly and time-consuming laboratory tests in some classifications. Voight (1970) has pointed out that data should not be collected only for purposes of classifying rock materials. A classification and the data required for it must be functional. Classifications are functional only if they contribute to the solution of real problems.

The present trend toward standardization of test procedures and gathering of data useful in rock classification enhances the evolution of functional classification schemes. The efforts of committees, commissions, and working parties of the ASTM, the ISRM, and the Geological Society (London) have done much to provide the needed standardization.

Intact Rock Classification

The geotechnical classification of intact rock is inherently simpler than that of rock masses because the variables introduced by discontinuities are not factors to be considered. The strength and deformability of intact rock specimens long have been index properties used in evaluating rock characteristics (Krynine and Judd, 1957). They are not representative of rock mass properties, being only contributors to the more complex characterization and classification of rock masses. However, the classification of intact rock is important because intact strength is basic to most modern rock mass classifications, and it increasingly becomes important as joint spacing decreases.

Uniaxial, or unconfined, compressive strength has been the standard strength parameter since its early use by John (1962) in his pioneering rock mass classification. The tangent modulus of elasticity at 50% of the uniaxial strength, E_{t50}, has been the accepted measure of intact rock deformability from the earliest classifications to the present (Coates and Parsons, 1966; Deere and Miller, 1966; Aufmuth, 1974).

The classification proposed by Deere and Miller in 1966 (Table 4.15) continues to be the standard because of its simplicity, although modifications have been made to compensate for rock anisotropy, use of elastic modulus, and different class limits. Deere and Miller divided strength into five classes based on uniaxial compressive strength. Strength ranges used by others are shown on Table 4.3.

Intact rock deformability may be measured by the tangent modulus,

Table 4.15 Intact Rock Strength Classification

Strength Classes	Uniaxial Compressive Strength	
Very high strength	>221 MPa	>32,000 psi
High strength	110–221	16,000–32,000
Medium strength	55–110	8,000–16,000
Low strength	28–55	4,000–8,000
Very low strength	<28	<4,000

Source: Deere and Miller, 1966.

E_{t50}, or by the modulus ratio, E_{t50}/σ_1, where σ_1 is the uniaxial compressive strength. Deere and Miller (1966) suggested three classes of modulus ratio for classification purposes (Table 4.16). More recently, investigators have used the tangent modulus as the measure of intact rock deformability.

The uniaxial compressive strength is used alone by many for intact rock classification (Franklin et al., 1971; Müller and Hofmann, 1971; Bieniawski, 1973; Golder Associates and J. F. MacLaren, Ltd., 1976; Geological Society, 1977; ISRM, 1978f). Others combine uniaxial compressive strength and the tangent modulus of elasticity (Deere and Miller, 1966).

The uniaxial compressive strength and deformability may be estimated from tests other than the definitive laboratory tests for strength and elastic modulus. The most commonly used are the point load test, Schmidt rebound hammer test, Shore hardness test, and determination of compressional wave velocity of the rock. The various relationships among these tests and with strength and deformability are described and illustrated in the discussion on intact rock properties. When testing specimens, anisotropy must be recognized and the testing should be conducted so that the classification reflects the anisotropy and anticipated rock use.

Qualitative field tests of rock strength have been devised that are based on rock hardness (Geological Society, 1977; Williamson, 1980). Comparison of the estimates with laboratory tests permits assignment of uniaxial compressive strength ranges to the hardness categories. One such classification is reproduced in Table 4.17. The field strength estimates obtained are very approximate and should be followed by more definitive testing procedures.

Geotechnical classification of the argillaceous and fine-grained sedimentary rocks presents special problems. The geological classification of these materials stresses grain size and distance between partings. Although these

Table 4.16 Intact Rock Modulus Ratio[a]

Ratio Classes	Ratio
High modulus ratio	>500
Average modulus ratio	200–500
Low modulus ratio	<200

[a] Where the ratio equals the tangent Modulus, E_{t50}, divided by UCS, σ_a

Source: Deere and Miller, 1966.

Table 4.17 Intact Rock Classification Based on Empirical Field Tests

Term	Unconfined Compressive Strength MN/m² (MPa)	Field Estimation of Hardness
Very strong	>100	Very hard rock—more than one blow of geological hammer required to break specimen.
Strong	50–100	Hard rock—handheld specimen can be broken with single blow of geological hammer.
Moderately strong	12.5–50	Soft rock—5 mm indentations with sharp end of pick.
Moderately weak	5.0–12.5	Too hard to cut by hand into a triaxial specimen.
Weak	1.25–5.0	Very soft rock—material crumbles under firm blows with the sharp end of a geological pick.
Very weak rock or hard soil	0.60–1.25	Brittle or tough, may be broken in the hand with difficulty.

Source: Reproduced by permission of the Geological Society of London from Geol. Soc. Eng. Group Working Party, The Description of Rock Masses for Engineering Purposes, Q. J. Eng. Geol., Vol. 10, 1977.

characteristics have a direct bearing on the geotechnical properties of the material, they do not indicate changes in strength that result from these variables or from changes in kind and amount of cementation and different degrees of compaction.

Both the slaking (disintegration) of argillaceous materials, with alternating wetting and drying cycles, and the softening, when immersed in water, have been recognized as indicators of strength (durability) (Franklin et al., 1971; Aufmuth, 1974; Morgenstern and Eigenbrod, 1974; Geological Society, 1977). Classifications using the slake-durability tests (described in the discussion of intact rock) evaluate materials too soft or weak for use of the Schmidt rebound hammer and/or the point load test (Tables 4.8, 4.9, and 4.10).

Rock Mass Classification

The geotechnical literature contains many references to the influence of discontinuities on the strength or quality of rock masses. In addition, the recognition that the geotechnical properties of a rock mass are of greater importance than those of intact rock has been widely accepted. As a result, most rock mass classifications that have found acceptance include some measure of the geometric and physical characteristics of discontinuities. The stability of rock masses in underground openings and in rock cuts and open-pit mines is the primary concern in most contemporary classifications. The deformability characteristics of a rock mass, for instance, from foundation and abutment loads from a thin-arch dam, may be reflected in a classification scheme.

Some rock mass classifications have been developed and are of primary value only for a specific purpose. Such is the case of the RSR (Rock Structure Rating) classification of Wickham et al. (1972), which was designed for tunnel-support applications. Other classifications, typified by Bieniawski's Geomechanics Classification (1974), have relatively wide application ranges.

A few classifications are based on a minimum of variables. For example, the classification of Franklin et al. (1971) utilizes only fracture spacing and intact rock strength (Figure 4.62). Others, such as the Geomechanics Classification of Bieniawski and the Q-system classification of Barton et al. (1974), are based on six or more parameters. Examples from these and

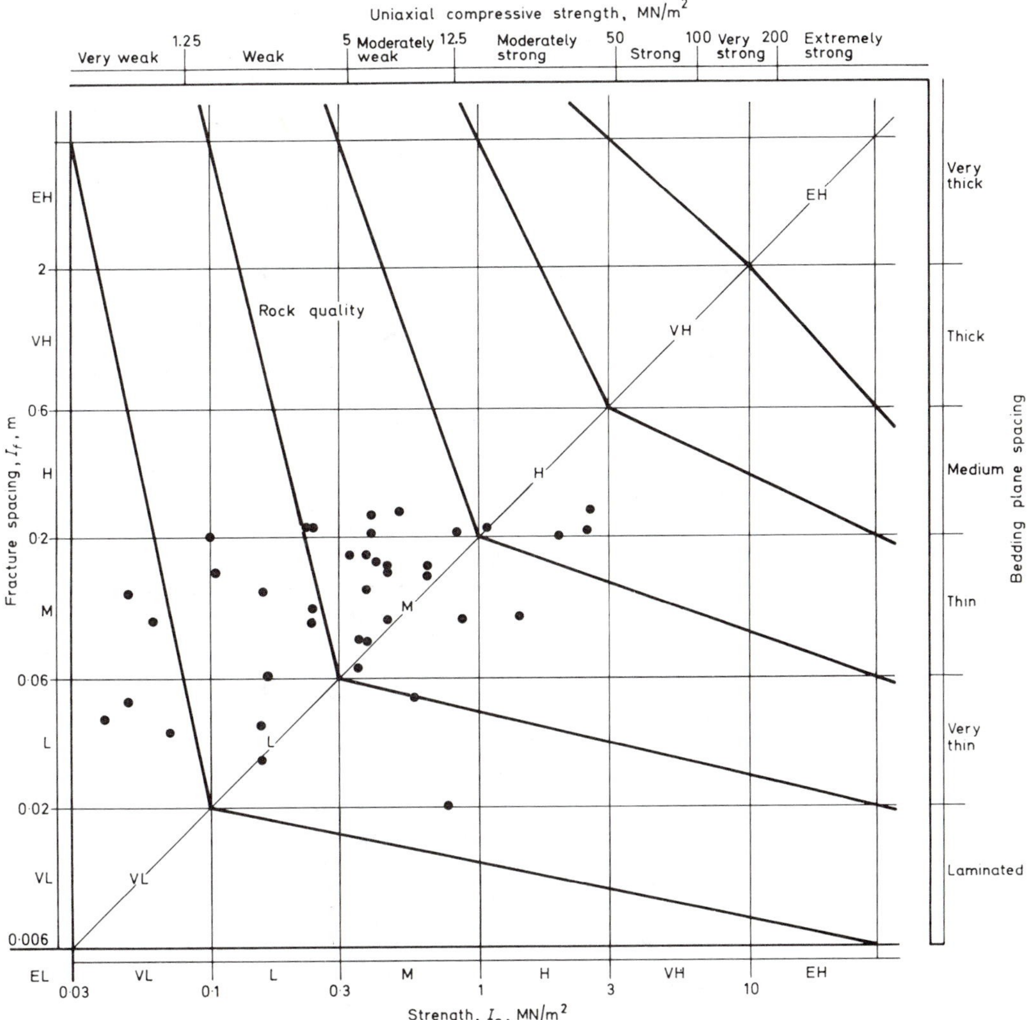

Figure 4.62 Rock mass quality classification based on UCS and discontinuity spacing, where I_s = point load index and classes range from very low (VL) to extremely high (EH). (Reprinted with permission from Trans. Inst. Min. Metall., Vol. 80, Sec. A, Bull. 770, J. A. Franklin, E. Broch, and G. Walton, 1971, Logging the Mechanical Character of Rock.)

other classifications will serve to illustrate the complex problem of classifying rock masses, whether for limited or more universal use.

Early classifications, understandably, were quite descriptive compared to some currently in use. In 1946, Terzaghi devised a rock mass classification for estimating tunnel-support requirements that, though qualitative, included the salient parameters of modern quantitative classifications: the presence, number, and character of discontinuities and the degree of chemical weathering (Table 4.18). In 1962, John quantified joint spacing and degree of weathering, and he added the intact uniaxial compressive strength of the rock (Figure 4.63). This classification has been modified successively by Müller and Hofmann (1971) and Bieniawski (1973).

The greatest modifications were made by Bieniawski, as shown in Figure 4.64. In this classification, the influence of weathering is implicit in the intact rock strength column. The addition of cohesion and friction-angle values provides for the influence of discontinuity surface roughness on rock mass quality. The influence of increased jointing on each class is indicated by the curve separating each adjacent class.

Degree of weathering and discontinuity spacing, orientation, continuity, and surface roughness are among the rock mass properties that are either indirectly or inadequately measured or that are not utilized at all by simple classifications (Figures 4.62 and 4.64). The more-comprehensive classifications correct these oversights.

Bieniawski's Geomechanics Classification (1974) uses RQD (see discussion of rock masses) in conjunction with uniaxial intact rock strength, joint spacing, joint separation, joint continuity, joint orientation, and groundwater flow (Table 4.19). Bieniawski recommended use of the point load test in the field to estimate intact rock strength. Each of these parameters and its five ranges are weighted as shown in Table 4.19*a*. Joint orientation is weighted relative to orientation of construction in the rock (Table 4.20). The total number of weighted points, the rock mass rating (RMR), is the basis for classifying the rock mass into one of five new classes from very good to very poor (Table 4.19*b*). The RMR has been related to the *in situ* modulus of deformation, using test data from a number of different engineering projects in South Africa (Figure 4.65). Table 4.21 is an example of the determination of RMR.

Table 4.18 Terzaghi's Descriptive Rock-mass Classification for Estimating Tunnel-Support Requirements

Intact rock—rock with no joints.
Stratified rock—rock with little strength along bedding surfaces.
Moderately jointed rock—rock mass jointed but cemented or strongly interlocked.
Blocky and seamy rock—jointed rock mass without any cementing of joints and weakly interlocked blocks.
Crushed rock—rock that has been reduced to sand-sized particles without any chemical weathering.
Squeezing rock—rock containing a considerable amount of clay.
Swelling rock—rock that squeezes primarily from mineral swelling.

Source: Terzaghi, 1946.

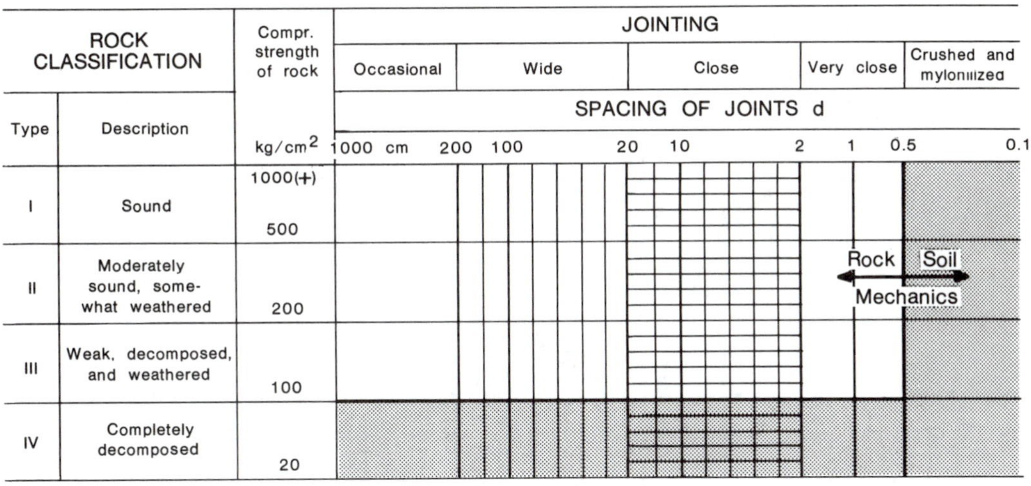

ROCK CLASSIFICATION		Compr. strength of rock	JOINTING				
			Occasional	Wide	Close	Very close	Crushed and mylonitized
			SPACING OF JOINTS d				
Type	Description	kg/cm²	1000 cm 200	100 20	10 2	1 0.5	0.1
I	Sound	1000(+) 500					
II	Moderately sound, somewhat weathered	200				Rock ← Mechanics	→ Soil
III	Weak, decomposed, and weathered	100					
IV	Completely decomposed	20					

Figure 4.63 Rock mass strength classification of John (1962). (Modified and reprinted with permission from Proc. Am. Soc. Civ. Eng., J. Soil Mech. and Found. Eng. Div., Vol. 88, K. W. John, An Approach to Rock Mechanics, 1962.)

The RSR classification developed by Wickham et al. (1971) combines the effects of rock type and structure, joint spacing and orientation, and joint conditions and anticipated water inflow for use in designing tunnel-support systems. Weighted values are assigned to various combinations of each pair of factors. In the case of rock type and structure, rocks are simply igneous, sedimentary, and metamorphic and structure ranges from massive (undeformed) to intensely faulted or folded. Joint spacing and orientation specifically relate to tunneling problems by having weighted values assigned for driving the tunnel through rock with different joint spacings relative to the direction of strike, direction and amount of dip, and direc-

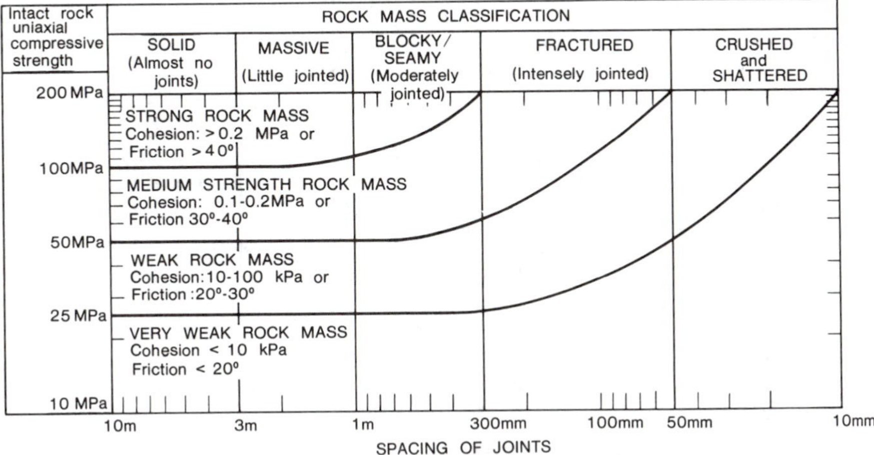

Figure 4.64 Bieniawski's rock mass classification utilizing joint spacing, intact strength, cohesion, and friction angle. (Reprinted by permission from Trans. S. Afr. Inst. Civ. Eng., Vol. 15, Z. T. Bieniawski, Engineering Classification of Jointed Rock Masses, 1973.)

Table 4.19 Geomechanics Classification Parameters, Ranges, Ratings, and Classes

a. Classification Parameters and Their Ratings

	Parameter					
1	USC of intact rock	>200 MPa	100–200 MPa	50–100 MPa	25–50 MPa	<25 MPa
	Rating	10	5	2	1	0
2	Drill-core quality RQD	90% to 100%	75% to 90%	50% to 75%	25% to 50%	<25% or highly weathered
	Rating	20	17	14	8	3
3	Spacing of joints	>3 m	1–3 m	0.3–1 m	50–300 mm	<50 mm
	Rating	30	25	20	10	5
4	Strike and dip orientations of joints	Very favorable	Favorable	Fair	Unfavorable	Very unfavorable
	Rating	15	13	10	6	3
5	Condition of joints	Very tight: separation <0.1 mm Not continuous		Tight: <1 mm and continuous No gouge	Open: 1–5 mm Continuous Gouge <5 mm	Open >5 mm Continuous Gouge >5 mm
	Rating	15		10	5	0
6	Groundwater inflow (per 10 m of tunnel length)	None		<25 l/min	25–125 l/min	>125 l/min
	Rating	10		8	5	2

b. Rock-Mass Classes and Their Ratings

Class No.	I	II	III	IV	V
Description of class	Very good rock	Good rock	Fair rock	Poor rock	Very poor rock
Total rating	100 ← 90	90 ← 70	70 ← 50	50 ← 25	<25

Source: Modified from Bieniawski, 1974.

Table 4.20 Influence of Tunnel Axis, Strike and Dip of Joints, and Relative Direction of Tunneling on the Rating of Jointing for Use in the Geomechanics Classification

Strike perpendicular to tunnel axis				Strike parallel to tunnel axis	
Drive with dip		Drive against dip			
Dip 45°–90°	Dip 20°–45°	Dip 45°–90°	Dip 20°–45°	Dip 45°–90°	Dip 20°–45°
Very favorable	Favorable	Fair	Unfavorable	Very unfavorable	Fair
Dip 0°–20°: Unfavorable, irrespective of strike					

Source: Modified from Bieniawski, 1974.

tion of tunnel construction. Joint-condition categories range from tight or cemented to severely weathered or open for different anticipated water inflow rates. Weighted values for each of the three pairs of factors are summed with high totals being desired.

One of the most comprehensive rock mass classifications to date is the Q-system, or NGI classification, developed in 1974 by Barton et al. of the Norwegian Geotechnical Institute (NGI) specifically for tunneling. The classification includes three-dimensional aspects of rock mass quality that are of critical importance in tunnel-roof stability. As in the Geomechanics Classification, the Q-system requires the determination of RQD. To this are added weighted numerical factors that define the number of joint sets (J_n), joint roughness (J_r), joint alteration (J_a), and two factors based on water flow in joints, (J_w) and an appraisal of rock-stress reduction (SRF) from weathering and tectonic action. The joint set number factor, J_n, is reproduced in Table 4.22 to illustrate a typical factor and its weighting.

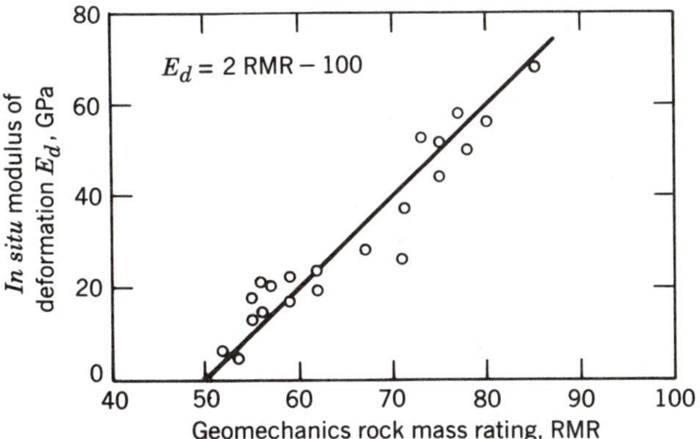

Figure 4.65 Relationship between *in situ* modulus of deformation, E_d, and RMR for the Geomechanics Classification. (Modified and reprinted with permission from Intl. J. Rock Mech. Min. Sci., Vol. 15, Z. T. Bieniawski, Determining Rock Mass Deformability: Experience from Case Histories. Copyright © 1978, Pergamon Press.)

Table 4.21 Example of Data Input to Obtain the RMR, Using the Geomechanics Classification

Parameter	Value	Rating	
UCS	153 MPa		5
Core quality RQD	90–94%		20
Spacing of joints	Set 1: 0.3–1 m	(20)	
	Set 2: 0.3–0.6 m	(20)	22
	Set 3: 2 m	(25)	
Orientations of joints	Set 1: Horizontal	(6)	
	Set 2: Vertical; parallel to tunnel axis	(3)	8
	Set 3: Vertical; perpendicular to tunnel axis	(15)	
Condition of joints	Separation < 1 mm Continuous joints		10
Ground water inflow	None		10
	TOTAL—RMR		75

Source: Bieniawski, 1974.

The ratio RQD/J_n is a measure of block size. Interblock shear strength is represented by J_r/J_a and J_w/SRF defines the active stress at the site. The product of the three ratios is used to define the rock quality index, Q. The higher the value of Q, the better the tunneling conditions of the rock mass.

The Q-system is too lengthy to reproduce here. It is sufficient to note that although such common classification parameters as intact strength and joint spacing are not included, the classification contains this information implicitly in factors RQD and J_n. A total of 60 subfactor descriptions and ratings are defined, with additional modifying multipliers and additive or subtractive values for further refinement for specific tunneling or geologic conditions.

Most of the factors used in even the most complex rock mass classifications may be obtained either visually or by rapid measurements of rock

Table 4.22 Descriptions and Ratings for Q-System Joint Set Number Parameter, J_n

Joint Set Number	(J_n)
Massive, no or few joints	0.5–1.0
One joint set	2
One joint set plus random	3
Two joint sets	4
Two joint sets plus random	6
Three joint sets	9
Three joint sets plus random	12
Four or more joint sets, random, heavily jointed, "sugar cube", etc.	15
Crushed rock, earthlike	20

Source: Reprinted with permission from Rock Mechanics, Vol. 6, N. Barton, R. Lien, and J. Lunde, Engineering Classification of Rock Masses for the Design of Tunnel Support, 1974, Springer-Verlag.

characteristics such as joint spacing and orientation. Strength is an intact rock characteristic that requires more definitive data. The use of tests such as the point load test for indirectly determining intact strength and the portable direct shear box for obtaining the friction angle along discontinuities are sufficiently well documented to permit their use in lieu of more extensive laboratory tests.

Although only a few investigators have incorporated the slake-durability test in a classification or rock mass description scheme (Aufmuth, 1974; Geological Society, 1977), its importance in defining rock mass quality cannot be overemphasized where silt- and clay-rich sedimentary rocks are left unprotected and exposed to the environment. In addition, the determination of the velocity index as a measure of rock mass quality is recommended by some (Geological Society, 1977) for rock mass classification in conjunction with RQD and fracture frequency.

The use of classification parameters has required that each, in turn, is classified as to whether it is RQD, joint spacing, surface roughness, strength, or others. Some evidence of this may be seen in Figure 4.64. Summaries of these secondary classifications as well as the numerous rock mass classifications have been prepared by Golder Associates and J. F. MacLaren, Ltd., (1976), Geological Society (1977), and Johnson (1979). Bieniawski (1980) has prepared a state-of-the-art review of rock classifications.

SUMMARY

Rock plays an important part in the design, construction, and performance of all projects that involve structures built on or in rock. Strength and deformation properties are primary factors in the utilization of both intact rock and rock masses in engineering projects. In most cases, rock mass characteristics are more important than those of intact rock. However, the contribution of intact rock properties to rock mass performance is so great that testing of intact rock cannot be ignored in problems that involve rock masses.

Geologic factors such as mineralogy, texture, grain size, and cementing material significantly affect intact rock strength and deformability. For instance, rocks that have interlocking textures typically are stronger than those that have clastic textures. Chemical weathering alters the engineering properties of all rocks, the degree being dependent primarily on rock type, climate, and time. The end result of chemical weathering is weakening of the rock.

Tests have been devised to characterize rock strength and deformability under different conditions, including those that simulate the *in situ* state. These include static and dynamic tests for determination of elastic moduli and index tests that provide estimates of strength and resistance to exposure by the atmosphere.

Strength and deformability of rock masses and susceptibility to chemical weathering are controlled by the presence of discontinuities such as bedding surfaces and joints. The elastic properties of intact rock contribute to the development of joints. The presence of intersecting sets of discontinuities significantly reduces rock mass strength compared to that of intact rock. Additional loss of strength from chemical weathering is enhanced by water movement through discontinuity paths. Characteristics of discontinuities such as orientation, frequency of occurrence, continuity, and surface roughness have an important role in the way a rock mass deforms and the stability of natural and manmade slopes, excavations, and tunnels.

As in the case of intact rocks, there are tests for determining the response of rock masses to loads. These may be conducted by large- or small-scale tests by means of variations in hydraulic jacking devices and by dynamic means. The deformation moduli obtained provide input for estimates of rock deformation under construction loads. In all cases, the presence of discontinuities plays a critical role.

The classification of rocks is as important to the engineer as it is to the geologist. For the engineer, the classification schemes devised provide for assignment of a rock mass to a certain performance category. The classification and its categories may be designed specifically for a given application, as in tunnel construction. To be of value, required data must characterize the rock, address the intended use, and be easily obtained.

REFERENCES

ASTM, 1979, Standard method of test for triaxial compressive strength of undrained rock core specimens without pore pressure measurements: Am. Soc. Test. Mater., ASTM Designation D 2664-67, Annual Book of ASTM Standards, Pt. 19, 632 pp.

Attewell, P. B., and Farmer, I. W., 1976, Principles of engineering geology: Halsted Press Div. of John Wiley & Sons, New York, 1,045 pp.

Aufmuth, R. E., 1974, Site engineering indexing of rock: Am. Soc. Test. Mater. Spec. Tech. Publ. 554, pp. 81–99.

Barton, N., and Choubey, V., 1977, The shear strength of rock joints in theory and practice: Rock Mech., Vol. 10, Nos. 1–2, pp. 1–54.

Barton, N., Lien, R., and Lunde, J., 1974, Engineering classification of rock masses for the design of tunnel support: Rock Mech., Vol. 6, No. 4, pp. 189–236.

Bernaix, J., 1974, Properties of rock and rock masses: Proc. 3rd Cong. Intl. Soc. Rock Mech., Denver, Colo., Vol. 1, Pt. A, pp. 9–38.

Bieniawski, Z. T., 1973, Engineering classification of jointed rock masses: Trans. S. Afr. Inst. Civ. Eng., Vol. 15, No. 12, pp. 335–343.

———, 1974, Geomechanics classification of rock masses and its application to tunneling: Proc. 3rd Cong. Intl. Soc. Rock Mech., Denver, Colo., Vol. II–A, pp. 27–32.

———, 1975, The point-load test in geotechnical practice: Eng. Geol., Vol. 9, 1975, pp. 1–11.

———, 1978, Determining rock mass deformability: Experience from case histories: Intl. J. Rock Mech. Min. Sci. & Geomech. Abstr., Vol. 15, pp. 237–247.

———, 1980, Rock classification: state of the art and prospects for standardization: Transp. Res. Rec. 783, pp. 2–9.

Birch, F., 1966, Compressibility; elastic constants: *in* Handbook of physical constants, S. P. Clark, Jr. (ed.), Geol. Soc. Am., Mem. 97, pp. 97–173.

Brekke, T. L., and Howard, T. R., 1972, Stability problems caused by seams and faults: Proc, 1st No. Am. Rapid Excavation and Tunneling Conf., Am. Inst. Min. Metall. & Pet. Eng., Chicago, Ill., Vol. 1, pp. 25–41.

Broch, E., and Franklin, J. A., 1972, The point-load strength test: Intl. J. Rock Mech. Min. Sci., Vol. 9, pp. 669–697.

Bukovansky, M., 1970, Determination of elastic properties of rocks using various on-site and laboratory methods: Proc. 2nd Cong. Intl. Soc. Rock Mech., Belgrade, Yugo., Vol. 1, pp. 329–332.

Clark, G. B., 1966, Deformation moduli of rocks: Am. Soc. Test. Mater., Spec. Tech. Publ. 402, pp. 133–174.

Clark, S. P., Jr. (ed.), 1966, Handbook of physical constants, rev. ed.: Geol. Soc. Am., Mem. 97, 587 pp.

Coates, D. F., 1964, Classification of rocks for rock mechanics: Intl. J. Rock Mech. Min. Sci., Vol. 1, pp. 421–429.

———, 1970, Rock mechanics principles, rev. ed.: Can. Dept. Energy, Mines and Res. Mines Branch Monograph 874, 363 pp.

Coates, D. F., and Parsons, R. C., 1966, Experimental criteria for classification of rock substances: Intl. J. Rock Mech. Min. Sci., Vol. 3, pp. 181–189.

Coon, R. F., and Merritt, A. H., 1970, Predicting in situ modulus of deformation using rock quality indexes: Am. Soc. Test. Mater., Spec. Tech. Publ. 477, pp. 154–173.

Cording, E. J., Hendron, A. J., Jr., MacPherson, H. H., Hansmire, W. H., Jones, R. A., Mahar, J. W., and O'Rourke, T. D., 1975, Methods for geotechnical observations and instrumentation in tunneling, Vols. 1 & 2: Rep. No. UILU–ENG 75 2022, Dept. of Civ. Eng., Univ. of Illinois at Urbana-Champaign, 566 pp.

Cording, E. J., and Mahar, J. W., 1974, The effect of natural geologic discontinuities on behavior of rock in tunnels: *in* Proc. 2nd No. Am. Rapid Excavation and Tunneling Conf., Vol. 1, Am. Inst. Min. Metall. & Pet. Eng., San Francisco, Calif., pp. 107–138.

Cording, E. J., and Mahar, J. W., 1978, Index properties for design of chambers in rock: Eng. Geol., Vol. 12, pp. 113–142.

Cruden, D. M., and Krahn, J., 1973, A reexamination of the geology of the Frank Slide: Can. Geotech. J., Vol. 10, pp. 581–591.

D'Andrea, D. V., Fischer, R. L., and Fogelson, D. E., 1965, Prediction of compressive strength from other rock properties: U.S. Bur. Mines Rep. Invest. 6702, 23 pp.

Dearman, W. R., 1974, Weathering classification in the characterization of rocks for engineering purposes in British practice: Bull., Intl. Assoc. Eng. Geol. No. 9, pp. 33–42.

Dearman, W. R., Baynes, F. J., and Irfan, T. Y., 1978, Engineering grading of weathered granite: Eng. Geol., Vol. 12, pp. 345–374.

Deere, D. U., Hendron, A. J., Jr., Patton, F. D., and Cording, E. J., 1967, Design of surface and near-surface construction in rock: Proc. 8th Symp. Rock Mech., Am. Inst. Min. Metall. & Pet. Eng., Minneapolis, Minn. pp 237–302.

Deere, D. U. and Miller, R. P., 1966, Engineering classification and index properties for intact rock: Tech. Rep. No. AFWL–TR–65–116, Univ. of Illinois, Urbana, 299 pp.

Deere, D. U., and Patton, F. D., 1971, Slope stability in residual soils: Proc. 4th Pan American Conf. on Soil Mech. and Found. Eng., San Juan, P.R., Vol. 1, pp. 87–111.

de la Cruz, R. V., 1977, Jack fracturing technique of stress measurement: Rock Mech., Vol. 9, No. 1, pp. 27–42.

de la Cruz, R. V., and Goodman, R. E., 1969, Theoretical basis of the borehole deepening method of stress measurement: Proc. 11th Symp. Rock Mech., Am. Inst. Min. Metall. & Pet. Eng., Berkeley, Calif. pp. 353–373.

Dodds, D. J., 1974, Interpretation of plate loading test results: Am. Soc. Tes. Mater., Spec. Tech. Publ. 554, pp. 20–34.

Donath, F. A., 1961, Experimental study of shear failure in anisotropic rocks: Geol. Soc. Am. Bull., Vol. 72, No. 6, pp. 985–990.

———, 1970, Some information squeezed out of rock: Am. Sci., Vol. 58, No. 1, pp. 54–72.

Fleming, R. W., Spencer, G. S., and Banks, D. C., 1970, Empirical study of behavior of clay shale slopes: U.S. Army Engineer Nuclear Cratering Group Tech. Rep. NCG No. 15, Vols. 1 and 2, 397 pp.

Fookes, P. G., Dearman, W. R., and Franklin, J. A., 1971, Some engineering aspects of rock weathering with field examples from Dartmoor and elsewhere: Q. J. Eng. Geol., Vol. 4, pp. 139–185.

Franklin, J. A., Broch, E., and Walton, G., 1971, Logging the mechanical character of rock: Inst. Min. and Metall., Trans., Sec. A, Vol. 80, Bull. 770, pp. A1–A9.

Franklin, J. A., and Chandra, R., 1972, The slake-durability test: Intl. J. Rock Mech. Min. Sci., Vol. 9, pp. 325–341.

Geological Society, 1970, The logging of rock cores for engineering purposes: Geol. Soc. (London) Eng. Group Working Party, Q. J. Eng. Geol., Vol. 3, pp. 1–24.

———, 1977, The description of rock masses for engineering purposes: Geol. Soc. (London) Eng. Group Working Party, Q. J. Eng. Geol., Vol. 10, pp. 355–388.

Golder Associates and J. F. MacLaren, Ltd., 1976, Tunnelling technology, an appraisal of the art for application to transit systems: Ontario Ministry of Transp. and Communication, Toronto, Can. 166 pp.

Goodman, R. E., 1976, Methods of geological engineering in discontinuous rocks: West, St. Paul, Minn., 472 pp.

Goodman, R. E., Van, T. K., and Heuze, F. E., 1972, Measurement of rock deformability in boreholes: Proc. 10th Symp. Rock Mech., Am. Inst. Min. Metall. & Pet. Eng., Austin, Tex., 1968, pp. 523–555.

Griggs, D. T., 1936, Deformation of rocks under high confining pressures: J. Geol., Vol. 44, No. 5, pp. 541–577.

Hadley, J. B., 1974, Madison Canyon (1959) landslide: *in* Rock Mechanics: The American Northwest, 3rd Cong. Expedition Guide, Intl. Soc. Rock Mech., pp. 134–137.

Haimson, B. C., 1976, Preexcavation deep-hole stress measurements for design of underground chambers—case histories: Proc. 3rd No. Am. Rapid Excavation and Tunneling Conf., Am. Inst. Min. Metall. & Pet. Eng., Las Vegas, Nev., pp. 699–714.

Hamel, J. V., 1974, Rock strength from failure cases—Left bank slope stability study, Libby Dam and Lake Koocanusa, Montana: Missouri River Div., U.S. Army Corps of Eng., Omaha, Neb., Tech. Rep. MRD–1–74, 215 pp.

Hamrol, A., 1961, A quantitative classification of the weathering and weatherability of rocks: Proc. 5th Intl. Conf. Soil Mech. Found. Eng., Paris, France, Vol. 2, pp. 771–774

Handin, J., 1966, Strength and ductility: *in* Handbook of physical constants, S. P. Clark, Jr. (ed.), Geol. Soc. Am., Mem. 97, pp. 223–289.

Hawkes, I., 1971, Photoelastic strain gauges and in situ rock stress measurements: Intl. Symp. on Determination of Stresses in Rock Masses, LNEC, Lisbon, Port., 1969, pp. 359–375.

Hawkes, I., and Moxon, S., 1965, The measurement of in situ rock stress using the photoelastic biaxial gauge with the core-relief method: Intl. J. Rock Mech. Min. Sci., Vol. 2, pp. 405–419.

Heuze, F. E., 1974, Analysis of geological data for the design of rock cuts: Proc. 3rd Cong. Intl. Soc. Rock Mech., Denver, Colo., Vol. 2, Pt. B, pp. 798–802.

Hoar, R. J., and Stokoe, K. H., 1978, Generation and measurement of shear waves in situ: Am. Soc. Tes. Mater., Spec. Tech. Publ. 654, pp. 3–29.

Hobbs, D. W., 1964, The tensile strength of rocks: Intl. J. Rock Mech. Min. Sci., Vol. 1, pp. 385–396.

Hoek, E., and Bray, J., 1974, Rock slope engineering: Inst. Min. Metall., London, Eng., 309 pp.

———, 1981, Rock slope engineering, rev. 3rd ed.: Inst. Min. Metall., London, Eng., 358 pp.

Hoek, E., and Brown, E. T., 1980, Underground excavations in rock: Inst. Min. Metall., London, Eng., 527 pp.

Hoskins, E., 1966, An investigation of the flat jack method of measuring rock stress: Intl. J. Rock Mech. Min. Sci., Vol. 3, pp. 249–264.

Irfan, T. Y., and Dearman, W. R., 1978, Engineering classification and index properties of a weathered granite: Bull., Intl. Assoc. Eng. Geol., No. 17, pp. 79–90.

ISRM, 1978a, Suggested methods for determining hardness and abrasiveness of rocks: Intl. Soc. Rock Mech. Comm. on Standardization of Laboratory and Field Tests, Intl. J. Rock Mech. Min. Sci. & Geomech. Abstr., Vol. 15, pp. 89–97.

———, 1978b, Suggested methods for determining sound velocity: Intl. Soc. Rock Mech. Comm. on Standardization of Laboratory and Field Tests, Intl. J. Rock Mech. Min. Sci. & Geomech. Abstr., Vol. 15, pp. 53–58.

———, 1978c, Suggested methods for determining tensile strength of rock materials: Intl. Soc. Rock Mech. Comm. on Standardization of Laboratory and Field Tests, Intl. J. Rock Mech. Min. Sci. & Geomech. Abstr., Vol. 15, pp. 99–103.

———, 1978d, Suggested methods for determining the strength of rock materials in triaxial compression: Intl. Soc. Rock Mech. Comm. on Standardization of Laboratory and Field Tests, Intl. J. Rock Mech. Min. Sci. & Geomech. Abstr., Vol. 15, pp. 47–51.

———, 1978e, Suggested methods for petrographic description of rocks: Intl. Soc. Rock Mech. Comm. on Standardization of Laboratory and Field Tests, Intl. J. Rock Mech. Min. Sci. & Geomech. Abstr., Vol. 15, pp. 41–44.

———, 1978f, Suggested methods for the quantitative description of discontinuities in rock masses: Intl. Soc. Rock Mech. Comm. on Standardization of Laboratory and Field Tests, Intl. J. Rock Mech. Min. Sci. & Geomech. Abstr., Vol. 15, pp. 319–368.

———, 1979a, Suggested methods for determining the uniaxial compressive strength and deformability of rock materials: Intl. Soc. Rock Mech. Comm. on Standardization of Laboratory and Field Tests, Intl. J. Rock Mech. Min. Sci. & Geomech. Abstr., Vol. 16, pp. 135–140.

———, 1979b, Suggested methods for determining water content, porosity, density, absorption and related properties and swelling and slake-durability index properties: Intl. Soc. Rock Mech. Comm. on Standardization of Laboratory and Field Tests, Intl. J. Rock Mech. Min. Sci. & Geomech. Abstr., Vol. 16, pp. 141–156.

———, 1985, Suggested method for determining point load strength: Comm. on Testing Methods, Intl. J. Rock Mech. Min. Sci. & Geomech. Abstr., Vol. 22, pp. 51–60.

Jaeger, J. C., and Cook, N.G.W., 1976, Fundamentals of rock mechanics, 2nd ed.: Chapman & Hall, London, Eng., 585 pp.

Jennings, J. E., and Robertson, A. M., 1969, The stability of slopes cut into natural rock: Proc. 7th Intl. Conf. Soil Mech. Found. Eng., Mexico City, Vol. 2, pp. 585–590.

Jesch, R. L., Johnson, R. B., Belscher, D. R., Yaghjian, A. D., Steppe, M. C., and Fleming, R. W., 1979, High resolution sensing techniques for slope stability studies: Rep. No. FHWA-RD-79-32, U.S. Dept. of Commerce, Natl. Bur. Standards, Boulder, Colo., 138 pp.

John, K. W., 1962, An approach to rock mechanics: Proc. Am. Soc. Civ. Eng., J. Soil Mech. Found. Eng. Div., Vol. 88, No. SM4, pp. 1–30.

———, 1968, Graphical stability analysis of slopes in jointed rock: Proc. Am. Soc. Civ. Eng., J. Soil Mech. Found. Eng. Div., Vol. 94, No. SM2, pp. 497–526.

Johnson, R. B., 1979, Factors that influence the stability of slopes: Rep. No. FHWA-RD-79-54, U.S. Geol. Surv., Denver, Colo., 129 pp.

Judd, W. R., and Huber, C., 1962, Correlation of rock properties by statistical methods: Intl. Symp. Min. Res., Vol. 2, Pergamon, Oxford, Eng., pp. 621–648.

Knill, J. L., 1974, Engineering geology related to dam foundations: Proc. 2nd Intl. Cong., Intl. Assoc. Eng. Geol., Vol. 2, Paper VI-PC-1, 7 pp.

Knill, J. L., and Price, D. G., 1972, Seismic evaluation of rock masses: Proc. 24th Intl. Geol. Cong., Montreal, Can., pp. 176–182.

Kohlbeck, F., and Scheidegger, A. E., 1977, On the theory of the evaluation of joint orientation measurements: Rock Mech., Vol. 9, No. 1, pp. 9–25.

Krynine, D. P., and Judd, W. R., 1957, Principles of engineering geology and geotechnics: McGraw-Hill, New York, 730 pp.

Kulhawy, F. H., 1975, Stress deformation properties of rock and rock discontinuities: Eng. Geol., Vol. 9, pp. 327–350.

Lane, K. S., and Heck, W. J., 1964, Triaxial testing for strength of rock joints: Proc. 6th Symp. Rock Mech., Am. Inst. Min. Metall. & Pet. Eng., Rolla, Mo., pp. 98–108.

Lee, F. T., Smith, W. K., and Savage, W. Z., 1976, Stability of highwalls in surface coal mines, western Powder River basin, Wyoming and Montana: U.S. Geol. Surv., Open-File Rep. No. 76-846, 52 pp.

Leeman, E. R., 1964, Absolute rock stress measurements using a borehole trepanning stress-relieving technique: Proc. 6th Symp. Rock Mech., Am. Inst. Min. Metall. & Pet. Eng., Rolla, Mo., pp. 407–426.

———, 1968, The determination of the complete state of stress in rock in a single borehole-laboratory and underground measurements: Intl. J. Rock Mech. Min. Sci., Vol. 5, pp. 31–56.

Lindner, E. N., and Halpern, J. A., 1978, In situ stress in North America: A compilation: Intl. J. Rock Mech. Min. Sci. & Geomech. Abstr., Vol. 15, pp. 183–203.

Lutton, R. J., 1977, Slaking indexes for design, *in* Design and construction of compacted shale embankments, Vol. 3: Report No. FHWA-RD-77-1, U.S. Army Eng. Waterw. Exp. Sta., Vicksburg, Miss., 94 pp.

Maidl, B., 1972, Classification of rocks according to their drillability: Rock Mech., Vol. 4, No. 1, pp. 25–44.

McMahon, B. K., 1968, Indices related to the mechanical properties of jointed rock: Proc. 9th Symp. Rock Mech., Am. Inst. Min. Metall. & Pet. Eng., Golden, Colo., pp. 117–128.

Merrill, R. H., 1964, In situ determination of stress by relief techniques: Proc. Intl. Conf. on State of Stress in the Earth's Crust, Elsevier, New York, pp. 343–380.

———, 1967, Three-component borehole deformation gage for determining the stress in rock: U.S. Bur. Mines, Rep. Invest. 7015, 38 pp.

Misterek, D. L., Slebir, E. J. and Montgomery, J. S., 1974, Bureau of Reclamation procedures for conducting uniaxial jacking tests: Am. Soc. Tes. Mater., Spec. Tech. Publ. 554, pp. 35–51.

Morgenstern, N. R., and Eigenbrod, K. D., 1974, Classification of argillaceous soils and rocks: Proc. Am. Soc. Civ. Eng., J. Geotech. Eng. Div., Vol. 100, No. GT 10, pp. 1,137–1,156.

Muir Wood, A. M., 1972, Tunnels for roads and motorways: Q. J. Eng. Geol., Vol. 5, pp. 111–126.

Müller, L., 1964, The rock slide in the Vajont Valley: Rock Mech. Eng. Geol., Vol. 2, pp. 148–212.

———, 1974, Rock mass behavior-determination and application in engineering practice; advances in rock mechanics: Proc. 3rd Cong. Intl. Soc. Rock Mech., Denver, Colo., Vol. 1, Pt. A, pp. 205–215.

Müller, L., and Hofmann, H., 1971, Selection, compilation and assessment of geological data for the slope problem: S. Afr. Inst. Min. Metall., Symp. on the Theoretical Background to the Planning of Open Pit Mines, Johannesburg, pp. 153–170.

Nichols, T. C., Jr., Abel, J. F., Jr., and Lee, F. T., 1968, A solid inclusion borehole probe to determine 3-dimensional stress changes at a point in a rock mass: U.S. Geol. Surv. Bull. 1258C, pp. C1–C28.

Nichols, T. C., and Lee, F. T., 1966, Preliminary appraisal of applied rock mechanics research on Silver Plume Granite, Colorado: U.S. Geol. Surv. Prof. Paper 550–C, pp. C34–C38.

Olsen, O. J., 1957, Measurement of residual stresses by the strain relief method: Colo. Sch. of Mines Q., Vol. 52, No. 3, pp. 183–204.

Onodera, T. F., 1963, Dynamic investigation of foundation rocks in situ: Proc. 5th Symp. on Rock Mech., Am. Inst. Min. Metall. & Pet. Eng., Univ. of Minnesota, Minneapolis, pp. 517–533.

Patton, F. D., 1966, Multiple modes of shear failure in rock: Proc. 1st Cong. Intl. Soc. Rock Mech., Lisbon, Port. Vol. 1, pp. 509–513.

Peltier, L., 1950, The geographic cycle in periglacial regions as it is related to climatic geomorphology: Ann. Assoc. Am. Geogr., Vol. 40, pp. 214–236.

Piteau, D. R., 1971, Geological factors significant to the stability of slopes cut in rock: S. Afr. Inst. Min. Met., Symp. on the Theoretical Background to the Planning of Open Pit Mines, Johannesburg, pp. 33–53.

———, 1973, Characterizing and extrapolating rock joint properties in engineering practice: Rock Mech., Supp. 2, 31 pp.

Priest, S. D., and Hudson, J. A., 1976, Discontinuity spacings in rock: Intl. J. Rock Mech. Min. Sci. & Geomech. Abstr., Vol. 13, pp. 135–148.

Roberts, A., Hawkes, I., Williams, F. T., and Murrell, S.A.F., 1964, The determination of the strength of rock in situ: Trans. 8th Intl. Cong. on Large Dams, Edinburgh, Scot., Vol. 1, pp. 167–186.

Rocha, M., 1970, New techniques for the determination of the deformability and state of stress in rock masses: Proc. Intl. Symp. Rock Mech., Madrid, Spain, pp. 289–302.

Rocha, M., Lopes, J.J.B., and Silva, J. N., 1966, A new technique for applying the method of the flat jack in the determination of stresses inside rock masses: Proc. 1st Cong. Intl. Soc. Rock Mech., Lisbon, Port., Vol. 2, pp. 57–65.

Saunders, M. K., and Fookes, P. G., 1970, A review of the relationship of rock weathering and climate and its significance to foundation engineering: Eng. Geol., Vol. 4, pp. 289–325.

Sowers, G. B., and Royster, D. L., 1978, Field investigation: *in* Landslides-analysis and control, Spec. Rep. 176, Transp. Res. Board, Natl. Acad. Sci., Washington, D.C., pp. 81–111.

Stapledon, D. H., 1968, Discussion-classification of rock substances: Intl. J. Rock Mech. Min. Sci., Vol. 5, pp. 371–373.

Steffen, O.K.H., and Jennings, J. E., 1974, Definition of design joints for two-dimensional rock-slope analyses: Proc. 3rd Cong. Intl. Soc. Rock Mech., Denver, Colo., Vol. 2, Pt. B, pp. 827–832.

Stephenson, B. R., and Murray, K. J., 1970, Application of the strain rosette relief method to measure principal stresses throughout a mine: Intl. J. Rock. Mech. Min. Sci., Vol. 7, pp. 1–22.

Stimpson, B., 1982, A rapid field method for recording joint roughness profiles: Intl. J. Rock Mech. Min. Sci. & Geomech. Abstr., Vol. 19, pp. 345–346.

Tarkoy, P. J., 1975, A study of rock properties and tunnel boring machine advance rates in two mica schist formations: Proc. 15th Symp. Rock Mech., Custer St. Park, S. Dak., pp. 415–445.

Terzaghi, K., 1946, Introduction to tunnel geology: *in* Rock tunneling with steel supports by R. Proctor and T. White, Youngstown Printing Co., Youngstown, Ohio, pp. 19–99.

Tse, R., and Cruden, D. M., 1979, Estimating joint roughness coefficients: Intl. J. Rock Mech. Min. Sci. & Geomech. Abstr., Vol. 16, pp. 303–307.

Voight, B., 1970, On the functional classification of rocks for engineering purposes: Proc. Intl. Symp. Rock Mech., Madrid, Spain, pp. 131–135.

Wallace, G. B., Slebir, E. J., and Anderson, F. A., 1972, Radial jacking test for arch dams: Proc. 10th Symp. Rock Mech., Am. Inst. Min. Metall. & Pet. Eng., Austin, Tex., pp. 633–660.

Wantland, D., 1964, Geophysical measurement of rock properties in situ: Proc. Int. Conf. on State of Stress in the Earth's Crust, Elsevier, New York, pp. 409–450.

Wawersik, W. R. and Fairhurst, C., 1970, A study of brittle rock fracture in laboratory compression studies: Intl. J. Rock Mech. Min. Sci., Vol. 7, pp. 561–575.

Wickham, G. E., Tiedemann, H.R.T., and Skinner, E. H., 1972, Support determinations based on geologic predictions: Proc. 1st No. Am. Rapid Excavation and Tunneling Conf., Chicago, Ill., Vol. 1, pp. 43–64.

Williamson, D. A., 1980, Uniform rock classification for geotechnical engineering purposes: Transp. Res. Rec. 783, pp. 9–14.

Wood, L. E., and Deo, P., 1975, A suggested system for classifying shale materials for embankments: Bull. Assoc. Eng. Geol., Vol. 12, pp. 39–55.

Wuerker, R. G., 1956, Annotated tables of strength and elastic properties: Am. Inst. Min. Metall. & Pet. Eng., Paper 663–G, 22 pp.

5

Subsurface Water

OCCURRENCE AND INFLUENCE OF SUBSURFACE WATER

Subsurface water is often a critical factor in various engineering works. It has impaired the function of dams, levees, and canals. It has shortened the design life of paved highways and airport runways. Subsurface water has contributed to the failure of slopes and building foundations, thus causing loss of life and property. These effects make understanding the occurrence, engineering significance, and control of subsurface water of vital interest to engineering geologists. There are other important aspects of subsurface water (also termed *groundwater*). Entire books and courses are devoted to understanding groundwater exploration and development. The emphasis here will be on dealing with subsurface water in the context of engineering works—thus, leaving the exploration and development aspects to others.

Precipitation is the principal source for the subsurface water found in soil or rock. Another source is subsurface flow from streams or bodies of water. Usually, subsurface water forms two distinct zones. Below the ground surface, the voids within soil or rock contain both water and air. This is the zone of aeration (Figure 5.1). Subsurface water in this zone is termed *vadose water.* At some depth, the void space will contain only water. This is the zone of saturation. The water in this zone is termed either *groundwater* or *phreatic water.* The boundary between the zones of aeration and saturation is the water table. The water table is sometimes referred to as the phreatic surface. The water table rises or falls as the amount of water increases or decreases within the soil or rock. Most engineering geology considerations dealing with subsurface water are based on the water within the zone of saturation.

The bounding conditions of the saturated zone define different water-bearing units called *aquifers* (Figure 5.2). Where the zone of saturation is

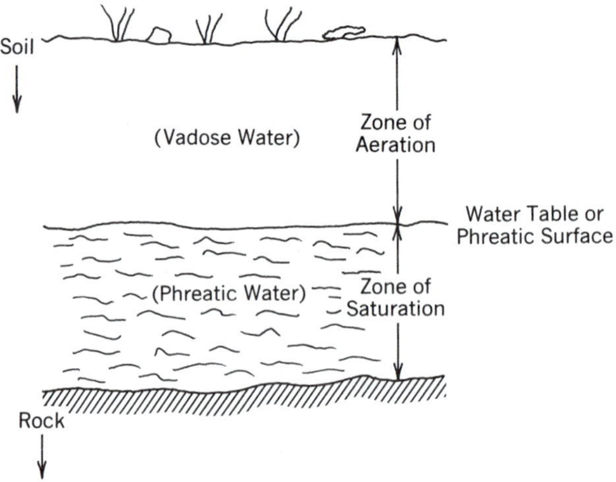

Figure 5.1 Representation of subsurface water conditions and related terms applied to them.

bounded by an impermeable layer below and the zone of aeration above, it is termed an *unconfined aquifer.* The slope of the water table defines the hydraulic gradient. Impermeable barriers may exist locally within the zone of aeration to produce zones of saturation with limited areal extent. These are known as *perched aquifers.* The *confined aquifer* is a zone of saturation bounded both above and below by impermeable layers. The water in a confined aquifer is usually under greater-than atmospheric pressure by the overlying layers. An imaginary line that connects the points to which water would rise in wells penetrating a confined aquifer is called the *piezometric surface.* It is analogous to the water table of an unconfined aquifer.

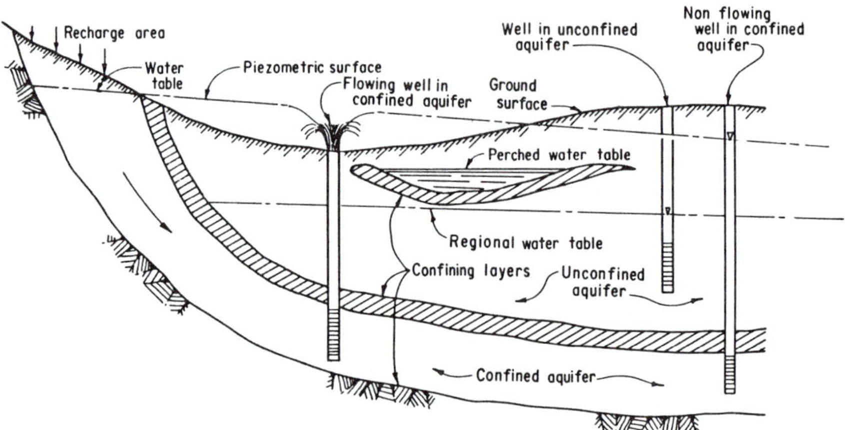

Figure 5.2 Diagram showing the conditions associated with different types of aquifers. (From USBR, 1981).

Permeability

Subsurface water is found in permeable materials. The space between solid particles can be occupied by either water or water and air. When only air fills voids in a soil mass that originally contained some water, the soil mass is considered to be in a drained condition. A soil mass with all voids filled with water, a saturated soil, represents the opposite case. It is an undrained soil. The partially drained condition in which the soil mass contains both water and air in the voids is a common condition. Permeability deals with changes between these conditions. It depends on both the size and amount of voids present in the soil mass and the degree to which these voids are interconnected. Permeability allows vertical movement of subsurface water, thus forming the zones of aeration and saturation. It also permits lateral movement or flow within a soil mass.

The coefficient of permeability, k, is computed to represent the permeability of a soil. The following equation, known as Darcy's Law, is used to determine this coefficient:

$$k = \frac{Q}{A}\frac{L}{h} = \frac{v}{i} \tag{5.1}$$

Where:

k = the coefficient of permeability

Q = discharge (the quantity of water per unit of time)

A = the cross-sectional area through with Q flows

L = the distance through which the head is lost

h = the pressure head lost

v = the discharge velocity

i = the hydraulic gradient, or ratio, of the head lost to the distance over which the loss occurs

The above formulation of Darcy's Law is difficult to apply because permeability is the most variable of the material properties commonly measured in engineering geology. Laboratory testing typically involves apparatus that applies water with a constant or falling head to determine the coefficient of permeability for a sample. Typical modification of Darcy's Law for using laboratory results would include the elapsed time during measurement and the substitution of a constant head for the pressure head lost (USBR, 1974). Laboratory results suffer from problems associated with segregation of particles, trapped air, or soluble mineral content (Cedergren, 1977).

More reliable permeability values can be obtained by well-performed field tests. Each of these tests procedures involves a well or a borehole. A commonly used procedure is the constant head test. A well is pumped at a constant rate and the drop in water level is measured in nearby observation wells. Difficulties in obtaining results may arise for materials with either very high or very low permeability. Variable head tests such as those described in the appendixes of USBR (1974) and in NAVFAC (1982) may permit field measurement for conditions unfavorable to constant head testing. Well and borehole tests require introducing a shape factor into Darcy's Law to reflect the effect of the circular-shaped cross-sectional area and the changing water level.

Cedergren (1977) notes an additional approach to field estimation of permeability that requires no pumping. This procedure employs the relationship of permeability to seepage velocity. Two boreholes are installed intersecting the water table. The seepage velocity is measured by injecting dye, radioactive materials, or similar detectable substances into the upgradient well and determining the time it takes to reach the second well. Effective porosity can be estimated or measured by testing the in-place soil. The hydraulic gradient is measured between the two wells. These values are then substituted into the following formulation of Darcy's Law to compute permeability:

$$k = \frac{v_s n_e}{i} \tag{5.2}$$

Where:

k = permeability of the soil

v_s = seepage velocity

n_e = effective porosity

i = hydraulic gradient

The coefficient of permeability is commonly expressed in cm/sec; ft/day or ft/yr and in./sec are also used to express permeability. Table 5.1 shows some representative permeabilities for different soils. *Pervious, semipervious,* and *impervious* are terms often applied to soils to characterize their degree of permeability. They represent a general range of permeabilities from less than 1 ft/yr for impervious soil to permeabilities exceeding 100 ft/yr for pervious soil.

Seepage and Hydrostatic Pressures

A series of piezometers can be inserted at different depths in saturated sediments under a lake (Figure 5.3*a*). The energy level is identical for each piezometer. The water levels coincide with the water level in the lake, showing that water within the sediments is motionless (Cedergren, 1977). A series of piezometers in a hillslope may also record identical pressure (Figure 5.3*b*). This means the energy, or pressure head, is the same at these points. Again, the energy line would be level. The pore pressure in the soil when the water is motionless is known as *hydrostatic pressure.* When a series of piezometers record different levels, the energy line is sloped (Figure 5.3*c*). This slope is called the *hydraulic gradient.* The water in the soil will flow toward the lower energy point. This produces seepage pressure.

Subsurface water influences the response of a soil mass to an applied

Table 5.1 Range of Permeability Values for Different Soil Material

Permeability

Ft^3/Ft^2/Day (ft/day)

10^5	10^4	10^3	10^2	10^1		10^{-1}	10^{-2}	10^{-3}	10^{-4}	10^{-5}

Ft^3/Ft^2/Min (ft/min)

10^1		10^{-1}	10^{-2}	10^{-3}	10^{-4}	10^{-5}	10^{-6}	10^{-7}	10^{-8}

Gal/Ft^2/Day (gal/ft^2/day)

10^5	10^4	10^3	10^2	10^1		10^{-1}	10^{-2}	10^{-3}	10^{-4}

Ms^3/M^2/Day (m/day)

10^4	10^3	10^2	10^1		10^{-1}	10^{-2}	10^{-3}	10^{-4}	10^{-5}

Relative Permeability

Very High	High	Moderate	Low	Very Low

Representative Materials

Clean gravel	— Clean sand and sand and gravel	— Fine sand	— Silt, clay, and mixtures of sand, silt, and clay	— Massive clay

Source: Modified from USBR, 1981.

load. In an unsaturated soil mass, the particles resist this compressional force. This resistance is based on the friction at points of contact between individual particles of the soil mass. It is termed *intergranular pressure.* The air within the voids has no practical effect. In a saturated soil, the particles respond to compressional force in the same manner as in an unsaturated state. However, water filling the voids is an incompressible material and resists compression.

The presence or absence of water in the voids leads to a distinction between effective stress and total stress (Wu and Sangrey, 1978). Effective stress is the resistance of the individual particles to the applied load. Under a load applied normal to an unsaturated soil mass, intergranular pressure resisting compression represents both effective stress and total stress. Under a saturated state, the resistance of the particles is the same as under an unsaturated state. The pore, or hydrostatic, pressure resists compression. Resistance to the applied load is both the pressure from grain-to-grain contacts and pressure from water in the voids. Stated another way, total stress is the sum of effective stress and pore pressure.

Effective stress and pore pressure are important because most force applied to a soil mass includes both a compressional and shear component (Bowles, 1979; Kenney, 1984). Water having no shear strength provides resistance only to the compressional component. Therefore, response of a soil mass to the shear component is dependent on effective stress. Because effective stress is determined by subtracting pore pressure from total stress, changes in subsurface water alter effective stress. Raising the water level within a soil mass increases pore pressure. Because effective stress in a

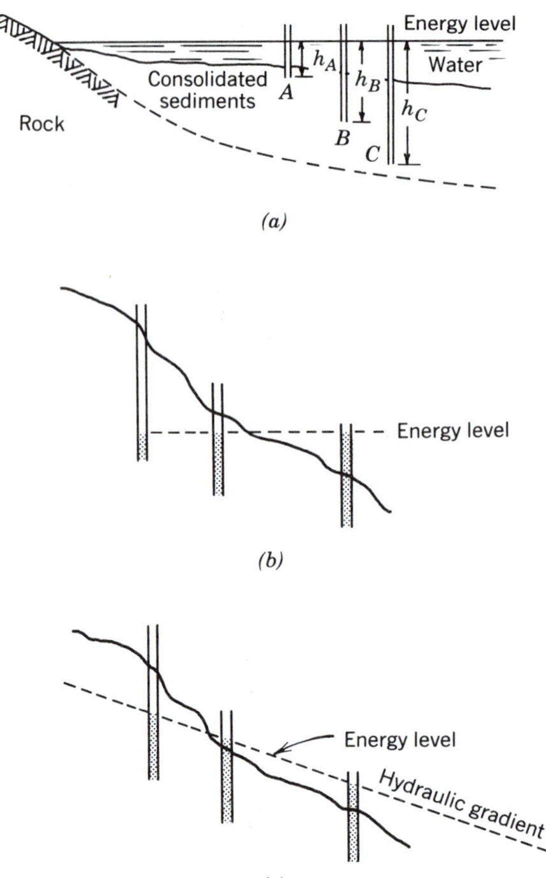

Figure 5.3 (a) Representation of the energy level for saturated sediments under a lake. (Reprinted with permission from H. R. Cedergren, Seepage, Drainage, and Flow Nets, 1977, John Wiley & Sons). (b) Conditions on a slope experiencing hydrostatic pressures. (c) Conditions on a slope experiencing seepage pressure.

saturated soil mass is total stress minus pore pressure, effective stress decreases with increased pore pressure. This means the ability of the soil mass to resist shear force is less. If the pore pressure increases to the point that effective stress approaches zero, it means the force of the water within the voids is so great that the particles are being pushed away from each other and barely touching. This defines a state in which the soil mass has the least resistance to applied shear forces.

A practical illustration of this principle is the short- and long-term stability of cuts and fills in a clay soil (Wu and Sangrey, 1978). The low permeability of clay soils means that they are often in an undrained or saturated state. A road fill placed on this soil applies both compressional and shear forces. The compressional load reduces the volume of the soil, producing the same effect as raising of the water level: hydrostatic pressure is increased. If the soil mass is able to resist the shear force applied, it will remain stable in the short term. Over time, the water may drain, returning to the same pore pressure that was present prior to the load being applied. This would represent conditions similar to a decrease in water level. As a result, the effective stress increases and the long-term stability is greater than short-term stability.

The margin of safety for maintaining a stable fill is greater than at the time the fill is placed. The opposite is true for the cut slope. Gravity will apply both compressional and shear components to the cut slope face. The removal of confining soil produces conditions similar to a decrease in water level. A decreased or negative pore pressure is the result. This creates a greater effective stress in the soil than in its preconstruction state. The initial cut slope is more stable in the short term than in its original state. When water returns, restoring the original pore pressure, it represents a decrease in effective stress. As a result, short-term stability of the cut slope is likely to be greater than long-term stability. If the margin of safety is small initially, there is a good chance that the cut slope will be unable to resist shear force as time passes.

Pore pressure as seepage pressure is a consequence of water flowing through a soil mass. Movement is due to a difference in total energy between two points rather than any pressure difference. Because flow velocity is so slow in most soil, it can usually be ignored. This energy can be expressed as *head*, where head is energy per unit weight of water. A field method for determining total head employs a device called a piezometer (see chapter 6). In its simplest form, the piezometer is an open tube inserted into the soil (Figure 5.4). The static height of water within the tube represents the pressure head at the bottom of the tube. The surface of the water in the tube is called the *piezometric level.* The elevation of the point at the bottom of the tube, the length of tube below the ground surface, and the depth of the piezometric surface from the ground surface are used to determine pressure head. This means that it can be expressed as:

$$h = z + h_p \tag{5.3}$$

Where:

h = total head

z = elevation head

h_p = pressure head

Pore pressure can be expressed in similar terms:

$$\mu = h_p \gamma_w \tag{5.4}$$

Where:

μ = pore pressure
h_p = pressure head
γ_w = the unit weight of water

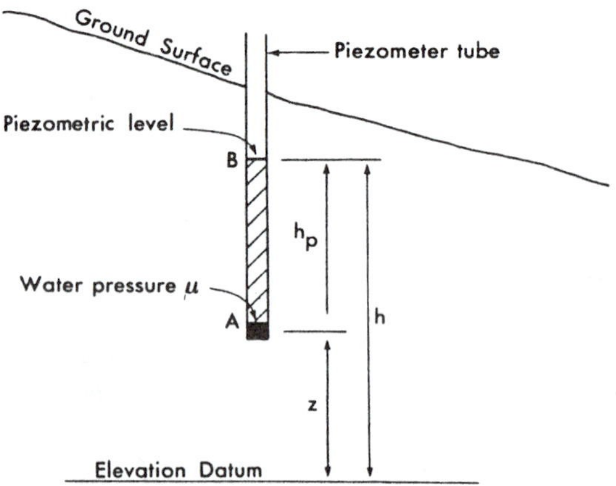

Figure 5.4 Cross section of a slope showing a piezometer installation and the relationship of measurements used to compute total head, pressure head, and pore pressure. (Reprinted with permission from C. Kenney, Properties and Behaviour of Soils Relevant to Slope Instability, *in* Slope Instability, D. Brunsden and D. Prior (eds.). Copyright © 1984, John Wiley & Sons. Reprinted by permission of John Wiley & Sons.)

Determining total head and pore pressure is necessary to forecast pore pressure changes and their possible effect on engineering works. An example of computing pore pressure from measurement of total head is given in Module 5.1.

The quantity of subsurface water flowing through a soil may need to be determined. The quantity of seepage under or through a levee is an example. This can be computed by rearranging Darcy's Law to solve for discharge:

$$Q = k\,A\,\frac{h}{L} \tag{5.5}$$

Where:

Q = discharge
k = the coefficient of permeability of the material
A = the cross-sectional area through which Q flows
h = the pressure head lost
L = the distance through which the head is lost

Constructing a flow net is a graphical method used to compute seepage quantity. Cedergren (1977) notes that the following assumptions are made:

1. The soil is homogeneous.
2. The voids are completely filled with water.

MODULE 5.1

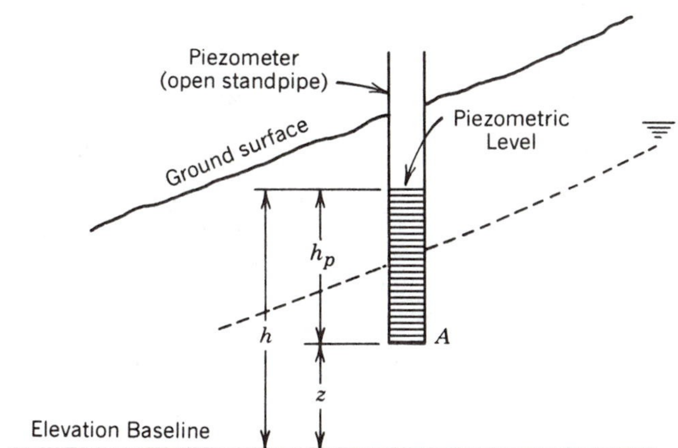

Compute total head (h) and pore-water pressure (μ) at point A when

$h_p = 5.32$ meters
$z = 246.6$ meters
$\gamma_w = 9.8$ kilonewtons/cubic meter

1. Total head

$h = z + h_p$
$h = 246.6$ meters $+ 5.32$ meters
$h = 251.92$ meters

2. Pore-water pressure

$\mu = h_p \gamma_w$
$\mu = 5.32$ meters $\times 9.8$ kilonewtons/cubic meter (kN/m^3)
$\mu = 52.1$ kilopascals (kPa)

3. No consolidation or expansion of the soil takes place.
4. The soil and water are incompressible.
5. Flow is laminar and Darcy's Law is valid.

A flow net is composed of flow lines and equipotential lines. Flow lines represent a few of the many paths that water flow may take through the soil mass. Equipotential lines are drawn at right angles to flow lines and represent points of equal energy or pressure. To start drawing a flow net, it is expedient to divide the total head by some whole number. Equipotential lines will extend perpendicularly from the water table connecting points representing the same amount of head. Flow lines are then drawn perpendicular to the equipotential lines to form equal squares (Figure 5.5). In practice, flow nets include curved lines that produce curvilinear rather than true squares (Cedergren, 1977). Seepage quantity is determined from the completed flow net using the equation:

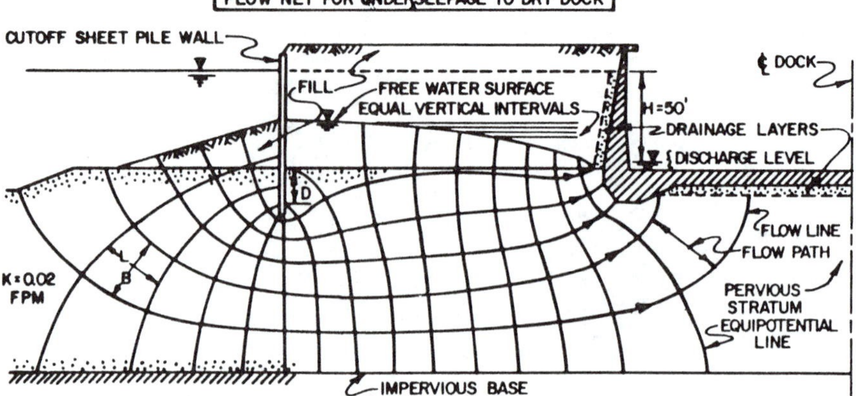

Figure 5.5 A simple flow net for a drydock. (From NAVFAC, 1982)

$$Q = k\frac{n_f}{n_d}H \tag{5.6}$$

Where:

Q = discharge
k = permeability of the material
n_f = number of flow paths
H = total head
n_d = number of equipotential drops

Module 5.2 provides an example that shows the calculation of seepage quantity from a flow net.

Lambe and Whitman (1969), Cedergren (1977), and NAVFAC (1982) contain more details on flow net construction. These references include instructions on constructing a flow net for a mass consisting of two materials with differing coefficients of permeability as well as other problems that may be encountered.

Another factor that may need to be determined is seepage force. Called *seepage pressure,* it is usually represented in terms of applying the force to a unit volume of soil. At a free face, seepage pressure may be great enough to dislodge individual particles. This alters flow lines, increases velocity, and causes greater seepage pressure to develop. In time, a pipe, or conduit, may form in the soil, thus threatening the integrity of the structure. Seepage force acts at right angles to the equipotential lines in a flow net. It is computed by multiplying the hydraulic gradient and the unit weight of water:

$$P_s = i\,\gamma_w \tag{5.7}$$

Where:

P_s = seepage pressure
i = hydraulic gradient, $\frac{h}{L}$
γ_w = unit weight of water

MODULE 5.2

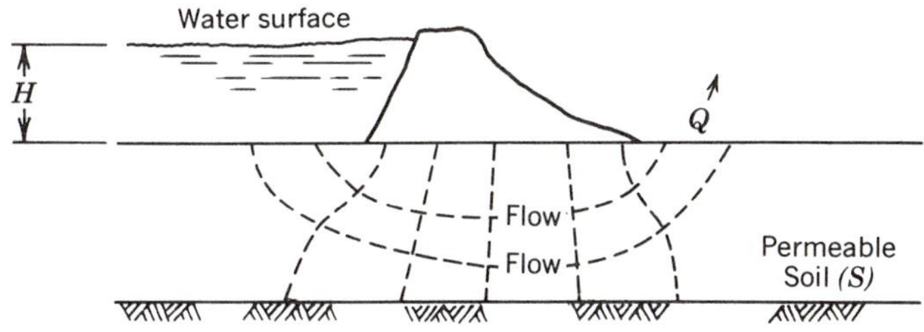

Determine the seepage quantity under the dam per meter of dam width per day when:

Head (H) = 20 meters and soil permeability (k) = 3.2×10^{-3} centimeters per second

1 day = 86,400 seconds

1 meter of width

$$Q = k\frac{n_f}{n_d} \text{H} \times \text{width}$$

$$Q = \left(3.2 \times 10^{-1} \text{ meters/second}\right)\left(\frac{2}{5}\right)\left(20 \text{ meters}\right)\left(\frac{86{,}400 \text{ seconds}}{1 \text{ meter}}\right)$$

$Q = 2.6$ cubic meters/day/meter of dam width

A seepage pressure example is given in Module 5.3. Piping is more likely to occur in finer-grained soils. Open voids, poorly compacted backfills, and ground cracks may serve as initiators of piping. A related concern is boiling. Boiling occurs when the force of soil acting downward is exceeded by the seepage pressure upward. A critical hydraulic gradient (i_c) defines this condition:

$$i_c = \frac{\gamma_T - \gamma_w}{\gamma_w} = \frac{\gamma_b}{\gamma_w} \tag{5.8}$$

Where:

γ_T = total unit weight of soil

γ_w = unit weight of water

γ_b = buoyant weight of soil

Common practice is to set a minimum factor of safety (FOS) of 2 for this condition (NAVFAC, 1982):

$$\text{FOS} = \frac{i_c}{i} \tag{5.9}$$

Where:

i = actual hydraulic gradient

MODULE 5.3

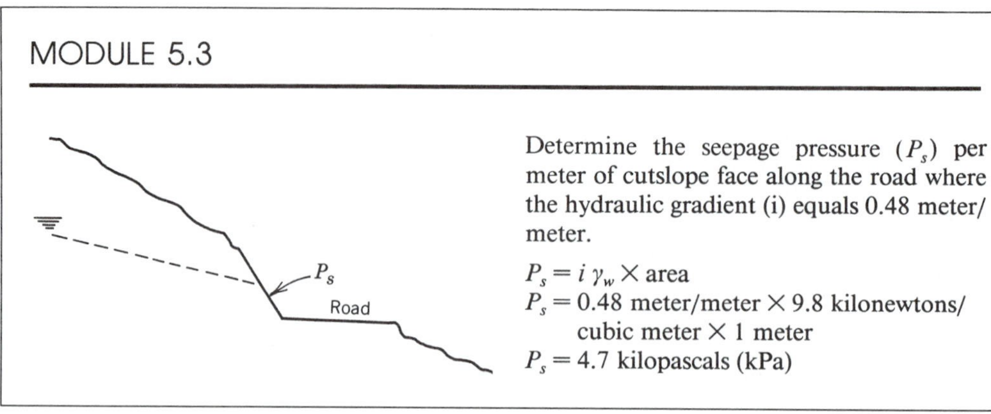

Determine the seepage pressure (P_s) per meter of cutslope face along the road where the hydraulic gradient (i) equals 0.48 meter/meter.

$P_s = i\,\gamma_w \times$ area

$P_s = 0.48$ meter/meter $\times$ 9.8 kilonewtons/cubic meter $\times$ 1 meter

$P_s = 4.7$ kilopascals (kPa)

ENGINEERING SIGNIFICANCE

Subsurface water is significant for three major aspects of engineering works: (1) subsurface water may pose a problem to construction, (2) it may be an erosional agent that degrades a structure, (3) subsurface water may be critical to the functioning of a structure.

Subsurface water as a medium for pollutant contamination is an increasingly significant aspect of engineering geologic investigation. Geologic information is critical to understanding contaminant plume migration and how best to monitor for detection or measurement of contamination in subsurface water.

Problems Created in Construction

Construction may be made difficult or even impossible by subsurface water. Neither excavation nor construction can be properly carried out with excessive water present. This is true even for seemingly minor projects. Construction of a small dam illustrates this point. During excavation of the keyway for a small earthen dam, subsurface water was intercepted. Flow did not hinder operation by the small bulldozer used in excavating the keyway. However, water flowed into the excavation too quickly after pumping ceased to permit the proper compaction of the impervious material forming the dam core. A second excavation upgradient was made to intercept the subsurface flow. By pumping this excavation, the keyway remained dewatered long enough to permit material to be placed and compacted by a hand-operated vibrator.

The Divide Cut of the Tennessee–Tombigbee Waterway serves as an example for a major project. This section of the project required excavation of over 150 million yd^3 of sand and clay. Site exploration determined that the 175-ft deep cut would intercept subsurface water through much of the alignment (Simmons, 1985). A dewatering system was installed to lower the water level an average of 50 ft to enable excavation to occur. This required an extensive series of dewatering trenches and wells to be installed in step with construction of the cut.

Water alone is not always the problem. The influence of subsurface water on soil may prove equally troublesome. Construction of an underpass for Interstate Highway 94 in Fargo, North Dakota, encountered subsurface water with a silty sand (Bell, 1968). The water caused the silty sand to have a "quicksand" character, making excavation extremely difficult. Dewatering at this site reduced the water level and stabilized the soil to allow the excavation and highway construction to be completed.

Water and water-related soil conditions may be encountered when tunneling in soft ground. Construction of the Bay Area Rapid Transit (BART) in San Francisco included a total of 64,576 ft of soft-ground tunneling (Taylor and Conwell, 1981). Excessive inflow of water and potential collapse of the tunnel heading in saturated materials was countered in two ways. Dewatering of the area adjacent to the tunnel was sometimes employed. In other areas, compressed air was used to produce pressures that ranged as high as 36 $lb/in.^2$ to counter the water inflow and the squeezing of saturated sediment into the tunnel heading.

Acting as an Erosional Agent

Uncontrolled subsurface water can act as an erosional agent. Cedergren (1977) states that structural failure caused by this agent can be divided into two groups: (1) boiling and piping where soil particles move and escape through an exit and (2) landslides, uplift pressures, and seepage pressures.

The Teton Dam failure in southeastern Idaho on June 5, 1976, is a good example of failure attributable to piping. Piping requires some initial discontinuity in the material to start the process. Bedrock fractures, animal burrows, unplugged drillholes, and similar features have provided such discontinuities. Open fractures and joints in the rock of the right abutment were the avenues for piping at Teton Dam (Flagg, 1979). In about 6 hr, the initial seepage led to a complete breach of the dam and loss of 40% of the dam embankment (150,000 yd^3).

The Teton Dam disaster illustrates another way subsurface water can be an agent of erosion. Landslides occurred along the reservoir as a consequence of rapid drawdown (Schuster, 1979; Schuster and Embree, 1980). The failure of the dam drained the reservoir quickly, leaving slope materials with the following conditions: (1) high saturated weights, (2) low shear strengths, and (3) pore pressures at disequilibrium for the saturated masses. Hundreds of shallow failures in unconsolidated materials occurred within the weeks immediately following the dam failure. Considerable landslide activity also occurred within welded tuff units owing to the influence of pore pressure changes on existing joints and fissures. Schuster and Embree (1980) note nearly all landslides are located below the maximum reservoir shoreline.

Landslides occurring on the edge of reservoirs may also result from filling that leads to increased pore pressure. The surface of the impounded water alters the water table in the adjacent slopes. These slopes experience a higher degree of saturation than under natural conditions. Just as some natural slopes become unstable owing to higher pore pressures induced by placing a fill, higher pore pressures resulting from this alteration of the local water table can lead to instability. Flagg (1979) notes that bedrock conditions at the edge of the Vaiont Reservoir in Italy were barely stable. The effect of the impoundment was to tip the balance in favor of slope failure.

The source of subsurface water leading to landsliding need not be artificial impoundments. Campbell (1975) related shallow landsliding in the Santa Monica Mountains in southern California to the soil, its moisture content, and slope steepness. He demonstrated that once the field capacity of the soil is reached, high rates of rainfall will often initiate shallow slope failures. However, predicting the response of subsurface water to rainfall is difficult (Sangrey et al., 1984). Increased subsurface water during heavy rainfall contributes to failure conditions. Subsurface water flows parallel to the slope. Flow moves more horizontally than vertically through the soil. As Cedergren (1977) notes, water moving in a generally horizontal direction destabilizes slopes. Horizontal flow creates seepage forces acting parallel to the slope, thus adding to the component of gravity acting in that direction. It is possible to have a highly saturated slope without destabilization occurring. If the material is sufficiently permeable or unsaturated that seepage is essentially vertical, the resulting forces acting on the slope are little different than under a dry state.

Subsurface water can act as an erosional agent through uplift pressure. For example, excessive uplift pressure led to the failure of the Malpasset Dam in southeastern France (Flagg, 1979). Pore pressure built up along the discontinuities defining an impermeable rock block in the left abutment. This pressure increase was first evident when excessive uplift pressure produced cracking in the concrete spillway apron. Transmission of these unrelieved, high pore pressures along the fractures initiated displacement of a wedge-shaped rock block from the left abutment. This resulted in the first total failure of a concrete arch-type dam.

Affecting the Functioning of a Structure

Subsurface water can be a factor that influences the functioning of a structure. Tunnels intended for rail or motor traffic, ponds for retaining water, or facilities isolating waste can be rendered less effective by subsurface water. For example, Davoren (1983) describes efforts to characterize subsurface flow in the Kaimai rail tunnel in New Zealand. Construction required the installation of drains to remove excess water. Study results showed that grouting to plug inflow was not needed. Tunnel effectiveness was adequately maintained by the previously installed drains.

Seepage control of surface impoundments is the key to maintaining proper function. For irrigation canals and surface water impoundments, seepage control reduces water losses. For wastewater lagoons, solid-waste disposal pits, and mine-tailings ponds, seepage control attempts to prevent or reduce contaminants reaching the natural subsurface water regime.

Bouwer (1982) examined the use of earth linings such as clay or soil with low permeability for both types of use. He states that prediction of seepage is needed to ensure that the right earth liner is selected, the appropriate thickness is used, and the best construction method for emplacement is employed.

Structures isolating waste from the subsurface environment are becoming increasingly important. This includes domestic and industrial waste. Foose and Hess (1976) provide a comprehensive case history of a landfill site in Pennsylvania. Determining subsurface water levels and flow were major aspects of the engineering geology work leading to successful design and operation of this facility. In contrast, White and Gainer (1985) describe the consequences of not controlling seepage at a waste site. Their study describes contamination resulting from disposal of uranium mill-tailings in unlined ponds at a site in Utah. This problem required not only the redesign of the ponds to prevent further contamination, but also the interception of contaminated subsurface water migrating off-site.

As a Medium for Contaminant Movement

Subsurface water can serve as the medium for dispersal of pollutants from a point or nonpoint source. Although this aspect of subsurface water overlaps with exploration and development of groundwater resources, engineering geologists are increasingly involved with characterization of subsurface water and conditions controlling its movement. The geologic information needed to address subsurface water pollution differs little from that needed to address subsurface water as it affects the construction or the functioning of a structure. The main difference is the focus of how this information is used. For the previously described aspects of subsurface water, the focus is on removing the water or preventing it from affecting the materials at the site or the functioning of the structure. With subsurface water as a medium for pollution, the concern is with the extent and movement of contamination. The geologist is concerned with detecting, monitoring, and measuring pollution in subsurface water.

Bedient et al. (1984) describe efforts to address contamination at an abandoned creosote facility in Conroe, Texas. The preliminary investigation included installing wells and boreholes to characterize the materials, identifying subsurface geologic conditions important to subsurface flow, determining subsurface water levels, and permitting water quality sampling. Water quality sampling is the only aspect notably different from investigation of subsurface water for other engineering geologic work.

Detection may involve identifying the location and depths for installing a series of wells in the vicinity of a newly constructed sewage-treatment plant. The intent of such a system would be to detect leakage of effluent should the plant malfunction in the future.

Monitoring and measurement are generally needed to characterize the extent of contamination from a source. A typical case could be determining the geologic and subsurface water conditions for monitoring the migration of a contaminant plume from a hazardous waste site.

CONTROL OF SUBSURFACE WATER

A variety of methods exist for controlling subsurface water. Choosing the method or combination of methods to use depends on the type of engineering work involved, site conditions, and the purpose for controlling the water.

Control methods can be reasonably grouped into four categories: (1) barriers, (2) liners, (3) wells, and (4) drains. Barriers are generally employed to reduce both quantity and velocity of subsurface water; liners are used to prevent the movement of subsurface water; drains reduce the quantity of water and direct its movement. Dewatering, the elimination of subsurface water, is accomplished though the use of wells. Each method has advantages and disadvantages that require careful attention when choosing the control method or methods for a specific project.

Barriers and Liners

Sheet pile cutoff walls, impervious cutoff trenches, and grouted or injected cutoff curtains are common types of barriers. Barriers are often part of structures. For example, earthen dams may incorporate a cutoff wall or grout curtain in their foundation to reduce the amount and speed of water flowing under the structure. Table 5.2 describes types of barriers, their applicability, and their characteristics.

The type of barrier to use at a site is dictated as much by subsurface conditions as by the type of structure involved. For instance, coarse-grained material and stratified soil with alternating layers of pervious and impervious material are ideal for use of sheet pile barriers. This is not the case when boulders or obstructions might be present. Conditions affecting the ability to excavate a trench to the necessary depth and maintain its shape will determine whether placement of impervious soil barriers are appropriate.

Liners are used to prevent seepage. This may involve preventing loss of water retained in a canal or impoundment or keeping water out of an area such as a waste-disposal site to prevent migration of contaminants. Table 5.3 contains brief descriptions of different types of liners and their applicability. Bentonite and related clays are often favored as liners to prevent water from reaching isolated waste. Some caution in relying on this method, however, should be exercised due to questions concerning their long-term efficiency (Lundgren and Soderblom, 1985).

Drains and Wells

Effective drains depend on the relative permeability of the drain materials and surrounding materials. Figure 5.6 shows an example of drains placed to maintain a road. Subsurface water enters the more-permeable drain material and is carried away before it can flow into the less-permeable road materials. The relative sizing of drain materials is critical to proper operation. It must have a sufficiently greater permeability than the soil it is protecting in order to intercept the subsurface flow. At the same time, it

Table 5.2 Descriptions of Different Types of Barriers Used to Control Subsurface Water and Factors Affecting Their Applicability and Performance

Method	Applicability	Characteristics and Requirements
Sheet pile cutoff wall	Suited especially for stratified soils with high horizontal and low vertical permeabiity or pervious hydraulic fill materials. May be easily damaged by boulders or buried obstructions. Tongue and groove wood sheeting utilized for shallow excavation in soft to medium soils. Interlocking steel sheetpiling is utilized for deeper cutoff.	Steel sheeting must be carefully driven to maintain interlocks tight. Steel H-pile soldier beams may be used to minimize deviation of sheeting in driving. Some deviation of sheeting from plumb toward the side with least horizontal pressure should be expected. Seepage through interlocks is minimized where tensile force acts across interlocks. For straight wall sheeting an appreciable flow may pass through interlocks. Decrease interlock leakage by filling interlocks with sawdust, bentonite, cement grout, or similar material.
Compacted barrier of impervious soil	Formed by compacted backfill in a cutoff trench carried down to impervious material or as a core section in earth dams.	Layers or streaks of pervious material in the impervious zone must be avoided by careful selection and mixing of borrow materials, scarifying lifts, aided by sheepsfoot rolling. A drainage zone downstream of an impervious section of the embankment is necessary in most instances.
Grouted or injected cutoff	Applicable where depth or character of foundation materials make sheetpile wall or cutoff trench impractical. Utilized extensively in major hydraulic structures. May be used as a supplement below cutoff sheeting or trenches.	A complete positive grouted cutoff is often difficult and costly to attain, requiring a pattern of holes staggered in rows with carefully planned injection sequence and pressure control.
Slurry-trench method	Suited for construction of impervious cutoff trench below groundwater or for stabilizing trench excavation. Applicable whenever cutoff walls in earth are required. Is replacing sheetpile cutoff walls.	Vertical-sided trench is excavated below groundwater as slurry with specific gravity generally between 1.2 and 1.8 is pumped back into the trench. Slurry may be formed by mixture of powdered bentonite with fine-grained material removed from the excavation. For a permanent cutoff trench such as a foundation wall or other diaphragm wall, concrete is tremied to bottom of trench, displacing slurry upward. Alternatively, well-graded backfill material is dropped through the slurry in

(continued)

Table 5.2 (*continued*)

Method	Applicability	Characteristics and Requirements
		the trench to form a dense mixture that is essentially an incompressible mixture; in working with coarser gravels (that may settle out), to obtain a more reliable key into rock, and a narrower trench, use a cement-bentonite mix.
Impervious wall of mixed in-place piles.	Method may be suitable to form cofferdam wall where sheet pile cofferdam is expensive, or cannot be driven to suitable depths, or has insufficient rigidity, or requires excessive bracing.	For a cofferdam surrounding an excavation, a line of overlapping mixed in-place piles are formed by a hollow shaft auger or mixing head rotated into the soil while cement grout is pumped through the shaft. Where piles cannot be advanced because of obstructions or boulders, supplementary grouting or injection may be necessary.
Freezing—ammonium brine or liquid nitrogen	All types of saturated soils and rocks. Forms ice in voids to stop water. Ammonium brine is better for large applications of long duration. Liquid nitrogen is better for small applications of short duration where quick freezing is needed.	Gives temporary mechanical strength to soil. Installation costs are high and refrigeration plant is expensive. Some ground heave occurs.

Source: NAVFAC, 1982.

Table 5.3 Descriptions of Different Types of Liners Used to Form an Impermeable Layer and Their Applicability

Method	Applicability and Procedures
Buried Plastic Liner	Impervious liner formed of black-colored polyvinyl chloride plastic film. Where foundation is rough or rocky, place a layer 2- to 4-inches thick of fine-grained soil beneath liner. Seal liner sections by bonding with manufacturer's recommended solvent with 6-inch overlap at joints. Protect liner by 6-inch min. cover of fine-grained soil. On slopes add a 6-inch layer of gravel and cobbles ¾- to 3-inch size. Anchor liner in a trench at top of slope. Avoid direct contact with sunlight during construction before covering with fill and in completed installation. Usual thickness range of 20 to 45 mils (.020" to .045"). Items to be specified include Tensile Strength (ASTM D412), Elongation at Break (ASTM D412), Water Absorption (ASTM D471), Cold Bend (ASTM D2136), Brittleness Temperature (ASTM D746), Ozone Resistance (ASTM D1149), Heat Aging Tensile Strength and Elongation at Break

(continued)

Table 5.3 *(continued)*

Method	Applicability and Procedures
	(ASTM D412), Strength—Tear and Grab (ASTM D751).
Buried Synthetic Rubber Liner	Impervious liner formed by synthetic rubber, most often polyester reinforced. Preparation, sealing, protection, anchoring, sunlight, thickness, and ASTM standards are same as Buried Plastic Liner.
Bentonite Seal	Bentonite placed under water to seal leaks after reservoir filling. For placing under water, bentonite may be poured as a powder or mixed as a slurry and placed into the reservoir utilizing methods recommended by the manufacturer. Use at least 0.8 pounds of bentonite for each square foot of area, with greater concentration at location of suspected leaks. For sealing silty or sandy soils, bentonite should have no more than 10 percent larger than 0.05 mm; for gravelly and rocky materials, bentonite can have as much as 40 percent larger than 0.05 mm. For sealing channels with flowing water or large leaks, use mixture of ⅓ each of sodium bentonite, calcium bentonite, and sawdust.
Earth Lining	Lining generally 2- to 4-feet thick of soils having low permeability. Used on bottom and sides of reservoir extending to slightly above operating water levels. Permeability of soil should be no greater than about 2×10^{-6} fpm for water supply linings and 2×10^{-7} fpm for pollution control facility linings.
Thin Compacted Soil Lining with Chemical Dispersant	Dispersant is utilized to minimize thickness of earth lining required by decreasing permeability of the lining. Used where wave action is not liable to erode the lining. Dispersant such as sodium tetraphosphate is spread on a 6-inch lift of clayey silt or clayey sand. Typical rate of application is 0.05 lbs/sf. Chemical and soil are mixed with a mechanical mixer and compacted by sheepsfoot roller. Using a suitable dispersant, the thickness of compacted linings may be limited to about 1 foot; the permeability of the compacted soil can be reduced to ⅒ of its original value.

Source: NAVFAC, 1982.

must not permit excessive passage of the smaller particles in the adjacent materials to avoid piping of the soil or clogging of the drain. Cedergren (1977) notes that the following design criteria are commonly used to address this problem:

$$\frac{D_{15}\text{(of filter)}}{D_{85}\text{ (of soil)}} < 4 \text{ to } 5 < \frac{D_{15}\text{(of filter)}}{D_{15}\text{ (of soil)}} \tag{5.10}$$

Where:

D = the diameter of particles at the percentile of gradation curve indicated by the subscript for the soil and filter medium

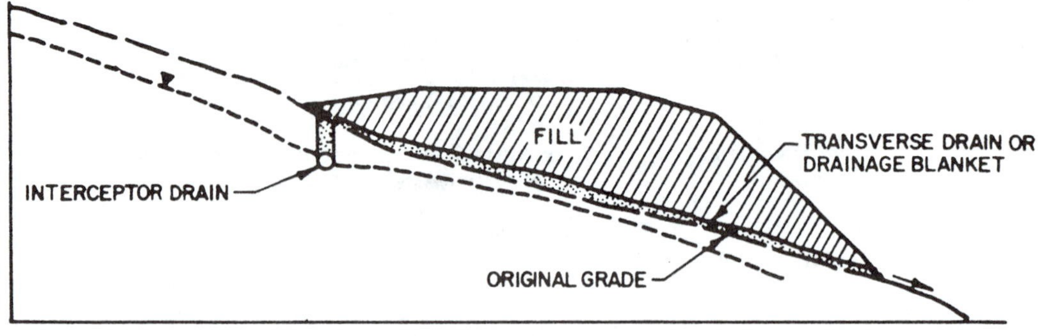

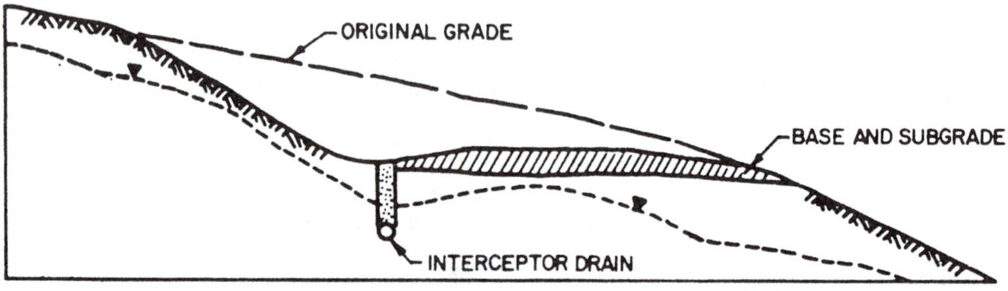

Figure 5.6 Examples of drains used to control subsurface water for cuts and fills along a road. (From NAVFAC, 1982)

This relationship was originally developed for use with graded aggregate. The growing use of geotechnical fabrics or filter cloth has modified this problem. Sizing aggregate is less critical for a drain where a filter cloth is included to prevent smaller particles in the soil from migrating into the drain. Fabrics may be ordered with differing effective-opening size (EOS) to permit the right fabric to be chosen for the gradation of a particular soil. The intent is not to totally prevent migration of particles. Instead, the filter cloth permits migration of very small particles capable of passing through the drainage aggregate. In this way, the actual filtering gradation is formed in the adjacent soil (Figure 5.7).

Drains are basically either blanket or interceptor types. Blanket drains are a narrow layer of drainage material placed between the source of subsurface flow and the area or structure being protected. Interceptor drains are some type of excavation filled with drainage material placed to catch and redirect the flow of subsurface water before it reaches the area of concern. Figure 5.8 shows several examples of blanket and interceptor drains.

The effectiveness of a blanket drain depends on its ability to carry away water from the site. Different designs for a particular site may have differing capacities. The drainage capacity of a blanket drain can be estimated by using a variation of equation 5.5:

$$Q = k\ i\ A \tag{5.11}$$

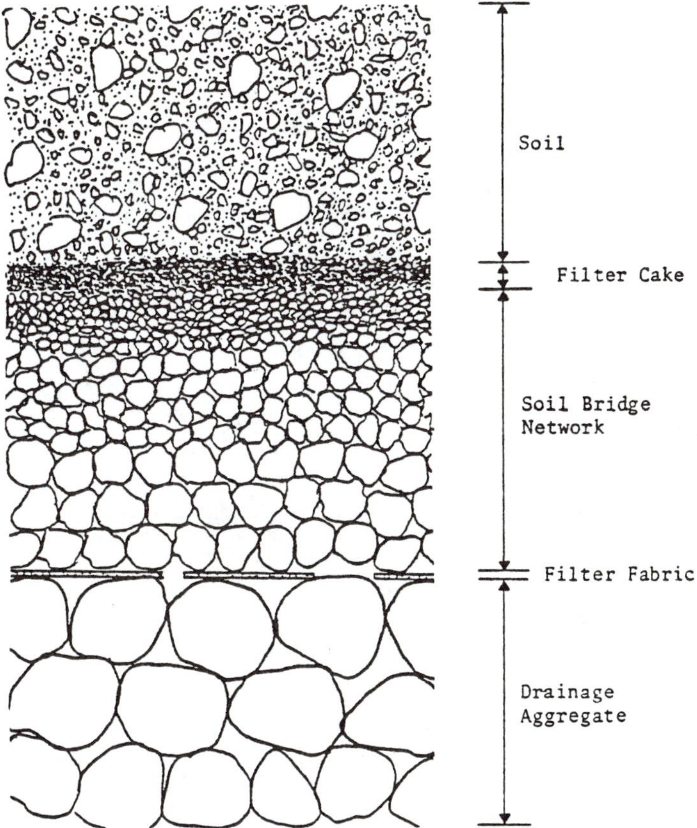

Figure 5.7 A diagram showing the rearrangement in soil particles next to a drain using filter fabric. (From Bell and Hicks, 1980)

Where:

Q = discharge
k = the coefficient of permeability
i = the average gradient in flow direction
A = cross-sectional area of blanket

Module 5.4 provides an example of calculating capacity for a blanket drain.

Common types of wells used to dewater a site are: sumps, well points, pumping wells, relief wells, and horizontal drains (or wells). A sump is merely a collection trench, or hole, deeper than the area being protected from subsurface water (Figure 5.9*a*). As the water collects, it is pumped away from the area, effectively drawing down the water table. A sump was used to dewater the keyway for the small earthen dam described on p. 224.

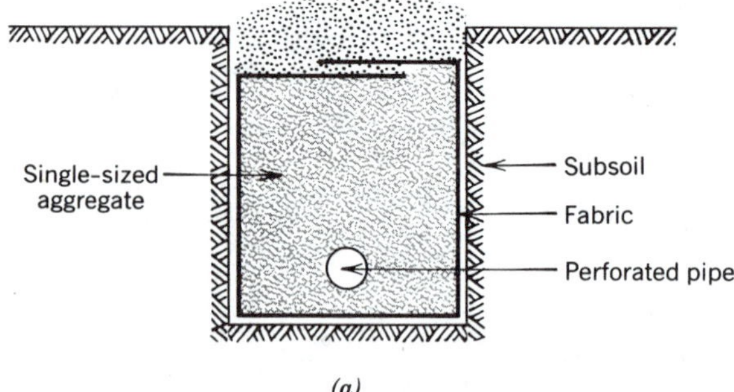

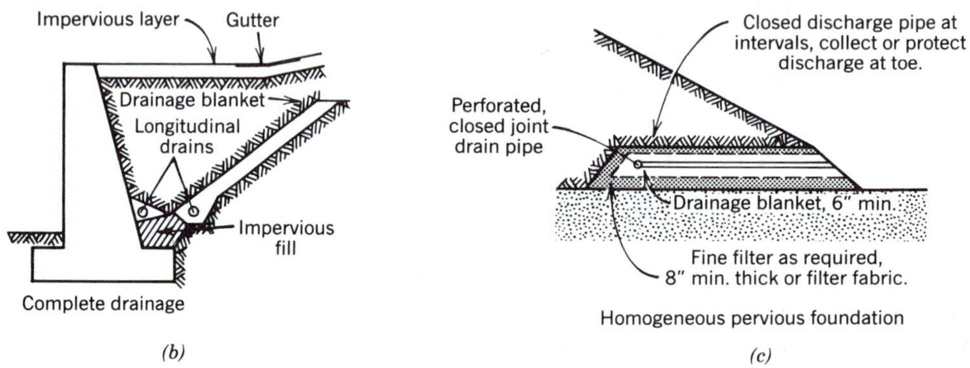

Figure 5.8 Examples of types of drains. (*a*) Trench-style drain. (*b*) Blanket drain behind a retaining wall. (c) Blanket drain protecting the toe of a foundation. (From Bell and Hicks, 1980; NAVFAC, 1982)

Well points accomplish the same purpose by being drilled to a shallow depth and attached to a suction pump to draw out the water (Figure 5.9*b*). Pumping wells are necessary to achieve dewatering to a greater depth owing to the limitations of the suction pumping used with well points. Pumping wells are drilled and equipped in a manner nearly identical to a well being developed as a water source. Relief wells are employed to decrease water in a confined aquifer. The artesian pressure forces the water upward through the relief well. The relief well is, in essence, a vertical interceptor drain. It is often used for dewatering during construction and to relieve uplift pressure on foundations.

Horizontal wells depend on gravity to move water through them (Figure 5.10). Royster (1980) provides a good overview of this type of well. They have proven effective in maintaining cut slope stability and in stabilizing landslides. Figure 5.11 shows a typical drill rig used in this type of work, placing one of nine horizontal wells used to stabilize a landslide. Smith (1980) studies the effectiveness of horizontal wells or drains in a variety of settings along California highways. He notes that determining the source and direction of groundwater movement in the area as well as the strata along which it moves are critical factors. The geology and topography of a location will dictate the most desirable location to intercept the subsurface

MODULE 5.4

A road crosses a slope where the water table is close to the surface. A blanket drain is incorporated into the road design. The blanket will have a gradient of .25 meter/meter, be 6 meters long and 0.3 meter thick. The coefficient of permeability for the drain will be 1.2×10^2 meters per second.

What is the drainage capacity, per meter of road, of this drain?

$Q = k\,i\,A \qquad A = 6 \text{ meters} \times 0.3 \text{ meter}$

$A = 1.8 \text{ square meters}$

$Q = (1.2 \times 10^2 \text{ meters per second})$
$(.25 \text{ meters/meter})\,(1.8 \text{ square meters})$

$Q = 54 \text{ cubic meters/second}$

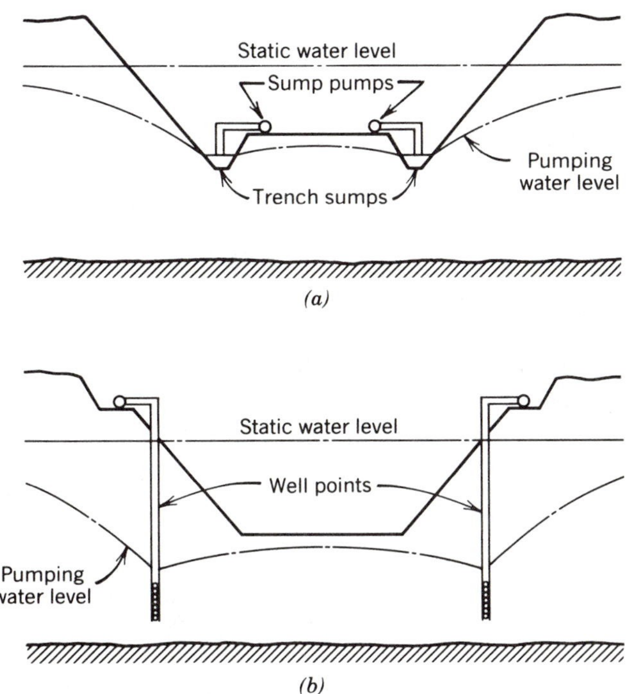

Figure 5.9 Examples of wells. (*a*) A system of sumps to dewater an excavation. (*b*) A set of well points protecting an excavation in a similar setting. (From USBR, 1981)

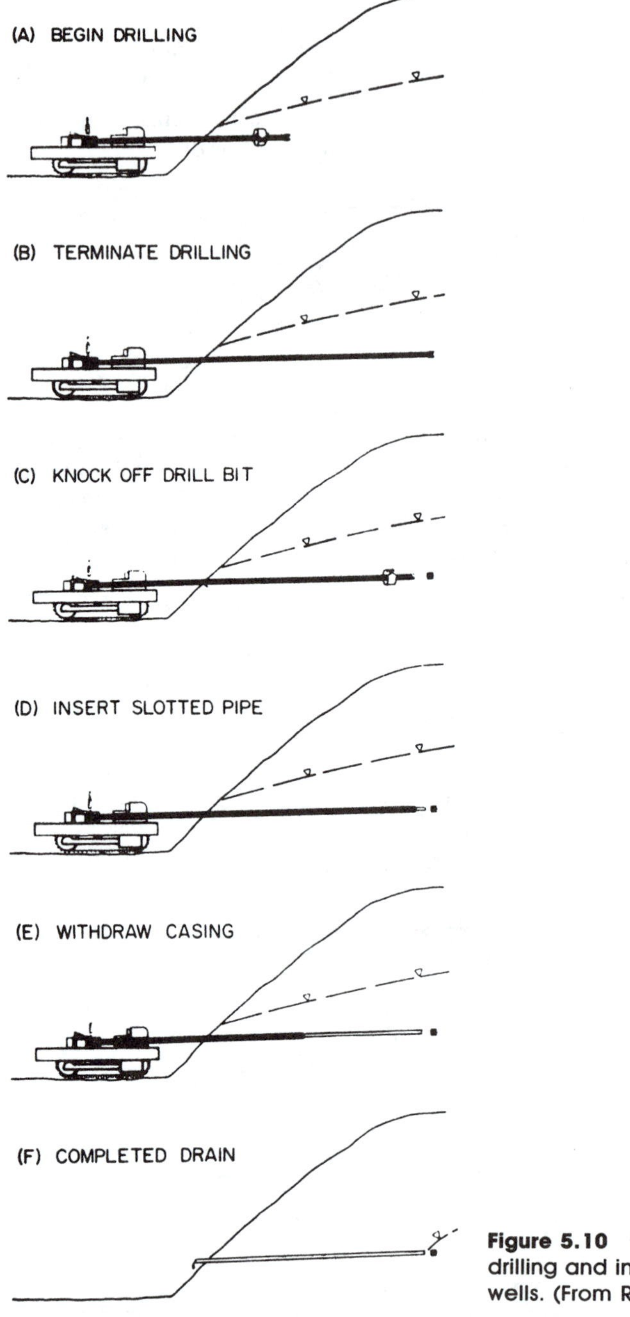

Figure 5.10 Main steps in the drilling and installation of horizontal wells. (From Royster, 1980)

Figure 5.11 A drill rig placing a horizontal well, following the steps shown in Figure 5.10. Nine horizontal wells are being installed to stabilize a landslide that failed twice between 1983 and 1986 and blocked the road.

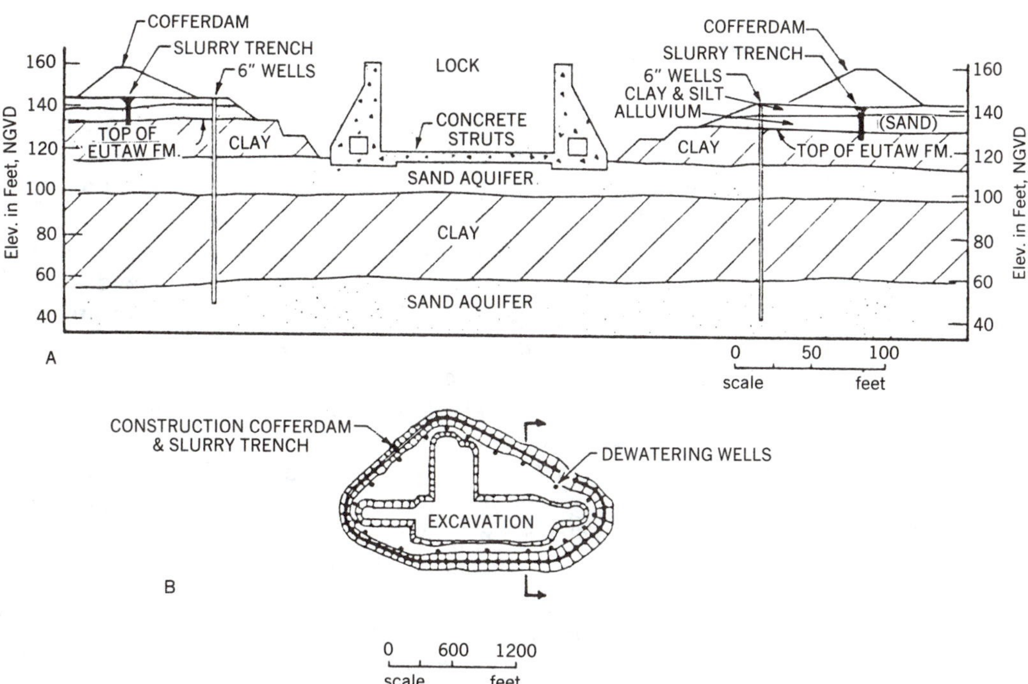

Figure 5.12 Cross section of the Columbus Lock segment of the Tennessee–Tombigbee Waterway showing the combining of barriers and wells to achieve control of subsurface water. (Reprinted with permission from Environmental Geology and Water Science, Vol. ½, J. H. Bryan, Hydrogeological and Geotechnical Aspects of the Tennessee–Tombigbee Waterway, 1985, Springer-Verlag.)

water before it reaches an area of possible instability. Attention should be paid to lateral as well as vertical changes in lithology when analyzing a site. In some instances, near-vertical barriers may exist such as gouge along shear zones; in other areas, confining beds may cause lateral flow by preventing downward percolation. By knowing these factors, spacing of wells and their length can achieve a better result than equal spacing and predetermined lengths.

Control of subsurface water in large projects or areas with extensive subsurface water may require a system that employs more than one method. Construction of the Columbus Lock on the Tennessee–Tombigbee Waterway illustrates this point (Bryan, 1985). Subsurface water was encountered in the alluvial sand and gravel in which the lock was excavated. Artesian pressures in the underlying Eutaw and Gordo aquifers needed to be relieved to prevent rupture from uplift pressures on the completed structure. Figure 5.12 shows the completed system used to achieve this control. Slurry trenches were used to cut off subsurface water from entering the excavation from the alluvial sands and gravels. Deep wells penetrating the underlying artesian aquifers accomplished relief of this underlying pressure. A total of 22 wells were placed at Columbus Lock. After construction, the only deep wells maintained were embedded in the lock walls. The other deep wells were abandoned and grouted.

SUMMARY

Subsurface water along with soil and rock are the three natural materials of most interest to the engineering geologist. Subsurface water influences how soil and rock respond under different stresses. Some engineering works are simply affected by the presence of subsurface water.

Precipitation and leakage from bodies of water are the source of subsurface water. It fills the voids or spaces within soil and rock. The degree of this saturation often governs the character of soil and rock. An unsaturated or dry soil mass resists a static load with the pressure derived from grain to grain contact. A saturated mass resists with intergranular pressure and pressure derived from the incompressible water filling the voids.

The significance of subsurface water to engineering encompasses: (1) problems posed to construction, (2) its action as an erosional agent, (3) its influence on functioning of a structure, and (4) its being a medium for pollution. Subsurface water often enters excavations to interfere with construction operations. In other instances, its presence causes materials used in construction to be less than optimum for their purpose. Uncontrolled water acts as an erosional agent to both engineered structures and natural features. It is especially important for dams, levees, and other structures impounding or conveying water. It can cause piping and slope failures that lead to severe impairment or destruction of these facilities. The function of other structures is affected by subsurface water control. Tunnels for roadways and railways need to be kept dry for proper operation. Wastewater lagoons and mine-tailings ponds need to retain water without uncontrolled seepage from subsurface sources. This is especially important where subsurface water can serve as a medium for pollution migration from controlled sites.

To achieve engineering objectives, it is necessary to control subsurface water. Basic control methods are: (1) barriers, (2) drains, (3) wells, and (4) liners. Barriers and liners are employed to prevent water from moving into undesirable locations. Drains and wells are intended to remove water prior to reaching a place where it is not wanted. Characterization of the geologic environment, especially of subsurface water, is essential to determining the best method or combination of methods for controlling subsurface water.

REFERENCES

Bedient, P. B., Rodgers, A. C., Bouvette, T. C., Tomson, M. B., and Wang, T. H., 1984, Ground-water quality at a creosote waste site: Ground Water, Vol. 22, pp. 318–329.

Bell, G. L., 1968, Engineering geology of interstate highway 94 underpass at Northern Pacific Railway, Fargo, North Dakota: *in* Engineering Geology Case Histories No. 6, Geol. Soc. Am., pp. 49–53.

Bell, J. R., and Hicks, R. G., 1980, Evaluation of test methods and use criteria for geotechnical fabrics in highway applications: Rep. No. FHWA/RD-80/021, Federal Hwy. Admin., Washington, D.C., 202 pp.

Bouwer, H., 1982, Design considerations for earth linings for seepage control: Ground Water, No. 20, pp. 531–537.

Bowles, J. E., 1979, Physical and geotechnical properties of soils: McGraw-Hill, New York, 478 pp.

Bryan, J. H., 1985, Hydrogeological and geotechnical aspects of the Tennessee–Tombigbee waterway: Environ. Geol. Water Sci., Vol. 1/2, pp. 25–50.

Campbell, R. H., 1975, Soil slips, debris flows, and rainstorms in the Santa Monica Mountains and vicinity, southern California: U.S. Geol. Surv. Prof. Paper 851, 51 pp.

Cedergren, H. R., 1977, Seepage, drainage, and flow nets: John Wiley & Sons, New York, 534 pp.

Davoren, A., 1983, Ground-water inflow in the Kaimai rail tunnel, New Zealand: Assoc. Eng. Geol. Bull., Vol. 20, pp. 387–391.

Flagg, C. G., 1979, Geological causes of dam incidents: Bull., Intl. Assoc. Eng. Geol., Vol. 20, pp. 196–201.

Foose, R. M., and Hess, P. W., 1976, Scientific and engineering parameters in planning and development of a landfill site in Pennsylvania: *in* Geomorphology and engineering, Dowden, Hutchinson & Ross, New York, pp. 289–312.

Kenney, C., 1984, Properties and behaviour of soils relevant to slope instability: *in* Slope instability, D. Brunsden and D. Prior (eds.), John Wiley & Sons, New York, pp. 27–65.

Lambe, T. W., and Whitman, R. V., 1969, Soil mechanics: John Wiley & Sons, New York, 553 pp.

Lundgren, T., and Soderblom, R., 1985, Clay barriers—a not fully examined possibility: Eng. Geol., Vol. 21, pp. 201–208.

NAVFAC, 1982, Design manual: soil mechanics: U.S. Dept. of Defense, NAVFAC–DM 7.1, Dept. of the Navy, Washington, D.C., 360 pp.

Royster, D. L., 1980, Horizontal drains and horizontal drilling: an overview: Transp. Res. Rec. 783, Transp. Res. Bd., Natl. Acad. Sci., Washington, D.C., pp. 16-20.

Sangrey, D. A., Harrop-Williams, K. O., and Klaiber, J. A., 1984, Predicting ground-water response to precipitation: J. Geotech. Eng., Vol. 110, pp. 957-975.

Schuster, R. L., 1979, Reservoir-induced landslides: Intl. Assoc. Eng. Geol. Bull., No. 20, pp. 8-15.

Schuster, R. L., and Embree, G. F., 1980, Landslides caused by rapid draining of Teton Reservoir, Idaho: Eng. Geol. Soils Eng. Symp. Proc., Vol. 18, pp. 1-14.

Simmons, M. D., 1985, Unwatering the Divide Cut of the Tennessee-Tombigbee waterway: a major challenge to construction: Environ. Geol. Water Sci., Vol. 1/2, pp. 51-67.

Smith, D. D., 1980, The effectiveness of horizontal drains: Rep. No. FHWA/CA/TL—80/16, Federal Hwy. Admin., Washington, D.C., 79 pp.

Taylor, C. L., and Conwell, F. R., 1981, BART—influence of geology on construction conditions and costs: Assoc. Eng. Geol. Bull., Vol. 18, pp. 195-205.

USBR, 1974, Earth manual, 2nd ed.: U.S. Bur. Reclamation, Denver, Colo., 810 pp.

———, 1981, Ground water manual: U.S. Bur. of Reclamation (Water and Power Resources Service), Denver, Colo., 480 pp.

White, R. B., and Gainer, R. B., 1985, Control of ground water contamination at an active uranium mill: Ground Water Monitoring Rev., Vol. 5, pp. 75-82.

Wu, T. H., and Sangrey D. A., 1978, Strength properties and their measurement: *in* Landslides, analysis and control, Spec. Rep. 176, Transp. Res. Bd., Natl. Acad. Sci., Washington, D.C., pp. 139-154.

6

Instrumentation

Instrumentation has an important role in the preconstruction, construction, and postconstruction phases of many civil engineering projects. Instrumentation involves obtaining *in situ* measurement data that have a variety of uses to the civil engineer. Examples of uses are: measurement of deformation of earth material surrounding underground openings, movement in natural or manmade slopes, earth-induced loads on structures, structural loads on earth materials, and pore-water pressure in soil and rock.

Some investigators include the determination of rock mass deformability and *in situ* state of stress in rock under instrumentation. These subjects as well as the instrument types used and their applications were discussed in chapter 4. The application of surveying techniques for monitoring of movement, although an important engineering consideration, is not treated here because it is a refinement of standard surveying techniques rather than an aspect of instrumentation. Papers by Gould and Dunnicliff (1971), Cording et al. (1975), and Wilson and Mikkelsen (1978) that describe engineering surveying instruments and their applications are recommended supplemental reading.

Although specific applications and results of instrumentation will be included with the instrument descriptions that follow, there are generalities applicable to all civil engineering projects that require instrumentation. One that must not be overlooked is the interaction of site geology with planned construction. Instruments placed without regard to estimates of the orientation of tectonically caused stress fields, distribution of discontinuities, differences in rock types to be encountered during construction, and groundwater conditions are but a few of the geologic factors to be considered.

In addition, instruments should be designed and constructed for ease of installation and use, should operate reliably and be easily accessible under adverse physical conditions, and should provide output data rapidly. The latter is of importance because design changes or remedial actions often need to be made rapidly for economy or safety. Underwood (1972), Cording et al. (1975), and Franklin (1977) have dealt with the geologic aspects of instrumentation and the criteria governing instrument design and use. Only those instruments should be employed that will provide data needed for the specific project.

INSTRUMENT COMPONENTS

Several equipment components are used in a sufficient number of instrument systems to justify independent examination. These all measure deformation or movement, a basic requirement of many types of instruments. The commonly used components operate electrically by resistance or inductance. In certain applications, some may measure rock deformation directly, as in the use of bonded resistance strain gauges for *in situ* stress determination described in chapter 4.

Resistance Devices

Electrical resistance devices operate on the principle that the electrical resistance of a wire is inversely proportional to its cross-sectional area. A change in wire length (strain) will, in turn, change the cross-sectional area and, therefore, the resistance. Determination of resistance changes is accomplished by ohm meter or Wheatstone bridge measurements. Temperature influence on resistance may be compensated for by comparison of an "active" unit with an "inactive" unit at the same temperature.

Bonded Strain Gauges

A commonly used resistance strain gauge utilizes a fine wire grid or etched-metal foil grid mounted (bonded) on a backing (Figure 6.1). The backing is cemented to the surface for which strains are to be measured. These strain gauges are calibrated by the manufacturer for change in gauge resistance for a given change in length. Strain indicators or readout units can be calibrated, in turn, to read strain directly in micro units from resistance changes. Strain indicators are Wheatstone bridge circuits or variations thereof.

Multidirectional planar strains may be measured simultaneously by a strain gauge rosette. The typical manufactured rosette consists of three overlapping, equally spaced, bonded resistance strain gauges. The angles between adjacent gauges in a rosette may not be equal for special applications such as for three-dimensional borehole deformation measurements described in chapter 4. Whether a rosette or a single gauge, the bonded resistance strain gauges require great care in mounting and protection from moisture. They are usually good for short-term use only if used independently of some other instrument.

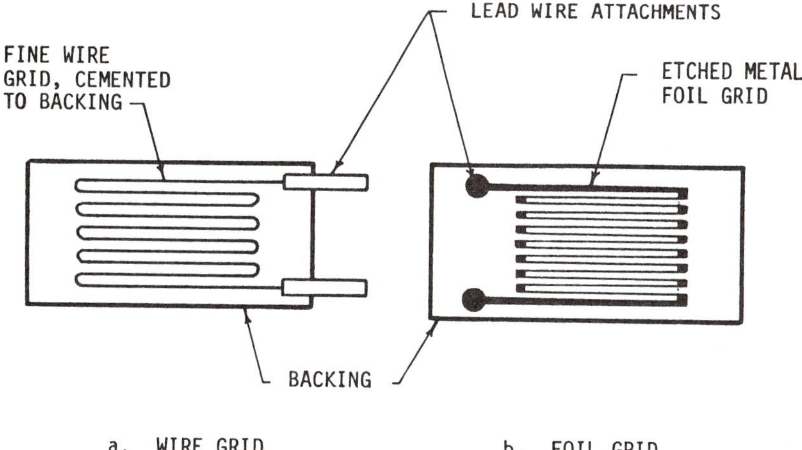

Figure 6.1 Bonded resistance strain gauges. (From Cording et al., 1975)

Unbonded Strain Gauges

The unbonded resistance strain gauge, typified by the Carlson strain meter, employs a fine resistance wire suspended under tension between two insulators so as to form a coil similar to that of the bonded strain gauge (Figure 6.2). The insulators are mounted on a deformable frame and the unit is encapsulated in a waterproof case and mounted on or within the strain-susceptible material. Any strain is transferred to the frame, which will lengthen or shorten the wire coil, altering its resistance. The Carlson units have a remarkable service record because of their design. Some that have been embedded in concrete dams for over 20 years are still functioning.

Potentiometers

Two styles of potentiometers are used in instruments. One is the linear potentiometer, which measures linear movement of a wiper-bearing shaft on a wire-wound cylindrical core. The other is a rotary potentiometer, which measures rotation of a wiper on a circular wire-wound core. The operation principle of potentiometers is standard.

Inductance Devices

Vibrating-Wire Strain Gauges

The vibrating-wire, or acoustical, strain gauge operates on the principle that the natural frequency of a tensioned wire is proportional to the tension. The tension, in turn, is a function of applied strain. Thus, if the natural frequency can be monitored, the strain can be determined, given the strain-frequency calibration of the particular gauge. The tensioned wire shown in Figure 6.3 is caused to vibrate by a direct current pulse sent to one of the electromagnets. The other electromagnet has a permanent magnet pole piece. The vibrating wire induces a current in the second electromagnet, which is transmitted to a readout unit.

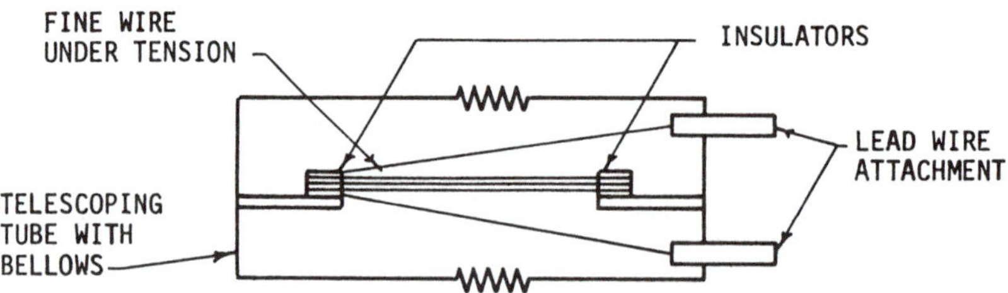

Figure 6.2 Unbonded resistance or Carlson strain gauge. (From Cording et al., 1975)

The induced current will have the frequency of the wire. This is a distinct advantage over the resistance-type strain gauge because frequency can be transmitted by wire over great lengths without significant distortion. In contrast, the voltage changes initiated by resistance gauges can be severely affected by lead-wire lengths and poor connections. A readout unit counts the frequency, which is converted to strain by the calibration factor of the gauge. Vibrating-wire gauges are always encapsulated to protect them from moisture and damage.

Linear Variable Differential Transformers (LVDTs)

The LVDT measures linear movement of a shaft. It does so by altering induced output from two secondary coils relative to the position of a ferromagnetic core when the primary coil is energized (Figure 6.4). Variations in the output signal of an LVDT are calibrated to the linear displacement of the core. The absence of any friction-resistance measuring mechanism is a distinct advantage over the linear potentiometer. The LVDTs are also less sensitive to moisture as the coils can be easily waterproofed.

Other Induction Devices

In addition to vibrating-wire strain gauges and LVDTs, induction in conjunction with gravity is the basis for the operation of servo-accelerometers utilized in some instruments. The servo-accelerometer transducer consists of either a pendulum or a cantilever with an attached coil that operates in a magnetic field. Movement from a null (vertical) position in the field is detected by current in the coil. The movement activates a servo unit that

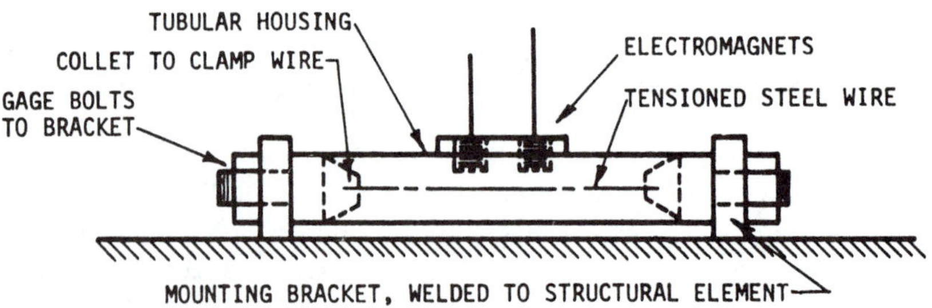

Figure 6.3 Tube-type vibrating-wire strain gauge. (From Cording et al., 1975)

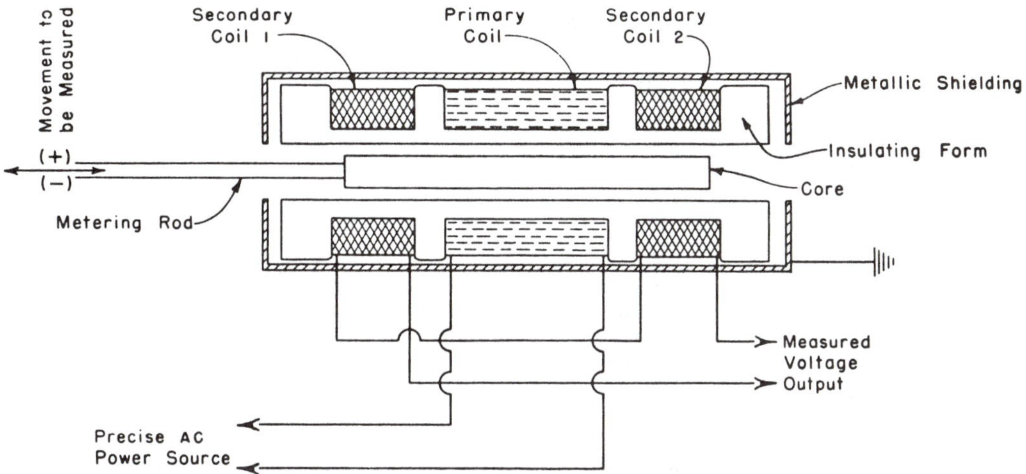

Figure 6.4 Linear variable differential transformer. (Copyright ©, ASTM, 1916 Race Street, Philadelphia, PA 19103. Reprinted with permission from Wallace et al., 1970.)

applies a restoring current to the pendulum or cantilever coil to return it to its original position. The current required to do this is a measure of the inclination from the vertical. Servo-accelerometers are used where exceptionally precise instruments are needed.

INSTRUMENT TYPES AND APPLICATIONS

For the sake of simplicity, instruments can be grouped into use categories, that is, measurement of loads and pressures, mass deformation, and pore pressure. Here, the more important instruments and what they measure will be examined for each of these categories. Some instrumentation systems are designed specifically for use in either rock or soil, whereas others may be equally adaptable to both materials. Examples of applications and results of use will be given where helpful. The integrated use of a variety of instrument types is common on most civil engineering projects of sufficient size to require the benefits of instrumentation.

Load and Pressure Measurement

The measurement of loads and pressures is an important role of instruments, especially during the construction phase of many civil engineering projects. For example, structural loads on steel mine and tunnel supports, rock bolts, and similar supportive structures are measured with load cells to monitor conditions and for comparison with design loads. Pressures developed in earth embankments during construction, whether as vertical pressures or lateral pressures on retaining structures, are measured by pressure cells. Pressures exerted on concrete tunnel linings in the form of external rock loads and internal pressures found in water-pressure tunnels also need to be monitored. A basic requirement of all pressure-measuring devices is that the compressibility or deformability of a cell must be matched to that of the material with which it is in contact. A rigid cell would inaccurately measure earth pressures, just as too flexible a cell would be inappropriate for measuring rock loads on steel sets in a tunnel.

Load Cells

The principal application of load cells is to measure loads in steel-set tunnel supports. The variety of cell-placement locations for horseshoe ribs and invert struts is shown in Figure 6.5. Load cells are recoverable and are replaced with shims of equal size after load monitoring has been completed. Load-cell measurement of the load history of a steel tunnel support is illustrated in Figure 6.6. The abutment load zone is the time period during the advancement of the tunnel face away from the monitored steel set. During that time period, vertical loads on the roof of the freshly excavated tunnel opening are being distributed to the tunnel walls by means of the steel supports. The cyclic nature of the loading history during this stress-distribution period is a product of support deformation, with resultant temporary reduction of load on the monitoring system.

Load cells also are used to monitor tension in selected rock bolts between the rock face and the rock bolt anchor. Rock bolt load cells are cylindrical in shape with an open axial portion to fit over the rock bolt (Figure 6.7). Dependence on use of torque wrenches to monitor rock bolt performance is reduced by the selective use of load cells. The cells are not affected by dirt and rust on bolt threads that may cause misleading high torques when wrenches are used on rock bolt nuts, even though anchorage effectiveness may have diminished.

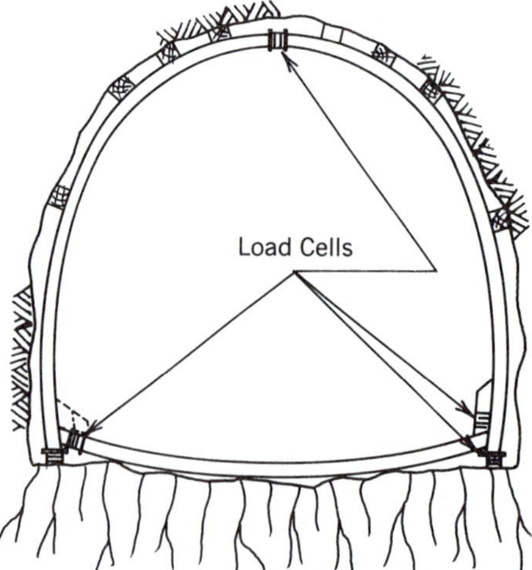

Figure 6.5 Load cell installation to measure rib and invert loads in tunnel supports. (Reprinted courtesy Slope indicator Co./Terrametrics.)

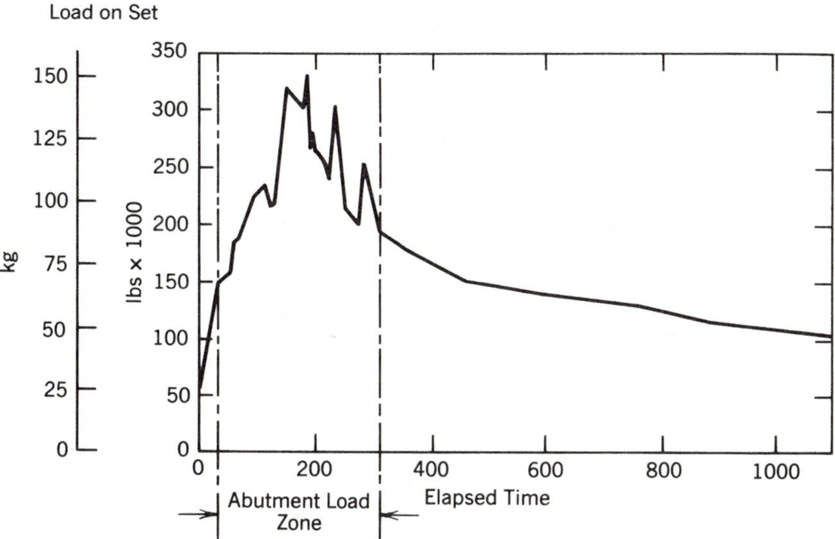

Figure 6.6 Load history curve obtained from load cells placed under footblocks of a steel set in a tunnel. Elapsed time is in hours for advancing tunnel force from zero time at location of monitored steel set. (Reprinted by permission from Hartmann, 1967.)

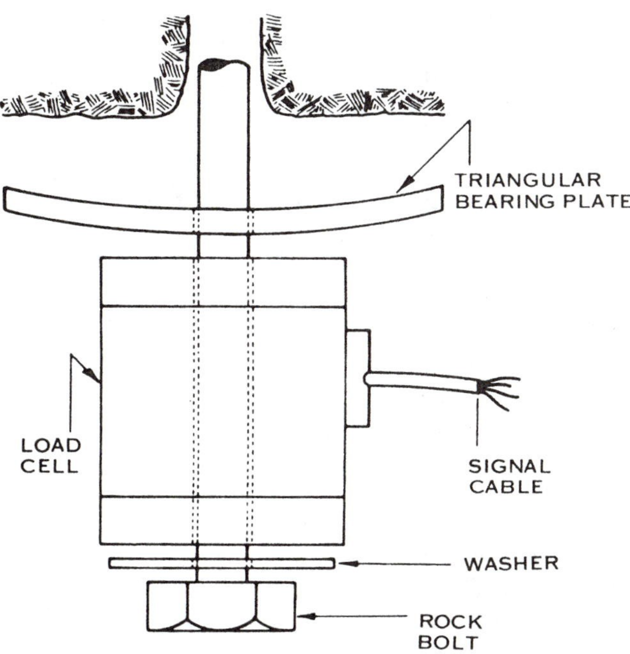

Figure 6.7 Resistance strain gauge-type rock bolt load cell. (Reprinted courtesy Slope indicator Co./Terrametrics.)

Resistance Type. Several designs of load cells are in use. The most common consists of a steel cylinder—either solid or hollow, depending on its application—to which are cemented resistance strain gauges of the type illustrated in Figure 6.1. Strain gauge orientation is parallel to the axis of the cylinder so that changes in length of the cylinder under load can be monitored. The unit is then encased in a protective cover that provides for waterproofing and, if necessary (e.g., in some tunnel applications), blast-damage protection. All load-cell strain gauge outputs for various loads are calibrated before use. Field measurements are made by connecting the load-cell cable to a strain gauge readout unit and corrected to load units by the calibration factor.

Induction Type. Other load cells utilize the vibrating-wire strain gauge. A typical application is to mount several gauges to the cell at equal spacings around the circumference. The unit is then covered, as in the case of the resistance strain gauge cells. Like other vibrating-wire strain gauge applications, the units are ideal for installations that require remote reading capability and long-term use.

Photoelastic Type. The principle of photoelasticity has been applied to load cells similar to that of the uniaxial stress meter described in chapter 4. The axis of a glass cylinder is oriented normal to the axis of the applied load within a steel cylinder. Polarized light transmitted through the cell is viewed with a handheld analyzer. Fringe measurement is accomplished as with the stress meter, with load values obtained from a load fringe number calibration chart for the particular load cell. Photoelastic load cells are used for monitoring rock bolt tension and loads on mine roof props. On-site reading of the load cell is required.

Mechanical Type. Purely mechanical load cells are also in use. One is the disk rock bolt load cell in which cup-type springs are compressed between two plates. On-site measurement is made by a readout dial gauge calibrated for this purpose.

Pressure Cells

Pressure cells are similar to flat jacks in their design and operation. They are installed in earth dams and embankments, concrete tunnel linings, and at the contact between earth fill and retaining or tunnel walls and surrounding material. Multiple installations for monitoring pressures at various points under earth dams and embankments, within tunnel linings, and along retaining walls are common.

The Gloetzl-type pressure cell is typical of those that operate hydraulically (Hartmann, 1967; Franklin, 1977). External pressure is transmitted to the fluid between two parallel plates (Figure 6.8). The fluid ranges from hydraulic fluid for soil applications to mercury for placement in concrete. The system is closed by a pressure diaphragm. The pressure applied to the cell is obtained by fluid pressure injected through the pressure line. When the pressures on opposite sides of the diaphragm are equal, the fluid bypasses and returns, resulting in a constant pressure that is recorded. The pressure may be converted to unit or total area values.

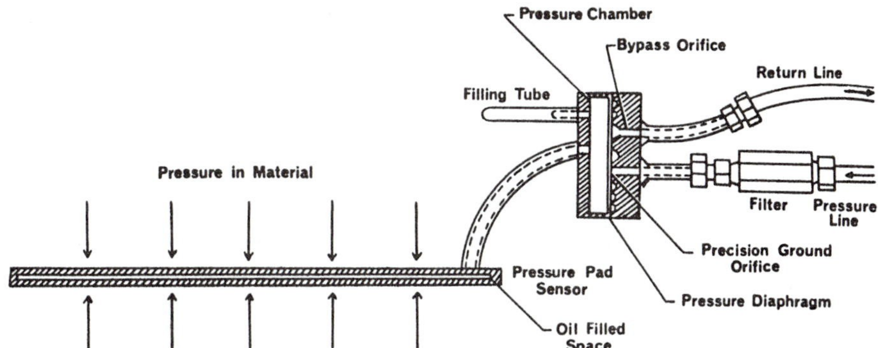

Figure 6.8 Gloetzl-type hydraulic pressure cell. (Reprinted courtesy Slope Indicator Co./Terrametrics.)

Tubing to the cells must be carefully placed and protected during construction. The tubes usually are terminated at a central reading site for convenience. Care must be taken in matching cells to material as well as being aware of the unique installation problems that arise, as in concrete when contraction occurs on hardening. The cells are not recoverable.

Axial Deformation Measurement

Measurements of deformation in earth materials are of considerable importance in many projects. In chapter 4, the measurement of rock mass deformation at selected distances along boreholes while conducting jacking tests was noted (Figure 4.44). Monitoring of axial deformation (or strain) along a borehole within the material surrounding a tunnel during construction, differential axial strains in unstable slopes, and vertical settlement of surface materials and structures are three other common installations.

Extensometers

The term *extensometer*, unless otherwise defined, applies to a device that measures changes in length along the axis of the device. Strain measurements may be made in other than boreholes. The simplest are tape and bar extensometers used, for example, to measure changes in the interior diameter of a tunnel, closure or convergence from stress relief, and relative movements in unstable soil and rock masses and slopes. Measurement points, or monuments, for such devices may be cemented or grouted into place for permanence.

Borehole Extensometers. Measurements of deformation in boreholes provide data on differential strain within a material rather than just total deformation at an exposed surface relative to an adjacent surface, as in the case of tunnel interior dimensions. In its simplest form, a borehole extensometer consists of a rod or tensioned wire that extends from an anchor point in a borehole to a measuring station at the drillhole collar mounted on the rock face (Figure 6.9). Measurement of changes in length from the anchor to the drillhole collar will be a measure of either compressional or

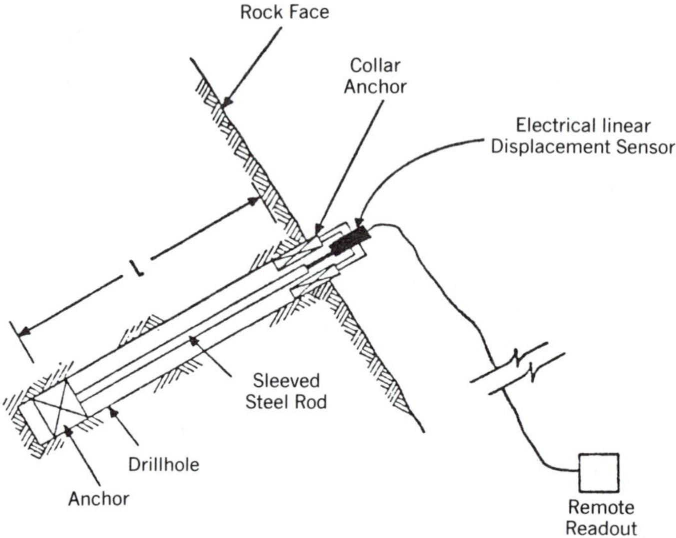

Figure 6.9 Borehole extensometer with remote electrical readout. (From Dunnicliff, 1982.)

tensional strain over the zone measured along the axis of the borehole. Strain gradients along the borehole are usually obtained by introducing multiple anchors, each with its mechanical connection to the monitoring point at the rock face. A typical multiposition, or multiple position borehole extensometer (MPBX), installation is illustrated by Figure 6.10.

Several anchoring techniques are employed, depending on the material in which the extensometer is installed. Mechanically or hydraulically expanding anchors that wedge against hard material or seat in softer material are most commonly used. A less-common practice is to grout the anchors and a cable-carrying plastic tube into the borehole. This is a useful installation method when blast-induced vibrations may loosen mechanical anchors.

Movement of anchor positions relative to the drillhole collar is measured mechanically and electrically in the typical MPBX. Mechanical measurements can be made at a sensing head situated at the drillhole collar. A typical design involves connecting the wire from an anchor to a spring-tensioned rod that is free to move axially within the sensor head as strain occurs. Strain is obtained from a dial gauge that measures changes in position of the rod relative to the sensor head plate. A major drawback of such measurements is that they cannot be obtained from a remote location but must be individually made for each anchor point by manual measurements.

More practical readout systems permit remote measurement of strains within the MPBX. Remote measurements are usually electrical in opera-

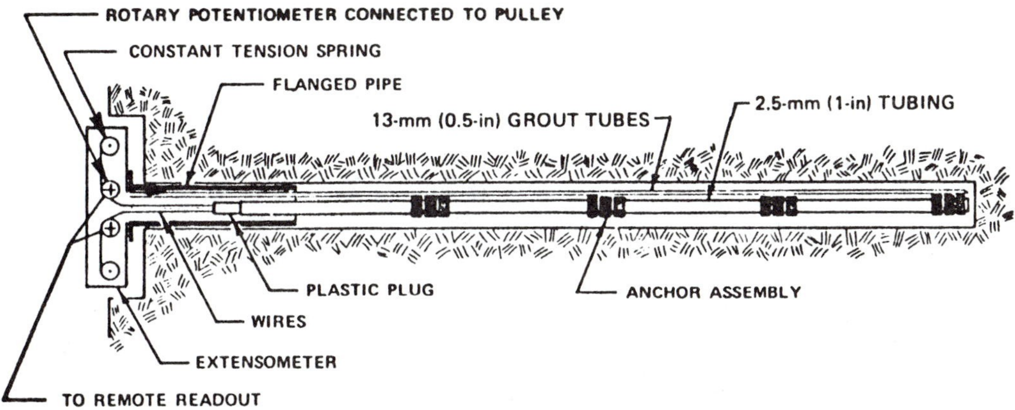

Figure 6.10 Installation of a four-point MPBX. (From Wilson and Mikkelsen, 1978)

tion, utilizing resistance strain gauges, rotary and linear potentiometers, LVDTs, and vibrating-wire strain gauges in the sensor head to measure rod or wire movement.

Borehole extensometers have many civil engineering applications where axial strain along a borehole must be monitored. A common installation is in underground openings—whether tunnels, shafts, or chambers—where closure on an opening must be measured (Figure 6.11). Axial compression and extension, or tension, may be graphically portrayed, as in Figure 6.12.

The stability of slopes may be monitored by borehole extensometers: Strains to various anchor points provide an indication not only of differential strains, or strain gradients, but also of strain accelerations when plotted

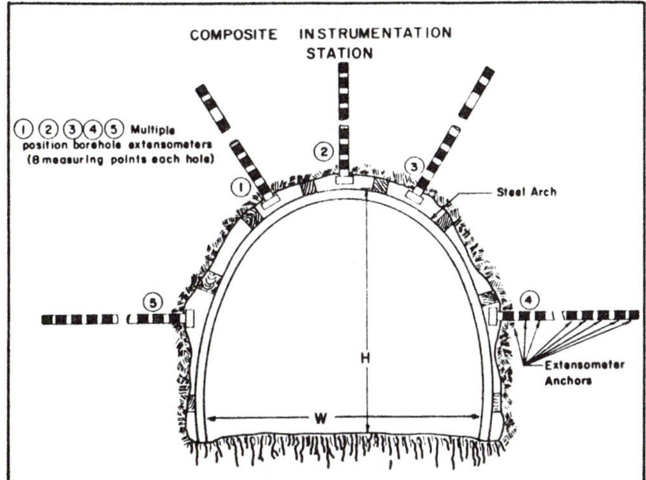

Figure 6.11 Extensometer arrangement for measurement of rock-mass strain gradients to evaluate roof tension, arch dimensions, and wall stability. (Reprinted courtesy Slope Indicator Co./Terrametrics.)

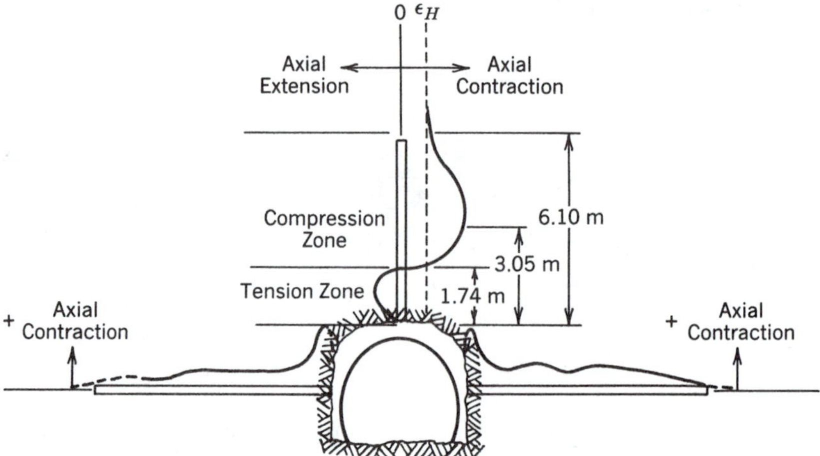

Figure 6.12 Strain distribution curves in an advancing tunnel measured by three extensometer installations at a single tunnel station. (Reprinted by permission from Hartmann, 1967.)

against time. Figures 6.13 and 6.14 illustrate a typical slope installation and measurements, respectively, obtained at the Cabin Creek Pumped Storage Project near Georgetown, Colorado. Figure 6.14 graphically indicates a factor to be considered in the use of borehole extensometers, that is, the sensor head usually is mounted in an unstable location with the deepest anchor ideally being in the most stable location. Thus, the greatest measured axial deformation typically will be at the sensor head, with decreasing amounts measured relative to the deepest anchor.

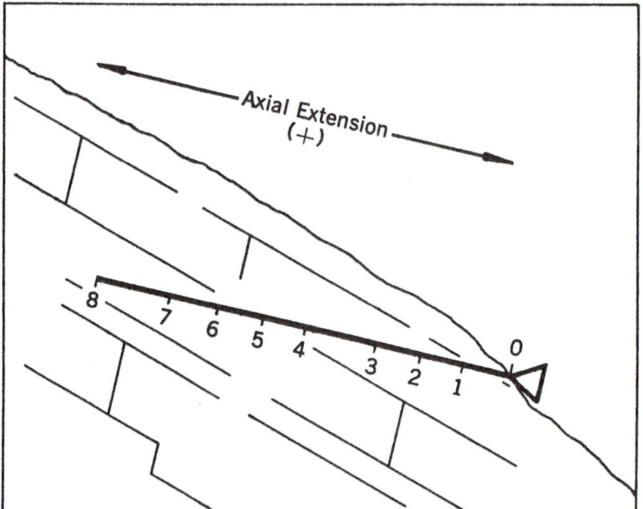

Figure 6.13 Multiple position borehole extensometer installation in a rock slope. (Reprinted courtesy Slope Indicator Co./Terrametrics.)

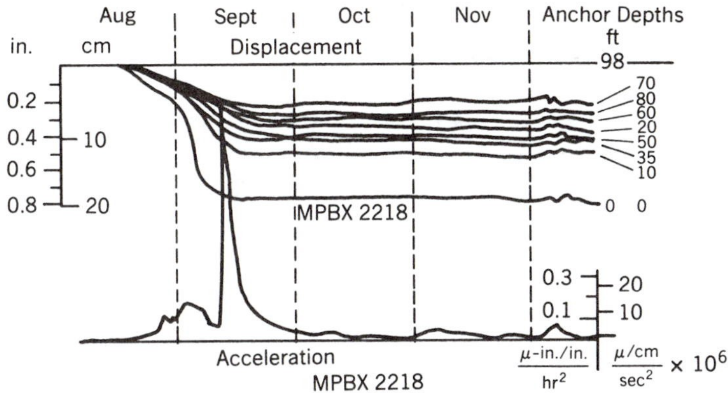

Figure 6.14 Displacement curves for eight-position borehole extensometer and accompanying movement acceleration graph for installation in an unstable slope. (From Dutro and Dickinson, 1974)

Settlement of surface structures constructed on material subject to loss of volume because of structural load can be monitored with vertically oriented extensometers utilizing boreholes. All borehole extensometers do not operate with anchor points and connecting rods and wires to the drillhole collar. One such type, a probe extensometer, makes use of ring magnets that are placed in a vertical borehole at selected distances from the surface by means of a plastic guide tube (Figure 6.15). A probe containing a magnetically activated switch is introduced into the borehole with provision for precise depth measurements (Burland et al., 1972; Smith and Burland, 1976). Settlements in the borehole result in variations in depths to the borehole magnets compared to original spacings. As a vertical measuring system, the probe extensometer is useful in determining settlement.

Settlement amounts with changing depth under a large silo structure over a two-month period are illustrated in Figure 6.16. A magnetic or

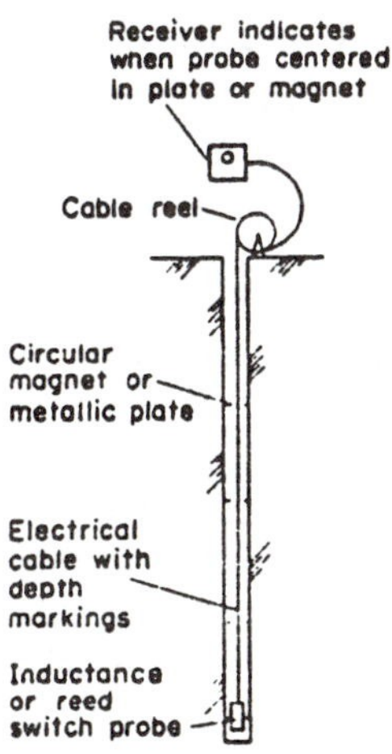

Figure 6.15 Probe-type extensometer with magnetic markers in borehole. (From ISRM, 1978. Reprinted with permission from Intl. J. Rock Mech. Min. Sci. & Geomech. Abstr., Vol. 15, Anon., Suggested methods for monitoring rock movements using borehole extensometers, 1978, Pergamon Press.)

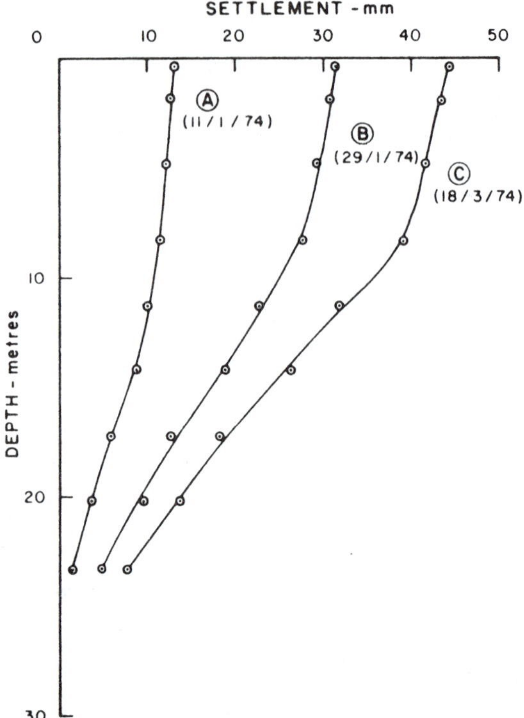

Figure 6.16 Distribution of settlements with depth and time in soft rock beneath a silo raft-type foundation. (From Smith and Burland, 1976. Reproduced by permission of the National Research Council of Canada from the Can. Geotech. J., Vol. 13, 1976.)

probe extensometer was used in this example. The magnetic extensometer is ideally suited to installations in soft materials where anchor points required in mechanical extensometers may be difficult to secure. Grouting of anchors may be employed if mechanical or standard MPBXs are installed for measurement of settlement. Where settlement results from collapse into a subsurface opening, precise leveling of the extensometer head must be conducted at times of settlement measurement. Only in this way will the indicated strain gradients be referenced to a constant surface elevation.

The International Society for Rock Mechanics Commission on Standardization of Laboratory and Field Tests has prepared suggested methods for use of borehole extensometers (ISRM, 1978). In addition, Franklin's paper on monitoring structures in rock (1977), the work of Cording et al. (1975), and vendor literature provide comprehensive application and construction details of extensometers, ranging from simple bar or tape units to the more sophisticated MPBX instruments. Several types of vertical borehole settlement instruments not included here are described by Cording et al. (1975) and Dunnicliff (1982).

Lateral Deformation Measurement

Measurement of deformation perpendicular to the axis of a borehole or of deflection of a borehole from the vertical is often a critical aspect of monitoring the stability of materials and associated structures. The need to monitor such transverse deformation occurs in tunnels, shafts, embankments, dams, and foundations—to list the more common applications. The terms *deflectometer* and *transverse extensometer* often are applied to the type of instrument that may have any orientation in space, as is required in some underground applications. The term *inclinometer* is applied to instruments that are either lowered down or installed in vertical or inclined boreholes from the surface for the purpose of measuring deviations from the vertical or original borehole inclination.

Deflectometers

Deflectometers or transverse extensometers are ideally suited to monitoring of shear deformation in a variety of engineering applications in which expected movement cannot be measured by lowering instruments down boreholes. These applications range from open-pit mines to dam abutments, encompassing many soil and rock excavations and a number of engineering structures and associated soil and rock foundations.

Deflectometers are articulated tubes that are installed in boreholes of any orientation. Lateral movement is detected at the joints in the tube, which are typically variable in number—as are the lengths of tubing or pipe between joints. Most deflectometers are designed to measure lateral deformation in only one plane and must be installed with knowledge of predicted shear movement direction. Instruments are also available that detect movement in two orthogonal planes, thus providing greater versatility in monitoring movement directions normal to the borehole.

Transverse movement can be measured electromechanically. Sensing elements form the joint assemblies in the deflectometer. Two approaches to measuring movement will serve to illustrate deflectometer operation. A taut wire may be used that extends the full length of the deflectometer tube assembly. Any movement of one tube segment relative to the adjacent segment serves to move the wire from its original central location in the tube. The deflection of the wire to a new position is converted to an angular change from the original orientation in space at a given joint. Measurement is made by using the potentiometer principle. The wire rests on a resistance element and slides from one position to another as lateral movement occurs. The resistance changes at each measurement point, as indicated by applying voltage to the wire and the resistance element. A readout unit permits switching to successive resistance elements for measurement of deflections at all joints. This design permits monitoring of movement in only one transverse plane. Accurate determination of hole deflections is dependent on the anchoring of the distal end of the deflectometer in material that will remain fixed in space, otherwise all measurements will be relative to an unknown base.

The other measurement system consists of articulated joint assemblies with cantilever transducers at each joint. Each cantilever has a resistance strain gauge mounted on it. Deflection at each joint is measured by a calibrated strain gauge readout unit that may be switched to each of the deflectometer joints. The design of the cantilever-type deflectometer permits installation of two mutually orthogonal cantilevers if desired. Deflections in each of the two planes may be obtained. Installation of this unit is reserved for sites at which lateral movement directions are not known prior to installation. Most installations of deflectometers in slopes, in locations adjacent to excavations, and in embankments are at sites that usually have highly predictable movement directions. These require only the single-cantilever–type or taut-wire–type deflectometers that measure deformation in one plane.

Inclinometers

Vertically oriented inclinometers have many engineering applications, including measuring differential movement across active landslide slip surfaces and measurement of lateral displacement in earth dams and embankments. Construction-related lateral movement that accompanies surface and underground excavations and monitoring of bulkhead performance are other examples.

Typical installations involve lowering a sensing probe down plastic or aluminum tubing installed in a borehole. Orientation of the probe is achieved by utilizing guide wheels on the probe that run in either of two pairs of grooves, or keyways, in the tubing (Figure 6.17). If square cross-section tubing is used, the opposite corners serve the same purpose as the keyways for guiding the probe. Lateral displacement readings are made that are perpendicular to the borehole axis in the two keyway planes. Each plane requires two measurement runs separated by 180°. Only one probe is required for obtaining deflection readings in numerous borehole installations equipped with inclinometer tubing, as the probe is not an integral part of the borehole casing. The guide tubing for each installation is grouted in place with material having rigidity comparable to that of the surrounding soil or rock.

As with the deflectometer, the sensing unit of the inclinometer reacts to changes in lateral deflection. In the typical inclinometer, the sensing element or transducer is enclosed in the probe that traverses the special tubing. Inclinometers typically utilize the pendulum principle, which provides for measurement of the probe axis relative to the vertical at selected measurement depths. The transducers employed include rotary potentiometers, LVDTs, bonded resistance strain gauges, vibrating-wire strain gauges, and servo-accelerometers. All but the servo-accelerometer utilize a simple pendulum that activates the transducer with varying degrees of precision in the range of +/− 1.25 to 2.54 cm/30.5 m. Precisions of +/− 0.13 to 0.25 cm/30.5 m are obtained with servo-accelerometer measuring systems.

The servo-accelerometer transducers are ideally suited for providing digital output that may be recorded on paper or magnetic tape. The tape may be used directly as computer input for rapid solution of lateral deflection values. Output from other transducers and their readout systems usually requires manual recording of dial readings or digital displays.

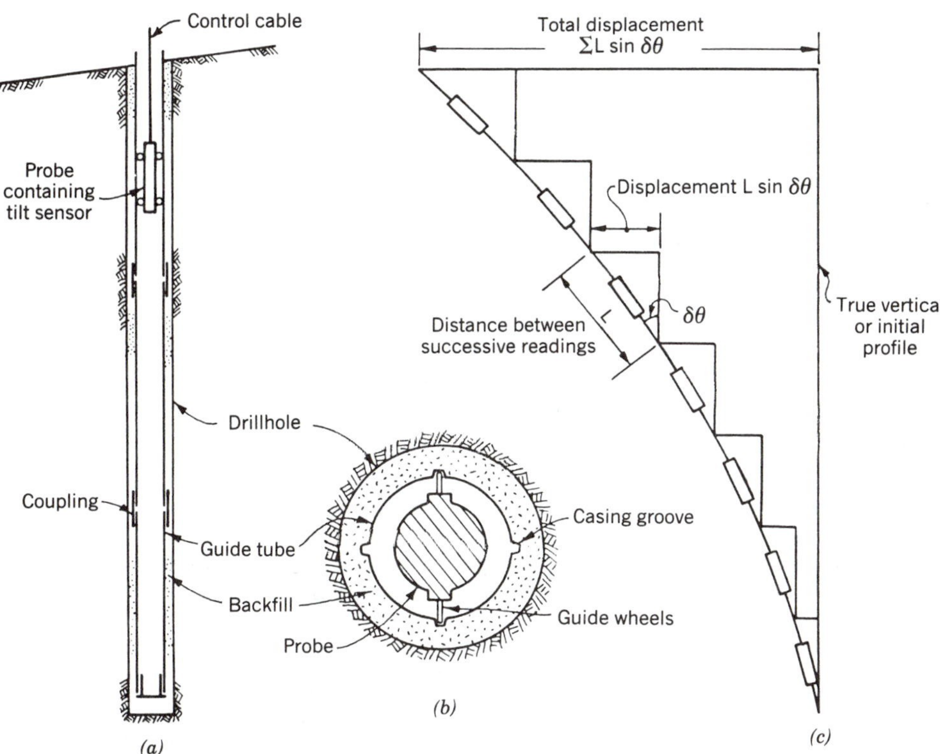

Figure 6.17 Probe-type inclinometer. (*a*) Probe being lowered in guide tube installed in drillhole. (*b*) Cross section of a drillhole, guide tube, and probe. (*c*) Incremental and total displacement diagram. (From Wilson and Mikkelsen, 1978)

Regular users of inclinometers have computer software designed to relieve the tedious job of reducing data from surveys.

Where continuous monitoring of lateral strain in a borehole installation is required, the typical probe-type inclinometer is of no value. In its place, a multisensor unit may be installed in standard grooved casing. With a variable number of servo-accelerometer sensors spaced at selected depths, movements can be monitored continuously from a remote site. A major advantage in addition to the remote monitoring of data is the availability of an alarm system. Maximum allowable deflection may be set and an alarm activated when that amount is reached. The advantages of this capability at failure-susceptible sites such as landslides, excavations, and open-pit mines are obvious. The remote readout feature also permits data transmission on demand over a telephone line.

Angular deflection measurements are made from the bottom of the tubing at successively chosen intervals as shown in Figure 6.17. As with the deflectometer, angular measurements and calculation of lateral displacements are made with the assumption that the bottom of the tube is fixed in space and not influenced by lateral movement. Borehole location, depth, and orientation of keyways should be compatible with the expected movement of depth zones and the direction of lateral displacement. The results of an inclinometer survey in a landslide are shown in Figure 6.18.

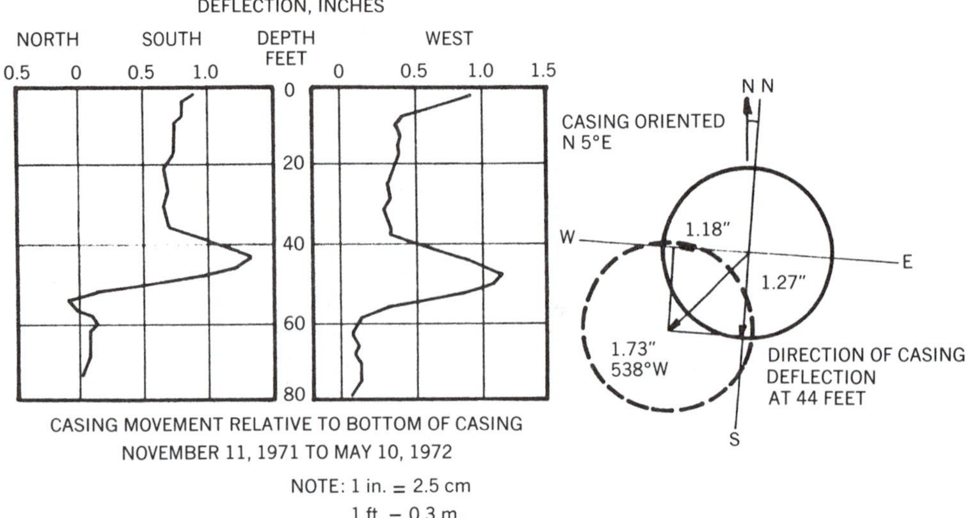

Figure 6.18 Graphical representation of casing deflections derived from inclinometer data in a landslide over a 6-month period. (From Tice and Sams, 1974)

In addition to readily available descriptive information from inclinometer manufacturers, detailed supplemental information may be found in Wilson (1962), Cording et al. (1975), ISRM (1977), and Wilson and Mikkelson (1978).

Shear Strips

A simple, economical device for localizing the position or positions of lateral deformation with respect to a drillhole is available. It is the shear strip, which is a strip of brittle material that breaks easily under shearing conditions. A typical shear strip has two parallel conductors printed the length of strip (Figure 6.19). At selected intervals, resistors of equal value are connected in parallel with the two conductors. The resistance of the strip, R_T, measured by an ohmmeter from either end will be

$$1/R_T = 1/R_1 + 1/R_2 + \ \ldots \ + 1/R_n.$$

A simple break in the shear strip permits localizing of the exact point of breakage relative to a given pair of resistors by simply measuring the postbreak resistances from either end. Multiple breaks permit determination of the total failure zone width as shown in Figure 6.20.

Shear strip installations are used as warning devices where the occurrence of shearing is more important to monitor than the amount of lateral displacement. They are either grouted in drillholes or enclosed by surrounding material during construction. Concrete tunnel linings, materials surrounding surface and subsurface excavations, earth dams and embankments, and dam foundations and abutments are examples of instances where shear strips are installed in addition to failure-susceptible slopes.

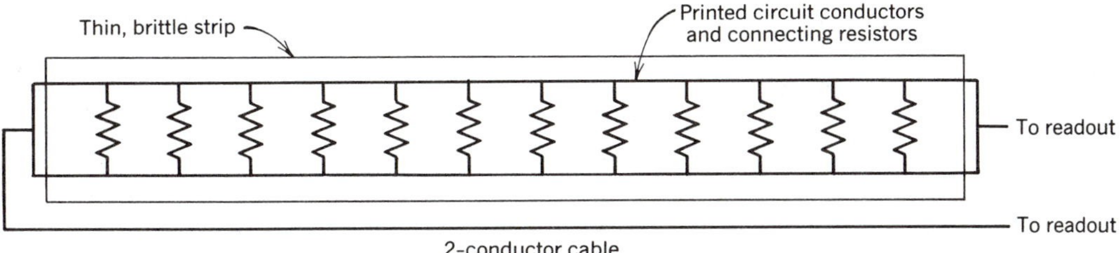

Figure 6.19 Diagram of shear strip construction.

Changes in strip resistance may be set to activate remote alarm systems in the event of shear failure of a strip.

Rotational Deformation Movement

The tilt or rotational (angular) movement of rock relative to a point can be measured by a device called a *tiltmeter.* Tiltmeters are gravity-activated instruments that commonly are designed to measure angles of inclination from the horizontal (tilt) along two orthogonal axes rather than from the vertical. They usually are mounted directly on rock faces at the surface or in underground openings that are subject to rotational movement. Measurements of tilt can be made periodically or continuously, depending on the system used.

Portable tiltmeters determine the tilt of a permanently mounted plate to which the tiltmeter sensor is fitted for measurements along the two horizontal axes. Tiltmeters have much in common with the probe-type inclin-

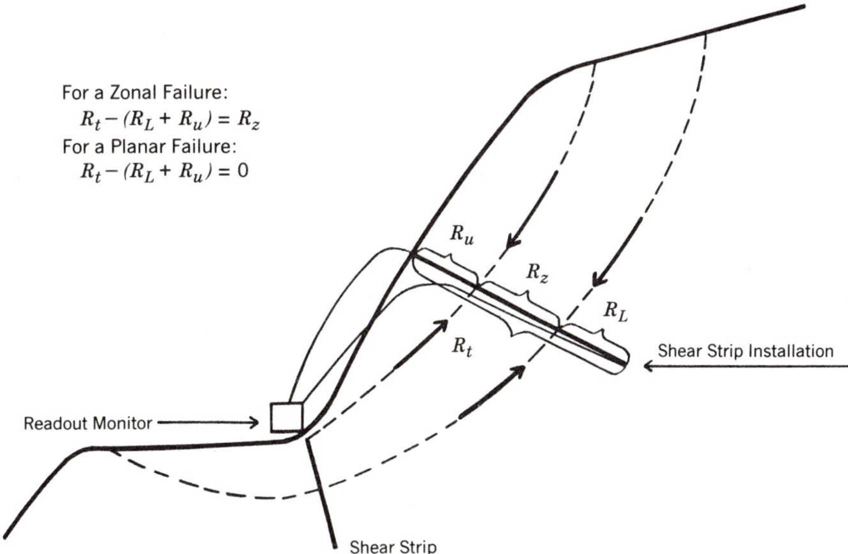

Figure 6.20 Localization of lateral movement by a shear strip installation in a slope. (Reprinted courtesy Slope Indicator Co./Terrametrics.)

ometer, in that the same instrument is used for measurements at many sites. They are, of necessity, periodic measurement devices. Data from permanently installed, or in-place, tiltmeters can be obtained at the site or remotely, periodically or continuously, depending on the design and need.

Most tiltmeters utilize electrolytic spirit level sensors or pendulums that actuate servo-accelerometers, vibrating-wire strain gauges, or photoelastic devices. In the electrolytic spirit level sensors, changes in the spirit level bubble position in the electrolytic liquid is monitored electrically (Sherwood and Currey, 1974). Instrument output is converted to axial rotation in minutes of arc.

The tiltmeter can be used as an independent, long-term source of rock mass deformation data relative to each measurement plate. As such, it is used to monitor any rotational movement in unstable natural and man-made rock slopes and excavations, subsidence over tunnels and mines, deformation changes in dam abutments, and deformation rates and directions during and following tunneling and mining operations (Figure 6.21). Tiltmeters can also be used to obtain movement directions for efficient placement of other rock mass-movement sensing instruments such as borehole extensometers. When used as monitoring instruments, acceleration of movement usually is more important than magnitude from the safety standpoint. Supplemental information on tiltmeter design, installation, and use can be found in ISRM (1977) and Wilson and Mikkelsen (1978).

Pore Pressure Measurement

Pore water pressure is a measure of the water pressure exerted at a given point in a soil or rock mass by the hydraulic head at that point. In an unconfined aquifer condition, it can be defined by the position of the water table relative to the point of measurement. The piezometric surface defines the head where measurement is taken in a confined state. Pore pressures in soils that have relatively uniform permeabilities are most easily sampled by

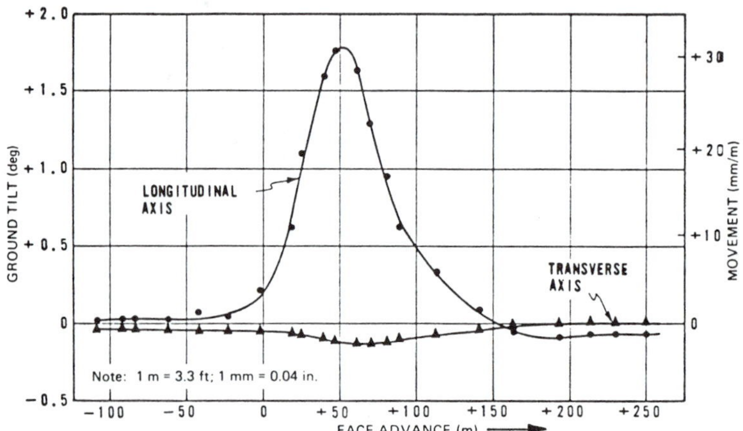

Figure 6.21 Tilt measured at ground surface owing to advance of a long-wall mining face. (From Wilson and Mikkelsen, 1978)

instrumentation. Sensing elements can be placed in such materials with the emphasis being on the obtaining of representative pressure samples at a particular site. However, changes in permeability that occur across a clay-sand boundary in a soil or a rock-joint boundary in rock require great care in the locating of the sensors. Effective stress calculations and determination of buoyant, or uplift, forces in soil masses or along discontinuities in rock masses are dependent on pressure data that correctly sample the local conditions.

Open Standpipe Piezometers

The simplest type of pore pressure instrument, or piezometer, consists of an open standpipe installation. An example of a widely used open piezometer is the Casagrande piezometer (Figure 6.22). This piezometer is installed in a borehole surrounded with clean, uniform sand to maintain support and constant-flow characteristics. Because many piezometers are installed to measure pore pressures in specific zones, they must be isolated from the rest of the borehole. This is accomplished by use of clay or cement grouts in the annulus above the piezometer as shown in Figure 6.22.

In the Casagrande piezometer, fluctuations in the free-water level in the tubing are measured from the surface. This may be done by lowering a two-conductor cable with a gap at the measuring probe. In contact with water, the open circuit in the downhole probe is completed. An ohmmeter at the surface registers the point of contact with water and the length of cable down the hole is measured. Pore pressure at the measuring point at a given depth is a function of the distance between that point and the free-water surface and the density of water.

The Casagrande piezometer is useful when rapid changes in pore pressure need not be monitored and when access for drilling and on-site monitoring of water levels is acceptable. Common well points and variations of them also are used as open piezometers. These also must be backfilled with sand and isolated from the rest of the hole by an impermeable seal.

Closed Hydraulic and Pneumatic Piezometers

Hydraulic and pneumatic piezometers are closed varieties of the standpipe design. In some hydraulic piezometers, a water-filled tubing extends to a surface pressure meter such as a Bourdon gauge or manometer. A second tube is used to clear bubbles from the system (Figure 6.23). Water is transferred from the surrounding material across sand to a porous plastic, ceramic, or sintered metal plate. The plate permits only water to cross, keeping the system free from sediment. Pressure changes in the closed part of the system are measured directly at the surface. Measurement depths usually are less than 8 m below the surface for this type of piezometer.

The most sensitive piezometers utilize a membrane, or diaphragm, to separate the pressure heads in the material and the piezometer. Porous materials are again used to permit passage of water into the sensing element to prevent sediment contamination. The external or *in situ* pore pressure acts against a flexible diaphragm or spring-loaded piston that causes movement. The external pore pressure may be determined either by

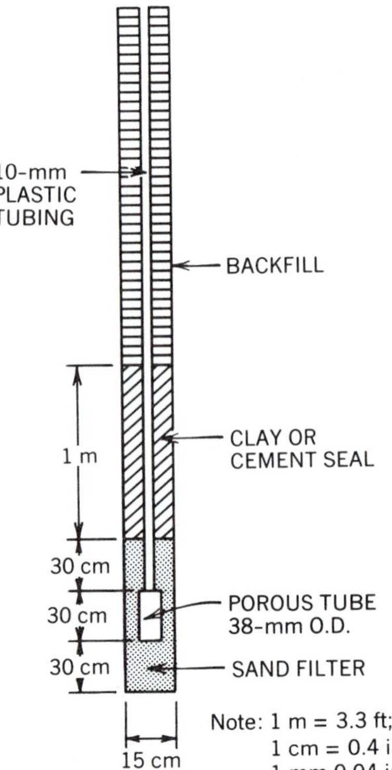

Figure 6.22 Casagrande-type piezometer installation. (From Wilson and Mikkelsen, 1978)

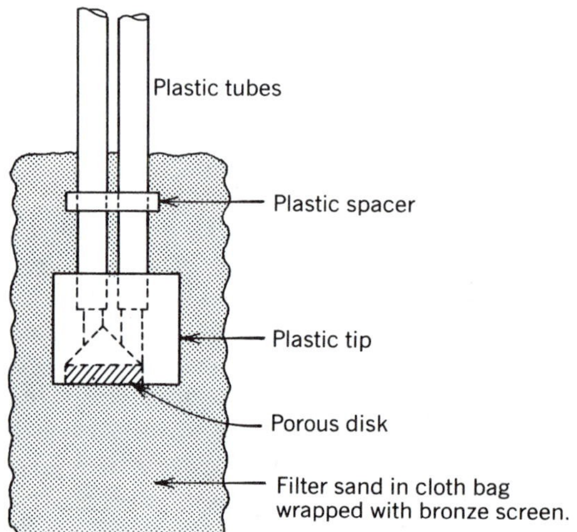

Figure 6.23 USBR- or twin-tube–type of hydraulic piezometer. (From Cording et al., 1975)

balancing the external pressure with liquid (hydraulic) or air (pneumatic) pressure in a closed system to a null position (Figure 6.24) or by measuring the deflection of the diaphragm. The balancing of *in situ* pressures may be accomplished by applying hydraulic or pneumatic pressure on the diaphragm or piston. When the pressures equalize, a bypass valve or vent maintains the pressure balance. At that time, the applied pressure is measured and converted to the unit *in situ* pore pressure. Other piezometers electrically measure the deflection of a diaphragm by means of a strain gauge attached to the diaphragm.

The measurement of pore pressures in soil and rock has many geotechnical applications. These range from monitoring the effectiveness of dewatering operations during surface excavations and tunnel construction to use in slope-stabilization procedures. Monitoring of pore pressures in earth dams and in foundation and abutment zones of all types of dams is a recognized necessity for maintaining dam safety. Monitoring of pore pressure differentials across impermeable barriers may be critical to the calculation of the factor of safety of soil and rock slopes with high potential for failure.

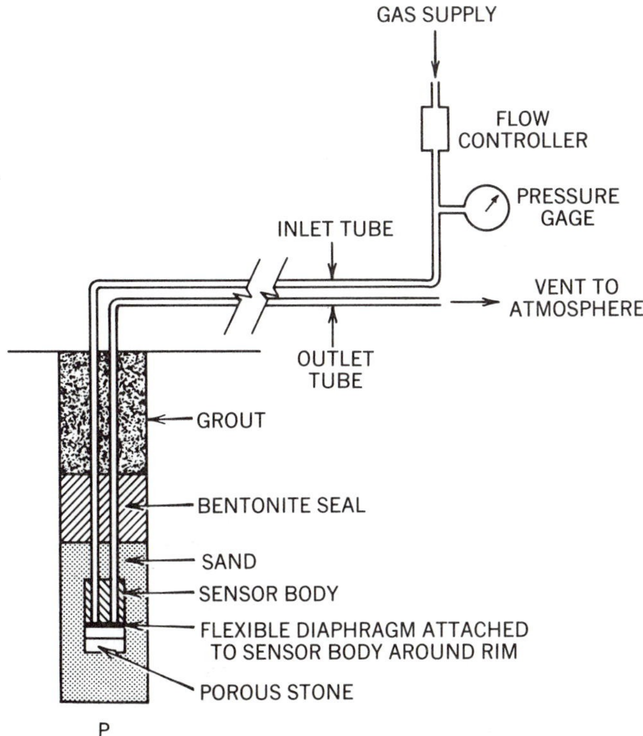

Figure 6.24 Typical pneumatic piezometer. (From Dunnicliff, 1982)

PLANNING AN INSTRUMENTATION PROGRAM

The variety of available instruments and the physical states that are measured provide opportunities for developing data bases of static and dynamic conditions at sites before, during, and after construction. Armed with the knowledge of what can be measured and the instrumentation types available for such tasks, it becomes the responsibility of the engineering geologist/geological engineer to plan properly for procurement and installation of the optimum kinds of instrumentation required for a project. In addition, it is necessary to schedule site preparation for, and installation of, instruments relative to planned construction progress to eliminate costly delays or possible deletion of one or more instrument systems because of poor planning and scheduling.

Because each project has relatively special instrumentation needs, it is not possible to do other than generalize in the area of planning an instrumentation program. The planning steps given by Dunnicliff (1982) serve as a useful planning framework. They are:

1. Define project conditions.
2. Define purpose of instrumentation.
3. Select variables to be monitored.
4. Make predictions of behavior.
5. Devise solutions to problems that may be disclosed by observations.
6. Assign contractual tasks and responsibilities.
7. Select instruments.
8. Plan recording of factors that may influence measured data.
9. Establish procedures for ensuring reading correctness.
10. Select instrument locations.
11. List specific purpose of each instrument.
12. Write instrument procurement procedures.
13. Write contractual arrangements for field instrumentation services.
14. Plan procedures subsequent to installation.

Dunnicliff should be consulted for more information about each listed planning step.

The various instruments described herein are not all equal in such characteristics as precision, accuracy, reliability, ease of installation, and cost. These and other factors have been addressed and summarized by Wilson and Mikkelsen (1978) and Dunnicliff (1982) for a variety of instrument types.

As noted earlier, a project typically requires more than one kind of instrumentation. Familiarity with summaries and case histories that involve instrument types, applications, installations, and the collection, reduction, and use of data are basic to an individual's effective utilization of instruments on a project. The following references in addition to those cited previously provide for such comprehensive familiarization: Wilson (1972); Franklin and Denton (1973); ASTM (1974); British Geotechnical

Soc. (1974); Hanna (1974); TRB (1974); ASTM (1975); Fukuoka (1980); Russell (1981); and Londe (1982).

SUMMARY

Instrumentation involves the use of specialized equipment to measure such things as deformation about underground openings, movement in slopes, geologic and structural loads, and pore pressures in soils and rocks. Measurements may be needed before, during, and following construction.

Most of the applications listed herein require measurement of strain. Various resistance, induction, and photoelastic devices are used for strain measurements, which, in turn, are basic to the design and operation of the specific instruments. Fluid-pressure measurement instruments permit determination of loads as well as pore pressures. Each instrument has its own application limitations, precision, accuracy, and installation requirements.

The type or types of instruments used on a project is dictated by the project and the site geology. Instruments selected must provide the data required at a given site. An instrument such as an extensometer that measures only axial strain is inappropriate for sites having the potential for lateral strain. Site geology involves such factors as whether construction is in soil or rock, the tectonic history in the case of a project involving rock, soil and rock mass heterogeneity, and groundwater conditions. The instruments used must function in the natural and construction environments specific to the site.

Planning of an instrumentation program requires coordinating the selection of proper equipment for the specific site, with project goals, available funds, construction schedule, and provision for placement of instruments and collection of output data.

REFERENCES

ASTM, 1974, Field testing and instrumentation of rock: Am. Soc. Test. Mater., Spec. Tech. Publ. 544, 188 pp.

———, 1975, Performance monitoring for geotechnical construction: Am. Soc. Test. Mater., Spec. Tech. Publ. 584, 194 pp.

British Geotechnical Society, 1974, Field instrumentation in geotechnical engineering: John Wiley & Sons, New York, 720 pp.

Burland, J. B., Moore, J. F. A., and Smith, P. D. K., 1972, A simple and precise borehole extensometer: Geotechnique, Vol. 22, pp. 174–177.

Cording, E. J., Hendron, A. J., Jr., MacPherson, H. H., Hansmire, W. H., Jones, R. A., Mahar, J. W., and O'Rourke, T. D., 1975, Methods for geotechnical observations and instrumentation in tunneling, Vols. 1 & 2: Rep. No. UILU-ENG 75 2022, Dept. of Civ. Eng., Univ. of Illinois at Urbana-Champaign, 566 pp.

Dunnicliff, J., 1982, Geotechnical instrumentation for monitoring field performance: Natl. Coop. Hwy. Res. Program, Synthesis of Hwy. Practice 89, Transp. Res. Bd., Natl. Acad. Sci., 46 pp.

Dutro, H. B., and Dickinson, R. O., 1974, Slope instrumentation using multiple position borehole extensometers: Transp. Res. Rec. No. 482, pp. 9–17.

Franklin, J. A., 1977, The monitoring of structures in rock: Intl. J. Rock Mech. Min. Sci. & Geomech. Abstr., Vol. 14, pp. 163–192.

Franklin, J. A., and Denton, P. E., 1973, The monitoring of rock slopes: Q. J. Eng. Geol., Vol. 6, pp. 259–286.

Fukuoka, M., 1980, Instrumentation: Its role in landslide prediction and control: Proc. Intl. Symp. on Landslides, Vol. 2, New Delhi, India, pp. 139–153.

Gould, J. P., and Dunnicliff, C. J., 1971, Accuracy of field deformation measurements: Proc. 4th Pan Am. Conf. on Soil Mech. and Found. Eng., Vol. 1, Am. Soc. Civ. Eng., San Juan, P.R., pp. 313–366.

Hanna, T. H., 1974, Foundation instrumentation: Trans Tech Publ. Ser. on Rock and Soil Mech., Vol. 1, No. 3, 372 pp.

Hartmann, B. E., 1967, Rock mechanics instrumentation for tunnel construction: Terrametrics, Golden, Colo., 154 pp.

ISRM, 1977, Suggested methods for monitoring rock movements using inclinometers and tiltmeters: Intl. Soc. Rock Mech. Comm. on Standardization of Laboratory and Field Tests, Rock Mech., Vol. 10, Nos. 1–2, pp. 81–106.

———, 1978, Suggested methods for monitoring rock movements using borehole extensometers: Intl. Soc. Rock Mech. Comm. on Standardization of Laboratory and Field Tests, Intl. J. Rock Mech. Min. Sci. & Geomech. Abstr., Vol. 15, pp. 305–317.

Londe, P., 1982, Concepts and instruments for improved monitoring: Am. Soc. Civ. Eng. Proc., J. Geotech. Eng. Div., Vol. 108, No. GT6, pp. 820–834.

Russell, H. A., 1981, Instrumentation and monitoring of excavations: Assoc. Eng. Geol., Bull., Vol. 18, pp. 91–99.

Sherwood, D. E., and Currey, B., 1974, Experience in using electrical tiltmeters: *in* Field instrumentation in geotechnical engineering, British Geotech. Soc., John Wiley & Sons, pp. 382–395.

Smith, P. D. K., and Burland, J. B., 1976, Performance of a high precision multipoint borehole extensometer in soft rock: Can. Geotech. J., Vol. 13, pp. 172–176.

Tice, J. A., and Sams, C. E., 1974, Experiences with landslide instrumentation in the southeast: Transp. Res. Rec. No. 482, pp. 18–29.

TRB, 1974, Landslide instrumentation: Transportation Research Board, Transp. Res. Rec. 482, 51 pp.

Underwood, L. B., 1972, The role of the engineering geologist in the instrumentation program: Bull. Assoc. Eng. Geol., Vol. 9, pp. 185–205.

Wallace, G. B., Slebir, E. J., and Anderson, F. A., 1970, *In situ* methods for determining deformation modulus used by the Bureau of Reclamation, Am. Soc. Test. Mater., Spec. Tech. Publ. 477, pp. 3–26.

Wilson, S. D., 1962, The use of slope measuring devices to determine movements in earth masses: *in* Field testing of soils, Am. Soc. Test. Mater., Spec. Tech. Publ. 322, pp. 187–197.

———, 1972, Instrumentation for Dams: Assoc. Eng. Geol. Bull., Vol. 9, pp. 143–157.

Wilson, S. D., and Mikkelsen, P. E., 1978, Field instrumentation: *in* Landslides-analysis and control, Spec. Rep. 176, Transp. Res. Bd., Natl. Acad. Sci., Washington, D.C., pp. 112–138.

7

Exploration

MAPS

Maps are a fundamental part of exploration. They serve two related functions. First, maps are a means for storing and transmitting information (Varnes, 1974). The location of observations or conditions found in an area could be recorded in written form. But this would be a cumbersome and inaccurate means to store and retrieve such information in comparison to a map. Second, maps convey specific information about the spatial distribution of factors or conditions (Varnes, 1974). Spatial relationships may be the type of data relevant to the purpose of an investigation: for example, a map showing the distribution of landslides over an area for an investigation of regional landslide hazards (Figure 7.1).

Varnes (1974) states that maps convey one or more of the three principal aspects of information. The first aspect is syntactic, that is, the statistical rarity of certain information. Infrequent or rare data are considered more important because of their rarity. In other words, anomalous aspects of a map are important. Figure 7.2 shows part of the earthquake epicenter map for California. The largest magnitude event that tends to stand out when viewing this map represents syntactic information. The second aspect is semantic. Every map is conveying information about something. For example, a geologic map displays different bedrock units in an area. It is a classification of rocks in order to form mappable units with an associated degree of certainty as to their description. For instance, a unit labeled as a conglomerate is expected to be a conglomerate as described, not a siltstone, nor a shale, nor other type of rock defined by any other unit on the map. This semantic aspect considers information contained in the map apart from who might use it or their purpose. The third aspect is pragmatic. This is simply the degree of response from the user. This may

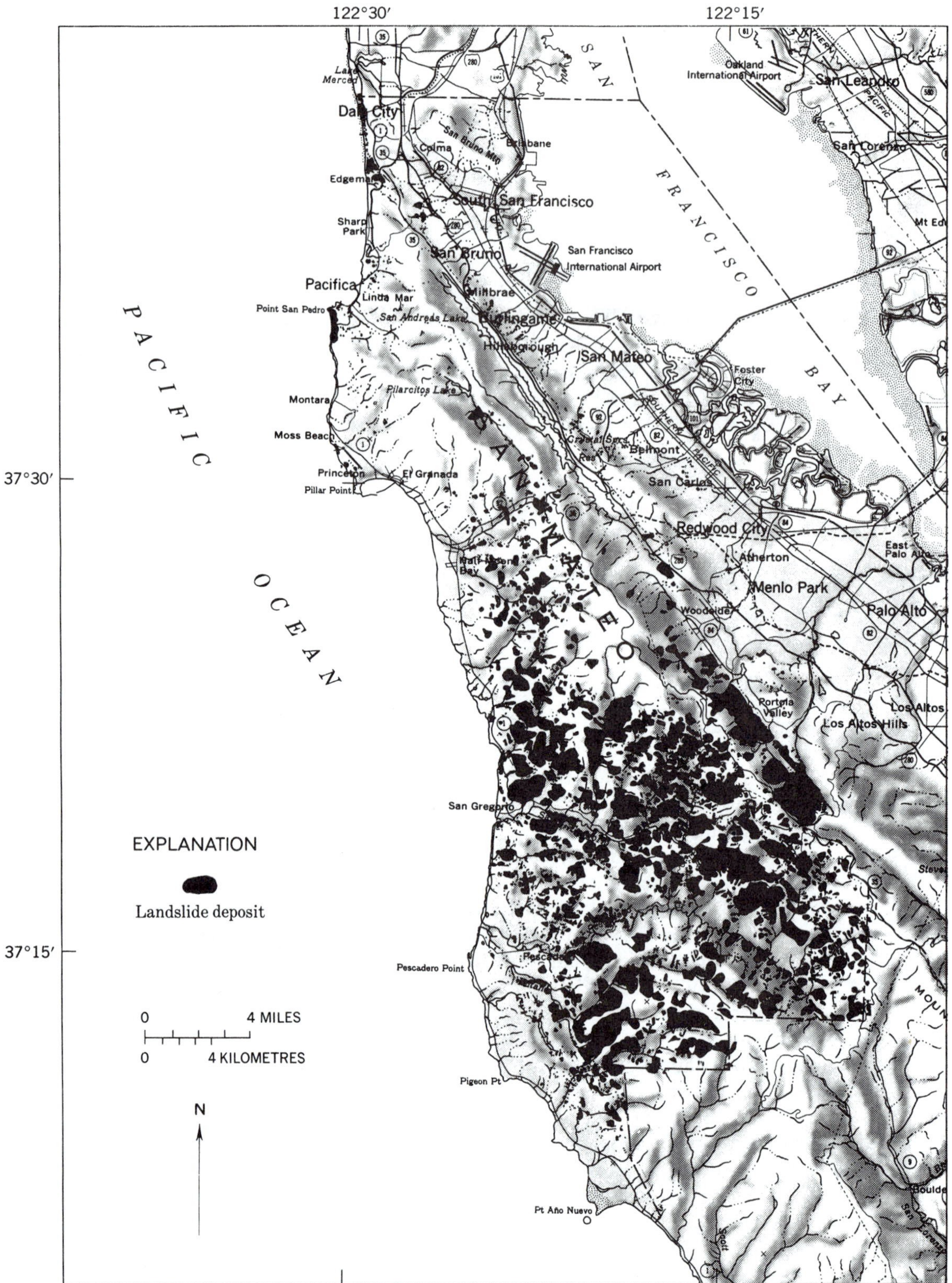

Figure 7.1 Map showing the location of identified landslides in San Mateo County, California. (From Nilsen and Brabb, 1975)

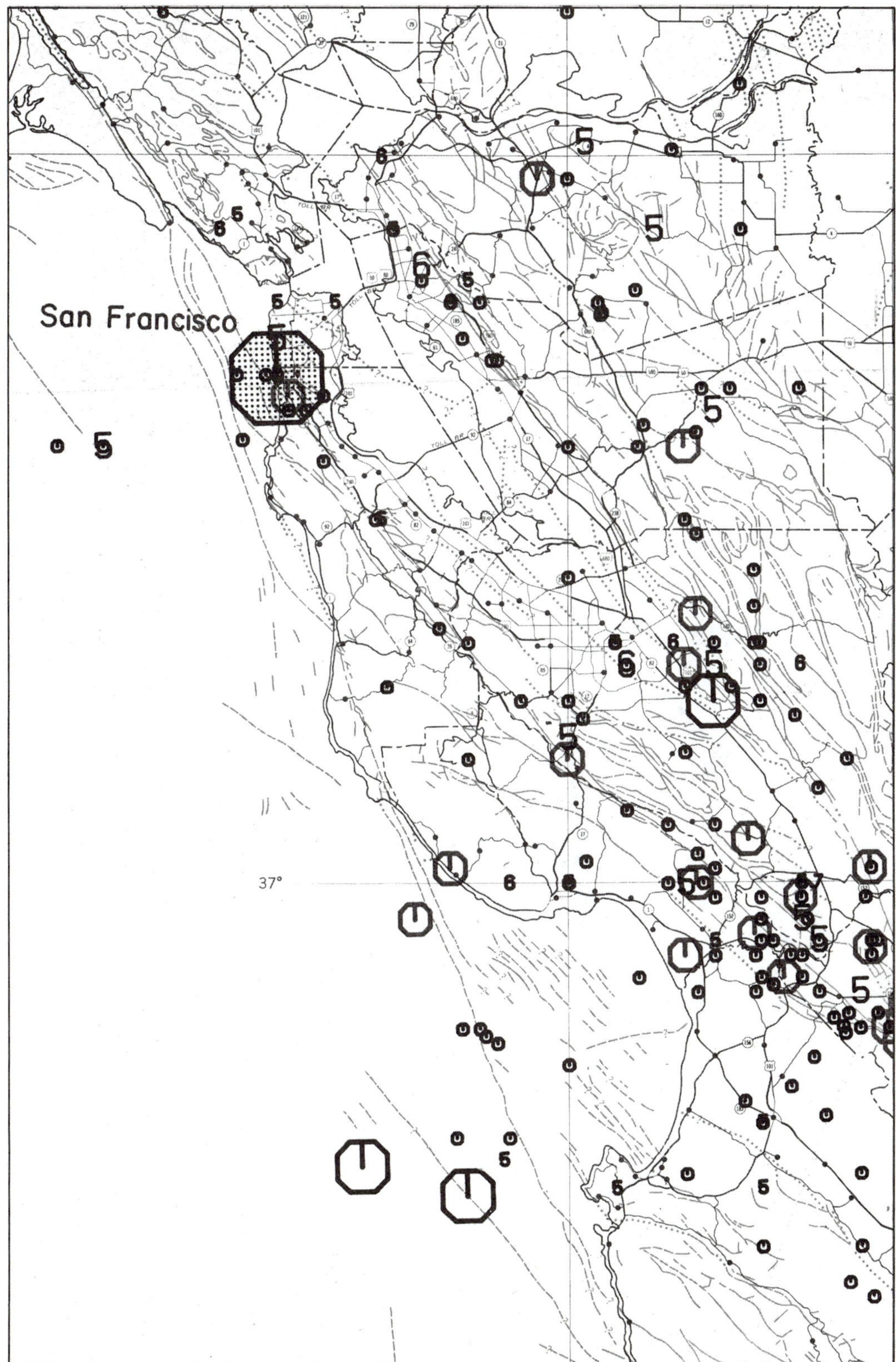

Figure 7.2 Part of the earthquake epicenter map of California showing the area near San Francisco. The magnitude of earthquakes is represented by the size of the epicenter symbol. Earthquakes occurring from 1900 through 1974 are represented. (From Real et al., 1978)

vary from no response to an intense response. It is strongly dependent on the individual receiving the information. A map that shows the variation in thickness of unconsolidated material over bedrock is unlikely to evoke much of a response from a petroleum geologist. This is in sharp contrast to the reaction of an engineering geologist involved with foundation studies.

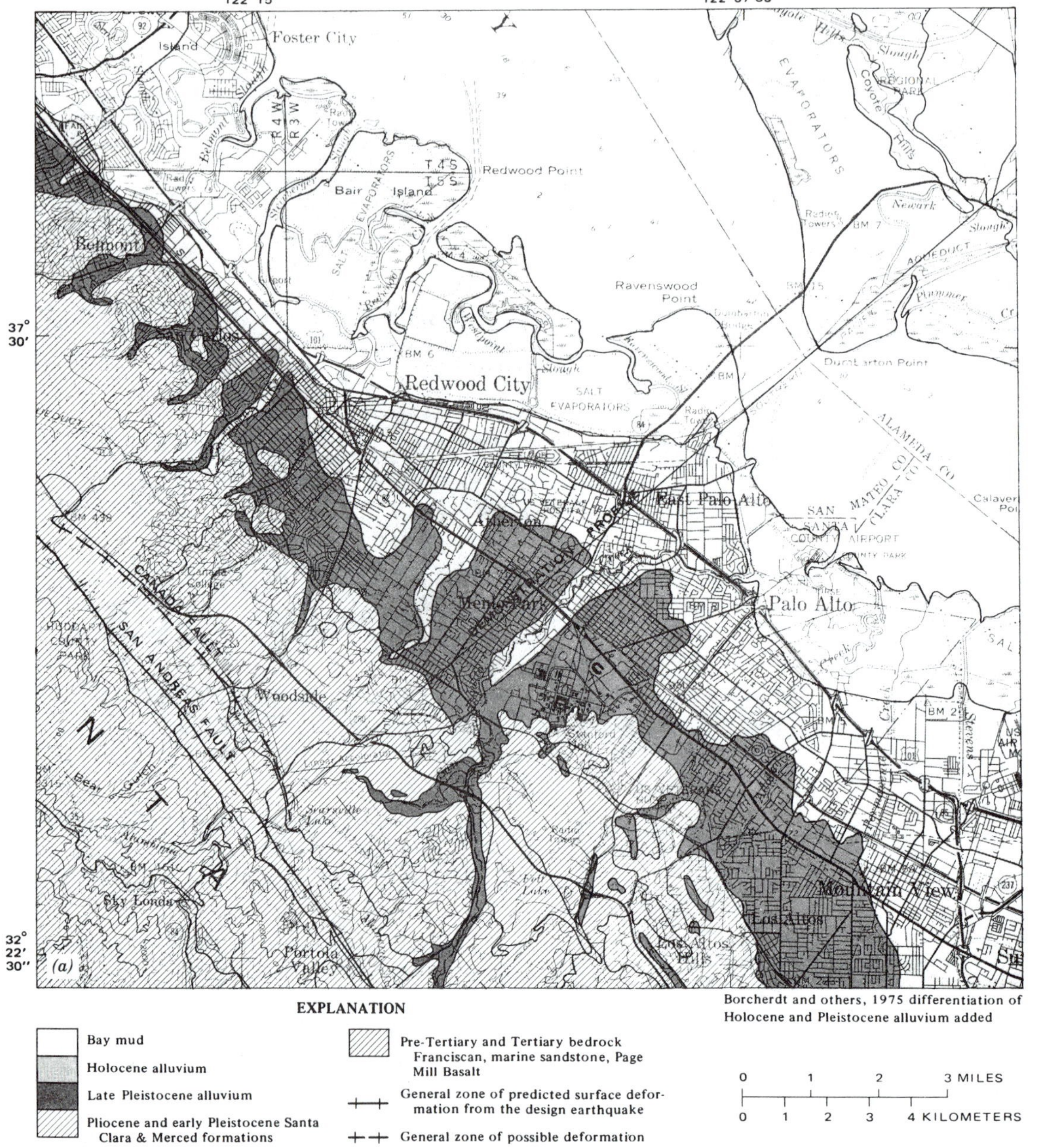

Figure 7.3 Two maps showing the region around Palo Alto, California. (*a*) A generalized geologic map. (*b*) Shows zones of relative potential for liquefaction and lateral spreading based on an interpretation of the response of geologic units to expected earthquake motion. (From Blair and Spangle, 1979)

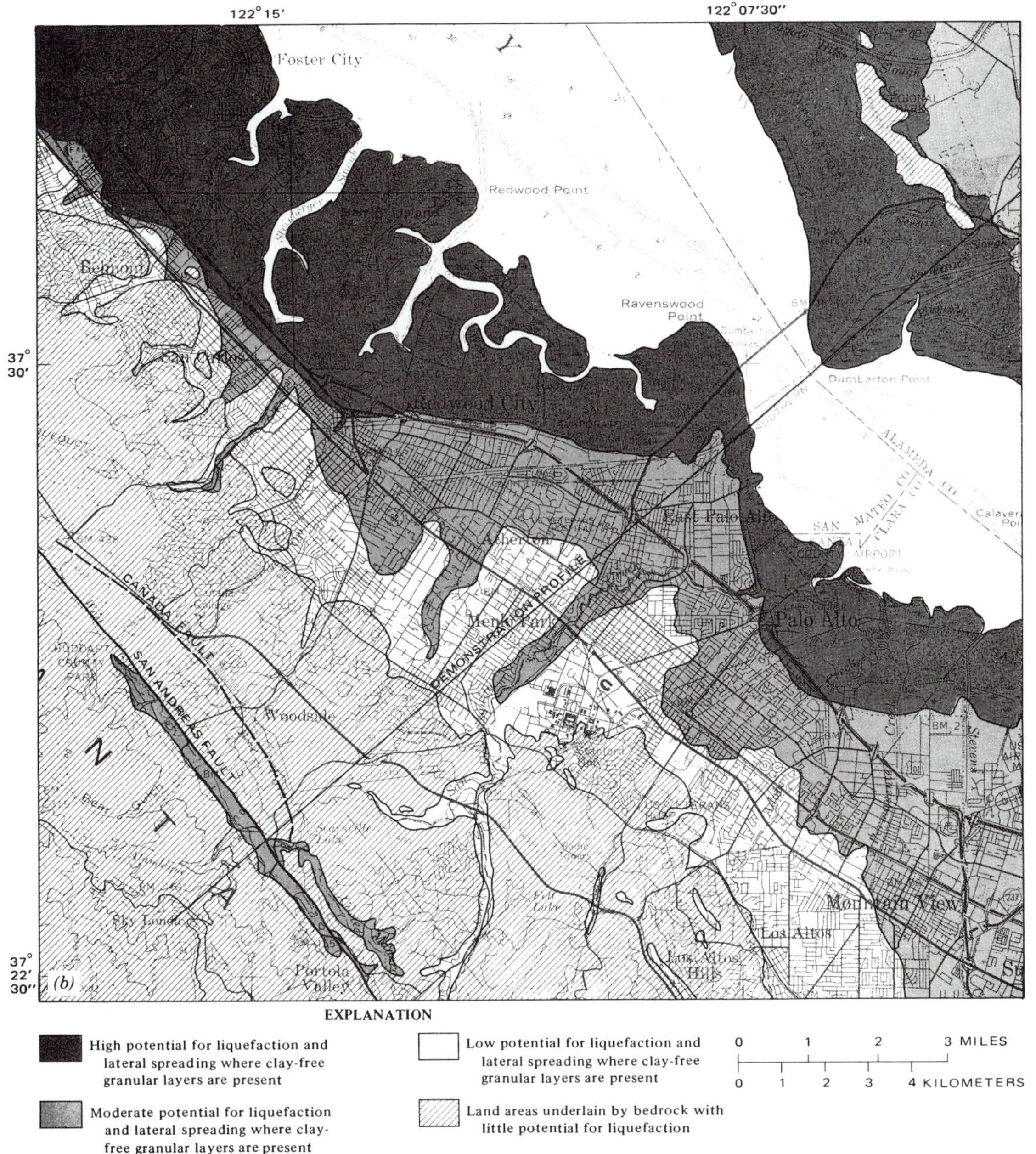

Mapping in engineering geology is an effort to convert semantic information to pragmatic information (Varnes, 1974). Blair and Spangle (1979) effectively demonstrate this point in their conversion of the generalized geologic map of the area near Palo Alto, California, to one that shows relative potential for liquefaction and lateral spreading, that is, a map useful for disaster planning (Figure 7.3).

Varnes (1974, p. 4) notes:

> The essence of mapping is to delineate areas that are homogeneous or acceptably heterogeneous for the intended purpose of the map. The resulting map consists of two parts that should never be considered separately: (1) the two-dimensional plan showing the outline of identified areas and (2) the explanation that tells in words and symbols what the essential attributes are that the enclosed area exhibit.

Soule (1980) provides a good illustration of this precept. The engineering-geologic maps he describes are prepared to assist in deciding land-use issues. Map units consist of various geologic hazard conditions present in the area (Table 7.1). These map units are generalized, but they are sufficiently homogeneous to describe hazard potential. Because such a map provides information for land-use decisions, their explanations include more than map-unit descriptions. A matrix defines how the elements in each map unit relate to expected land uses (Figure 7.4).

Engineering geologic maps can be categorized according to their purpose, content, or scale (UNESCO, 1976). More simply, engineering geo-

Table 7.1 Representative Map-Unit Descriptions for the Engineering Geologic Map of the Crested Butte–Gunnison Area, Colorado[a]

LANDSLIDE-EARTHFLOW AREA: area with demonstrably active natural movement of landslides and/or earthflows. Evidence for modern slope movement(s) includes distinctive physiography and disrupted vegetation or structures.

UNSTABLE SLOPE: slope with landslide-earthflow physiography but where modern slope movement is not apparent or is uncertain. Such areas have undergone slope movement in the recent geologic past (late Pleistocene-Holocene). Owing to climate changes and other factors, some of these areas have become stabilized in the natural state, whereas other places included in this category are metastable or possibly even slowly failing (moving) at the present time.

POTENTIALLY UNSTABLE SLOPE: slope with most attributes of an unstable slope, but where past or present slope failure is not apparent. The most significant of these attributes are composition of surficial and bedrock materials, proximity and geological similarity to slopes that have failed in the past or are failing now, slope angle and aspect, soil moisture conditions, and microclimate.

ROCKFALL AREA: area subject to rapid, intermittent, nearly unpredictable rolling, sliding, bounding, or free-falling of large masses of rock, rocks and debris, or individual rock blocks. Such areas are most commonly adjacent to unvegetated, barren, steep and/or fractured and jointed bedrock cliffs.

MUDFLOW-DEBRIS FAN AREA: area subject to rapid mud and debris movement after mobilization by heavy rainfall or snowmelt runoff. The essential elements of these areas are: (1) a source of mud and debris, usually in the upper reaches of a drainage basin or its contiguous sideslopes; (2) a drainageway or channel down which this mud and debris move; (3) a debris or alluvial fan formed by successive episodes of deposition of mud and debris.

HIGH WATER-TABLE AREA: area where groundwater is at or near the ground surface much of the year. These areas, shown only in places adjacent to major drainages, are evidenced by riparian vegetation and streambank physiography. Numerous such areas too small to be shown at this map scale are found contiguous to smaller drainages or associated with ancient and modern landslides and earthflows.

[a] See Figure 7.4.

Source: Soule, 1976.

GEOLOGIC HAZARDS FOR COMMON LAND USES

HAZARDS MAP UNIT	Residential Development: High density	Residential Development: Low density	Roads	Utilities	Open-space recreational complexes including ski areas, but not associated structures	Commercial and industrial development, including larger residential buildings such as condominiums and apt buildings	Low-value light-weight agricultural buildings	Agriculture uses grazing and similar
Landslide-earthflow area	3† A B C D E F H REMEDIAL ENGINEERING TYPICALLY IS PROHIBITIVELY EXPENSIVE	3 A B C D F H MAY BE POSSIBLE WITH CAREFUL ENGINEERING.	3† A B C D E F H TYPICALLY NOT FEASIBLE WITHOUT CAREFUL ENGINEERING.	2 A B C E F H COMPATIBLE WITH OPEN-SPACE LAND USE	1† A B COMMONLY FEASIBLE. MAINTENANCE COSTS MAY BE HIGH.	3† A B C F H MAINTENANCE COSTS PROBABLY WILL BE HIGH.	2† A F H MAINTENANCE COSTS MAY BE HIGH.	1† A B E USUALLY MINOR PROBLEMS EXCEPT FOR IRRIGATION DITCHES AND FENCES.
Unstable slope	3 A C D E F H REMEDIAL ENGINEERING USUALLY NECESSARY.	2† A C D E F H REMEDIAL ENGINEERING MAY BE NECESSARY.	3 A C D E F H REMEDIAL ENGINEERING USUALLY NECESSARY.	1† A C F H REMEDIAL ENGINEERING MAY BE NECESSARY.	1 A F COMMONLY FEASIBLE.	3 A C D E F H REMEDIAL ENGINEERING NECESSARY.	2 A F H REMEDIAL ENGINEERING NECESSARY.	1 A D E USUALLY MINOR PROBLEMS EXCEPT WHERE DITCH LEAKAGE CAUSES EARTHFLOWS.
Potentially unstable slope	3 C D E F REMEDIAL ENGINEERING MAY BE NECESSARY.	2 C D E F REMEDIAL ENGINEERING MAY BE NECESSARY.	2† A C D E F H MAY EXPERIENCE DIFFICULTIES WITHOUT CAREFUL PLANNING/ ENGINEERING.	1 C D E CAREFUL PLANNING CAN MINIMIZE HAZARD.	0 F TYPICALLY NO DIFFICULTIES.	2 A C F H REMEDIAL ENGINEERING NECESSARY.	2 A C F H REMEDIAL ENGINEERING NECESSARY.	0 E USUALLY MINOR PROBLEMS EXCEPT IN AREAS OF INTENSE CULTIVATION OF HILLSLOPES.
Rockfall area	3† A B C F H RARELY COMPATIBLE WITHOUT ELABORATE AND EXPENSIVE MITIGATION.	3 A B C F H CAREFUL SITING TYPICALLY NECESSARY TO MINIMIZE HAZARD.	3 A B C F H REMEDIAL ENGINEERING CAN MINIMIZE HAZARD.	2 A B C H CAREFUL PLANNING CAN MINIMIZE HAZARD.	1† B CAREFUL SITING CAN MINIMIZE HAZARD.	2 A B C H MAINTENANCE COSTS PROBABLY WILL BE HIGH.	2† A B C F H CAREFUL SITING CAN MINIMIZE HAZARD.	1 B USUALLY FEW OR MINOR PROBLEMS.
Mudflow-debris fan area	3† B C D F H RARELY COMPATIBLE WITHOUT ELABORATE AND EXPENSIVE MITIGATION.	3 B C D F H RARELY COMPATIBLE WITHOUT ELABORATE AND EXPENSIVE MITIGATION.	3 B D H COMPATIBLE ONLY WITH ELABORATE AND EXPENSIVE MITIGATION.	2 B F H POSSIBLY EXCESSIVE MAINTENANCE NECESSARY.	2 B COMMONLY FEASIBLE IF RISK IS ACCEPTABLE	2 B C D F H OCCASIONAL VERY HIGH MAINTENANCE COSTS CAN BE EXPECTED.	2 B D F H OCCASIONAL VERY HIGH MAINTENANCE COSTS CAN BE EXPECTED.	0 B USUALLY FEW OR MINOR PROBLEMS.
High water table area	3 G H BASEMENTS AND SEPTIC TANK SEWAGE DISPOSAL USUALLY NOT FEASIBLE.	2† G H BASEMENTS AND SEPTIC TANK SEWAGE DISPOSAL USUALLY NOT FEASIBLE.	3 G H USUALLY DIFFICULT - DEPENDS ON TYPE OF DEVELOPMENT. FLOOD PLAIN DETERMINATION MAY BE NECESSARY.	0 G SOME REMEDIAL ENGINEERING MAY BE NECESSARY IN UNUSUAL CASES.	1 G USUALLY LITTLE DIFFICULTY. POSSIBILITY OF FLOOD DAMAGE.	2† G H MAY REQUIRE SPECIAL CONSTRUCTION TECHNIQUES OR REMEDIAL ENGINEERING.	2 G H MAY REQUIRE SPECIAL CONSTRUCTION TECHNIQUES OR REMEDIAL ENGINEERING	0 G DESIRABLE FOR MANY KINDS OF AGRICULTURE.

EXPLANATION OF CHART SYMBOLS

- 3 HIGH HAZARD
- 2 MODERATE HAZARD
- 1 LOW HAZARD
- 0 VERY LOW, IF ANY HAZARD

EXAMPLE: TYPICAL POTENTIAL HAZARD FOR INDICATED LAND USE. / CONDITIONS AFFECTING ACTUAL DEGREE OF HAZARD. / COMMENTS APPLICABLE TO MOST CASES

MEANING OF LETTER SYMBOLS

- A ESPECIALLY SEVERE ON SLOPES GREATER THAN 30 PERCENT.
- B SLOPE MOVEMENT INTERMITTENT DEPENDENT ON VARIATION IN WEATHER OR OTHER FACTORS.
- C OVERSTEEPENING OR CUTTING OF SLOPES CAN INCREASE HAZARD GREATLY.
- D ARTIFICIAL OR NATURAL INCREASE IN GROUND MOISTURE CAN INCREASE HAZARD GREATLY.
- E REMOVAL OF NATURAL VEGETATION CAN INCREASE HAZARD GREATLY.
- F HAZARD MAY DECREASE CONSIDERABLY AS SLOPE DECREASES.
- G VARIES SEASONALLY.
- H DETAILED ENGINEERING GEOLOGY STUDIES NECESSARY DURING PRE-PLANNING STAGES OF DEVELOPMENT.

Figure 7.4 Matrix incorporated into the map explanation accompanying the engineering geologic map for the Crested Butte–Gunnison, Colorado area. (From Soule, 1976)

logic maps are either descriptive or interpretative in nature. Descriptive maps relate conditions in an area that may bear on some engineering geologic problem. They are measurable or observable characteristics obtained by modifying existing mapped data or by actual mapping of a specific characteristic of interest. The landslide-inventory map shown in Figure 7.5 is typical of descriptive maps (Wieczorek, 1984). Interpretative

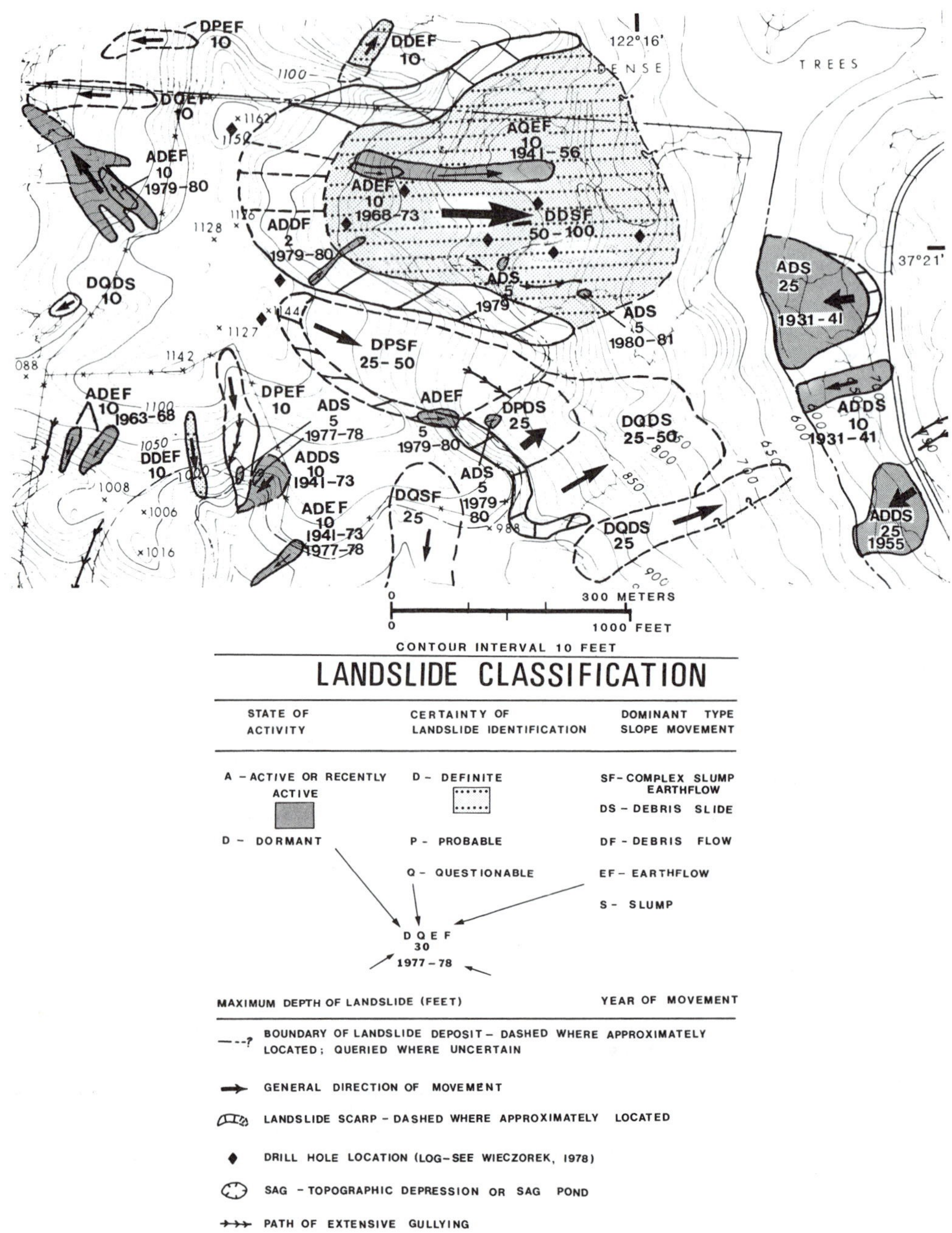

Figure 7.5 Part of a detailed landslide inventory map for an area near La Honda, California, (From G. F. Wieczorek, Preparing a Detailed Landslide-Inventory Map for Hazard Evaluation and Reduction, *Bulletin of the Association of Engineering Geologists*, Vol. 21, No. 3, August 1984, pp. 337–342. Used by permission.)

maps transform basic data into a form appropriate to address some engineering-geologic problem. They are often the product of manipulation of existing maps. Descriptive maps such as those that show thickness of overburden and rock rippability can be used with other information to make an interpretative map showing the suitability for subsurface installations such as sewers and electrical lines.

Four operations commonly used to generate a different map are: (1) generalization, (2) selection, (3) addition and superposition, and (4) transformation. Generalization requires use of data from a more detailed map to construct the new map. Subsidence in an area might be generalized simply by changing the contour interval (Figure 7.6). Selection involves discrimination in the choice of information to represent. Choosing to emphasize sand deposits over other deposits in a surficial geology map would involve selection (Figure 7.7). Addition is the combining of two or more maps or simply adding different data to existing map units. The term *superposition* more readily reflects this operation. Superimposing a map that shows different slope-steepness categories on a map of bedrock units is an example of addition. The fourth operation, transformation, is an actual changing of the character and meaning of boundaries, areas, or symbols to make the map more understandable and meaningful to the user. An isopleth map that shows the percentage of existing landslides can be transformed into a landslide-susceptibility map by assigning relative degrees of hazard to certain percentage values (DeGraff, 1985). Table 7.2 briefly describes how these four operations might influence the map and its accompanying explanation.

It should be recognized that maps are a selective representation. Consider a map that shows fault locations. The faults do not represent all the mappable information in that area. Faults were selected for representation as opposed to types of bedrock exposed, depth to the water table, or any other attribute, or characteristic, present within that area. As a consequence, map preparation involves a number of conscious decisions to arrive at what form this selective representation will take. First, the purpose of the map and its units must be identified. Varnes (1974) notes that the purpose includes deciding whether the map units will pertain to time; space; the inherent qualities, or properties, of real matter; the relationship among objects; or a combination of two or more of these characteristics. Keaton (1984) wanted maps that documented soils and geologic data with direct engineering significance. This required map units pertaining to the properties of real matter. To achieve this purpose, the Genesis-Lithology-Qualifier (GLQ) system is used instead of conventional geologic map symbols (Figure 7.8). The GLQ map units and their symbols represent the properties of surficial materials within an area. These properties are more easily recognized by symbols that represent the genetic origin of the deposit and the general particle size present, with qualifiers indicating landform. Second, a decision is needed on the essential attributes to be used in identifying a map unit. DeGraff and Romesburg (1980) choose to use the relative proportion of land disturbed by past landslides to define relative landslide susceptibility. The proportion represents the amount of landslides, in acres, that occur on land with certain bedrock, slope-steepness,

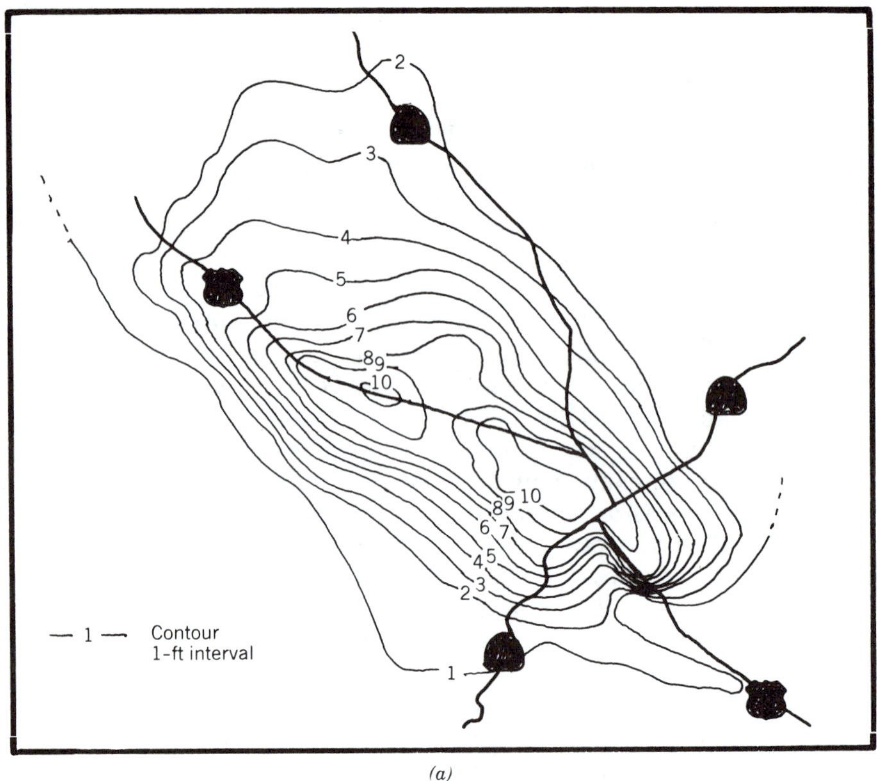

(a)

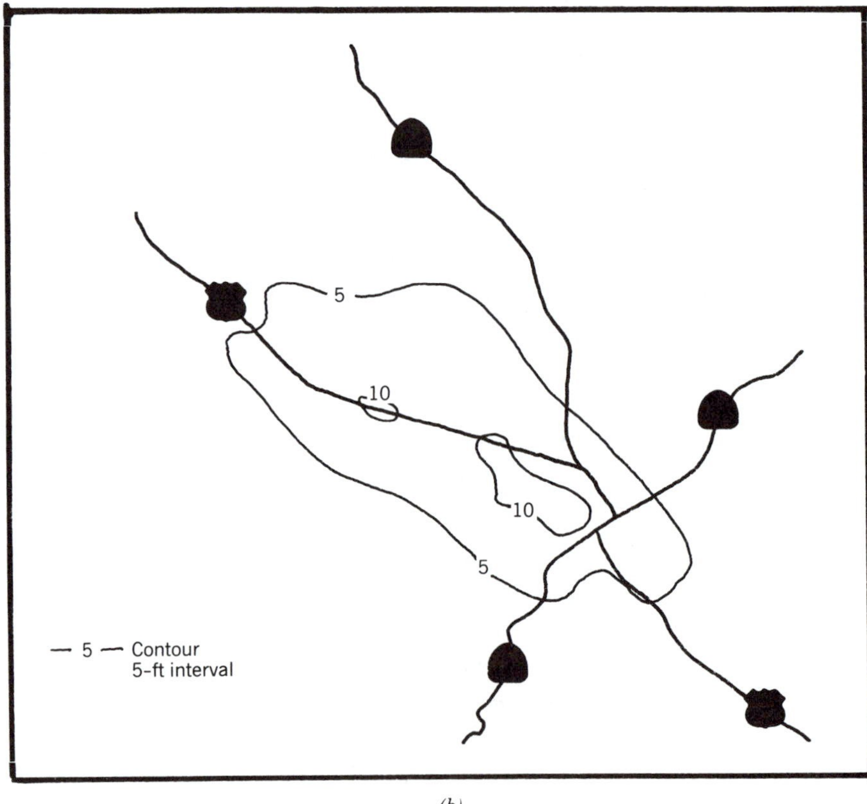

(b)

Figure 7.6 Two maps of land subsidence from 1934 to 1967 in the Santa Clara Valley, California. Map (*b*) is a generalized version of map (*a*). (Modified from Helley et al., 1979)

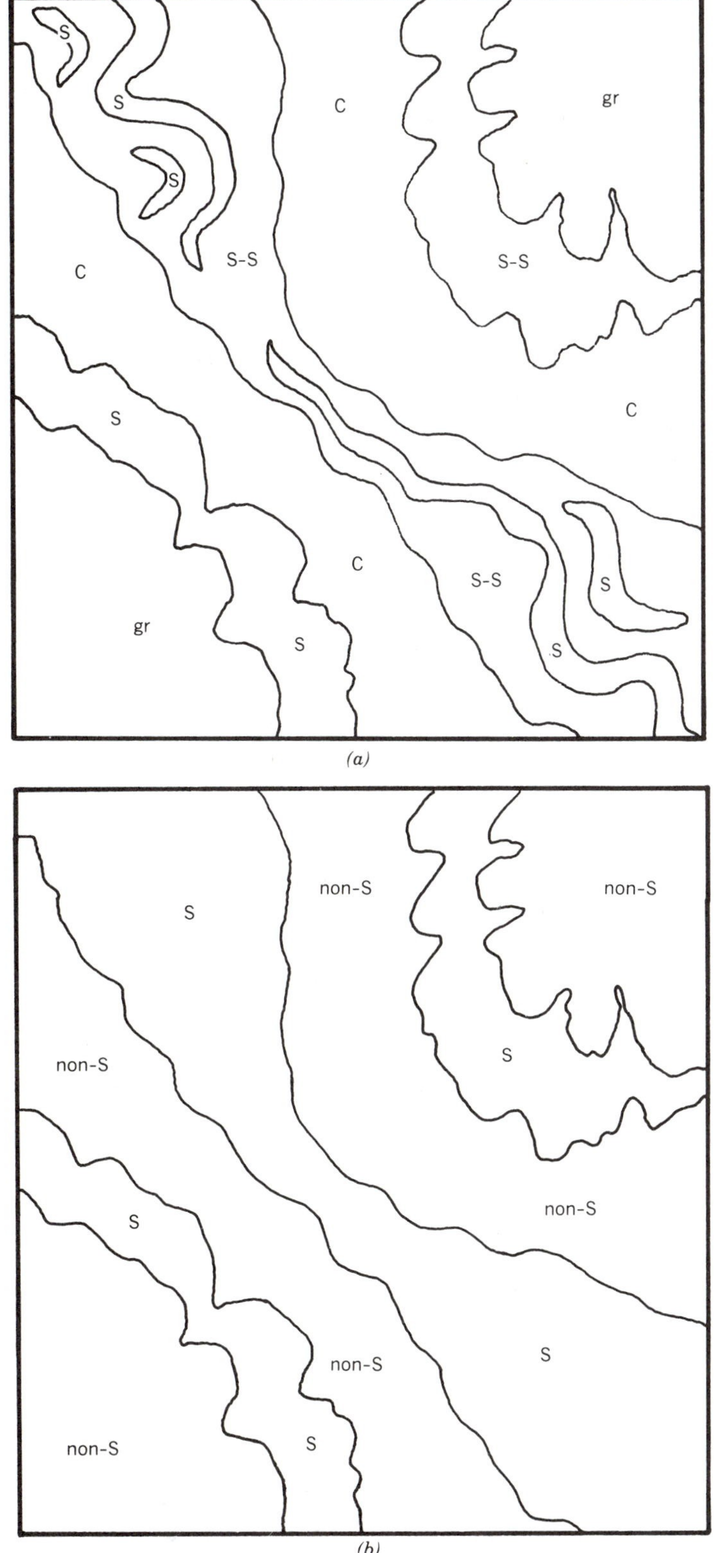

Figure 7.7 Two maps illustrating the process of selection. (*a*) Shows alluvial units: gr = gravel, s = sand, s-s = silty sand, and c = clay. After selection, (*b*) shows the same area with the units: s = sand and non-s = all materials other than sand.

Table 7.2 Summary of the Effects on Map Graphics and Explanation of the Four Common Operations Employed to Generate Maps

Operation	Cause, Reason, of Purpose	Effects on Graphic Portrayal	Effects on Language Statements
Generalization:			
Spatial	To achieve emphasis or clarity; may be required cartographically after reduction in scale.	Boundaries made straighter; inliers erased; several symbols in a given area replaced by one.	Changes generally not necessary; can be made less specific to fit increase in heterogeneity.
Typological (same as grouping)	To clarify concepts, add emphasis, or remove detail unimportant to purpose.	Lines erased; fewer symbols used.	Must be recast to make broader.
Selection:			
Spatial	To limit area of interest to user.	Units outside of boundaries deleted.	Some may require modification.
Typological	To emphasize or to fit particular needs.	Some units deleted.	Some statements deleted.
Division:			
Spatial	To divide area for examination, sampling, or scanning.	Lines added.	No effect on typological units; areal units defined.
Typological	Need for detail of new kind.	Lines added.	New units defined.
Addition:			
Spatial	To extend areal coverage or increase detail.	Map enlarged or information made denser.	None.
Typological			
Same map	To add related information.	None.	New attributes added.
Superposed maps	To add information of a different kind.	All map elements superposed.	Statements still refer to identifiable areas for each attribute.
Transformation:			
Cartographic	Change of medium.	None.	None.
Spatial	Change of scale or projection.	Size or shape of units altered; may require generalization.	None.

(continued)

Table 7.2 *(continued)*

Operation	Cause, Reason, of Purpose	Effects on Graphic Portrayal	Effects on Language Statements
Temporal	Use in future.	Depends whether attributes are constant or changing.	Depends whether attributes are constant or changing.
Typological	Change in actual or potential use for map.	Lines erased but not added without new field study.	Minor to complete change in definition of map units.

Source: Varnes, 1974.

and slope-orientation characteristics divided by the total amount of land with identical characteristics that is present in the study area. Map units are defined by different parts of the range of proportions. Thus, each susceptibility category is defined by specific combinations of bedrock units on slopes with a certain slope steepness and orientation. Third, the degree of internal heterogeneity that will be tolerated while still achieving the intent of the map must be determined. The delineation of potential volcanic hazards in the Long Valley–Mono Lake area of California reflects the degree of internal heterogeneity in the careful description of mapped units (Figure 7.9) (Miller et al., 1982).

Engineering geologists engaged in exploration activities will both read and use maps as well as prepare them. Whether using or making maps, the engineering geologist should always remain aware that some degree of uncertainty is present within a map. Varnes (1974, p. 42) gives four reasons for representing this uncertainty in both graphic and written terms that are worth keeping in mind:

1. Well-designed graphics yield more efficient transmittal of spatial information than do words.
2. Users of engineering geologic maps are generally more interested than other users of geologic data in the accuracy of both attribute-at-a-point and area-of-an-attribute information and in the homogeneity of map units.
3. Being usually outside the science of the mapmaker, the user of engineering geological maps has no way to assess the qualifications and doubts that attend the lines around the map units unless he is shown and told. Matters of probability that we geologists believe we comprehend almost instinctively need explication in both written and graphic language. Otherwise, the user may receive a false impression either of unwarranted security or of unwarranted doubt.
4. Adherence to a philosophy of "conservatism" such as that advocated by Wentworth, Ziony, and Buchanan (1970), in practice requires having a variety of means for showing uncertainty.

Innovations in engineering geologic mapping can be expected in the future. This will stem, in part, from a wider use of computers and digitized data bases in map production. Hasan and West (1982) illustrate how a

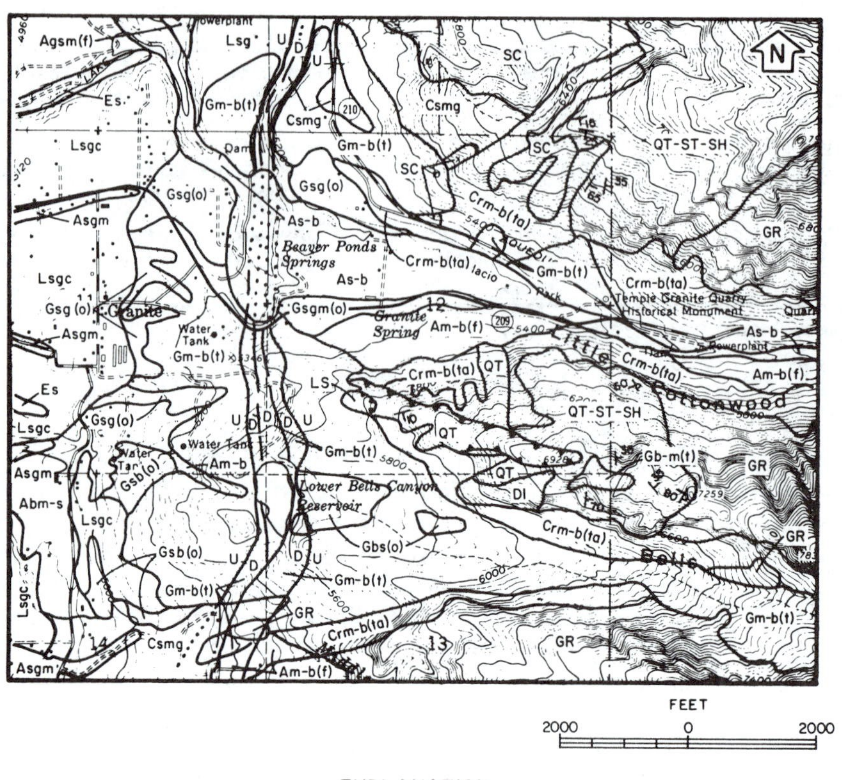

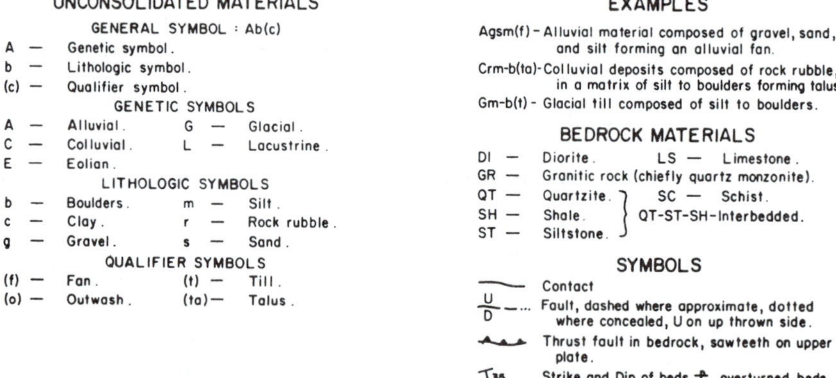

Figure 7.8 An engineering geologic map covering part of the Draper quadrangle, Utah. Map units represented on this map are based on the GLQ system. (From Keaton, 1982)

digitized data base can address regional land appraisal. Other developments may involve three-dimensional representations of engineering geologic data such as the mapping described by Kempton and Cartwright (1984). No matter how innovative these future approaches might be, the same questions of logic and basic map operations will apply. For more detailed discussion of the logic and application of engineering geologic maps, Varnes (1974) and UNESCO (1976) are recommended.

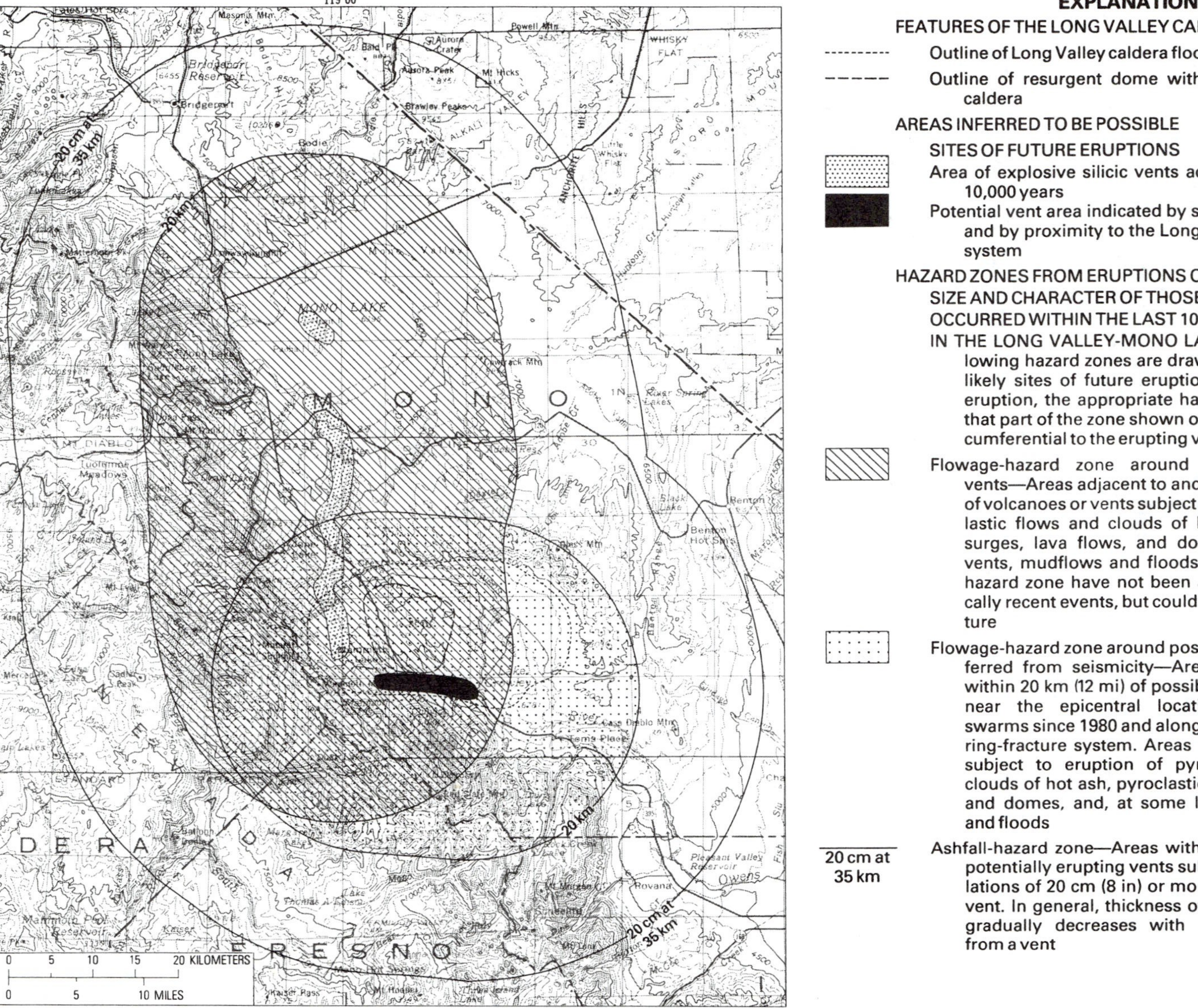

EXPLANATION

FEATURES OF THE LONG VALLEY CALDERA

- Outline of Long Valley caldera floor
- Outline of resurgent dome within the Long Valley caldera

AREAS INFERRED TO BE POSSIBLE

SITES OF FUTURE ERUPTIONS

- Area of explosive silicic vents active during the last 10,000 years
- Potential vent area indicated by seismicity since 1980 and by proximity to the Long Valley ring-fracture system

HAZARD ZONES FROM ERUPTIONS OF THE SIZE AND CHARACTER OF THOSE THAT HAVE OCCURRED WITHIN THE LAST 10,000 YEARS IN THE LONG VALLEY-MONO LAKE AREA—The following hazard zones are drawn around relatively likely sites of future eruptions. During a future eruption, the appropriate hazard zone would be that part of the zone shown on the map that is circumferential to the erupting vent or vents

- Flowage-hazard zone around existing explosive vents—Areas adjacent to and within 20 km (12 mi) of volcanoes or vents subject to eruption of pyroclastic flows and clouds of hot ash, pyroclastic surges, lava flows, and domes, and, at some vents, mudflows and floods. Some parts of the hazard zone have not been affected by geologically recent events, but could be affected in the future
- Flowage-hazard zone around possible future vents inferred from seismicity—Areas adjacent to and within 20 km (12 mi) of possible future vents at or near the epicentral location of earthquake swarms since 1980 and along a part of the caldera ring-fracture system. Areas within this zone are subject to eruption of pyroclastic flows and clouds of hot ash, pyroclastic surges, lava flows, and domes, and, at some locations, mudflows and floods
- 20 cm at 35 km — Ashfall-hazard zone—Areas within 35 km (22 mi) of potentially erupting vents subject to ash accumulations of 20 cm (8 in) or more downwind from a vent. In general, thickness of ash accumulations gradually decreases with increasing distance from a vent

Figure 7.9 Map and explanation of potential volcanic hazards in the central part of the Long Valley–Mono Lake area, California. (From Miller et al., 1982)

REMOTE SENSING

Remote sensing in broadest terms involves the collecting of data about an object, a surface, or material without physical contact regardless of the distance separating the observer and the feature. Unless restrictions are applied, remote sensing includes geophysical exploration methods. Remote sensing in this text refers to those methods that provide data about objects or features on the earth's surface as well as physical properties of surficial materials that are obtained in the absence of any matter in the intervening space (Rib and Liang, 1978).

Remote sensing in this restricted sense is used to directly or indirectly provide an image of the earth's surface from instruments that are on some platform such as an airplane or satellite. Aerial photographs from aircraft and Landsat imagery from satellites are examples. Terrestrial photogrammetry is a variety of remote sensing that utilizes images of features obtained from surface instruments rather than aircraft and satellites.

Remote sensing as defined here makes use of electromagnetic energy within the spectrum illustrated by Figure 7.10. Sensors detect and record energy that, for the most part, comes from the sun and is reflected or emitted from the earth. In addition, energy may be transmitted from the sensor platform and reflected back to the sensor from the surface. The methods employed are passive and active, respectively. Reflectance, or emission of energy from the surface, is not uniformly transmitted through the atmosphere for all of the spectral wavelengths or frequencies. Electromagnetic energy is selectively absorbed by the atmosphere. The resulting *atmospheric windows* through which energy may be transmitted are shown in Figure 7.10*a*.

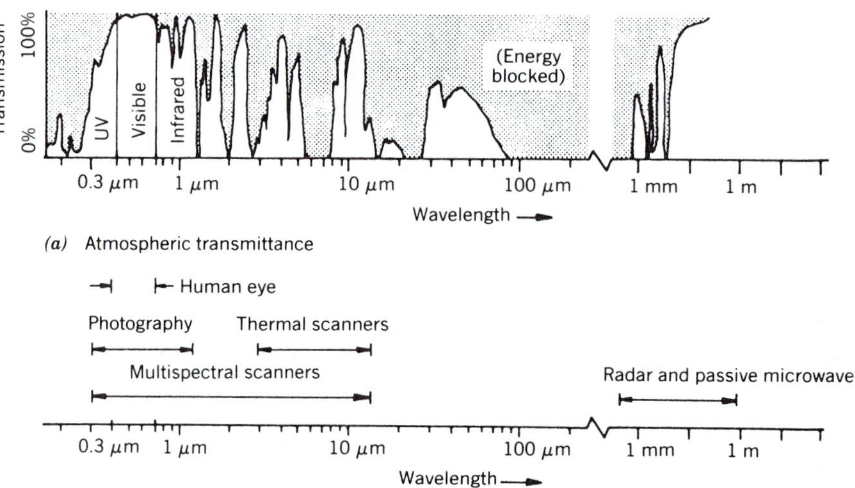

Figure 7.10 Electromagnetic spectrum. (*a*) Atmospheric transmission. (*b*) Wavelength ranges used for remote-sensing systems. (Reproduced with permission from Remote Sensing and Image Interpretation, T. M. Lillesand, and R. W. Kiefer, 1979. Copyright © John Wiley & Sons, Inc.)

Although the human eye has a spectral sensitivity range that coincides with an atmospheric window having maximum energy transmission, it has a very limited spectral range compared with the electromagnetic spectrum usable in remote sensing (Figure 7.10*b*). This limited range seriously restricts the frequencies that may be useful for interpreting surface features or materials. Figure 7.11 illustrates the problem. Within the visible spectrum, deciduous and coniferous trees have overlapping spectral reflectances and thus are undifferentiated by the eye on the basis of color alone. At longer wavelengths, however, the spectral reflectance ranges for the two types of trees exhibit a significant difference. The deciduous trees appear to be lighter in color from the higher reflectance. A remote-sensing imaging system sensitive to the longer wavelengths permits discrimination between them. The variety of remote-sensing systems available today allows selection of one or more systems as well as image-enhancing techniques tailored to the specific needs of the investigator. For the case illustrated by Figure 7.11, color infrared (IR) photography and multispectral scanning (MSS) provide coverage of the wavelength range in which the reflectances differ.

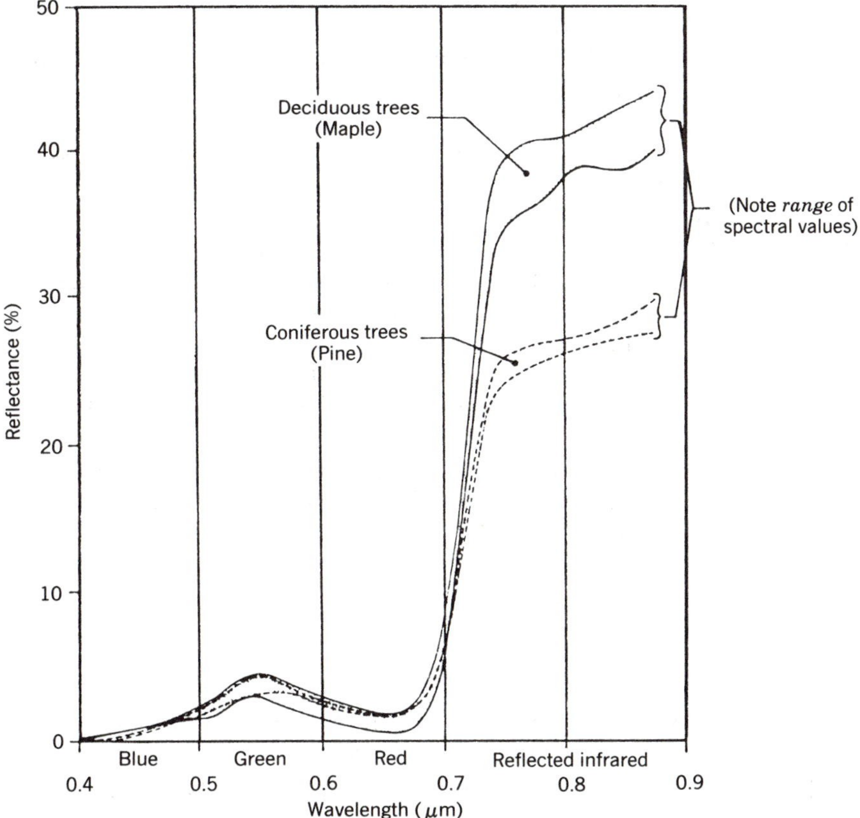

Figure 7.11 Generalized spectral reflectance envelopes for deciduous and coniferous trees. (Reproduced with permission from Remote Sensing and Image Interpretation, T. M. Lillesand, and R. W. Kiefer, 1979. Copyright © John Wiley & Sons, Inc., and courtesy of Canadian Aeronautics and Space Institute and Canadian Remote Sensing Society.)

Sensing systems of value in engineering geologic investigations include: black-and-white, color, and color IR photography; simultaneous digital recording of several spectral ranges, or bands (MSS); digital recording of the more-limited thermal range; and active sensing in the microwave range. Each has its uses singly or in combination with other systems. Most provide opportunities to selectively enhance or improve the image for given applications.

"Standard" aerial photography is available at low cost for many areas and requires simple viewing equipment for routine interpretation. Other sensor systems require costly contracted flying of an area, with recording of image data by sophisticated airborne scanning equipment or the purchase of digital tapes that have multispectral data received from satellites. Most enhancement techniques and many of the kinds of available imagery require computer facilities for preparation of the final products needed for interpretation. Thus, although there is a variety of sensors and data-handling procedures available for engineering-geologic purposes, the one or more remote-sensing methods selected will be dependent on several factors. These are the mating of system capabilities with the information requirements of the job, associated production costs, and available computer facilities. One must be prepared to make an assessment of needs and available resources that will permit the investigator to obtain the greatest benefits for the investigation with fiscal, time, and equipment limitations.

Engineering-geologic investigations benefit from the use of remote-sensing methods as noted in Stallard, 1972; Chandler, 1975; Norman and Watson, 1975; West et al., 1976; and Rengers and Soeters, 1980. A more complete inspection of an area is possible from a near-vertical vantage point, a view seldom available to the investigator on the ground. A set of imagery provides essentially instantaneous views of an area compared to longer time spans for surface mapping. Satellite imagery is periodic and may be used to monitor changes of surface features with time. A practical factor to be considered in site investigations is that remote sensing surveys will not cause property values to increase early in a project, as is the case when a field party works in an area. A disadvantage is the loss of detail that will differ in magnitude with the image scale and sensor system employed.

The efficient investigation of sites by remote-sensing methods does not preclude the need for selected data from the site—data known as *ground truth*. The amount of ground truth required is dependent on the sensing systems used, the experience of the interpreter, and the analogies that may be made with similar sites.

In addition to the references cited herein, the reader should become familiar with one or more of a number of publications containing overviews of the variety of sensors, enhancement techniques, and applications presently available. Representative of these are: Reeves (1975), Barrett and Curtis (1976), Sabins (1978), Lillesand and Kiefer (1979), Siegal and Gillespie (1980), and Paine (1981).

Aerial Photography

Since early engineering applications in the 1940s (Eardley, 1943; Jenkins et al., 1946; Frost and Woods, 1948), aerial photography has continued to be the most economical and useful of the various remote-sensing systems. As new sensing systems have been brought into the arena of civil engineering applications, aerial photography also has progressed in its versatility with the addition of color and color IR photography and a variety of enhancement techniques.

Photography in a range of scales from 1:15,000–1:40,000 is available for much of the United States in black-and-white and color prints and in positive color transparencies. Photos in this scale range are satisfactory for many applications. The USDA Forest Service and the U.S. Geological Survey are federal agencies that sell air photos in this range of scales. High-altitude, small-scale (<1:60,000) photography is also available for selected areas. Such smaller-scale photos are useful in inventorying large features such as major landslides and in recognizing large geologic structures and major drainage patterns. Where greater detail and small areal coverage are needed, large-scale (>1:15,000) coverage of a specific area may be obtained from private contractors. Desirable scales for various stages of highway location and construction are illustrated by Figure 7.12.

Photointerpretation

Photointerpretation has been defined as the art of examining photographic images for the purpose of identifying objects and judging their significance (Barrett and Curtis, 1976). It differs from photo analysis with respect to judgmental conclusions that are not a part of photo analysis. Photointerpretation inherently contains bias. The degree to which photointerpreta-

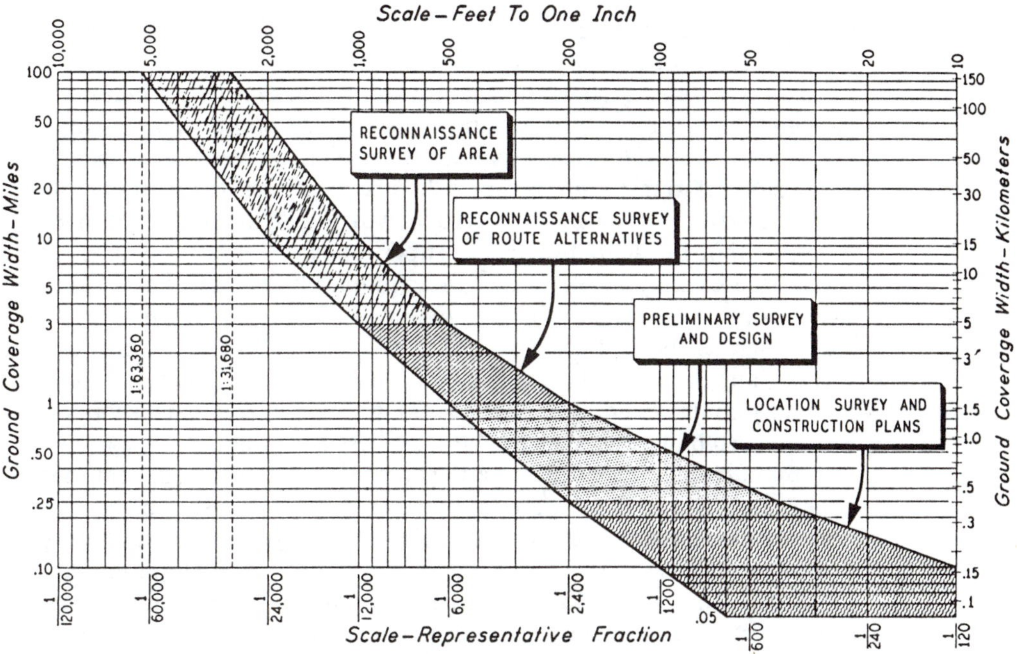

Figure 7.12 Ranges of scales of air photos for various stages of highway location and construction. (From W. T. Pryor, 1964. Reproduced with permission from "Photogrammetric Engineering," Vol. 30, No. 1. Copyright © 1964 by the American Society of Photogrammetry.)

tion is biased and effective is dependent upon the user's knowledge of the geologic and engineering characteristics as well as on the problems associated with features seen on the photos. An example is the identification by photo analysis of a landform such as an end moraine. The photointerpreter carries the identification process further—for instance, by making assessments of material type based on vegetation, drainage, and parent materials. How thorough and correct these assessments are depend on the interpreter's knowledge of, and experience with, similar features.

Photointerpretation is possible because of the three-dimensional view of the earth's surface obtained from photos by use of standard aerial photographic procedures and an evaluation of the two- and three-dimensional features, called *elements*, that are visible when viewed stereoscopically. Rib and Liang (1978) have pointed out that no other technique in the wide array of remote-sensing methods provides the three-dimensional overview of the terrain from which interrelations of topography, drainage, surface cover, geologic materials, and human activities on the landscape can be viewed and evaluated.

Stereovision. The three-dimensional, or stereoscopic viewing, of the earth's surface is uniquely a characteristic product of typical aerial photographic surveys regardless of scale or film type. Current standards call for aerial surveys to be obtained by taking photographs vertically downward from aircraft. Stereoscopic coverage is obtained by sequentially overlapping photographs along a flight line of predetermined length. Additional coverage is obtained by flying parallel and overlapping flight lines (sidelap). The standard overlap and sidelap are 60% and 30%, respectively. Photography obtained from federal agencies will have been flown along either N–S or E–W flight lines.

The third dimension of surface features is the product of being able to view the same surface object from two separated points of view along the flight line. This is the same method by which we perceive depth (third dimension) if our vision is normal. The adjacent, overlapping photos (stereo pairs) are arranged for stereo viewing by laying one photo over the other, so that the shared areas are superimposed. Surface features within this 60% area may be examined stereoscopically by use of a pocket-type stereoscope or the larger mirror stereoscope. Versions of the mirror stereoscope such as the Old Delft model permit changes in magnification and vertical and horizontal traversing of the photos by means of built-in, manually controlled optics. A disadvantage of all mirror stereoscopes is the lack of portability afforded by the lens stereoscopes.

If there are sloping surfaces on the area viewed, they will appear to slope at greater angles to the horizontal than is actually the case. This phenomenon is called vertical exaggeration. Vertical exaggeration results, in part, from the differences in distance between the aircraft when photographing the stereo pairs of the surface and the distance between the interpreter's eyes when viewing the same surface. Additional causes of exaggeration are altitude and camera lens focal length.

Another problem associated with the lens focal length and the altitude of the aircraft from which photos are being taken is the change in scale that occurs on a photo as one views radially outward toward the photo edges from the center. Changes in elevation (relief) relative to a datum elevation also change the scale within the photo (Paine, 1981). Therefore, unless photos have been rectified, that is, corrected by special optical plotting equipment for both geometric and relief distortion to produce an orthophotograph, one should not use photos for planimetric (two-dimensional mapping) purposes. By-products of the production of orthophotographs are orthophotoquads, which are the planimetric and area equivalence of standard 7½-minute topographic maps. Both are obtainable from the U.S. Geological Survey. Although the advantage of the third dimension is lost, orthophotoquads provide greater detail of surface features such as rock outcrop patterns, vegetative cover, and land use than do topographic maps.

Photo Elements. The reliability of photo interpretation is dependent on the recognition and assessment of a number of elements that are visible on the photos. Some elements are dependent on stereo viewing of photos, an interpretive advantage not available in most other kinds of imagery. Although the list of elements may differ among photointerpreters, several are consistently used by all. These are topography, drainage patterns, erosion, vegetation, land use, and either photo tone for black-and-white photos or color for color and color IR photos.

Topography includes the size, shape, and relative elevations of surface (topographic) features. Some photointerpreters consider each of these as separate elements. Recognition of surface relief and evaluation of the geologic or engineering importance of such features is almost totally dependent on the stereo viewing of photos. An example of the typically associated topographic factors of size, shape, and relative elevations may be found by comparing a hogback with an end moraine, landforms with significantly different materials and associated uses and/or engineering properties. Sizes of both may be similar. However, the cross-section shapes seen stereoscopically will differ greatly. In addition, the ridgelike form of the hogback usually will be more linear, with visual identification of rock units often possible. End moraines more commonly do not form well-defined ridges and are curved or sinuous. Relative elevations on hogbacks will be controlled by strata of differing thicknesses and resistances to weathering and erosion as drainage developed in the area. The elevations along and across a hogback will be predictable because of these factors.

Differences in elevation on end moraines, however, are quite random and will be dependent on many factors. These are variations in sediment load discharged at the melting ice front, random deposition around blocks of ice along the ice front, outflow drainage channels developed during melting episodes, and downstream deposition of poorly graded (well-sorted) materials downstream with evidence of multiple stream channels. When one considers the topographic expressions of the many landforms in addition to those noted, that is, dunes, volcanic cones, terraces, and so on, it is obvious that the interpreter must have considerable experience in recognizing and correctly interpreting topography.

Drainage patterns are, in part, inseparable from topography, as both implicitly involve elevation changes and related topographic patterns.

However, topographic features need not be erosional in origin as in the case of depositional or altered landforms (fans, moraines, dunes, landslides, etc.). Drainage patterns are of considerable value in assessing soil characteristics and thickness and rock lithologies and structure. Six of the commonly occurring patterns are illustrated by Figure 7.13.

Dendritic, rectangular, trellis, and radial patterns are primarily due to underlying rock conditions. Dendritic patterns indicate relatively flat-lying rocks of uniform composition. Jointing in rocks with thin soil cover causes rectangular patterns. Dipping sedimentary rock strata of differing resistance to erosion are the cause of trellis drainage patterns. Conical structures such as volcanoes and cinder cones and roughly circular exposures of resistant crystalline rocks cause radial drainage patterns. Eroded structural domes in sedimentary rocks will have a combined pattern of modified trellis and radial drainage, a pattern classically displayed in the Black Hills

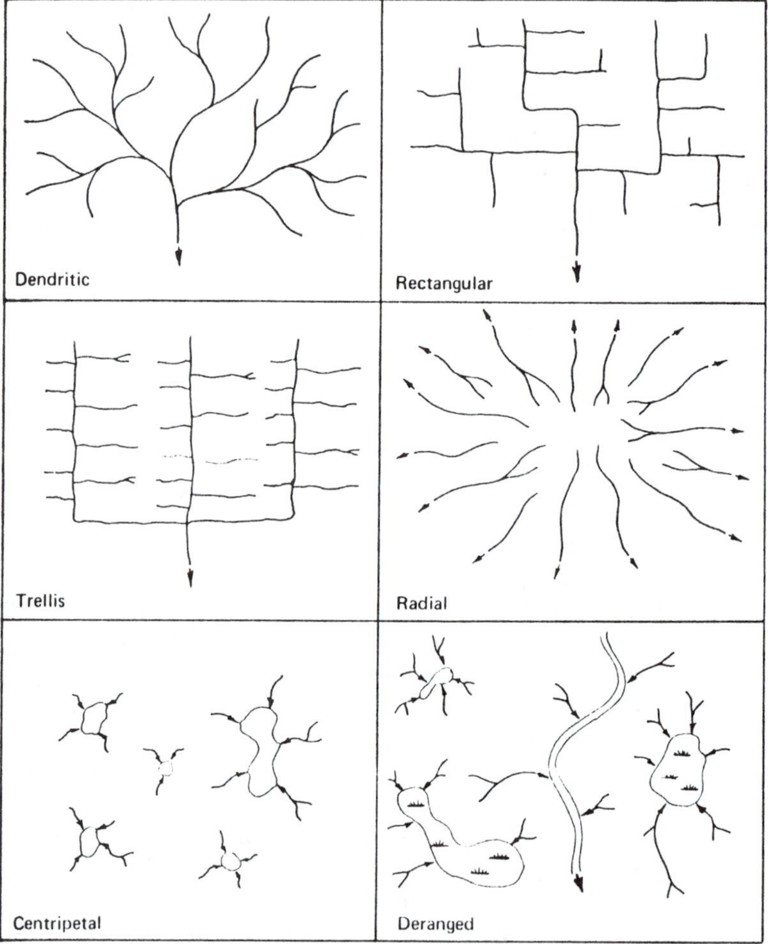

Figure 7.13 Six basic drainage patterns. (Reproduced with permission from Remote Sensing and Image Interpretation, T. M. Lillesand, and R. W. Kiefer, 1979. Copyright © John Wiley & Sons, Inc.)

of South Dakota. Centripetal drainage patterns may indicate the presence of sinkholes and karst conditions in the underlying sedimentary rock or internal drainage in thick, permeable soils that have a deranged drainage pattern. The deranged pattern, in turn, typically is produced by the disruption of original drainage by glacial deposition. Mass movements of soil and rock as in large landslides may cause the same disrupted pattern. The latter illustrates the need for knowledge of all processes and the ability of the interpreter to discern all visual characteristics for correct evaluation of an area.

The density of drainage patterns provides useful clues about local conditions. Impermeable soils normally have dense or closely spaced drainage channels. Permeable soils under similar circumstances have less dense, or widely spaced, channels. Before an assessment of soil permeability can be made, the interpreter must conclude that the soil thickness is sufficient to make the soil the controlling factor. This may be done under ideal conditions by confirming the absence of bedrock control on the drainage pattern, a condition implying thick soil cover. Geologic history of an area must be known before the densities of drainage patterns can be used confidently to imply differences in soil permeability. In areas subjected to continental glaciation, successive ice sheets during the Pleistocene Epoch may not have advanced as far as previous ones. In such cases, density of drainage is also controlled, in part, by the time available to reestablish drainage systems and by the subjection of older exposed surfaces to erosion by meltwater runoff. This is illustrated by Figure 7.14 of areas covered by older (dense pattern) and younger Wisconsinan (less-dense pattern) glacial soils.

Erosion as a photo element is intimately related to drainage pattern development and the evolution of erosional landforms (topography). Inasmuch as erosion plays such a major part in the previous two elements, its use in photointerpretation is often restricted to stereo examination of gully development. Assuming gully development to be restricted to easily eroded material, that is, soils or rocks such as shales having soil-like properties, interpretation makes use primarily of gully cross sections. Smoothly curved cross sections may develop in clays. Soils composed of silt may exhibit a U-shaped cross section and V-shaped cross sections are indicative of sands and gravels. As in all cases where interpretive guidelines have been developed, there will be exceptions that require experience and knowledge of geologic history and the influence of climatic factors. An example is the development of near-vertical walled, flat-floored channels called *arroyos* that develop in arid and semiarid areas as a result of episodic, relatively high-intensity storms. In western Colorado and eastern Utah, large areas underlain by shale have extensive arroyos. The more typical curved cross sections are developed in similar materials in the Dakotas under different climatic conditions.

Vegetation as a photo element is closely associated with climatic conditions and soil thickness and type, especially where not influenced by land-use practices. Although trees are not exclusive indicators of the above conditions, they provide a means for describing the use of vegetation in photointerpretation. Thick tree cover normally indicates sufficient water

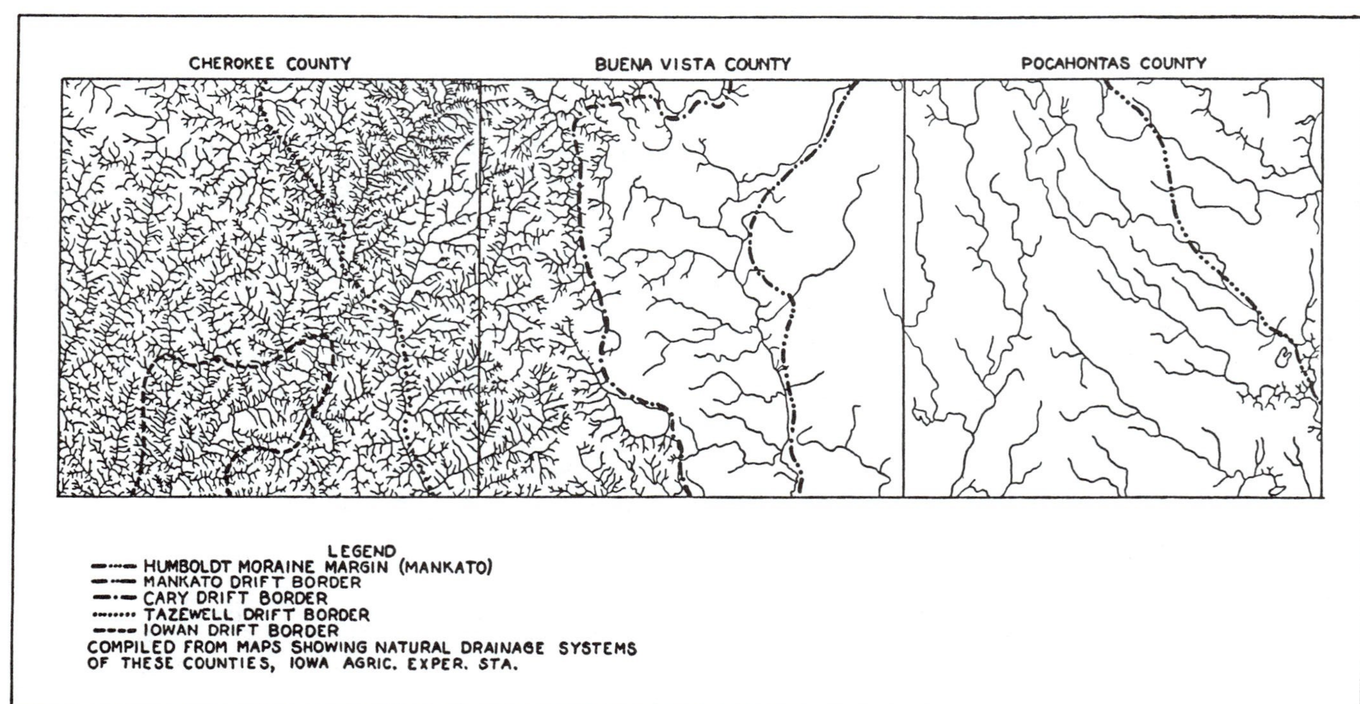

Figure 7.14 Drainage density difference between Early (Tazewell Substage) and Middle (Cary Substage) Wisconsinan drift in northwestern Iowa. (Reproduced with permission from R. V. Ruhe and Am. J. Sci., Vol. 250, 1952.)

and optimum soil type and thickness needed for growth. The dense cover of deciduous trees and/or conifers in many of the eastern, southeastern, and northwestern states is detrimental to detailed interpretation of geology. With reductions in available moisture and/or ideal soil types and thickness, natural as well as human-controlled patterns of tree cover, kinds of trees, and density of tree cover can be useful in interpretation. Patterns and density in general will be governed by available soil moisture, as shown by concentrations of trees along drainage ways in the Great Lakes and Great Plains states. Certain deciduous trees such as oaks prefer well-drained, sandy soils and may be used, for instance, in areas with soils deposited in conjunction with continental glaciation to define soil boundaries. Late fall and early winter photography may permit identification of oaks as they tend to retain their leaves longer than other trees. Orchards, identified by their ordered patterns, have similar soil and moisture requirements and interpretive use.

In the semiarid western states, slope aspect is an additional control on tree distribution. North-facing slopes retain moisture longer than the south-facing slopes that receive more sunlight. Although elevation is an additional control over the type of trees found, it is possible to make preliminary interpretations of moisture and soil thickness based on cover density and tree type. Unless an area is exceptionally dry, thin soils will be the cause of sparce tree cover. The speckled appearance caused by piñon-juniper tree cover in many western states is indicative of dry soil conditions as well as thin soil. The presence of deciduous cottonwoods and aspen in such an area defines the limits of higher soil moisture and thickness. Large, widely spaced ponderosa pines grow on slopes with more moisture than the piñon-juniper stands, but less than that required by spruce, lodgepole pine, aspen, firs, and cottonwoods. It is common to find slope aspect controlling the distribution of ponderosa pines on dryer south-facing slopes and mixtures of the other trees on the more moist north-facing slopes. The relatively easily recognized tall, conical spruce and deciduous aspen are useful in outlining areas of higher soil moisture.

In all cases in all areas, localization of vegetation, including tree cover and variations in vegetation type and density, are subject to local conditions. Although generalities provide a basis for interpretation of surficial conditions, interpreters must become familiar with the controls exerted locally by careful comparison of all vegetation types and distribution shown on photos and actual site conditions (ground truth).

Land use is an element based on human activities. As with the other elements, the kinds of land use are numerous and often are restricted regionally by geology and climate. Again, it is imperative that a photointerpreter is cognizant of both the land-use practices in the area being investigated and the visual characteristics of such practices. Several examples that involve interpretation of soil and rock conditions will serve to illustrate the diversity and interpretive aspects of land-use practice.

The presence of rock quarries may indicate rock types useful as crushed stone or aggregate, as raw materials for manufacture of Portland cement, or as dimension stone. Gravel pits and associated processing equipment indicate the presence of poorly graded (well-sorted) materials that probably pass aggregate soundness requirements as well as deposits of sufficient thickness to warrant the expense of setting up an extraction plant. Roughly rectangular excavations without obvious evidence of quarrying activity may have been used for fill or borrow material. The excavations are called borrow pits and, though they may not identify the soil gradation, they at least are evidence of the minimum thicknesses of soil cover. If filled with water, borrow pits are excellent indicators of the depth to the water table. Some gravel pits may be similarly useful, but caution must be exercised as pumping may reduce the water level to facilitate extraction of sand and/or gravel. Drainage ditches or soil moisture patterns showing the presence of buried drainage tile reveal the need for lowering a high water table for a given land use.

The value to engineering geology of such interpretations—and the many others that involve road cuts in soil and rock, agricultural practice, and so on—is dependent on the requirements of each individual case. Photointerpretation of land-use practice, as with other elements, provides the basis for more efficient and economical on-site investigation.

Photo tone and color are elements of photointerpretation that are associated with black-and-white and color photography, respectively. In general, a user of air photos is better acquainted with color differences and intensities related to surficial conditions than with black-and-white tone. Tonal differences usually are variations in shades of gray from black to white. They appear on black-and-white photos as a result of the reflective characteristics of surface materials. Interpretive applications are mainly concerned with variations in moisture content of soils. Dryer soils tend to reflect sunlight more than wet soils. Thus, both soil gradation and depth to the water table may interact to influence photo tones. Stereo vision permits the interpreter to identify localized areas of higher elevation that could be misinterpreted as changes in soil permeability rather than topographically controlled differential moisture content in a uniform soil gradation. The finer soils tend to drain less readily under such conditions, so that the sharpness or distinctness of the tonal boundaries may be useful in interpreting soil properties.

Photo tones are the basis for identifying buried drainage tiles, as noted previously under the land-use element. They also may be used to identify poorly drained areas that contain high organic content, a condition that causes minimal reflection or a dark gray to black tone on photos. The presence of a surface layer of permeable soil over impermeable soil or rock may cause subsurface or "phantom" drainage patterns that influence tonal variations. The time of day and, more important, the time of year may influence soil moisture. Tonal variations may be absent during times of high soil moisture. Experience with local conditions, both natural and manmade, must be the basis for correct interpretation of tones and their patterns.

Recognition of the various photo elements and understanding their application to engineering geology are of considerable importance. Summaries of photo elements, soil types, and engineering characteristics for different rock types and climatic conditions are given in Tables 7.3 through 7.5. Preliminary site evaluation may be made concerning such factors as

Table 7.3 Characteristic Topographic Features and Associated Inferences for Igneous Rocks

From Aerial Photos and Geologic Maps					Inferences			
Landform Climate	Topography	Drainage & Texture	Photo Tone	Gully Type	Soil Texture	Soil Drainage	Land Use	Engineering Characteristics
Granite								
Humid and intrusive	Bold and domelike	Dendritic Medium	Light (uniform)	Variable	Silty sand	Poor	Agriculture Forestry	Excavation difficult Poor aggregate Good base
Arid and intrusive	A-shaped hills	Dendritic Fine or internal	Light (banded)	None	Fine	Poor	Barren (rangeland)	
Basalt								
Extrusive flows	Flat to hilly	Parallel or internal	Dark (spotted)	None	Clay to rock	Good	Agriculture to barren	Blasting not difficult. Landslides common. Soil
Volcanic								
Extrusive	Cinder cone	Radial C to F	Dark	Variable	Silty clay	Poor (surface) Good (subsurface)	Barren	
Fragmented								
Tuff	Sharp-ridged hills (variable height)	Dendritic Fine	Light	[V-shaped gully profile symbol]	Noncoh. sand to dust	Excellent	Forestry or Grass	Blasting not required. Unstable soil. Septic systems easily contami-nated
Imbedded								
Flows	Terraced hills	Parallel dendritic	Light and dark (banded)	Variable	Variable	Variable	Agriculture to barren	Unstable soil

Source: Reprinted with permission from D. P. Paine, Aerial Photography and Image Interpretation for Resource Management, 1981, John Wiley & Sons, Inc., as adapted by permission from D. S. Way, Terrain Analysis, a Guide to Site Selection Using Aerial Photographic Interpretation, 1973, Van Nostrand Reinhold Co., Inc. (Hutchinson Ross Publishing Co.).

Table 7.4 Characteristic Topographic Features and Associated Inferences for Sedimentary Rocks

From Aerial Photos and Geologic Maps					Inferences			
Landform Climate	Topography	Drainage & Texture	Photo Tone	Gully Type	Soil Texture	Soil Drainage	Land Use	Engineering Characteristics
Shale								
Humid	Rounded hills	Dendritic Med.-fine	Light (mottled)		Fine Silt-clay	Poor	Agriculture Forestry	Excellent base Poor septic systems
Arid	Rough-steep	Dendritic Fine	Light (banded)		Med. Silty	Very poor	Barren	Excavation difficult. Poor aggregate
Sandstone								
Humid	Massive and steep	Dendritic Coarse	Light		Sandy	Excellent	Forestry	Excellent base Shallow to bedrock
Arid	Flat table	Dendritic Med.-coarse	Light (banded)	None	Fine	Poor	Barren	Poor septic systems
Limestone								
Humid	Flat or rough sinkholes	Internal	Mottled		Silt-clay	Poor to good	Agriculture	Shallow soil Poor septic systems
Arid	Flat table	Dendritic	Light	None	Fine	Poor	Barren	Poor septic systems
Interbedded								
Flat (humid)	Terraced	Dendritic Med.-coarse	Medium (banded)	Variable	Variable	Variable	Some agric. Some forest	Good base Variable to bedrock
Tilted (humid)	Parallel ridges	Trellis Med.	Medium (banded)		Fine	Fair	Agriculture Forestry	Good base Excavation difficult

Source: Reprinted with permission from D. P. Paine, Aerial Photography and Image Interpretation for Resource Management, 1981, John Wiley & Sons, Inc., as adapted by permission from D. S. Way, Terrain Analysis, a Guide to Site Selection Using Aerial Photographic Interpretation, 1973, Van Nostrand Reinhold Co., Inc. (Hutchinson Ross Publishing Co.).

Table 7.5 Characteristic Topographic Features and Associated Inferences for Metamorphic Rocks

From Aerial Photos and Geologic Maps					Inferences			
Landform Climate	Topography	Drainage & Texture	Photo Tone	Gully Type	Soil Texture	Soil Drainage	Land Use	Engineering Characteristics
Slate								
Humid and arid	Many small sharp ridges of same height	Rectangular Fine	Light		Coh. Fine		Unproductive natural veg. thin soil	Excavation difficult Rock slides
Schist								
Humid	Steep rounded hills	Rectangular Med.-fine	Light (uniform)		Mod. coh. sand-clay	Good to poor	Cultivated Forested	Seepage problems. Poor septic systems
Arid	Rugged	Rectangular Fine	Light (banded)		Mod. coh. sand-clay	Good	Thin soil (grass and scrub)	Little excavation needed
Gneiss								
Humid and Arid	Parallel, steep, sharp ridges	Ang. dend. Med.-fine	Light (uniform)		Mod. coh. sand-silt	Fair	Natural (forested, grass or scrub)	Much blasting Fair aggregate

Source: Reprinted with permission from D. P. Paine, Aerial Photography and Image Interpretation for Resource Management, 1981, John Wiley & Sons, Inc., as adapted by permission from D. S. Way, Terrain Analysis, a Guide to Site Selection Using Aerial Photographic Interpretation, 1973, Van Nostrand Reinhold Co., Inc. (Hutchinson Ross Publishing Co.).

rock type and structure, physical properties and thickness of soil, slope stability, and design of detailed site investigations following careful recognition and interpretation of photo elements.

Color and Color IR Photography

The availability of color prints and transparencies has added an important dimension to photo interpretation. Although the human eye can distinguish about two hundred gradations on the gray scale, some twenty thousand spectral hues and intensities can be recognized in color (Barrett and Curtis, 1976). As a result, one can detect more easily on color photos the subtle differences in the colors of soil and rock, the variations in soil moisture, and the related differences in vegetation type, distribution, and vigor that is needed for correct interpretation of conditions for engineering applications. Rib (1975) has reported that, for the same scale, color photography permits greater discrimination of small features and objects when compared with black-and-white photos. In addition, improved interpretation accuracy and increased speed of analysis are benefits of natural color photography.

Color IR photography is responsive to a different spectral range than natural color photography. Rather than the three visible primary additive colors (blue, green, and red) used in color photography, color IR responds to green, red, and near infrared (Figure 7.10). The green-sensitive layer develops to a yellow-positive image, whereas red produces magenta and infrared appears as cyan. Thus, the colors produced have resulted in the term *false color* for color IR photography. Healthy vegetation appears as various shades of red, with different kinds of vegetation having identifiable color signatures. Water and wet areas appear as blue and blue-gray.

Differences in soil moisture are more easily seen on color IR photos than with natural color photographs. Moisture differences are noted in exposed soil as well as in changes in vegetation type and vigor. Color IR photos also permit better recognition of boundaries between soil-covered and exposed rock areas. Table 7.6 summarizes the signatures of selected differences in vegetation and surface conditions for color and color IR film. Although the term *photo* has been used in our discussion, it should be noted that best interpretation results are obtained when positive transparencies are used in color and color IR photography. By comparison, positive prints are best when using black-and-white photography.

There is general agreement that for engineering-geologic applications of photointerpretation, natural color photography is the best single type of imagery to be found. If a combination of types of photography can be justified economically, natural color and color IR transparencies provide optimal information (Rib and Liang, 1978).

Other Imagery

Although aerial photography utilizing black-and-white, color, and color IR film is versatile and economical, there are limitations in its use that have resulted in the development of a variety of more sophisticated remote-sensing methods. As noted earlier (Figures 7.10 and 7.11), the photo-

Table 7.6 Terrain Signatures on Normal Color and Color IR Film

Subject	Signature on Normal Color Film	Signature on Color IR Film
Healthy vegetation:		
Broadleaf type	Green	Red to magenta
Needle-leaf type	Green	Reddish brown to purple
Stressed vegetation:		
Previsual stage	Green	Darker red
Visual stage	Yellowish green	Cyan
Autumn leaves	Red to yellow	Yellow to white
Clear water	Blue-green	Dark blue to black
Silty water	Light green	Light blue
Damp ground	Slightly darker	Distinct dark tones
Shadows	Blue with details visible	Black with few details visible
Water penetration	Good	Green and red bands; good. IR band: poor
Contacts between land and water	Poor to fair discrimination	Excellent discrimination
Red bed outcrops	Red	Yellow

Source: From Remote Sensing—Principles and Interpretation by F. F. Sabins, Jr., W. H. Freeman & Co. Copyright © 1978.

graphic methods described have limited spectral coverage. Images are composites of wavelengths reflecting from the earth's surface. The viewing of scenes that have limited wavelength ranges for optimum imaging, as in the case of deciduous versus coniferous trees (Figure 7.11), benefits from systems that record such limited wavelength ranges. Cloud cover may seriously limit photography. Thermal emission differences among surface materials and associated moisture contents may be masked by surface reflectance on the photographic films described earlier.

The MSS, thermal, and microwave (radar) systems have been developed to improve imaging and interpretive capabilities. All obtain image data by scanning the surface perpendicularly to the flight line. The MSS and thermal sensors receive energy from the surface (passive). Radar systems transmit energy to the surface for reflection back to the sensors (active). None of these systems substitutes for the interpretive advantages and greater detail offered by stereo aerial photography. However, the additional interpretation capability gained for engineering-geologic applications may outweigh the additional costs of procurement, data preparation, and more specialized personnel and equipment. Although multispectral and thermal data may be recorded on film, greater versatility in image production and enhancement is gained by digital recording of scene data. Photographic imagery may be digitized for data processing, however. Only digital or scanning techniques will be described as there are no conceptual differences involved.

Multispectral Scanning Imagery

As seen on Figure 7.10, MSS encompasses the wavelength ranges of photography and thermal scanners. As the name implies, more than one spectral range, or band, is simultaneously recorded by scanning procedures illustrated by Figure 7.15. Landsat-1 and -2 satellites have been the source

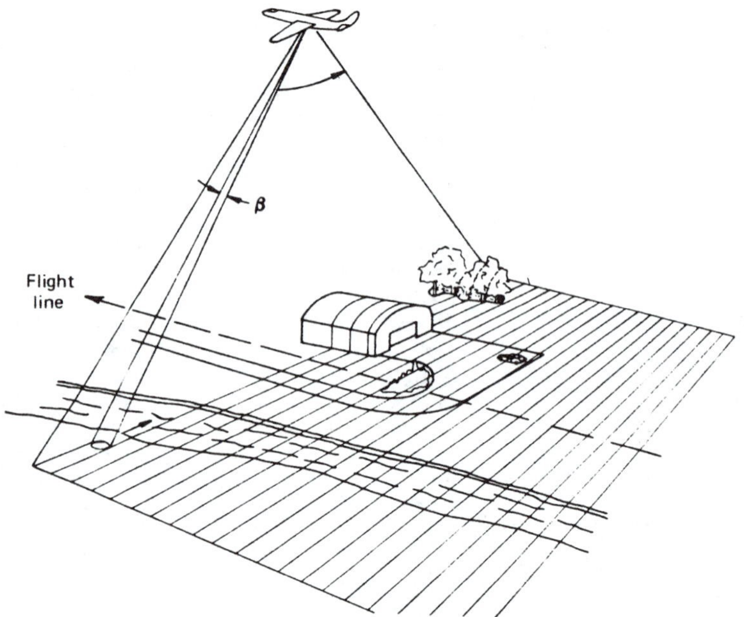

Figure 7.15 Scanning pattern employed in airborne MSS system. (Reproduced with permission from Remote Sensing and Image Interpretation, T. M. Lillesand and R. W. Kiefer, 1979. Copyright © John Wiley & Sons, Inc.)

of most MSS data available since 1972. Landsat has four spectral bands as shown on Table 7.7. Custom multispectral scanning with as many as 24 bands may be obtained using aircraft (Lowe, 1980). Scale and spectral ranges will differ from those of Landsat imagery. Landsat MSS data have the advantage of an 18-day cycle at a constant sun angle that permits monitoring of the surface under optimum conditions. A major drawback of Landsat imagery for detailed work is the approximately 80-m resolution in individual scenes covering approximately 10,000 mi^2. Landsat-D, with a seven-band sensing system, called a *thematic mapper*, was put in operation during 1982. Six of its bands have a resolution of 30 m, with the seventh band a thermal channel, being 120 m. Four of the MSS bands have spectral ranges similar to those in Table 7.7. Two additional bands provide improved recognition of vegetation and rock composition.

Landsat MSS band data may be obtained in several forms. Black-and-white prints and positive films for each channel can be obtained to examine spectral signatures and reflectances of various surface materials and objects. Differentiation of soil and rock types and degree of rock alteration is made by comparing tonal variations on different spectral bands although reflectance from vegetation may mask surface materials (Siegal and Goetz, 1977; Abrams, 1980; Siegrist and Schnetzler, 1980). Mapping of four soil types in one scene by use of airborne MSS data is illustrated by Figure 7.16.

The best directly available Landsat data come in the form of color-composite images. These are analagous to color IR photography and are made

Table 7.7 Landsat-1 and -2 MSS Bands and Related Information

Band	Color or Spectral Range	Wavelength (μm)
4	green	0.5–0.6
5	red	0.6–0.7
6	near IR	0.7–0.8
7	near IR	0.8–1.1

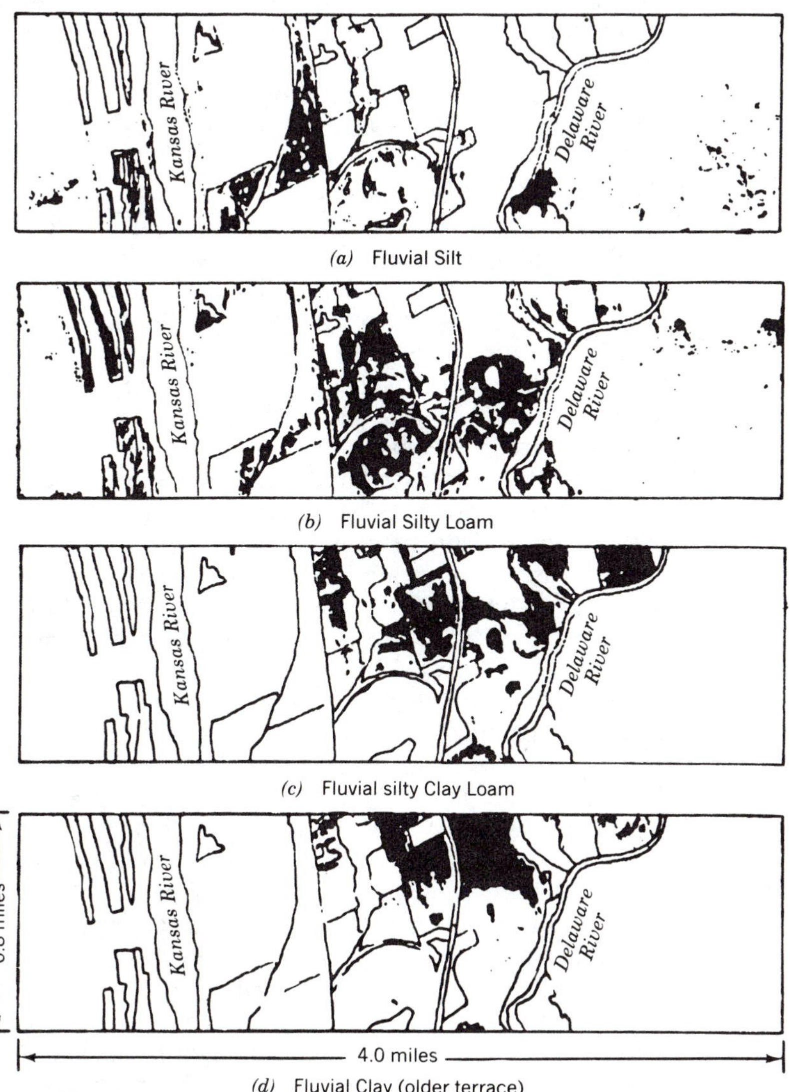

Figure 7.16 Multispectral soils recognition for Kansas River flood plain. (From Wagner, 1972)

from channels 4, 5, and 7. Positive color IR transparencies also may be made from black-and-white positives by use of filters.

Digital tapes of the four Landsat channels also may be obtained. Of the types of image data available, digital imagery provides the greatest flexibility through the playback capabilities of tape and the variety of enhancement techniques that may be applied. Images that enhance specific surface characteristics may be made by combining or ratioing selected band data (Lillisand and Kiefer, 1979).

In addition to the four spectral bands, Landsat imagery has numerous advantages. The broad overview permits better recognition of major tectonic features such as lineaments. Small-scale viewing of drainage patterns, major landforms, vegetation, and land use may improve the interpretation of structure and soil and rock types made from larger-scale aerial photography. Landsat images also are almost distortion free as opposed to larger-scale, lower-altitude sensing systems. They have the characteristics of orthophotographs. More-detailed information on Landsat satellites and their MSS imagery may be obtained from Sabins (1978) and Paine (1981).

Thermal Infrared Imagery

The thermal IR spectral wavelength range from 3 m to 14 m is an extension of the infrared portion of the spectrum utilized by color IR photography (Figure 7.10). Although this portion of the electromagnetic spectrum is included in some multispectral operations, it may be imaged independently. When this is done, the result is known as thermal or thermal IR imagery. The systems of greatest value utilize scanners (as with MSS systems) that similarly respond to energy being reflected or emitted from the earth's surface (passive).

The thermal wavelength range contains energy emitted or radiated from the earth, which previously has stored energy from the sun. Emissivity is a measure of the radiating efficiency of a surface. The tendency of a material to transfer as well as retain heat is dependent on its thermal properties, which are conductivity, specific heat, and density. These may be used to derive additional thermal properties such as thermal inertia and thermal diffusivity (Table 7.8). Of these, thermal inertia is of special interest as it is a measure of thermal response of material to temperature change, permitting differentiation of various surface materials.

All objects above 0°K (−273°C) emit electromagnetic radiation. It is fortuitous that the 8–14-m portion of the spectrum is both the best for geologic purposes and a window (Figure 7.10) permitting transmission through the atmosphere. Nighttime (predawn) thermal imagery eliminates reflected IR from sunlight and provides a period of stable temperatures that reduces the influence of terrain and slope aspect.

Thermal IR images do not have the resolution of photographic images for a given scale. Thus thermal data is used specifically to improve imaging of certain surface conditions that are not represented well by photography. An example is material with similar reflectance characteristics but different thermal properties. The principal uses of thermal IR are for mapping changes in lithology, soil composition, and anomalies in groundwater flow in conjunction with air photo interpretation. In each case, the various

Table 7.8 Thermal Properties of Common Rocks and Soils (Handbook Values)[a,b]

Rock/Soil Type	K	ρ	C	k	P
Basalt	0.0050	2.8	0.20	0.009	0.053
Clay soil (moist)	0.0030	1.7	0.35	0.005	0.042
Dolomite	0.012	2.6	0.18	0.026	0.075
Gabbro	0.0060	3.0	0.17	0.012	0.055
Granitic rocks	0.0075	2.6	0.16	0.016	0.052
	0.0065				
Gravel	0.0030	2.0	0.18	0.008	0.033
Limestone	0.0048	2.5	0.17	0.011	0.045
Marble	0.0055	2.7	0.21	0.010	0.056
Obsidian	0.0030	2.4	0.17	0.007	0.035
Peridotite	0.0110	3.2	0.20	0.017	0.084
Pumice, loose (dry)	0.0006	1.0	0.16	0.004	0.009
Quartzite	0.0120	2.7	0.17	0.026	0.074
Rhyolite	0.0055	2.5	0.16	0.014	0.047
Sandy gravel	0.0060	2.1	0.20	0.014	0.050
Sandy soil	0.0014	1.8	0.24	0.003	0.024
Sandstone, quartz	0.0120	2.5	0.19	0.013	0.054
	0.0062				
Serpentine	0.0063	2.4	0.23	0.013	0.063
	0.0072				
Shale	0.0042	2.3	0.17	0.008	0.034
	0.0030				
Slate	0.0050	2.8	0.17	0.011	0.049
Syenite	0.0077	2.2	0.23	0.009	0.047
	0.0044				
Tuff, welded	0.0028	1.8	0.20	0.008	0.032

[a] Values are given in cgs units for 20°C.
[b] K = thermal conductivity; ρ = density; C = specific heat; k = thermal diffusivity, and P = thermal inertia.

Source: From Janza, 1975, reproduced with permission from Manual of Remote Sensing, 1st Ed. Copyright © 1975 by the American Society of Photogrammetry.

material thermal inertias provide differences in emissivity, which are recorded.

Engineering-geologic applications include: improved delineation of fault boundaries, soil-rock boundaries and associated seepage zones, mapping near-surface drainage patterns, noting variations in rock weathering depths, and identifying potential sinkhole collapse sites. Rib and Liang (1978), Sabins (1978), Lillesand and Kiefer (1979), Warwick et al. (1979), and Paine (1981) should be consulted for more detailed information concerning thermal IR theory and engineering geologic applications.

Microwave (Radar) Imagery

Imaging of the earth's surface also may be conducted in the microwave range of the electromagnetic spectrum (Figure 7.10). Microwave imaging includes active and passive systems. The latter is an extension of thermal IR mapping into the microwave wavelengths, which is useful in measuring soil moisture and mapping karst and mined areas (Dedman and Culver,

1972; Estes et al., 1977). Greater utilization of microwave imaging of surface features is found in the active mode in which short bursts or pulses of electromagnetic energy are transmitted to the surface and then received as reflected energy. This active system is commonly refered to as radar. Its inherent distance-ranging characteristics, which have been responsible for its development for military purposes, are ideal for scientific applications.

Radar developed for imaging purposes scans the earth's surface by electronically viewing a swath of surface laterally from an aircraft as shown in Figure 7.17. The forward movement of the aircraft provides the second dimension of the image similar to that obtained from the mechanical scanning systems described previously. The technique is called side-looking airborne radar (SLAR) because of the less than 90° viewing angle of the surface. Overlapping images along the flight line, or azimuth, direction may permit stereo viewing of surface features.

Terrain features are illuminated by the energy pulses from the airborne transmitter. The intensity of the returned energy is dependent, in part, on the orientation of the reflecting surface to the transmitting antenna and surface characteristics. The SLAR scenes are photographed sequentially as they appear on the instrument's cathode-ray tube. A mountain front parallel to the flight line and relatively perpendicular to the pulse propagation direction will provide strong energy return or high image intensity. Bodies of water and other smooth, relatively horizontal surfaces will reflect the energy away from the receiving antenna, resulting in a dark image. Variations in reflecting angles, surface roughness, and vegetative cover influence the return intensities between the extremes noted. Surface objects that extend above the surface cast radar shadows, the sizes of which are dependent on the incidence angle, all other factors being equal (Figure 7.18). The shadows emphasize surface topography and structural trends.

Energy from the transmitter is polarized in either the horizontal (H) or vertical (V) plane. Most of the returned energy will be in the same plane. If the receiving antenna is in the same plane as the transmitter, the SLAR images are like polarized, that is, HH or VV. Cross-polarization to detect surface modification or polarization may be obtained by orienting the

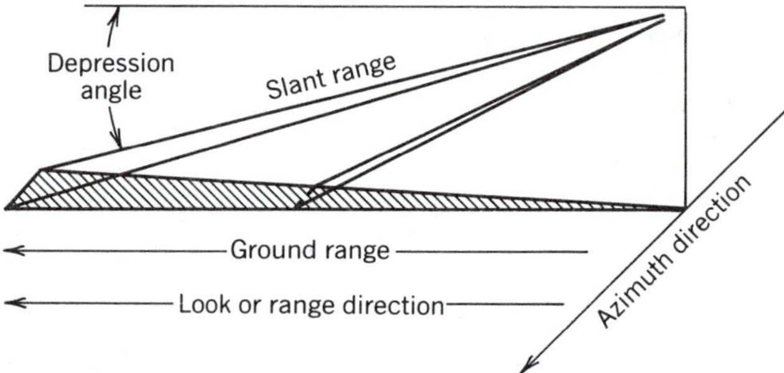

Figure 7.17 Schematic diagram of SLAR imaging pattern. (Reproduced with permission from Aerial Photography and Image Interpretation, D. P. Paine, 1981. Copyright © John Wiley & Sons, Inc.)

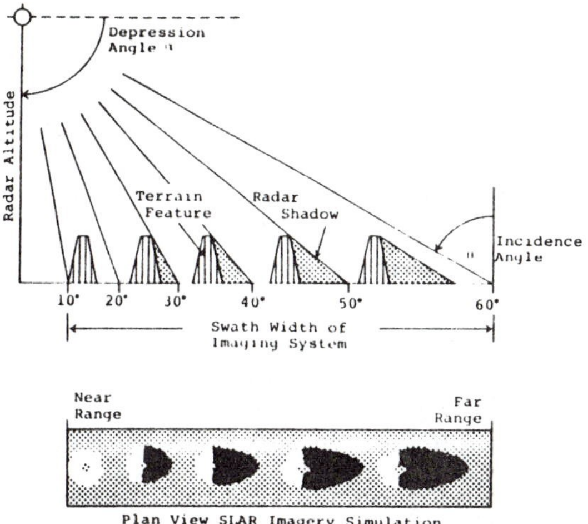

Figure 7.18 Shadowing characteristics of SLAR-imaging systems. (From Dellwig and Burchell, 1972)

receiving antenna perpendicular to the transmitting antenna (HV or VH). Cross-polarization may accentuate differences in surface moisture and vegetation, but it has limited geologic value.

Typically, SLAR imagery is complementary to aerial photography because of lower resolution and small scale (usually <1 : 125,000). In spite of both of these seeming handicaps, SLAR imagery is of great value in identifying surface structural and drainage features, patterns, and trends that might escape notice on air photos because of the enhancement of such terrain features by low-angle, oblique illumination. The loss in observable details compared with aerial photographs has the advantage of emphasizing major features that may be obscured by more surface detail.

By far the greatest advantage of SLAR is the penetration of cloud cover by certain wavelengths in the microwave range. Areas where clouds obscure the surface for extended periods of time may be imaged with photographic quality. Also, SLAR may be obtained at night, a factor of considerable value at the higher latitudes during winter months.

Additional information on SLAR principles, instrumentation, interpretation, and engineering geology case histories may be obtained from Lillesand and Kiefer (1979), McEldowney and Pascucci (1979), and MacDonald (1980).

Enhancement of Remote-Sensing Imagery

Optimal displaying of selected information from all types of images may be obtained by various methods of image enhancement. Most enhancement techniques require a computer and digitized image data. Image information contained on photographs (optical or analog images) may be enhanced directly from the film or may be converted to digital form by a variety of methods that are beyond the scope of this text. Following data manipulation, all digital imagery may be converted to optical or analog form for interpretation.

Enhancement techniques are too numerous to be treated in detail here. However, several types will illustrate applications to engineering geology. Images inherently have different densities related to reflectance levels from surface features. An image may be divided, or sliced, into arbitrarily chosen density ranges. Each slice may be assigned and displayed in a separate color chosen to enhance or alter its visual impact on the viewer. One application of this density-slicing-enhancement technique has been to improve the interpreter's ability to identify the boundaries of landslide masses, especially where vegetative cover obscures the view. Variations in soil-moisture distribution by slope failures and reflectance from vegetation abnormally stressed by movement or moisture anomalies may be identified, as shown in Figure 7.19.

The MSS images are especially amenable to enhancement because of the multiple images that are analogous to those obtained by density slicing. The digital data for each channel may be enhanced by adjusting, or stretching, the reflectance contrasts. Also enhanced or nonenhanced channels may be superimposed and assigned colors, as in density slicing. Various pairs of spectral bands or channels may be combined or ratioed to enhance desirable features.

Spectral ratioing has engineering-geologic applications, in that visually subtle differences in reflectivity and emissivity from different rock and soil types and altered versus unaltered rock masses may be enhanced by combining selected spectral bands. Further enhancement of desired features may be obtained by making color composites of ratioed images by using one or more of the contrast-stretching techniques. Terrain-caused differences in illumination that can obscure lithologic contacts may be removed by ratioing as illustrated by Figure 7.20.

The following references provide details of enhancement techniques and engineering geologic applications; McKean et al., 1977; Sabins, 1978; Lillesand and Kiefer, 1979; Abrams, 1980; Gillespie, 1980; and Zall and Michael, 1980.

Terrestrial Photogrammetry

As defined earlier, terrestrial photogrammetry utilizes images obtained by surface instruments. Horizontal rather than vertical photography is used to obtain stereo photographs of vertical or near-vertical surfaces. In engineering-geologic applications, the photos are used typically to map discontinuities exposed on rock surfaces of open-pit mines, dam abutments, and similar surfaces. Repetitive photography also may be used to monitor and document small- and large-scale movements of soil and rock slopes.

Terrestrial photogrammetry by definition involves measurements of features shown on the photos such as orientation of exposed discontinuities and their spacings as well as surface roughness. The advantages of obtaining accurate quantitative data for an inaccessible yet potentially unstable rock wall, as in open-pit mines, are obvious. Specialized equipment such as

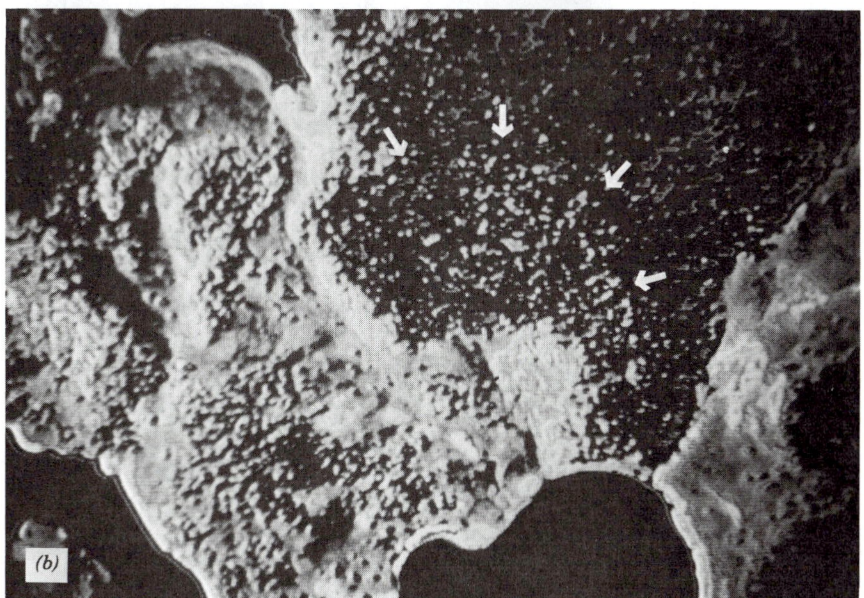

Figure 7.19 Application of density slicing to reveal potential landslide hazards in tree-covered (lodgepole pine) slopes. (*a*) Black-and-white reproduction of color IR photo showing prominent landslide outlined by arrows. (*b*) Selected density slice showing the outline resulting from stress on trees of a possible landslide (additional arrows). (From McKean et al., 1977)

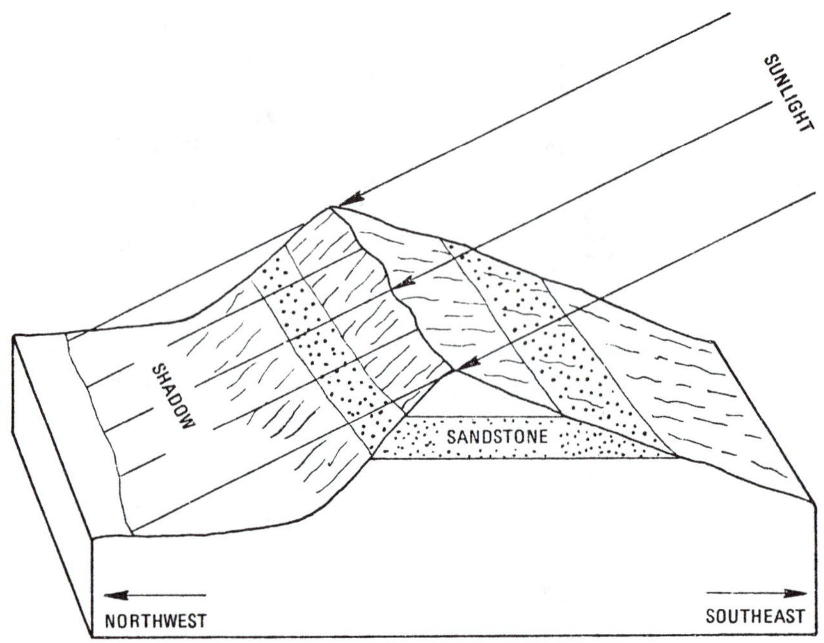

SANDSTONE REFLECTANCE

ILLUMINATION	BAND 4	BAND 5	RATIO 4/5
Sunlight	28	42	0.66
Shadow	22	34	0.65

Figure 7.20 Removal of illumination differences by band ratioing. (From Remote Sensing—Principles and Interpretation by F. F. Sabins, Jr., W. H. Freeman & Co. Copyright © 1978.)

the phototheodolite is required for obtaining overlapping stereoscopic images of a surface (Figure 7.21). Standardization has been developed for optimal stereo viewing of a surface. It has been suggested (Geological Society, 1977) that two overlapping photographs of the same surface be taken from positions parallel to the surface and separated by 5% of the distance to the surface.

Stereoscopic plotting instruments are needed for producing an image for measurement purposes, making use of targets painted or positioned on the surface. This equipment and the need for the services of a trained photogrammetrist often preclude the extensive use of terrestrial photogrammetry in engineering geology. Photogrammetric measurements usually are not required for photointerpretation, simple mapping, or the monitoring of surface features. The reader interested in pursuing this subject and related engineering applications is referred to the following representative litera-

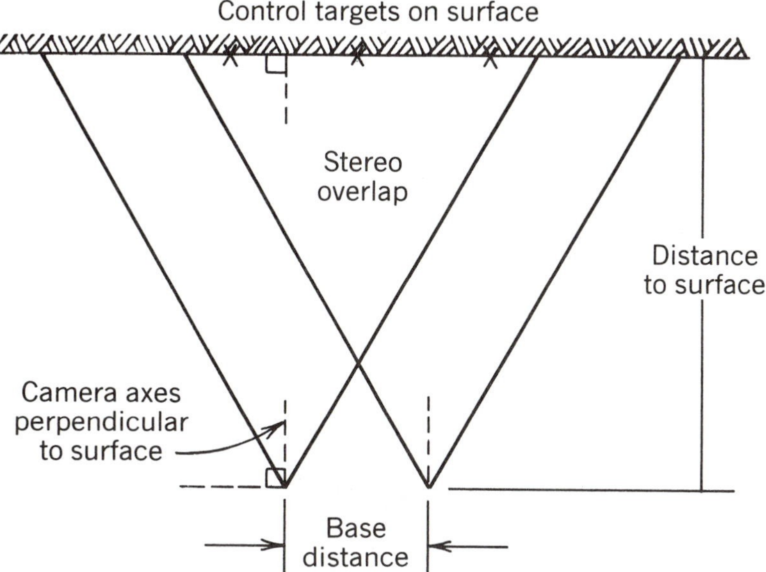

Figure 7.21 Camera positions for obtaining stereopair of terrestrial photos.

ture: Rengers, 1967; Wickens and Barton, 1971; Ross-Brown and Atkinson, 1972; and Ross-Brown et al., 1973.

The engineering geologist may find that less rigorous applications of the same principles used in terrestrial photogrammetry may be of value. A handheld or tripod-mounted camera may be used to obtain stereopairs for determination of discontinuity patterns, division of an exposed rock surface into structural zones, and monitoring of changes occurring over time. Camera positions are similar to those shown in Figure 7.21. In this simple application of terrestrial photogrammetry, it is not possible to make accurate measurements of features or to transfer data obtained to maps of exposed surfaces. Thus, technically it is not a photogrammetric procedure.

Simple forms of true terrestrial photogrammetry using handheld cameras have been described by Williams (1969) and Goodman (1976). With data such as camera-lens focal length, measured distances from the camera to the surface, and horizontal and vertical angles to identifiable features or targets that are obtained by use of a Brunton compass, it is possible to construct working maps or sketches of surface features.

SUBSURFACE EXPLORATION

Exploration in engineering geologic studies frequently includes subsurface investigation. This exploration defines subsurface conditions that influence the siting, design, and performance of a project. Surface investigation is often the starting point for subsurface exploration. Identification of fault traces, the need to know the thickness of soils with certain characteristics, the lack of exposures permitting identification of underlying bedrock, and uncertainty about the relation of springs and seeps to the regional water table are typical situations arising during surface investigation that define the need for subsurface exploration.

Subsurface exploration identifies underground features and conditions, characterizes their physical properties, and delineates their vertical and lateral extent. Both direct and indirect means are used to accomplish this work. Direct means include exploratory excavation and borehole exploration. Geophysical methods are the main form of indirect exploration.

Exploratory Excavation

Table 7.9 describes the more common types of exploratory excavation and their capabilities and limitations. Exploratory excavation provides a means for both sampling subsurface material and mapping subsurface features and conditions. Sampling from exploratory excavation differs little from sampling of surface exposures (USBR, 1974). Mapping the lateral extent of subsurface conditions is also little different from surface mapping. However, mapping the vertical extent of subsurface conditions involves logging to represent these findings.

Exploratory excavation may provide sufficient subsurface data so that drilling or geophysical methods need not be used. More often, exploratory excavation is one of several types of subsurface methods employed. The area of consideration in planning exploratory excavation will be dictated by the results of surface exploration and the needs of the particular project. Lund and Euge (1984) illustrate how the needs of the project influence subsurface exploration work. Their investigation for the Palo Verde Nuclear Generating Station addressed subsurface conditions for both Category I and other-than Category I structures. Category I structures are those critical to plant operation, especially the safe shutdown of the reactors. Category I excavations required mapping of the walls and floors at a scale of 1 in. to 10 ft, with more detailed mapping at 1 in. to 1 ft in areas requiring special attention. Other-than Category I excavations were visually inspected. At least one wall would be photographed with mapping at scales from 1 in. to 1–10 ft when some conditions seemed to warrant attention (Lund and Euge, 1984).

There are general considerations to remember in planning exploratory excavation. Effective use of exploratory excavation requires tying excavation sites to a survey of the area under study. This permits the information gained from each excavation to be related to the planned work and to other information sources. The logging of exploratory excavations requires vertical survey control, too. Stations 2-m apart or less are usually sufficient. Lines affixed to the trench wall from these stations transfer the vertical stationing into the trench or pit (Hatheway and Leighton, 1979). The method used to excavate trenches or pits should be the least disturbing to the soil or rock present at the site to ensure the least distribution to the features being exposed. Another consideration in using exploratory excavation is the safety of individuals conducting logging and sampling. Trench walls in some soils will require shoring to prevent collapse. Collapse can be avoided in bulldozer trenches by stepping the walls at less-steep angles.

Table 7.9 The Principal Methods for Exploratory Excavation and Their Associated Uses, Capabilities, and Limitations

Exploration Method	General Use	Capabilities	Limitations
Hand-Excavated Test Pits and Shafts	Bulk sampling, *in situ* testing, visual inspection.	Provides data in inaccessible areas, less mechanical disturbance of surrounding ground.	Expensive, time-consuming, limited to depths above groundwater level.
Backhoe Excavated Test Pits and Trenches	Bulk sampling, *in situ* testing, visual inspection, excavation rates, depth of bedrock and groundwater.	Fast, economical, generally less than 15-ft deep, can be up to 30-ft deep.	Equipment access, generally limited to depths above groundwater level, limited undisturbed sampling.
Drilled Shafts	Preexcavation for piles and shafts, landslide investigations, drainage wells.	Fast, more economical than hand excavated, min. 30-in. diameter, max. 6-ft diameter.	Equipment access, difficult to obtain undisturbed samples, casing obscures visual inspection.
Dozer Cuts	Bedrock characteristics, depth of bedrock and groundwater level, rippability, increase depth capability of backhoes, level area for other exploration equipment.	Relatively low cost, exposures for geologic mapping.	Exploration limited to depth above groundwater level.
Trenches for Fault Investigations	Evaluation of presence and activity of faulting and sometimes landslide features.	Definitive location of faulting, subsurface observation up to 30 ft.	Costly, time-consuming, requires shoring, only useful where datable materials are present, depth limited to zone above groundwater level.

Source: From NAVFAC, 1982.

Logging the vertical exposure in a trench or pit is sometimes called face mapping (Hatheway, 1982). Face mapping provides a record of rock and soil conditions. The relationship of contacts between subsurface units, their characteristics, and the spacing and orientation of discontinuities can be represented in face mapping. Hatheway and Leighton (1979) note that trench logging, or face mapping, can be either subjective or objective. Subjective trench logging portrays key factors. The result is a schematic representation that accurately shows the subsurface conditions while omitting subordinate or accessory details deemed unnecessary (Hatheway and Leighton, 1979). Figure 7.22 is a subjective face map relating the subsurface extent of faults to key stratigraphic features having a bearing on determining the age of movement. Objective trench-logging attempts to show all the physical features of the exposed face regardless of relative importance (Hatheway and Leighton, 1979). Figure 7.23 is an example of

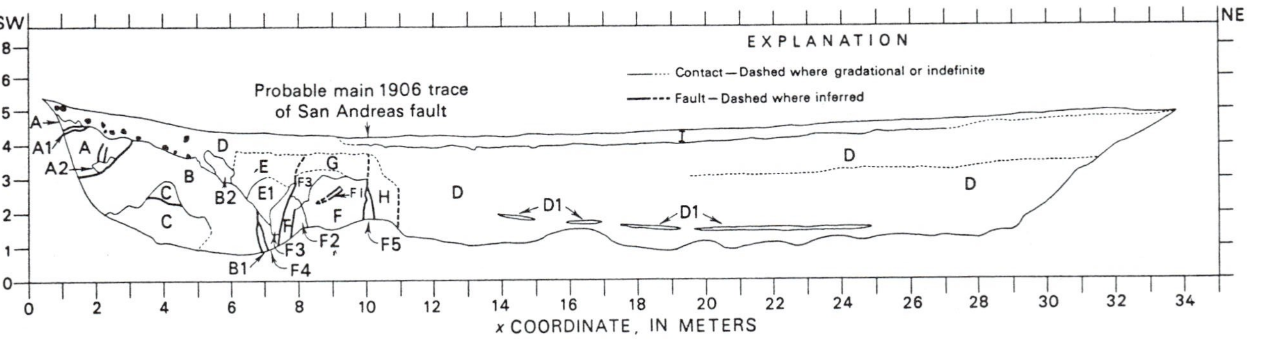

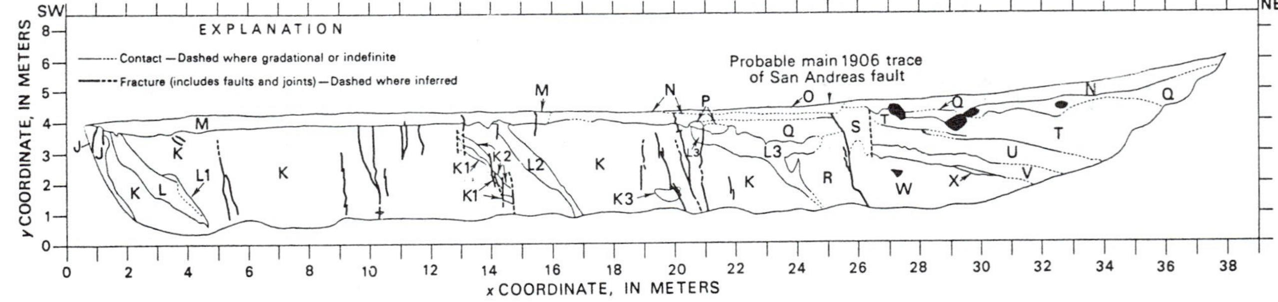

Figure 7.22 An example of subjective face mapping of a trench across the San Andreas fault in northern San Mateo County, California. Letters refer to key soil stratigraphic units identified during mapping. (From Bonilla et al., 1978)

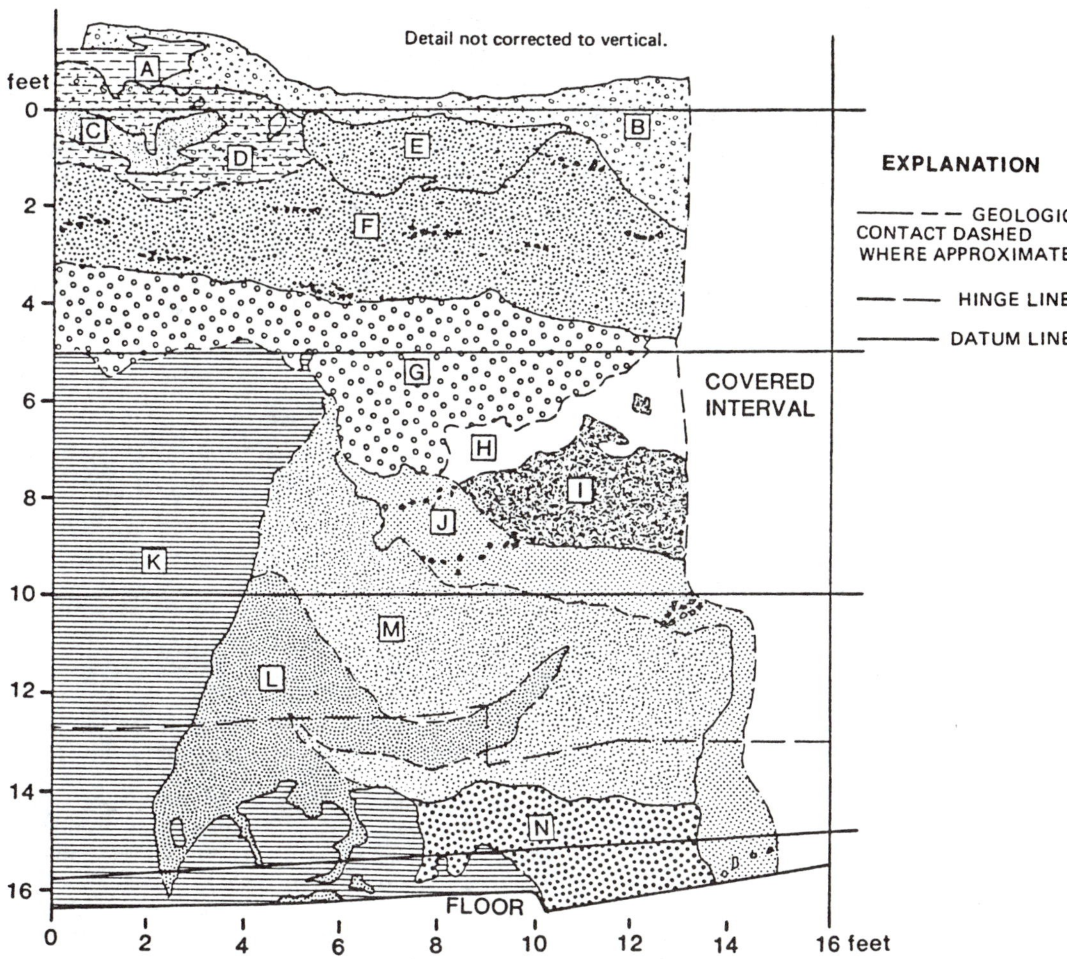

A - SILTY CLAY; (CL), dark brown, medium plasticity stiff, moist, noncalcareous.

B - SILTY GRAVELLY SAND; (SM), light brown, poor to moderately sorted, subangular to subrounded, noncalcareous.

C - SILT and SAND; (ML & SM), light reddish brown, well sorted, subangular, dense, slightly moist, noncalcareous.

D - GRAVELLY SILTY SAND; (SM), light brown to reddish brown, poorly sorted, angular to subangular clasts of varied compositions, noncalcareous, cobbles to 8" dia.

E - SAND with SILT (SP - SM); greenish to reddish brown; moderately sorted, subrounded to rounded, slightly moist, dense, noncalcareous, crossbedded and micaceous.

F - SILTY SAND; (SM), light brown, moderate to well sorted, medium to coarse, subangular to subrounded quartz and lithic sand, occasional gravel to 3" in diameter.

G - GRAVELLY CLAYEY SAND; (SC - CL), brown, stiff, dry, calcareous; sand and clay in near equal proportions.

H - SILT - CLAYEY SILT; (ML), light brown, low plasticity, dry, stiff.

I - SANDY CLAYEY SILT; (ML), brown, low to medium plasticity, dry, firm, occasional gravel to 1½" in diameter.

J - SAND; (SP), grey, moderate to well sorted, angular to subrounded, fine to medium grained quartz, mica and lithic sand, loose, dry, calcareous.

K - SILTY CLAY and CLAYEY SILT; (CL & ML), reddish brown, low to medium plasticity, very stiff to hard, calcareous, locally up to 10% very fine micaceous sand.

L - SILTY SAND; (SM), light greyish brown, very fine to medium, subangular quartz, mica, and lithic sand, medium dense, calcareous, crossbedded

M - SILTY GRAVELLY SAND; (SM - GM), brown, poorly sorted, subangular to subrounded, very fine to coarse grained quartz and lithic sand, loose, dry, calcareous, gravel and sand in near equal proportions, gravel to 1½" diameter.

N - SAND - SILTY SAND; (SP - SM), light brown, poor to moderately sorted, angular to subrounded, very fine to coarse lithic and quartz sand, dry loose to medium dense, calcareous, thinly bedded.

Figure 7.23 An example of objective face mapping from an excavation for the Palo Verde Nuclear Generating Plant, Arizona. (From W. R. Lund and K. M. Euge, Detailed Inspection and Geologic Mapping of Construction Excavations at Palo Verde Nuclear Generating Station, Arizona, *Bulletin of the Association of Engineering Geologists*, Vol. 21, No. 2, May 1984, pp. 179–189. Used by permission.)

an objective trench log. Zellmer et al. (1984) used objective face mapping of trenches in determining the repair needs to a damaged road. It revealed the extent of fissures present and the types of materials involved. The understanding gained from this work helped justify employing a nonstandard and somewhat novel repair to the damaged section of road. Without this information, more traditional repair methods costing more and not fully achieving the desired long-term result would have been instituted. Table 7.10 lists some uses of face mapping during different phases of a project. Hatheway and Leighton (1979) provide detailed discussions on recognizing different features and conditions in trench logging.

Borehole Exploration

Borehole exploration includes a variety of methods to drill holes for sampling materials in the subsurface and for mapping the extent of features and conditions. Table 7.11 describes the main drilling methods used in

Table 7.10 A Listing of Uses for Face Mapping of Exploratory Excavation Associated with Different Phases of Construction

Phases	Uses
Pre-Construction	• As a basis for prediction of stability factors in structural design; • To determine relative risk of fault activity from "capable" faults in safety-related structures such as nuclear power plants, hospitals, and schools; • Verification of foundation stability and permeability conditions beneath dams and other water-retention structures; • To evaluate optimal excavation methods and rates of performance of machinery; • To assist in modifying orientation of underground structural components to take advantage of controlling elements of rock structures; • Basis for design of internal support systems for underground structures; • Prediction of groundwater inflow and dewatering requirements; • Verification of physical modelling assessments of landslides and other forms of mass wastage.
As-Constructed	• As a general record of conditions encountered in construction; • As a basis for pay for some variable-rate items of construction; • Record alleged conditions underlying contractor/owner disputes; record rock quality, limits of over-excavation, presence/absence of contract-expected ground conditions.
Post-Construction	• A record of stress-concentration or earthquake-induced failure conditions in underground structures or in foundation components of surface structures, or in an earth or earth/rock dam; • As a record of natural conditions in the structure and surrounding exposed ground, at the time of proposed modifications to the structure.

Source: From A. W. Hatheway, Trench, Shaft, and Tunnel Mapping, *Bulletin of the Association of Engineering Geologists*, Vol. 19, No. 2, May 1982, pp. 173–180. Used by permission.

Table 7.11 The Common Drilling Methods Used in Borehole Exploration with Descriptions of the Procedures Used and Applicability

Boring Method	Procedure Utilized	Applicability
Auger boring	Hand- or power-operated augering with periodic removal of material. In some cases continuous auger may be used requiring only one withdrawal. Changes indicated by examination of material removed. Casing generally not used.	Ordinarily used for shallow explorations above water table in partly saturated sands and silts and soft-to-stiff cohesive soils. May be used to clean out hole between drive samples. Very fast when power-driven. Large diameter bucket auger permits examination of hole. Hole collapses in soft soils and soils below groundwater table.
Hollow-stem flight auger	Power operated, hollow stem serves as a casing.	Access for sampling (disturbed or undisturbed) or coring through hollow stem. Should not be used with plug in granular soil. Not suitable for undisturbed sampling in sand and silt.
Wash-type boring for undisturbed or dry sample	Chopping, twisting, and jetting action of a light bit as circulating drilling fluid removes cuttings from holes. Changes indicated by rate of progress, action of rods, and examination of cuttings in drilling fluid. Casing used as required to prevent caving.	Used in sands, sand and gravel without boulders, and soft to hard cohesive soils. Most common method of subsoil exploration. Usually can be adapted for inaccessible locations such as on water, in swamps, on slopes, or within buildings. Difficult to obtain undisturbed samples.
Rotary drilling	Power rotation of drilling bit as circulating fluid removes cutting from hole. Changes indicated by rate of progress, action of drilling tools, and examination of cutting in drilling fluid. Casing usually not required except near surface.	Applicable to all soils except those containing such large gravel, cobbles, and boulders. Difficult to determine changes accurately in some soils. Not practical in inaccessible locations because of heavy truck-mounted equipment, but applications are increasing since it is usually most rapid method of advancing borehole. Soil samples and rock cores usually limited to 6 in.
Percussion drilling (churn drilling)	Power chopping with limited amount of water at bottom of hole. Water becomes a slurry that is periodically removed with bailer or sand pump. Changes indicated by rate of progress, action of drilling tools, and composition of slurry removed. Casing required except in stable rock.	Not preferred for ordinary exploration or where undisturbed samples are required because of difficulty in determining strata changes, disturbance caused below chopping bit, difficulty of access, and usually higher cost. Sometimes used in combination with auger or wash bor-

(continued)

Table 7.11 (continued)

Boring Method	Procedure Utilized	Applicability
		ings for penetration of coarse gravel, boulders, and rock formations. Could be useful to probe cavities and weakness in rock by changes in drill rate.
Rock-core drilling	Power rotation of a core barrel as circulating water removes ground-up material from hole. Water also acts as coolant for core barrel bit. Generally hole is cased to rock.	Used alone and in combination with boring types to drill weathered rocks, bedrock, and boulder formations.
Wire-line drilling	Rotary-type drilling method where the coring device is an integral part of the drill rod string, which also serves as a casing. Core samples obtained by removing inner barrel assembly from the core barrel portion of the drill rod. The inner barrel is released by a retriever lowered by a wire-line through drilling rod.	Efficient for deep hole coring over 100 ft on land and offshore coring and sampling.

Source: From NAVFAC, 1982.

borehole exploration. The choice of drilling method will depend on availability of equipment, cost, accessibility, the kinds of materials likely to be encountered, and the type of information sought.

The information needs of particular projects will affect not only the choice of drilling method, but also the layout and depth of borings. Boyce (1982) suggests that the number and spacing of boreholes should be adequate for defining the lateral and vertical extent of any condition likely to influence the structure being planned. This includes being able to identify geologic structures such as faults, folds, and joints. Table 7.12 provides some general guidelines for boring layouts appropriate to investigation for different structures. Related guidelines for boring depths are shown in Table 7.13. The basic premise for deciding what depth to require is that all pertinent materials must be penetrated by the boreholes (Boyce, 1982).

Compared to exploratory excavation, boreholes have a major drawback. The investigator is unable to access the features or conditions encountered in the subsurface. One exception is large diameter boreholes that permit an individual to be lowered to the bottom of the hole. Although it is not a common form of borehole exploration, it does permit logging similar to face mapping. Sampling is usually accomplished by placing a sampler down the borehole as it is being drilled. A great variety of samplers are available to accommodate the different drilling methods and the types of materials that may be encountered. Table 7.14 lists some of the common

Table 7.12 General Guidelines for Appropriate Boring Layout for Different Structural Investigations

Areas for Investigation	Boring Layout
New site of wide extent.	Space preliminary borings 200- to 500-ft apart so that area between any four borings includes approximately 10% of total area. In detailed exploration, add borings to establish geological sections at the most useful orientations.
Development of site on soft compressible strata.	Space borings 100–200 ft at possible building locations. Add intermediate borings when building sites are determined.
Large structure with separate closely spaced footings.	Space borings approximately 50 ft in both directions, including borings at possible exterior foundation walls at machinery or elevator pits, and to establish geologic sections at the most useful orientations.
Low-load warehouse building of large area.	Minimum of four borings at corners plus intermediate borings at interior foundations sufficient to define subsoil profile.
Isolated rigid foundation, 2,500 to 10,000 sq ft in area.	Minimum of three borings around perimeter. Add interior borings depending on initial results.
Isolated rigid foundation, less than 2,500 sq ft in area.	Minimum of two borings at opposite corners. Add more for erratic conditions.
Major waterfront structures such as dry docks.	If definite site is established, space borings generally not farther than 50 ft, adding intermediate borings at critical locations such as deep pumpwell, gate seat, tunnel, or culverts.
Long bulkhead or wharf wall.	Preliminary borings on line of wall at 200-ft spacing. Add intermediate borings to decrease spacing to 50 ft. Place certain intermediate borings inboard and outboard of wall line to determine materials in scour zone at toe and in active wedge behind wall.
Slope stability, deep cuts, high embankments.	Provide three to five borings on line in the critical direction to provide geological section for analysis. Number of geological sections depends on extent of stability problem. For an active slide, place at least one boring upslope of sliding area.
Dams and water-retention structures.	Space preliminary borings approximately 200 ft over foundation area. Decrease spacing on centerline to 100 ft by intermediate borings. Include borings at location of cutoff, critical spots in abutment, spillway, and outlet works.

Source: From NAVFAC, 1982.

types with information on their use (Figure 7.24). Figure 7.25 shows the operation of a thin-walled fixed-piston sampler. The sampler is withdrawn from the hole so that the sample may be cataloged and sent to a laboratory for testing.

In sampling rock, the sampler is drilled rather than driven into the

Table 7.13 Suggested Borehole Depths for Borehole Exploration for Certain Types of Structures

Areas of Investigation	Boring Depth
Large structure with separate closely spaced footings.	Extend to depth where increase in vertical stress for combined foundations is less than 10% of effective overburden stress. Generally all borings should extend to no less than 30 ft below lowest part of foundation unless rock is encountered at shallower depth.
Isolated rigid foundations.	Extend to depth where vertical stress decreases to 10% of bearing pressure. Generally all borings should extend no less than 30 ft below lowest part of foundation unless rock is encountered at shallower depth.
Long bulkhead or wharf wall.	Extend to depth below dredge line between ¾ and 1½ times unbalanced height of wall. Where stratification indicates possible deep stability problem, selected borings should reach top of hard stratum.
Slope stability.	Extend to an elevation below active or potential failure surface and into hard stratum or to a depth for which failure is unlikely because of geometry of cross section.
Deep cuts.	Extend to depth between ¾ and 1 times base width of narrow cuts. Where cut is above groundwater in stable materials, depth of 4 to 8 ft below base may suffice. Where base is below groundwater, determine extent of pervious strata below base.
High embankments.	Extend to depth between ½ and 1¼ times horizontal length of side slope in relatively homogeneous foundation. Where soft strata are encountered, borings should reach hard materials.
Dams and water-retention structures.	Extend to depth of ½ base width of earth dams or 1 to 1½ times height of small concrete dams in relatively homogeneous foundations. Borings may terminate after penetration of 10 to 20 ft in hard and impervious stratum if continuity of this stratum is known from reconnaissance.

Source: From NAVFAC, 1982.

material. The bit normally used to drill the hole is replaced at certain depths with a coring bit. The rock core is extracted from the core barrel and treated in a manner similar to a soil sample. Continuous sampling in either soil or rock is not usually done. Drilling is much slower for continuous sampling and increases the cost of borehole exploration. There must be some specific circumstance to justify this increased time and cost. Bock (1981) provides a good example. Subsurface investigation for constructing the Washington Metro involved identifying conditions that might affect the advancing of the shield used in tunneling. Sandstone interbeds present along part of the tunnel alignment would affect shield progress. As Bock (1981) points out, the usual sampling interval on boreholes might miss some of these beds (Figure 7.26). Recognizing the potential for this problem resulted in requiring continuous sampling through the depth that tunneling would occur. Sampling the transition zone between soil and rock presents some unique problems in choosing the way to obtain samples. Table 7.15 indicates some sampling methods appropriate to the type of conditions that may be encountered in this zone.

In addition to sampling, penetration-resistance tests are often performed during the drilling of the borehole. The standard penetration test (SPT) is based on the number of blows required to drive a split tube sampler a distance of 12 in. This test has specific requirements on the size of the sampler and the manner in which it is driven. This number, or *N* value, is correlated to factors such as the relative density of fine-grained, granular soil; undrained shear strength; and shear modulus at very small strains. Cone penetrometers rather than samplers may be used for penetration-resistance testing. Figure 7.27 shows the operation of a Dutch Cone test and the typical results obtained. Results for the cone penetration test can be correlated to bearing capacity, relative density of sands, strength and sensitivity of clays, and overconsolidation of clays. USBR (1974) and NAVFAC (1982) contain detailed discussions on all aspects of borehole exploration. Included is information of the use of specific devices and tests in borehole exploration.

Logging of boreholes is as important as face mapping in exploratory excavation. Without this careful recording of data, there is little value to conducting borehole exploration. Like a face map, a log is a written record of materials and conditions encountered in the subsurface. The information contained in the log will influence the extent of additional investigation, siting, and design. Boyce (1982) provides a comprehensive listing of information required in a good log (Table 7.16). For correlation, individual logs are plotted as graphic columns and arranged to show the variation in subsurface conditions present in the study area. It is only through accurate logging that a clear picture of subsurface conditions can be obtained.

Geophysical Methods

Geophysical methods are useful in determining a variety of the physical properties of soil and rock. Of greatest value for engineering applications are those methods that measure the propagation of shock waves and conductance of electrical current through earth materials. These are the seismic and electrical methods of geophysical exploration, respectively.

Geophysical methods of exploration provide an indirect source of information about the subsurface. Their typical uses in civil engineering applications are to supplement data gained from surface mapping and shallow drillholes and to provide a bridge between test data obtained from small samples in the laboratory with *in situ* conditions at a site. Calculation of depths to bedrock between drillholes and determination of the *in situ* influence of discontinuities on rock mass physical properties compared to an intact sample in the laboratory are examples of the more common applications. Surface and downhole applications of seismic and electrical methods currently are in use for engineering purposes.

Table 7.14 Listing of the Types of Borehole Samplers with Brief Descriptions of Their Dimensions and Suggested Uses

Sampler	Dimensions	Best Results in Soil or Rock Types	Methods of Penetration	Causes of Disturbance or Low Recovery	Remarks
Single Tube		Primarily for strong, sound and uniform rock.		Fractured rock. Rock too soft.	Drill fluid must circulate around core—rock must not be subject to erosion. Single tube not often used for exploration.
Double Tube		Nonuniform, fractured, friable, and soft rock.		Improper rotation or feed rate in fractured or soft rock.	Has inner barrel or swivel that does not rotate with outer tube. For soft, erodible rock. Best with bottom discharge bit.
Triple Tube		Same as double tube.		Same as double tube.	Differs from double tube by having an additional inner split-tube liner. Intensely fractured rock core best preserved in this barrel.
Split Barrel	2-in. OD—1.375-in. ID is standard. Penetrometer sizes up to 4-in. OD/3.5-in ID available.	All fine-grained soils in which sampler can be driven. Gravels invalidate drive data.	Hammer driven	Vibration	SPT is made using standard penetrometer with 140# hammer falling 30-in. Undisturbed samples often taken with liners. Some sample disturbance is likely.
Retractable Plug	1-in. OD tubes 6-in. long. Maximum of six tubes can be filled in single penetration.	For silts, clays, fine and loose sands.	Hammer driven	Improper soil types for sampler. Vibration.	Light weight, highly portable units can be hand carried to job. Sample disturbance is likely.

Augers:					
Continuous Helical Flight	3-in. to 16-in. diameter. Can penetrate to depths in excess of 50 ft.	For most soils above water table. Will not penetrate hard soils or those containing cobbles or boulders.	Rotation	Hard soils, cobbles, boulders.	Rapid method of determining soil profile. Bag samples can be obtained. Log and sample depths must account for lag between penetration of bit and arrival of sample at surface.
Disk	Up to 42-in. diameter. Usually has maximum penetration of 25 ft.	Same as flight auger.	Rotation	Same as flight auger.	Rapid method of determining soil profile. Bag samples can be obtained.
Bucket	Up to 48-in. diameter common. Larger available. With extensions, depths greater than 80 ft are possible.	For most soils above water table. Can dig harder soil than above types, and can penetrate soils with cobbles and small boulders when equipped with a rock bucket.	Rotation	Soil too hard to dig.	Several type buckets available including those with ripper teeth and chopping buckets. Progress is slow when extensions are used.
Hollow Stem	Generally 6-in. to 8-in. OD with 3-in. to 4-in. ID hollow stem.	Same as bucket.	Same	Same	A special type of flight auger with hollow center through which undisturbed samples or SPT can be taken.
Diamond-Core Barrels	Standard sizes 1½-in to 3-in OD, ⅞-in to 2⅛-in core. Barrel lengths 5 to 10 ft for exploration.	Hard rock. All barrels can be fitted with insert bits for coring soft rock or hard soil.			

(continued)

Table 7.14 *(continued)*

Sampler	Dimensions	Best Results in Soil or Rock Types	Methods of Penetration	Causes of Disturbance or Low Recovery	Remarks
Shelby Tube	3-in. OD–2.875-in. ID most common. Available from 2-in. to 5-in. OD. 30-in. sample length is standard.	For cohesive fine-grained or soft soils. Gravelly soils will crimp the tube.	Pressing with fast, smooth stroke. Can be carefully hammered.	Erratic pressure applied during sampling, hammering, gravel particles, crimping tube edge, improper soil types for sampler.	Simplest sampler for undisturbed samples. Boring should be clean before lowering sampler. Little waste area in sampler. Not suitable for hard, dense, or gravelly soils.
Stationary Piston	3-in. OD most common. Available from 2-in. to 5-in OD. 30-in. sample length is standard.	For soft to medium clays and fine silts. Not for sandy soils.	Pressing with continuous, steady stroke.	Erratic pressure during sampling, allowing piston rod to move during press. Improper soil types for sampler.	Piston at end of sampler prevents entry of fluid and contaminating material. Requires heavy drill rig with hydraulic drill head. Generally less disturbed samples than Shelby. Not suitable for hard, dense, or gravelly soil. No positive control of specific recovery ratio.
Hydraulic Piston (Osterberg)	3-in. OD most common—available from 2-in. to 4-in. OD, 36-in sample length.	For silts-clays and some sandy soils.	Hydraulic or compressed air pressure.	Inadequate clamping of drill rods, erratic pressure.	Needs only standard drill rods. Requires adequate hydraulic or air capacity to acti-

					vate sampler. Generally less disturbed samples than Shelby. Not suitable for hard, dense, or gravelly soil. Not possible to limit length of push or amounts of sample penetration.
Denison Sampler	Samplers from 3.5-in OD to 7¾-in. OD (2.375-in to 6.3-in. size samples). 24-in. sample length is standard.	Can be used for stiff to hard clay, silt and sands with some cementation, soft rock.	Rotation and hydraulic pressure.	Improperly operating sampler. Poor drilling procedures.	Inner tube face projects beyond outer tube that rotates. Amount of projection can be adjusted. Generally takes good samples. Not suitable for loose sands and soft clays.
Pitcher Sampler	Sampler 4.125-in. OD uses 3-in. Shelby tubes. 24-in sample length.	Same as Denison.	Same as Denison.	Same as Denison.	Differs from Denison, in that inner-tube projection is spring controlled. Often ineffective in cohesionless soils.
Hand-cut Block or Cylindrical Sample	Sample cut by hand.	Highest quality undisturbed sampling in cohesive soils, cohesionless soil, residual soil, weathered rock, soft rock.		Change of state of stress by excavation.	Requires accessible excavation. Requires dewatering if sampling below groundwater.

Source: Modified from NAVFAC, 1982.

Figure 7.24 Shelby tube sampler being removed from truck-mounted, continuous-flight, hollow-stem auger.

Seismic Methods

All applications of the seismic methods, whether refraction, reflection, or acoustic, are based on the fact that the elastic properties of soils and rocks determine the velocities of wave propagation through them. The higher the elastic modulus, for example, the higher the velocity. The relationships between the elastic moduli and propagation velocities are covered in chapter 4.

Waves are propagated through the earth and along the surface of the earth as body waves and surface waves, respectively. Body waves are the faster and are used most commonly in engineering seismic applications although the slower surface waves have an important role in more sophisticated applications.

Body waves are propagated at different velocities by two kinds of particle motion. The faster body waves travel as a compression pulse through any medium, namely, solid, liquid, or gas. Particle motion is in the direction of wave propagation and the waves commonly are referred to as compressional waves. As the fastest of all waves, they also are known as primary or P-waves because they arrive first at a detector or receiver from a common source.

Compressional waves are used for most engineering seismic applications. They are the dominant body-wave form generated by the use of explosives in a shallow drillhole, by hammer blows to a metal plate on the surface, or by weight-dropping equipment. Compressional wave velocities for a variety of soil and rock types are listed by many literature sources such as Redpath (1973), Dobrin (1976), and Telford et al. (1976).

Shear waves are the second kind of body wave. They are propagated through a solid medium by particle motion that is perpendicular to the direction of travel. Shear waves arrive at a receiver from a common source after compressional waves (Figure 4.9) and before surface waves. For this reason, they are often referred to as secondary or S-waves. Shear waves travel only through solids, a characteristic of value in engineering applications. The usual sources of shock-wave energy do not generate strong shear waves, and specialized techniques are used to generate them.

As noted earlier, the velocity of wave propagation is controlled by the elastic properties of earth materials. This holds for both compressional and shear waves. In rocks, the dominant factors that influence velocity are crystallinity and porosity. Rocks having crystalline textures and low porosities have higher shock-wave velocities, just as they have high elastic moduli and compressive strengths. Porosity typically is altered in a given

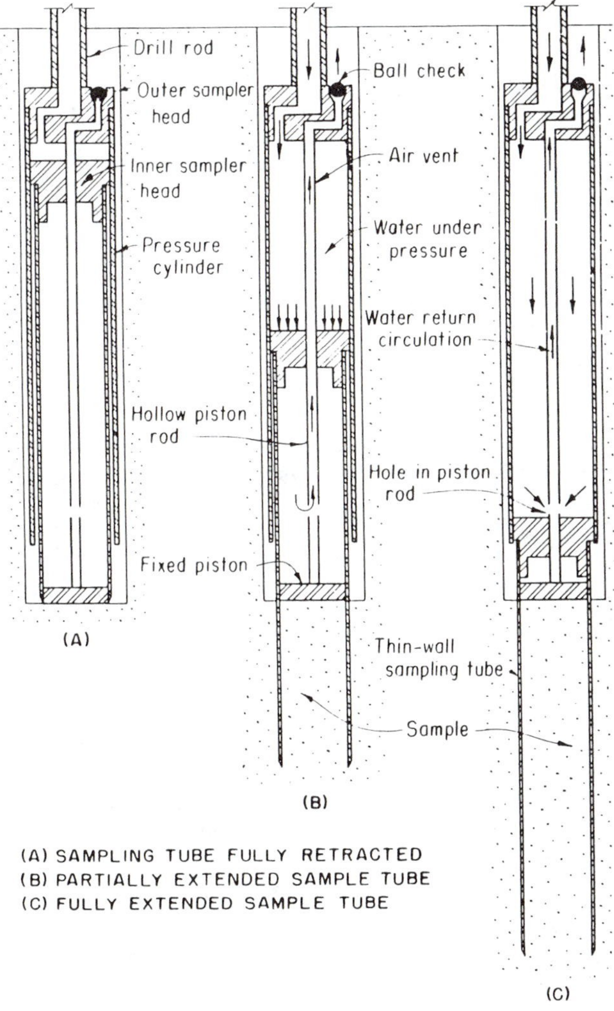

Figure 7.25 Diagram showing the operation of a thin-wall fixed-piston sampler. (From USBR, 1974)

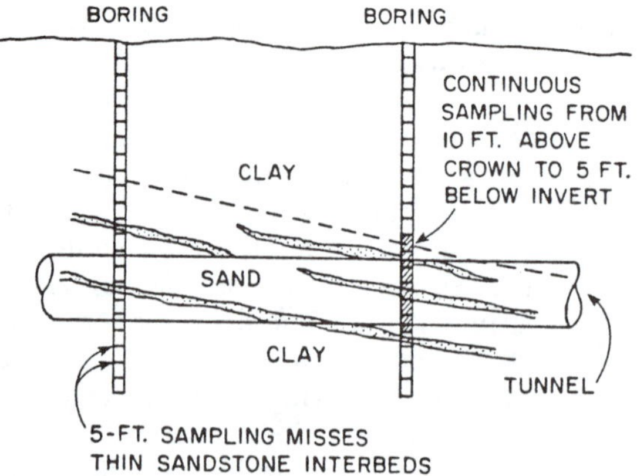

Figure 7.26 Diagram showing sampling in boreholes for the Washington Metro. Sandstone interbeds might be missed by traditional 5-ft interval sampling in borehole on the left. To ensure detection of these layers, continuous sampling in the borehole on the right is used through the depths where the tunnel is expected to pass (C. G. Bock, The Interrelationship of Geologic Conditions with Construction Methods and Costs, Washington Metro, *Bulletin of the Association of Engineering Geologists,* Vol. 18, No. 2, May 1981, pp. 187–194. Used by permission.

Table 7.15 Descriptions of Materials Likely to Be Encountered in the Transition Zone Between Soil and Bedrock with Recommended Sampling Method

Description of Material	Sampling Method
Colluvium—loosely packed, poorly sorted material.	Driven samples or triple tube core barrel. Double tube barrel is required for boulders. Denison sampler can be used if no boulders are present.
Structureless residual soil—the soil shows none of the fabric of the rock from which it is derived.	Driven samples or triple tube core barrel. Dennison sampler can be used. Hand-cut samples are best.
Decomposed rock containing rounded boulders that may be harder than surrounding material.	Driven samples or triple tube core barrel. Double tube barrel is required to sample boulders.
Decomposed rock containing angular boulders separated by thin seams of friable material.	Double tube core barrel with triple tube barrel in weak seams.
Slightly decomposed rock—friable material, if present, is limited to narrow seams.	Double tube core barrel.

Source: From NAVFAC, 1982.

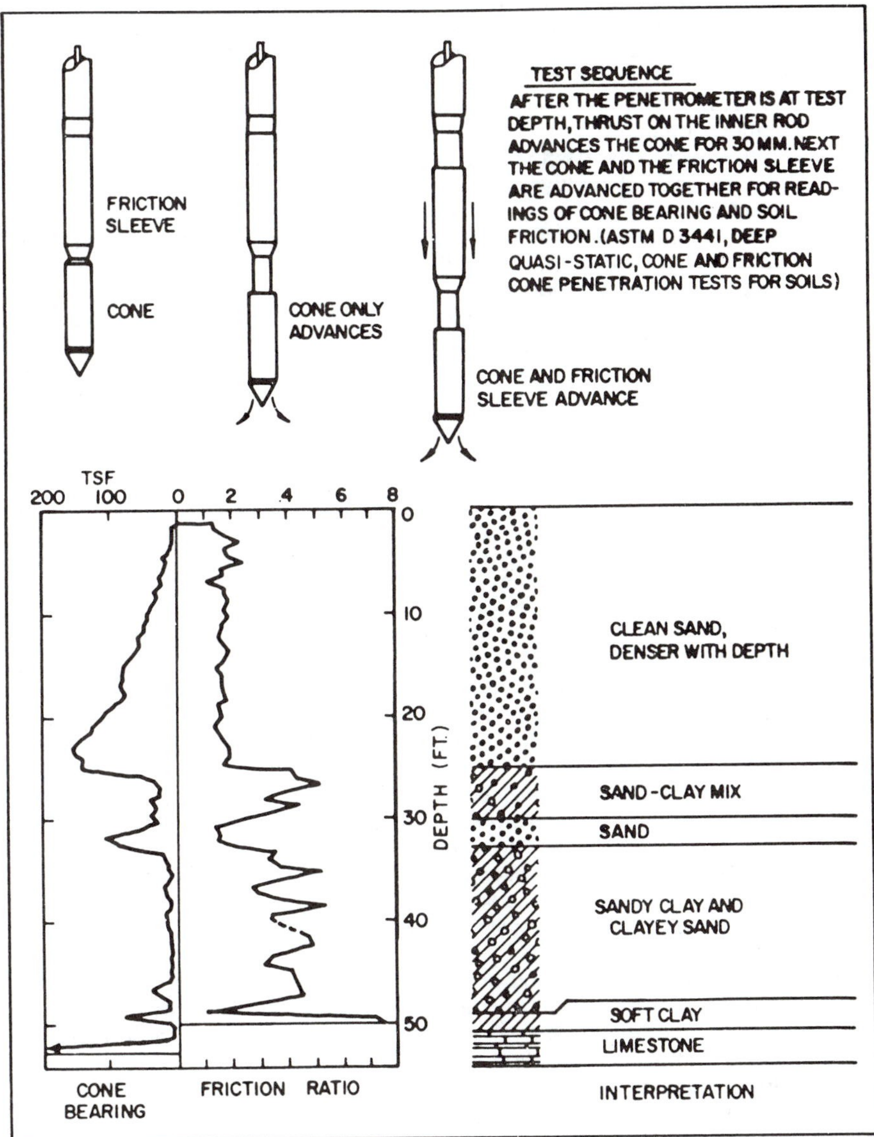

Figure 7.27 Diagram showing the operation of the Dutch Cone Penetrometer and an example of typical results. (NAVFAC, 1982)

rock type by cementation and/or grain size, both characteristics of soils and detrital sedimentary rocks. Mineralogy also has a velocity-controlling role. The presence of clay in limestone, for instance, will result in a reduction in velocity compared to a pure limestone. Aside from these factors, the major contributors to reduction in velocity in rocks are discontinuities such as bedding surfaces, joints and foliation and chemical weathering, or decomposition, of the rock mass.

The velocity of compressional waves in porous soils may be increased to

Table 7.16 Listing of Items That Should Be Included in a Borehole Log

(1) Hole Number, Location, and Surface Elevation—Show location by station number or by reference to some base. Show elevation above mean sea level if it is known, otherwise elevation from an assumed datum.

(2) Depth—Record the depth of the upper and lower limits of the layer being described.

(3) Name—In unconsolidated materials, record the name of the primary constituent first and then, as a modifier, the name of the second most prominent constituent, for example: sand, silty. Usually two constituents are enough. If it is desirable to call attention to a third, use the abbreviation w/_____ after the name, for example: sand, silty w/cbls (with cobbles).

(4) Texture—Record size, shape, and arrangement of individual minerals or particles. In consolidated rock, descriptive adjectives are usually enough. In unconsolidated material, use descriptive adjectives for size and give an average maximum size in inches or millimeters. Record shape by such terms as equidimensional, tabular, and prismatic and by the degree of roundness. Record by estimated relative amounts.

(5) Structure—Describe any features of rock structure observed, such as bedding, laminations, cleavage, joining, concretions, or cavities. Where applicable, include information on size, shape, color, composition, and spacing of structural features.

(6) Color—Record color for purposes of identification and correlation. Color may change with water content.

(7) Moisture Content—Note whether the material is dry, moist, or wet.

(8) Mineral Content—Record identifiable minerals and the approximate percentage of the more abundant minerals. Describe any mineral that is characteristic of a specific horizon and record its approximate percentage even though it occurs in minor amounts. Record the kind of cement in cemented materials.

(9) Permeability—Estimate the relative permeability and record it as impermeable, slowly permeable, moderately permeable, or rapidly permeable. If a field permeability test is run, describe the test and record the results.

(10) Age, Name, and Origin—Record geologic age, name, and origin, for example: Jordan member, Trempeleau formation, Cambrian age; Illionian till; recent alluvium. For sediments, identify the genetic type of the deposit. Such identification helps in correlation and in interpreting data from test holes. Similarly, knowing that a material is of lacustrine or eolian origin or that it is a part of a slump, slide, or other form of mass movement helps in evaluating a proposed construction site.

(11) Strength and Condition of Rock—Record rock condition by strength, degree of weathering, and degree of cementation.

(12) Consistency and Degree of Compactness—Describe consistency of fine materials as very soft, soft, medium, stiff, very stiff, and hard. Describe degree of compactness of coarse-grained soils as very loose, loose, medium, dense, and very dense.

(13) Unified Soil Classification Symbol—For all unconsolidated materials give the unified soil classification symbol. If there is any doubt about the proper classification of material, record it as "CL or ML" and "SW or SM," not by the borderline symbols. Record the results of field-identification tests, such as dilatance, dry strength, toughness, ribbon, shine, and odor.[a]

(14) Blow Count—Where the standard penetration test is made, record the results and the test elevation or depth. This test shows the number of blows under standard conditions that are required to penetrate 12 in. or, with refusal, the number of inches penetrated by 100 blows. The latter is commonly recorded as 100/d, where d equals the number of inches penetrated in 100 blows.

(15) Other Field Tests—If other field tests are made, record the results and describe each test so that there is no doubt as to what was done. Examples are vane-shear test, pressure test, field-density test, tests for moisture content, acetone test, and the use of an indicator such as sodium fluorescein to trace the flow of ground water.

(16) Miscellaneous Information—Record any drilling difficulties, core and sample recovery and reasons for any losses, type and mixture of drilling mud used to prevent caving or sample loss, loss of drilling fluid, and any other information that may help in interpreting the subsurface condition.

(17) Water Levels—Record the static water level and the date on which the level was measured. Wait at least one day after the hole has been drilled to measure the water level to allow time for stabilization.

[a] See Table 3.4 and accompanying text.

Source: From R. C. Boyce, An Overview of Site Investigation, *Bulletin of the Association of Engineering Geologists*, Vol. 19, No. 2, May 1982, pp. 167–171. Used by permission.

that of water (approximately 1,525 m/sec) in the saturated state, provided the unsaturated soil has a velocity less than that of water (W. E. Johnson, 1976). Shear waves do not propagate through water and, as a result, shear-wave velocities in earth materials are not affected by pore water. Comparison of compressional- and shear-wave velocities obtained over a span of time for a sand will indicate changes in degree of saturation, as dry sand normally has a relatively low compressional wave velocity (<600 m/sec). An additional use of shear-wave velocities is in the calculation of *in situ* dynamic elastic moduli, that is, Young's modulus, shear modulus, and Poisson's ratio (Figure 4.10).

Shock waves in earth materials follow multiple paths from source to receiver. In the near surface, waves take a direct path from source to receiver and the measurement of elapsed travel time for a measured distance results in the wave velocity through that material. Waves moving downward into the earth may be refracted and reflected at velocity interfaces. The ray-path diagram is a convenient method of showing wave propagation by direct, refracted, and reflected paths (Figure 7.28). Unless otherwise noted, all velocities refer to compressional waves.

Seismic data are obtained by recording shock-wave travel time between a source and a receiver, or geophone, for various chosen distances. Two general types of seismographs are used in engineering applications. One type permits the simultaneous, multichannel recording of shock-wave arrivals at a number of geophone locations from a single energy source such as a buried explosive charge or a weight-dropping system. The output from the geophones may be recorded in analog and digital forms in a variety of ways, all having a time base to permit extraction of elapsed travel times to each geophone location. Two important advantages of multichannel recording are (1) single energy source for all geophone channels and (2) more sophisticated filtering, recording, data processing, and printout capabilities than simpler single-channel seismographs. Disadvantages include higher first cost, greater operating cost, larger size, and often the need for greater peripheral support such as computer hardware and software.

The second type of seismograph is a single-channel instrument that records shock-wave travel time from a source to a single geophone location. As a result, the operation must be repeated for different geophone distances until a suitable number of travel times have been obtained. Single-channel seismographs may record geophone output in several ways such as on an oscilloscope; on a chart, as in the multichannel unit; or only as a first-arrival pulse time digitally (or by some other means). Timing systems are inherently a part of each system. Single-channel seismographs

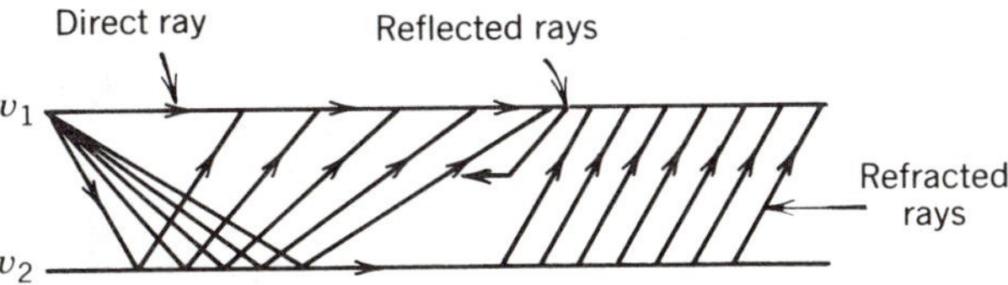

Figure 7.28 Direct, refracted, and reflected ray paths for a two-layer case where $V_1 < V_2$.

usually employ the hammer-plate energy source. The advantages of single-channel seismographs are low initial cost, small size, and the provision for tailoring of geophone recording distances to a given site as work progresses. Disadvantages include time required for obtaining data, lack of a single energy source for the entire geophone spread, and usually a restriction of use to refraction seismic surveys.

Some single- and multichannel seismographs provide for signal enhancement. This important feature permits the stacking or enhancement of shock waves from multiple hammer blows or weight drops at the same spot. Not only will weak signals be strengthened, but random noise present during recording will be partially canceled, improving the signal-to-noise ratio at site where unwanted vibrations create problems.

Refraction Method. Of the three shock-wave travel paths, refracted energy is the most widely used in engineering applications. Among the advantages of the refraction method are surface measurement of wave velocities in buried units, determination of velocity (material) interface dip and irregularities, low-cost equipment, and rapid data computation. Disadvantages are the need for a surface distance of roughly three to five times the depth of investigation for data collection, the need for an increase in velocity of each unit with increasing depth, and increasing minimum layer thicknesses with increasing depth.

The wave-front diagram is a convenient means of illustrating the refraction method. This diagram displays the position of the advancing compression pulse, or first arrival, for a selected time interval. The refraction method utilizes first-arrival energy as opposed to later-arrival energy used by the reflection method. Figure 7.29*a* is a wave-front diagram for two parallel layers, with the higher velocity layer underlying the slower layer to meet the refraction requirements imposed by Snell's law. Wave-front arrivals along the surface can be converted into a travel-time graph having distance and time axes as shown in Figure 7.29*b*. The surface location at which waves from adjacent layers arrive simultaneously is the critical distance. It is shown on the travel-time graph by a change in slope. Each layer velocity is represented on the graph by a linear segment having a slope ideally equal to the inverse velocity of each respective layer. This is a unique attribute of the refraction method that provides characterizing velocities for buried soil and rock layers in their natural *in situ* state of consolidation, fracturing, moisture content, and confinement. Equations for calculating depth to the V_2 layer are shown on Figure 7.30. A calculation example is shown in Module 7.1.

Because the velocity of wave travel through material is an indication of the physical nature of the material, it is imperative that the velocities obtained in the field are representative of the material. If a velocity interface at depth is not parallel with the surface, the velocities recorded at the surface are apparent rather than the true velocities. The velocities recorded and plotted will be less than or greater than true velocity when energy is traveling down dip or up dip, respectively (Figure 7.31). The shallower down-dip end will also exhibit a lower intercept time and critical distance compared to the deeper or up-dip end.

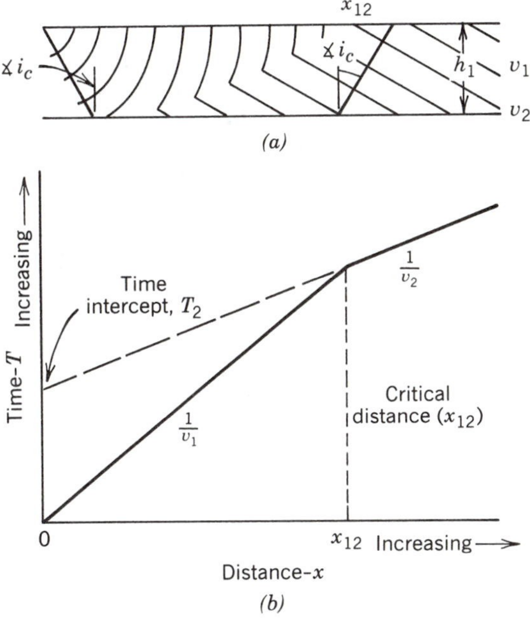

Figure 7.29 Refraction wave-front diagram (*a*) and resulting travel-time graph (*b*) for a two-layer case where $V_1 < V_2$. Angle i_c is the critical angle of incidence and h_1 is the thickness of the V_1 layer.

Recording data in up- and down-dip directions, or reverse profiling, is a necessity in engineering applications as the corrected, or true, velocities obtainable are necessary for correct depth calculation and evaluation of the physical properties of *in situ* material. Reversed profiles also indicate interface irregularities that appear as departures from the ideal linearity of the travel-time velocity lines (Figure 7.32*a*). These time irregularities or departures permit calculation of depths along the irregular interface as shown in Figure 7.32*b* (Hagedoorn, 1959; Hawkins, 1961; Redpath, 1973; Palmer, 1981).

Equations for depth calculations involving more than two layers may be found in the geophysics sources cited earlier. In addition, Redpath (1973) and Mooney (1976a) are recommended for similar information as well as for detailed examples of the influence of subsurface conditions on travel-time graphs and subsequent calculations. These conditions include dipping beds; irregular layer interfaces; velocity reversals with increasing depth, or hidden-layer problem; and minimum layer thickness, or blind-zone problem (Figure 7.33). The velocity-reversal problem often is encountered in engineering applications where dense, high-velocity soils overlie less-dense, low-velocity soils. This situation precludes accurate interpretation and depth calculation. If the V_2 layer velocity in Figure 7.33 is changed to less than that of the V_1 layer, the travel-time graph would also "hide" the V_2 layer and just indicate the V_1 and V_3 layers, with an increase in the critical distance caused by the "lost" time in the V_2 layer.

Snell's law:

$$\frac{v_1}{v_2} = \frac{\sin i_i}{\sin i_r}$$

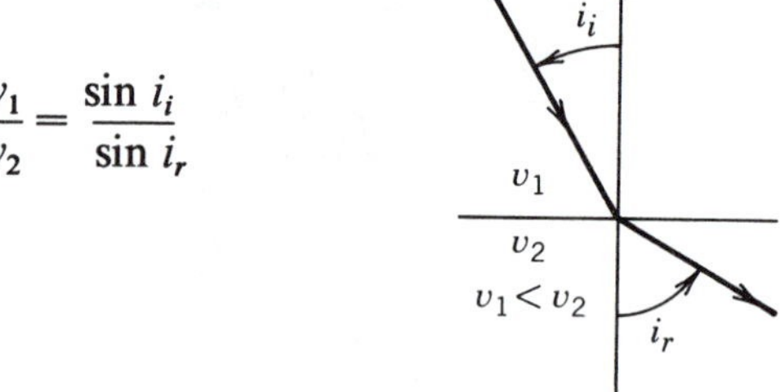

For 90° refraction,

$$\sin i_c = \frac{v_1}{v_2}, \tag{7.1}$$

Where:

i_c = critical angle of incidence

Thickness, h_1, equations:
Time-intercept method

$$h_1 = \frac{T_2 v_1}{2 \cos i_c} \tag{7.2}$$

Critical-distance method

$$h_1 = \frac{x_{12}(1 - \sin i_c)}{2 \cos i_c} = \frac{x_{12}}{2}\sqrt{\frac{v_2 - v_1}{v_2 + v_1}} \tag{7.3}$$

Figure 7.30 Equations required for calculation of upper-layer thickness for a two-layer refraction case. See Figure 7.29 for variables used.

Reflection Method. The use of reflected shock waves to calculate depths to lithologic changes has many advantages over the refraction method. Among the advantages are improved depth-calculation accuracy; better resolution of strata, irrespective of velocity changes and thicknesses with depth; and absence of any dependence of investigation depth on geophone spread length. Figure 7.34 illustrates the ray paths for reflected energy for a two-layer case. In a limiting case, the distance X may equal zero and reflections will still be obtained at the surface, including those from additional reflecting surfaces with increasing depth. The first-arrival refracted energy, by comparison, cannot arrive closer to the energy source than the critical distances for given refraction interfaces. These distances are controlled by the numbers of layers, their thicknesses, and the ratios of velocities from the adjoining layers.

The reflection method utilizes average velocity between the surface and a reflecting surface to calculate depths, thus eliminating the calculation

MODULE 7.1

Two-layer refraction seismic problem (refer to Figures 7.29 and 7.30).

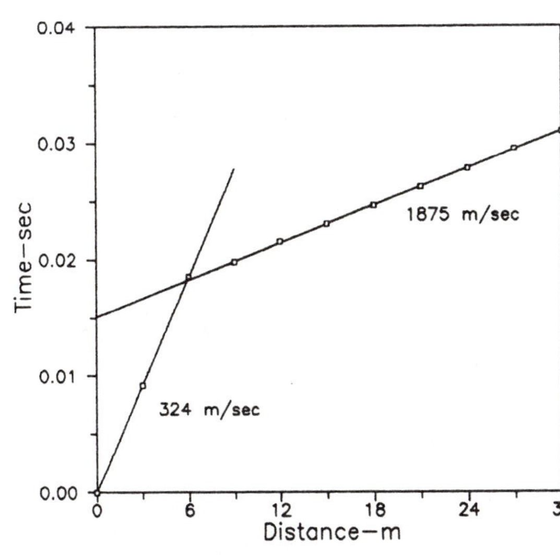

Critical angle, i_c, calculation:

$$\sin i_c = \frac{v_1}{v_2}$$

$$\sin i_c = \frac{324}{1875} = 0.173$$

$$i_c = 10°$$

Data shown by graph:

$v_1 = 324$ m/sec $\quad T_2 = 0.015$ sec

$v_2 = 1875$ m/sec $\quad x_{12} = 5.9$ m

Depth calculations, time-intercept method:

$$h_1 = \frac{T_2 v_1}{2 \cos i_c} = \frac{(0.015)(324)}{(2)(0.985)}$$

$$h_1 = 2.5 \text{ m}$$

Critical distance method:

$$h_1 = \frac{x_{12}(1 - \sin i_c)}{2 \cos i_c} = \frac{(5.9)(1 - 0.174)}{1.970}$$

$$h_1 = 2.5 \text{ m}$$

or

$$h_1 = \frac{x_{12}}{2}\sqrt{\frac{v_2 - v_1}{v_2 + v_1}} = \frac{5.9}{2}\sqrt{\frac{1551}{2199}}$$

$$h_1 = 2.5 \text{ m}$$

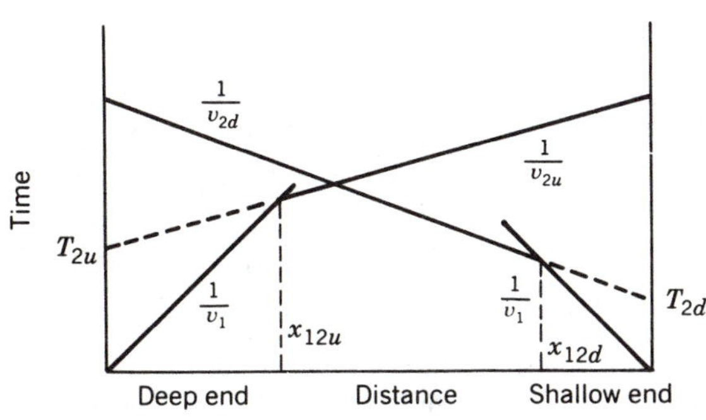

Figure 7.31 Reversed refraction travel-time graph for dipping two-layer cases. V_{2u}, T_{2u} and X_{12u} (updip) > V_{2d}, T_{2d}, and X_{12d} (downdip).

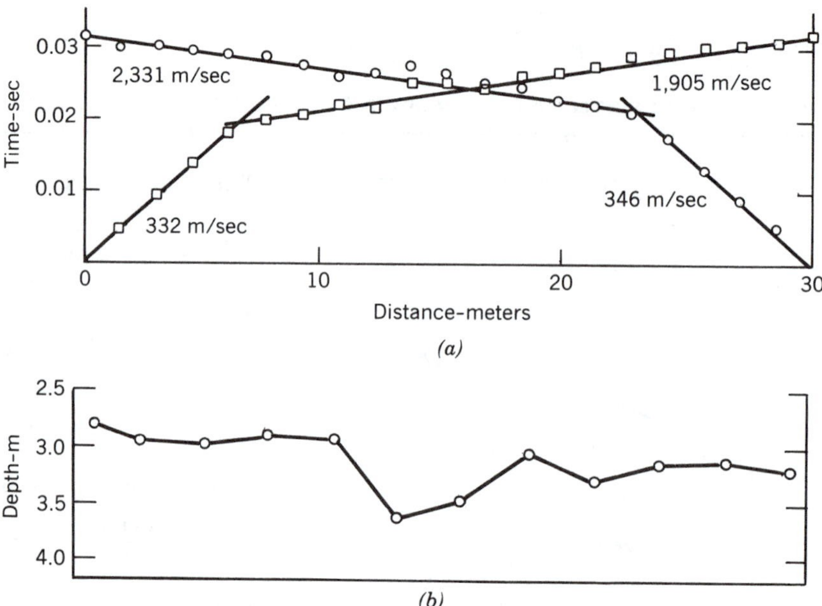

Figure 7.32 Refraction data for site with alluvium over irregular shale bedrock surface. (*a*) Reversed profile travel-time graph. (*b*) Calculated depths to bedrock for points on V_2 layer shared by reversed profiles (Hawkin's method used).

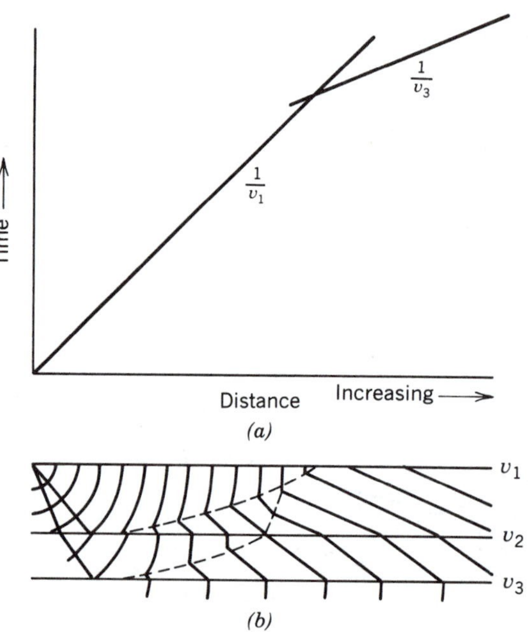

Figure 7.33 (*a*) Refraction wave-front diagram and (*b*) travel-time graph for a blind-zone, or minimum-layer-thickness problem. Note that wave fronts from V_2 do not appear at the surface as first arrivals.

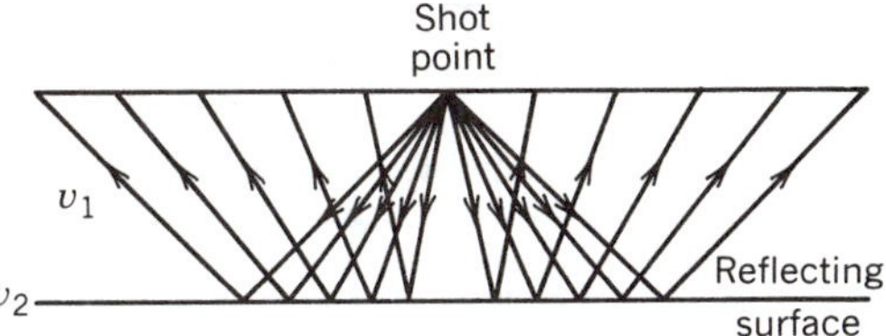

Figure 7.34 Ray-path diagram for reflected waves from a parallel surface. Split-spread profile shown.

errors from velocity reversals or thin layers found in refraction surveys. In reflection surveys, investigation depth is not dependent on horizontal distances, the need for increasing layer velocities with depth, or minimum layer thicknesses.

The average velocities used in reflection seismic surveys do not permit estimates to be made of the engineering properties of the discrete soil and rock units that lie between the surface and the reflecting surface that are included in the average velocity. Average velocities may be obtained by recording travel time from surface-energy sources to receivers at selected depths in a drillhole. In lieu of a drillhole of suitable depth and downhole geophones, reflection times from the source to given geophones may be squared and plotted against the squares of the surface distances from the source to the respective geophones. The average velocity is obtained from the slope of the line, $1/V^2$, as shown by Module 7.2*b* (Dobrin, 1976; Sheriff, 1978). Squaring of distances and times is required for linear plotting of reflection arrivals as noted in the following material.

Reflection data are converted to depth and dip values much differently than by the refraction method. A travel-time plot of reflected energy from a planar interface results in a hyperbolic curve (shown in Figure 7.35 and Module 7.2*a*) rather than the linear direct and refracted wave velocities generated by the same two layers. The reason for this is shown by the ray paths and associated travel times in Figure 7.34.

The hyperbolic curve is the plot of reflected energy from the same $V_1 - V_2$ interface. The reflection curve intersects the time axis at the time required for vertical reflection. The point of tangency with the V_2 line is where the reflection path is the same as that of the refraction critical angle of incidence. With increasing distance the reflected time-line approaches the V_1 velocity.

Depths are calculated by use of travel time to a given geophone along the curve, the surface distance to that geophone, and the average velocity to the reflecting surface. For vertical reflection where $X = 0$ (Figure 7.35 and Module 7.2*b*), this reduces to

$$H = \tfrac{1}{2}(V_{av}T) \tag{7.4}$$

Where:

V_{av} = average velocity to reflector
T = two-way travel time to reflector

MODULE 7.2

Reflection seismic-depth calculation at site having soil over shale.

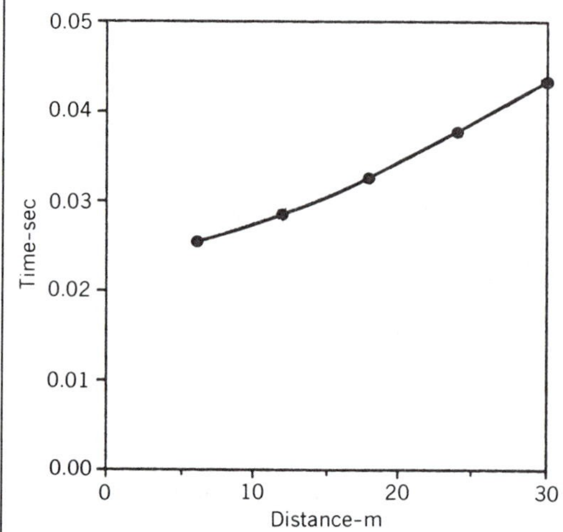

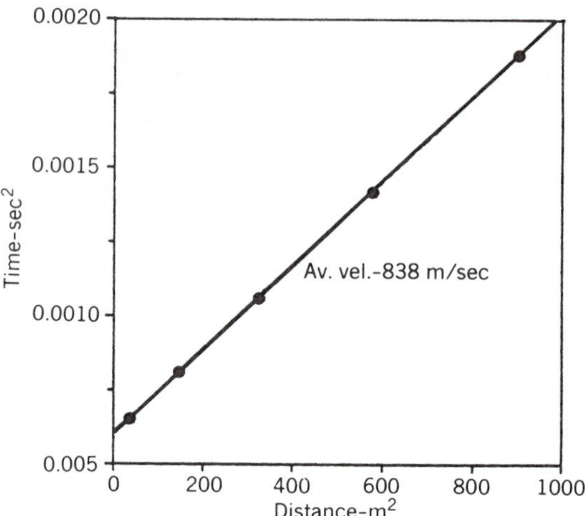

Module 7.2*a* is a travel-time plot of the following data:

Distance—m	Time—sec
6	0.0255
12	0.0285
18	0.0326
24	0.0377
30	0.0434

The plotted curve is a hyperbola, as noted in the text. It is not in the best form for depth calculation. Module 7.2*b* shows the transformation of these data for calculation of average velocity and depth.

Module 7.2*b* is the linear plot obtained from squaring the distance and time data in Module 7.2*a*. Linearity of the plot confirms the input data as reflection events. The slope of the line provides the average velocity of one or more soil layers overlying shale bedrock. The bedrock velocity below the reflection interface does not enter into the calculations. The square root of the time axis intercept (0.0006 sec) is the two-way vertical travel time to bedrock. The depth calculated is 10.3 m using equation 7.4.

Where the reflecting interface is parallel to the surface, reflection times to given distances on either side of the center will be equal as shown by Figure 7.34. The increases in reflection time to the surface outward from the center are known as move-out time and, for this case, normal move-out (NMO) time. For the case of a dipping interface, the move-out times will no longer be the same outward from the center, with down-dip times being greater than NMO and up-dip times being less than NMO. The dip in degrees may be calculated by comparing actual move-out times with the ideal NMO time (Dobrin, 1976).

In spite of its advantages in obtaining more accurate depths to velocity

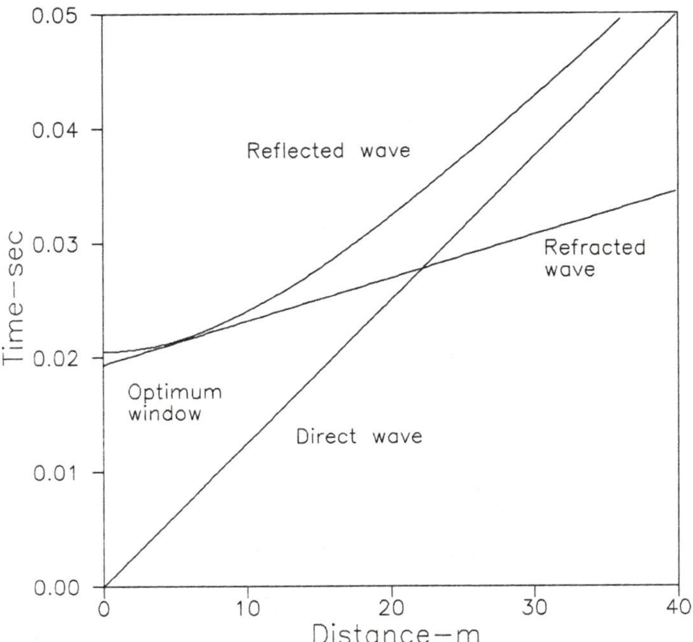

Figure 7.35 Travel-time graph of direct, refracted, and reflected waves for a two-layer case.

interfaces, the reflection method has not found widespread use in shallow engineering applications. A reason may be found by examining a travel-time graph (Figure 7.35) of first-arrival direct, refracted, and reflected energy from the same buried interface. At all points on the reflection curve, the reflected waves arrive later than direct and refracted waves. Normally, only reflection times from much deeper reflecting surfaces can be recorded in the surface distance shown—after all direct and refracted waves have passed. As a result, it is difficult to recognize and use shallow-depth reflection arrivals without some procedural or instrumental changes.

In order to record reflection arrivals from shallow interfaces, geophone spread lengths may be shortened to less than the refraction critical distance for a given buried interface, yet be long enough to record reflection arrivals before passage of surface waves. This recording length is known as the *optimum window* and is shown on Figure 7.35. Electronic filters may be used to attenuate direct and refracted waves and a variety of signal-enhancement procedures may be employed to accentuate reflected arrivals so that dependence on the optimum window is eliminated. Slow surface waves may complicate reception of reflected energy; special geophone arrangements may be used to eliminate their reception. These techniques coupled with new energy sources, recording of digital data, and computer-aided analysis of data are bringing the benefits of reflection seismology to the geotechnical investigation of near-surface materials and the calculation of depths (Mooney, 1976b; Hunter et al., 1982; Barbier, 1983; Hunter et al., 1984; King et al., 1986; Knapp and Steeples, 1986).

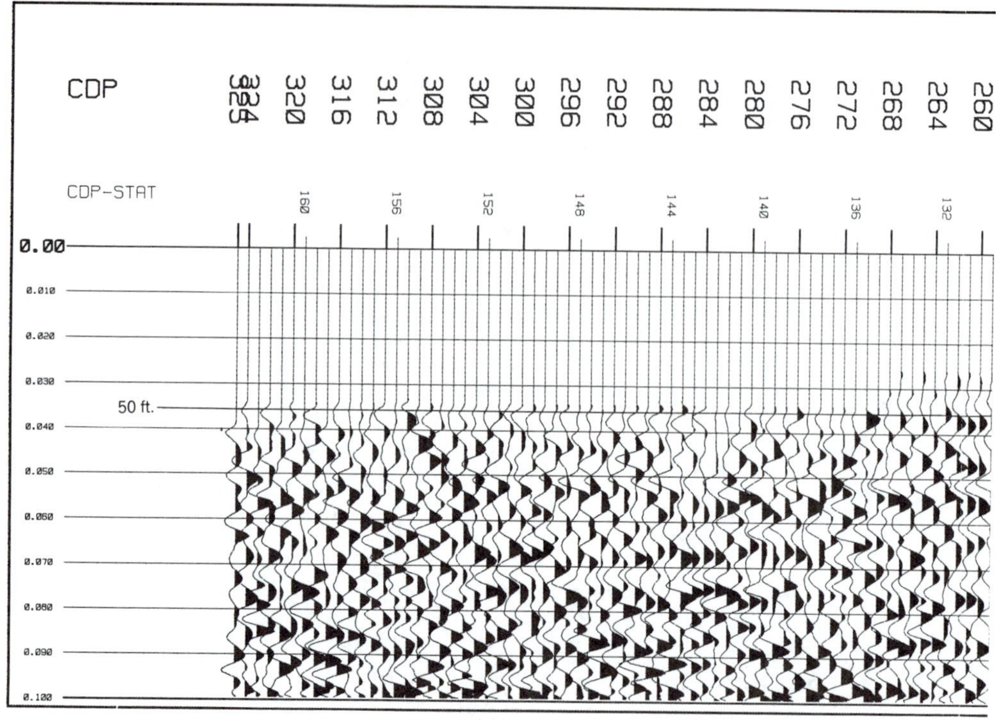

Figure 7.36 Shallow-reflection profile of site having alluvium over shale bedrock. Depth line shown is site specific for average (rms) velocity of materials overlying bedrock.

Figure 7.36 illustrates a computer-processed shallow-reflection profile obtained by using high-frequency geophones and common-depth-point (CDP) and roll-along field techniques (Dobrin, 1976). The soil-bedrock interface is shown at an average depth of 50 ft. This depth was calculated by using the two-way vertical reflection time to the interface and the average velocity of the soil layer. Each vertical trace is the composite of all reflections from the same point (CDP) on the interface at a given distance along the surface. This is an effective means of enhancing the reflection signals.

Optimum use of seismic principles involves a combination of refraction and reflection methods. The refraction method is easier and more economical to use and provides velocity data representative of material physical properties. However, it is limited by velocity and thickness constraints in many potentially useful engineering applications. The reflection method eliminates the velocity reversal and thin-layer problems when calculating depths, but it is more difficult and costly to use. Equipment is available for both methods. The greatest engineering application versatility is obtained from equipment that (a) is adaptable to both methods, (b) records geophone output on hard copy and on tape in digital form, and (c) permits signal enhancement.

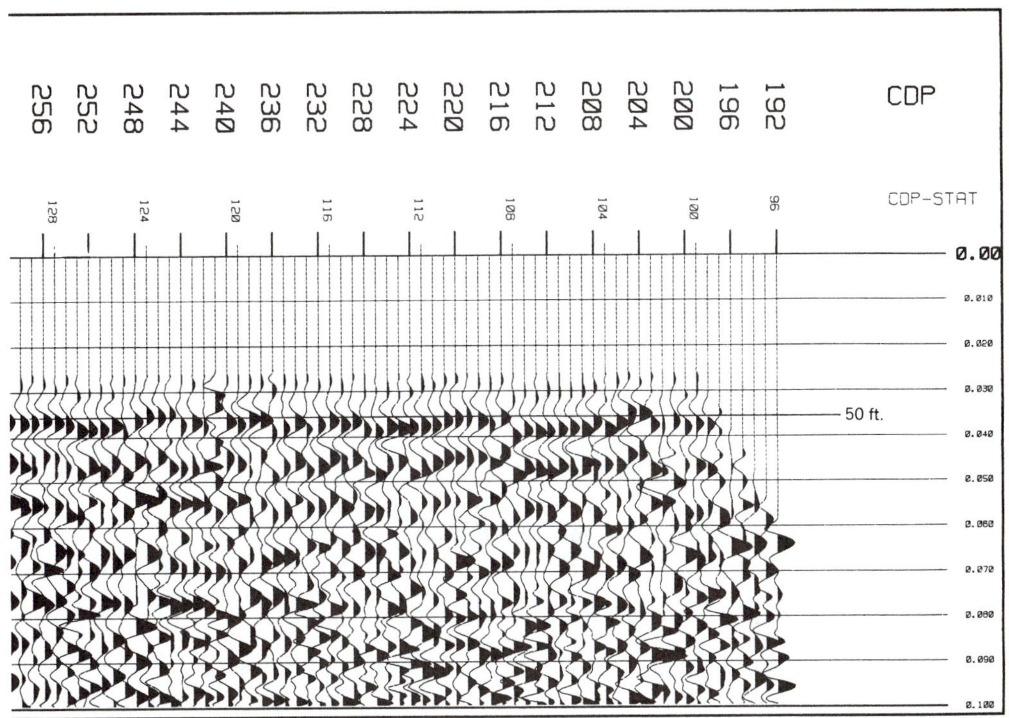

Engineering Applications. Calculation of depth to bedrock is the most common application of the seismic method, with use of the refraction technique the primary choice because both depth and individual velocity data are obtained. Figure 7.37 depicts a refraction travel-time graph from such an application. Site conditions are two alluvium layers over shale. The greater depths are associated with the higher apparent velocities, intercept times, and critical distances. The calculated depth of 7.4 m to the third layer at the A end is greater than the depth of 6.0 m at the B end. Depths were calculated using reversed profile technique illustrated by Figure 7.31.

This example describes the stock-in-trade of seismic-engineering applications. This is true to such a degree that if depth accuracy is poor from velocity reversals or thin beds, the refraction method may be rejected as a means of obtaining useful data. The reflection method, though more demanding technically and operationally, can be substituted if accurate depths are of primary concern. However, much useful information concerning soil and rock quality still may be obtained from recorded refracted velocities.

An example of the use of refraction velocities for other than depth calculation is illustrated by the relationship between rippability, or ease with which soil or rock can be excavated mechanically, and seismic veloc-

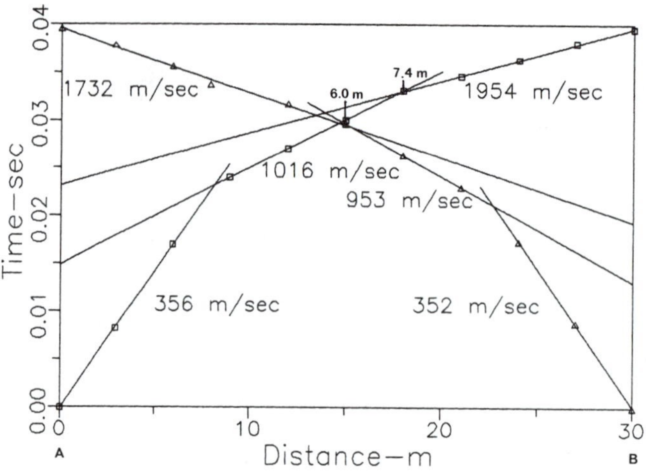

Figure 7.37 Refraction travel-time graph for a site having alluvium (V_1 and V_2) over shale bedrock $9V_3$.

ity. This relationship has been recognized and in use since the 1950s. Charts such as that in Figure 7.38 have been prepared for specific tractor and ripper combinations. Velocity data for determination of rippability may be obtained economically by use of single-channel seismographs. As a generalization, Knill (1974) concluded that most rocks with *in situ* velocities below 2,700 m/sec can be excavated mechanically. Case histories have been published by Bailey (1972).

The refraction method is useful in obtaining thickness data for calculating landslide volumes and slip surface geometry. Landslide movement in soil and/or rock disturbs the natural state, resulting in a reduction in compressional wave velocity in the disturbed materials. The velocity reduction may be sufficient to permit depth calculation to the disturbed/undisturbed material interface. Case histories by Trantina (1963), Carroll et al. (1972), Brooke (1973), Murphy and Rubin (1974), and Bogoslovsky and Ogilvy (1977) provide data for landslides in a variety of soil and rock types.

Reductions in shear-wave velocities in landslide material also have been reported. If shear-wave velocities can be obtained at a landslide site, the V_s/V_p ratios of disturbed and undisturbed materials are of considerable value in defining the slip surface and in seasonally monitoring the water table and degree of saturation in a slide mass.

Prior to slope failure, the opening of cracks in a rock mass will reduce the velocity of shock waves. This relationship between joint frequency and velocity was noted in chapter 4 in the discussion of velocity ratio and RQD. Periodic measurement of velocity changes has been used in areas surrounding open-pit mines to monitor the initiation and progression of tension fracturing that could lead to slope failure (Lacy, 1963; Dechman and Oudenhoven, 1976).

Similarly, attenuation of seismic energy typically is greater in landslide material, a factor related to velocity reduction. Using equipment which can record geophone output, progressive attenuation of energy at a site over a span of time may be suggestive of potential slope failure (Tamaki and Ohba, 1971).

The detection of subaudible rock noise (SARN), or acoustic emissions, associated with movement within a landslide mass has been developed from microseismic monitoring of rock bursts and mine-roof stability initiated by Obert and Duvall (1942, 1957). Typically, SARN is monitored by receivers in boreholes in conjunction with a slope indicator, a tilt meter, and surveying data. McCauley (1976) has emphasized that the noise rate, not the number of events, is the significant measurement. The rate increases as the slope stability decreases. This is illustrated (Figure 7.39) in a classic application of the method in predicting a major slope failure in an open-pit mine by Kennedy and Niermeyer (1971). There is evidence that stability changes can be detected earlier by rock noise rates than by other monitoring systems. Improved instrumentation has resulted in the three-dimensional locating of acoustic emission sources in addition to counting noise rates (Leighton and Steblay, 1977; Hardy, 1981).

Knill (1970) has reported the usefulness of compressional velocity in predicting the amount of cement grout needed to fill joints and fractures at

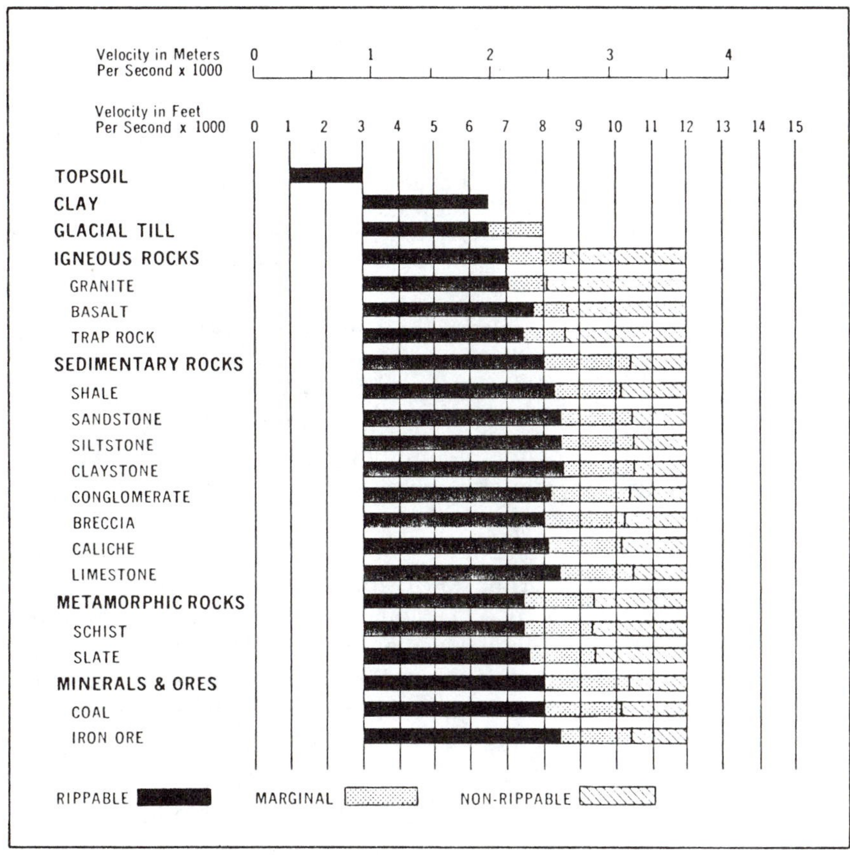

Figure 7.38 Seismic velocity-rippability chart for Caterpillar D9H tractor and No. 9 Series ripper. (Courtesy of Caterpillar Tractor Co.)

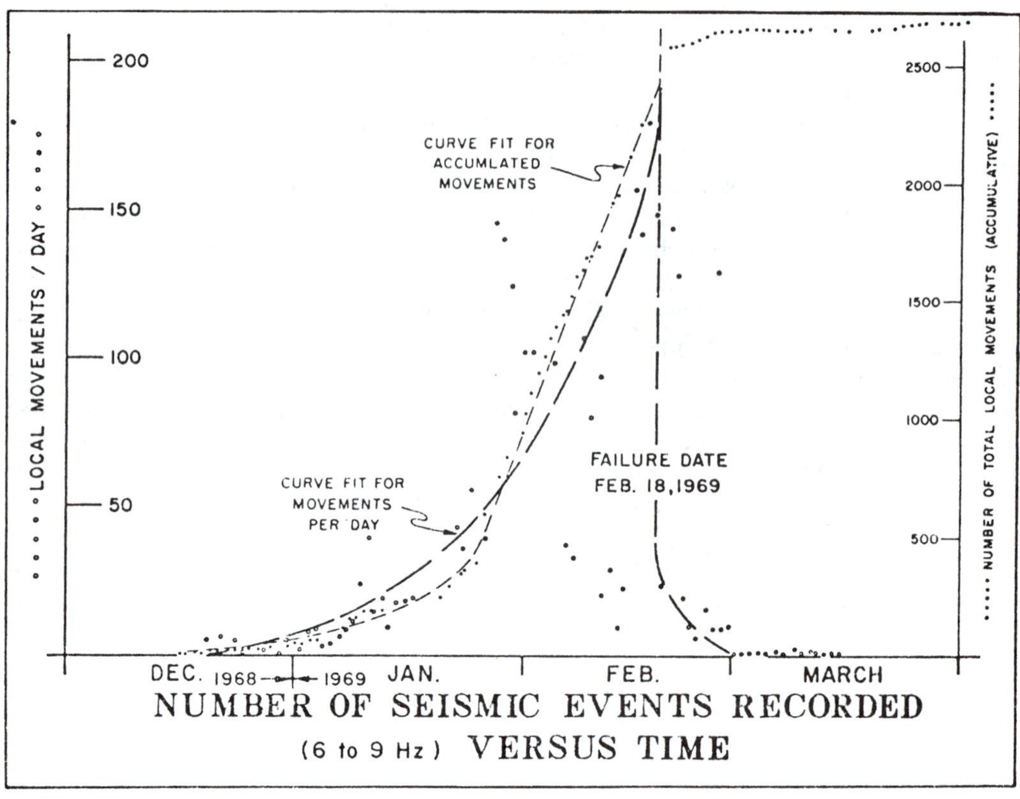

Figure 7.39 Number of seismic events recorded versus time prior to slope failure in the Chuquicamata Mine, Chile. (Reprinted with permission from Proc. Symp. on Theoretical Background to the Planning of Open Pit Mines, B. A. Kennedy and K. E. Niermeyer, Slope Monitoring Systems Used in the Prediction of a Major Slope Failure at the Chuquicamata Mine, Chile, 1971, A. A. Balkema.)

dam sites. The relationship in Figure 7.40 is based on the interactions between grout take, fracture index (velocity ratio, V_{field}/V_{lab}), and the *in situ* velocity, V_{field}. The assumption is made that velocity ratio is a measure of fracturing and serves as a fracture index. Field examination should indicate the validity of the assumption for a given site. The effectiveness of a grouting program also can be evaluated qualitatively by comparing *in situ* velocities before and after grouting, with a velocity increase indicating grout filling of fractures (Lane, 1964; Knill and Price, 1972).

A corollary of the above application is the use of P- and S-wave velocities to estimate rock mass quality and deformation characteristics (Sjøgren et al., 1979; Stephansson et al., 1979; Aikas et al., 1983). Rock mass quality is usually estimated from P-wave velocities obtained from direct- and refracted-wave surveys. Reductions in velocity of waves through rock masses are the result of increased joint frequency and/or weathering or decomposition, both decreasing the quality of the mass. Velocities and RQD values are directly proportional. Examples of this use of seismic velocities include the design of tunnel-support systems and rock-bolting programs (Price et al., 1970).

Uphole (or downhole) and crosshole measurements of seismic velocities (described briefly in chapter 4 relative to velocity or fracture index) are being used widely in more-sophisticated analyses of soil and rock mass dynamic elastic properties. The modulus of elasticity (Young's modulus) and shear modulus are the mass properties of greatest importance, with shear modulus assuming a dominant role with time. The reason for dynamic shear-modulus importance is its use in estimating or determining the response of soil and rock materials to earthquake waves. Applications include seismic response and liquifaction potential of tailings dams, stability analysis of earth dams, and the dynamic behavior of nuclear- and hydroelectric-power-plant foundations to earthquake excitation.

Crosshole measurement of P- and S-waves between two or more boreholes (Figure 4.51) has proven to be of greatest value in the determination and zonation of soil and rock mass dynamic properties. The following references provide a survey of typical engineering uses of downhole and crosshole shear-wave studies: Stokoe and Woods, 1972; Mooney, 1974; Viksne, 1976; Hoar and Stokoe, 1978; Marcuson and Curro, 1981; Baoshan and Chopin, 1983; and Bruce et al., 1983.

In concluding our discussion of engineering applications of seismic methods, it is important to note that the latest development in evaluating mass dynamic properties of soil and rock masses is derived from interdisciplinary research. The concepts and computer software developed for medical computer tomographic (CT) scanning of the human body have been applied to crosshole P- and S-wave propagation. Multiple energy source and receiver positions in adjacent boreholes result in multiple overlapping travel paths through the intervening material similar to those used with medical X-rays. The results are displayed graphically to isolate selected physical properties in the two-dimensional area between the boreholes (Figure 7.41). Papers by Johnson et al., 1979; Lytle, 1979; and Cosma,

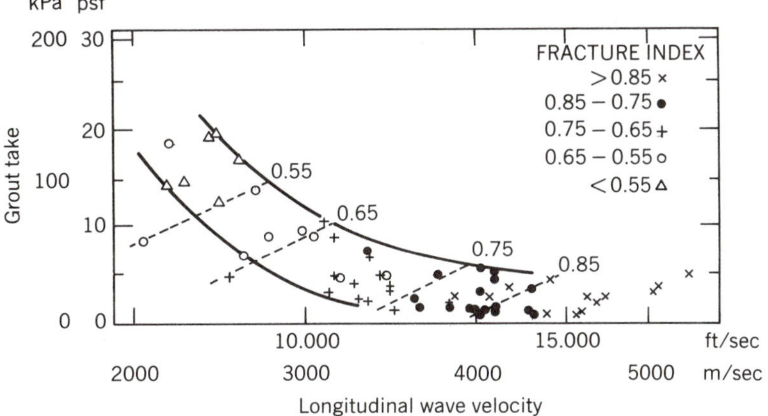

Figure 7.40 Relationships among P-wave velocity, fracture index (V_{field}/V_{lab}), and grout take. (Reprinted with permission from Proc. on In Situ Investigations in Soil and Rock, J. L. Knill, Application of Seismic Methods in the Prediction of Grout Take, 1970. Copyright © the British Geotechnical Society, 1970.)

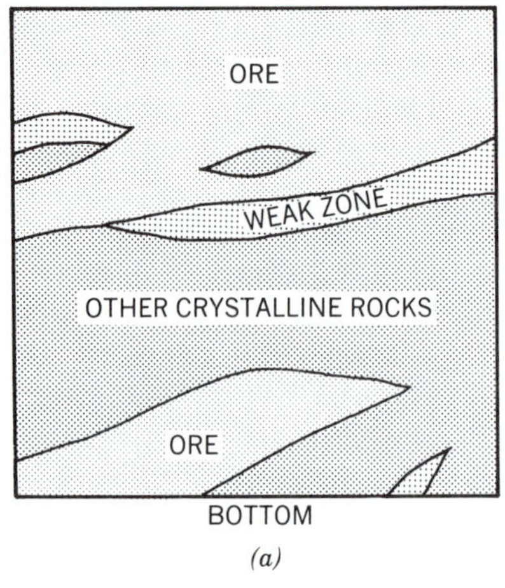

(a)

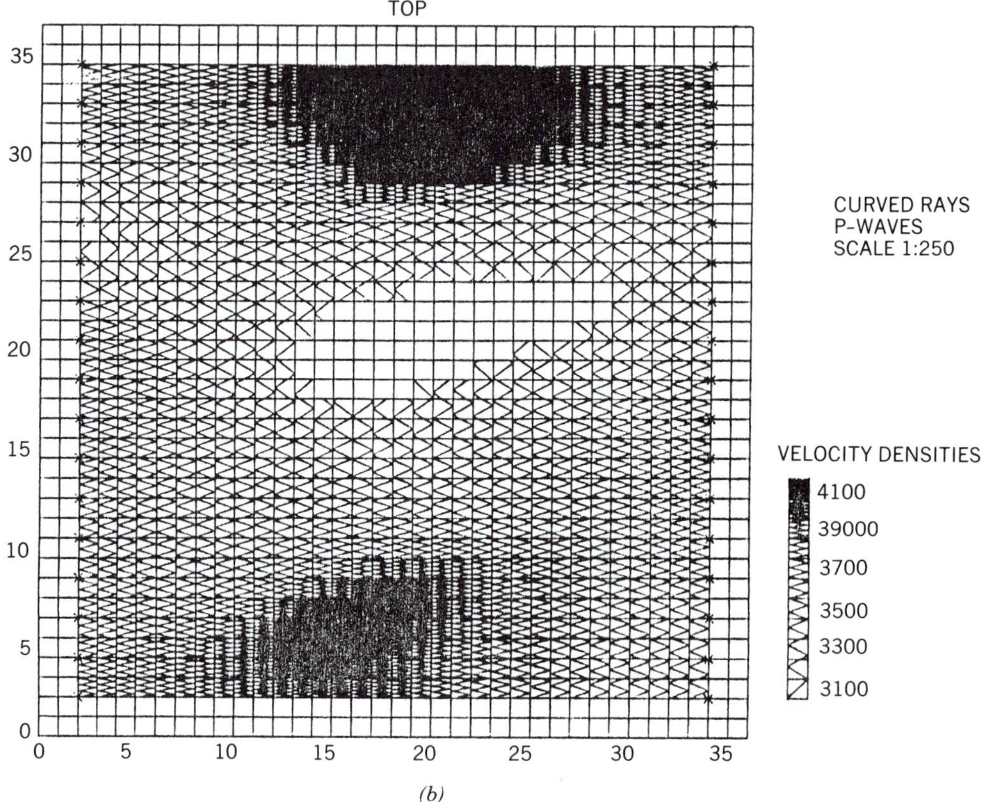

(b)

Figure 7.41 Utilization of computer tomography in crosshole seismic investigation of the subsurface. (a) Generalized geologic cross section where velocity of ore is highest and that of weak zone is lowest. (b) Output from tomographic analysis of cross section where computer velocity-density areas correspond with known geologic conditions. (Reprinted by permission from Bull. Int. Assoc. Eng. Geol., No. 26–27, C. Cosma, Determination of Rock Mass Quality by the Crosshole Seismic Method, 1983.)

1983 summarize this innovative use of another discipline's technology to subsurface investigations.

Electrical Methods

There are many geophysical exploration methods that make use of electrical and electromagnetic principles. In general, they operate by ionic conductance of current through earth materials and by induction of currents in materials by electromagnetic fields. Humanmade and natural sources of electrical energy are used in exploration. Determination of the electrical properties of earth materials may be determined from airborne, surface, and downhole surveys.

The electrical method most commonly utilized for engineering projects involves the surface measurement of resistance of earth materials to controlled current flow by ionic conductance. Its use is based on the fact that variations in mineralogy, texture, and contained water determine the resistance of a material to current flow. The composite influence of these factors governs the resistance of a given material. Because electrical resistance is also a function of the volume being measured, electrical prospecting employs resistivity, which is the unit-volume resistance to current flow. The method utilizing resistivity is the electrical earth resistivity method.

Earth Resistivity Method. In the several variations of the resistivity method, current is introduced to the earth through two electrodes that are pushed into the surface soil (Figure 7.42). The resistance of the material to current flow is obtained from Ohm's law by measuring the potential drop between two electrodes that are positioned within the field of the current electrodes. For a given current flow, the potential drop across the surface will vary with, and be proportional to, the resistance of the material to current flow. The conversion of resistance to resistivity is dependent on the position and spacing of the potential electrodes relative to the current electrodes.

The depth of investigation of a resistivity survey is proportional to the current electrode separation. Depth of investigation increases with increased electrode spacing. The penetration of current lines into the subsurface is, in part, dependent on the resistance of buried layers to current flow. The same electrode separations might result in greatly different current

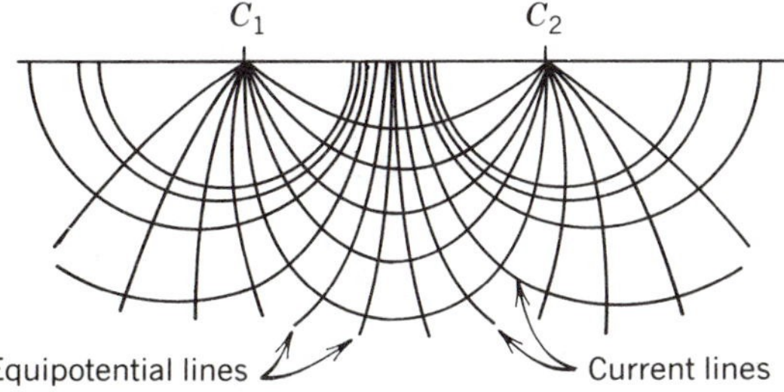

Figure 7.42 Current and equipotential line distribution in a vertical plane through the current electrodes, C_1 and C_2.

flow patterns for a case with a low-resistance surface layer over a buried highly resistive material versus the opposite situation. As a result, no simple proportionality exists between electrode spacing and depth of investigation. This is an inherent weakness of the resistivity methods.

The resistivity values obtained from a survey provide apparent rather than true resistivities. The reason for this is that every volume measured is a composite of all layers above a given current penetration. With increasing depth of penetration, individual layers with differing resistivities will have decreasing influence on calculated resistivity values or they may be masked completely by the overlying material. Thus, a minor change in apparent resistivity measured at the surface may represent a buried layer that is of greater subsurface importance than indicated at the surface. This problem is similar to that encountered in refraction seismology in which resolution of velocity layers with increasing depth requires ever-increasing thicknesses. Some of the interpretation methods to be discussed have been devised to provide improved definition of resistivity layers with increasing investigation depths.

In contrast to wave velocities in refraction seismology, resistivity values are not representative of specific physical properties of earth materials nor do they remain constant over time. The flow of current through soil and rock is by ion conduction, which is dependent on a combination of the conductivity of the fluid present, porosity, and percentage of saturation. Dissolved salts in water provide for ion conductance of the electrical current. The conductance that is the reciprocal of resistance is directly proportional to the amount of dissolved salts in water, or salinity. The amount of fluid regardless of its salinity that can be present is controlled by the porosity of the material. The more interconnected the pore spaces, the greater the ease of ion migration through the material. In addition, the degree of saturation that varies with the seasons, in turn, affects conductance or (in the context of this discussion) resistivity. Seasonal fluctuations in resistivity of as much as 200% have been reported (Brooke, 1973).

The rock-forming minerals normally are highly resistive to current flow. An exception, which complicates resistivity work, is the presence of clay minerals. The exchangeable ions in the clays may separate from the lattice and make the pore water conductive even though the formation water may not be saline. As a result, clays have low resistivities—whether occurring as clay-rich soils or as shales.

If we are certain that the groundwater in an area is fresh, low resistivities are representative of clay. Conversely, freshwater—a poor conductor—will cause high resistivities when present in the pore spaces of a clean or clay-free soil or in the pores or joints of a porous or dense, relatively clay-free rock. Note, however, that there is nothing distinctive about the kind of material that has high- or low-resistivity values, as is the case with seismic velocities. For instance, it would be possible on the basis of high resistivity to drill expecting to encounter a porous, freshwater-bearing sand and instead encounter a tight sandstone. Also, saline pore water in a sand or porous or highly fractured rock gives low resistivity values that are also indicative of clay or shale.

Because of these perplexing problems, there is the need for subsurface control of materials and thickness from either exposures or boreholes. It bears repeating that geophysical methods do not provide primary data and are best used to fill in between drillholes and to locate drill sites. Resistivity surveys are no exception.

Measurements of apparent resistivities in practice cause published tables of resistivity values for various soil and rock types to be misleading when raw field data are examined. Unless layer resistivities are obtained by interpretive procedures, one may find no correlation between field-derived values and tabulated values for a known material. For example, a dry, highly resistive sand overlying a low-resistivity clay will cause all field values to be high even though electrode-separation manipulation permits gathering apparent resistivity data from depths within the clay zone. Variations in pore water salinity also prevent uniqueness of resistivity values. Table 7.17 provides guidelines for estimating resistivity ranges that might be encountered for various materials in the field. Other sources of such data are Clark (1966), Dobrin (1976), and Telford et al. (1976).

The two electrode configurations or arrays most often used are the Wenner and Schlumberger configurations shown in Figure 7.43. Resistivity is calculated when the Wenner array is used by the equation:

$$\rho = 2\pi aR, \text{ where } R(\text{ohms}) = V(\text{voltage})/I(\text{current}) \tag{7.5}$$

The equation for the Schlumberger array is:

$$\rho = \pi(L^2/2l)R \tag{7.6}$$

The distances "a," "L," and "l" may be measured in English or SI units, with the resulting resistivities being ohm-ft or ohm-m.

Use of the Wenner configuration requires that the electrode spacing "a" is kept equal. The distance "L" for the Schlumberger configuration is varied while keeping (within certain limits) "l" constant. The "a" and "L"

Table 7.17 Generalized Resistivity Ranges for Rocks of Different Lithology and Age[a]

Age	Marine Sedimentary Rocks	Terrestrial Sedimentary Rocks	Extrusive Rocks (Basalt, Rhyolite)	Intrusive Rocks (Granite, Gabbro)	Chemical Precipitates (Limestone, Salt)
Quaternary and Tertiary age	1–10	15–50	10–200	500–2,000	50–5000
Mesozoic	5–20	25–100	20–500	500–2,000	100–10,000
Carboniferous Paleozoic	10–40	50–300	50–1,000	1,000–5,000	200–100,000
Early Paleozoic	40–200	100–500	100–2,000	1,000–5,000	10,000–100,000
Precambrian	100–2,000	300–5,000	200–5,000	5,000–20,000	10,000–100,000

[a] ohm-m

(*Source:* Reprinted with permission from G. V. Keller and F. C. Frischknecht, Electrical Methods in Geophysical Prospecting. Copyright © 1966, Pergamon Press.)

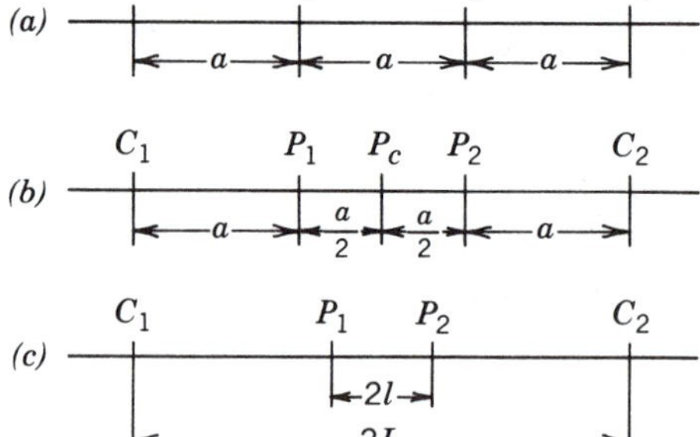

Figure 7.43 Commonly used resistivity electrode configurations. (*a*) Wenner. (*b*) Lee-partitioning. (*c*) Schlumberger. C = current electrodes, P = potential electrodes.

spacings determine depths of investigation for these two resistivity variations.

The apparent resistivities obtained for different electrode spacings are assumed to represent laterally uniform materials. This condition may not be met, especially where soils are involved. Knowledge of lateral changes can be of considerable value in interpreting subsurface conditions. For this reason, the Lee-partitioning configuration, a variation of the Wenner electrode configuration, employs a third potential electrode that is placed midway between the two potential electrodes (Figure 7.43).

Lateral variations in resistance to current flow beneath an electrode spread results in varying potential drops between the potential electrodes because of nonuniform distributions of current and equipotential lines. The third electrode permits detection of this nonuniform distribution by comparing potential drops of the two halves of the original potential electrode spacing. Conversion of these potential readings to resistivities is achieved by using the equation:

$$\rho = 4\pi aR \qquad (7.7)$$

The value of "a" remains the same as that used for the Wenner configuration. Equal calculated resistivity values imply lateral uniformity.

Resistivity surveys may be conducted by vertical and horizontal profiling. In vertical profiling, the center of the electrode spread is kept constant and the electrode spacings "a" and "L" are increased. The resulting resistivity values portray varying apparent resistivities from material and moisture changes as the depth of investigation is increased. For this reason, the method is often referred to as *electrical drilling*, or *sounding*. The resistivity data are plotted on the ordinate relative to increasing electrode spacings on the absissa (Figure 7.44). The use of linear or log-log paper depends on the interpretation technique. A common engineering application is determination of depth to bedrock, based on comparison of data with local control such as borings or exposures. The Wenner survey shown in Figure 7.44 was made at a site that had conductive soil over resistive sandstone bedrock. The influence of the sandstone can be seen, but this type of plot does not provide depth estimates.

Horizontal profiling, or electrical mapping, employs constant electrode spacings with data points at different map locations. This method permits

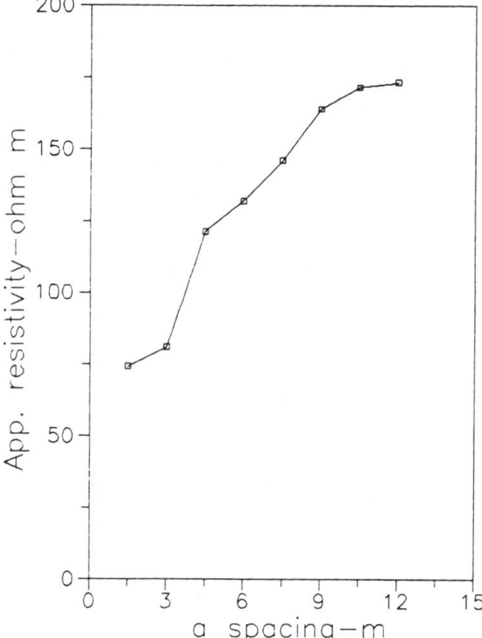

Figure 7.44 Plots of resistivity data—Wenner electrode configuration.

surveying differing resistivity values over an area for a chosen depth of investigation or, more properly, a chosen electrode separation. Contoured values may correspond to bedrock topography, soil changes, and variation in moisture content, depending on the local subsurface conditions, electrical material contrasts, and survey requirements. Figure 7.45 illustrates a

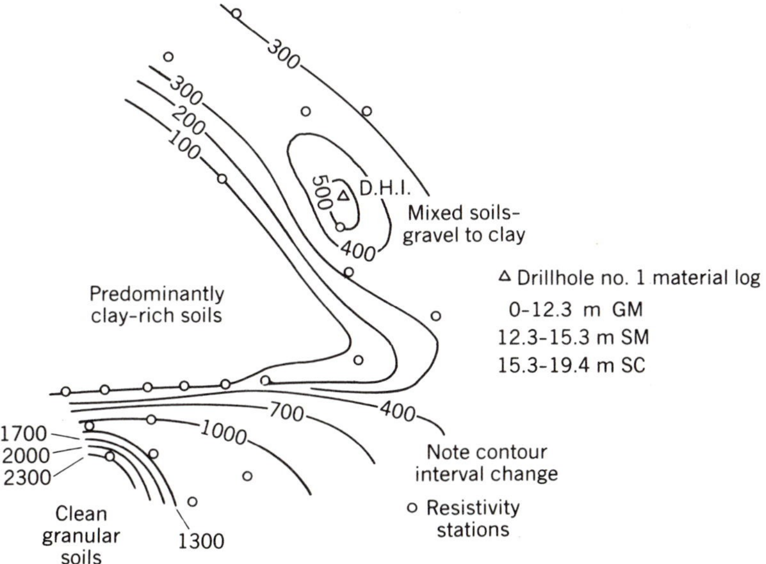

Figure 7.45 Resistivity contour map and associated soil types at a construction site. Resistivity values in ohm-m for an "a" spacing of 18 m.

survey for which the values differ in response to soil changes at an excavation site. The spacings used in horizontal profiling are usually chosen after conducting a vertical-profile survey at a subsurface control point. An investigator may choose to sample more than one electrode spacing in traversing a site.

Several interpretation techniques are available for use in vertical-profile resistivity surveys. The intent of each is to define the depth at which a resistivity change occurs. Two of the techniques to be discussed also provide layer resistivities for better interpretation of subsurface material and moisture conditions. Only in these cases can comparisons be made with tabulated resistivities as those in Table 7.17.

Regardless of the interpretation method employed, the goal of resistivity surveying is to determine the layered nature of resistive materials in the subsurface. Examination of apparent resistivity versus electrode spacing obtained from vertical profiling reveals by observation the layering and the relative resistivities of the layers as shown in Figure 7.46. The type H curve, for example, represents a three-layer case where $\rho_1 > \rho_2 < \rho_3$, where ρ_1 is the surface layer. Combinations of the letters H, K, A, and Q may be used to describe the sequence of resistivities when more than three layers are present. Bhattacharya and Patra (1968) have illustrated a variety of such combinations.

Of the two electrodes arrays in common use (Wenner and Schlumberger), the Wenner configuration and its Lee variation have been used most widely in engineering applications. This is, in part, the result of the availability of empirical interpretation techniques that facilitate rapid field interpretation of Wenner array data. In addition, the Wenner electrode arrangement is more sensitive to local resistivity variations at the shallow depths required for most engineering applications.

Prior to describing the various interpretation techniques, it must be emphasized that all interpretations are inherently ambiguous. Subsurface control must be used for comparison with results obtained from resistivity surveys. An exact match of depths is fortuitous.

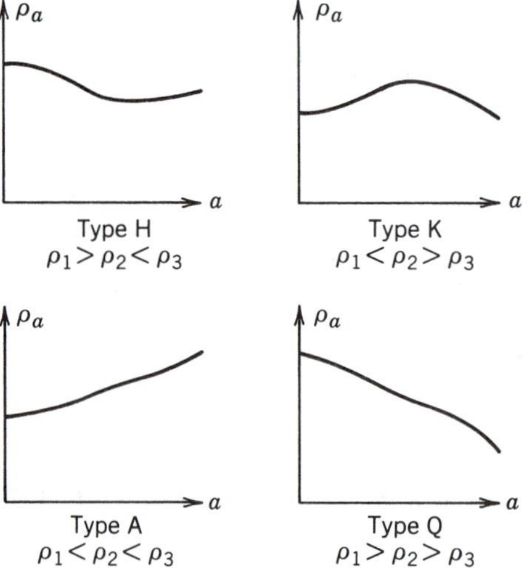

Figure 7.46 Sample curves for multilayered cases.

There is no agreement about the relation of electrode spacing to depth when using empirical interpretation techniques with the Wenner electrode configuration. In shallow-engineering-application practice, the "*a*" spacing is considered to be equal to depth. This relationship should be modified to conform with control data.

The simplest and least satisfactory technique for displaying Wenner and Lee data is that of plotting apparent resistivity versus "*a*" spacing, using linear scales as in Figure 7.44. The layered state of the subsurface is shown with accompanying increases the decreases in apparent resistivity. Lateral variations may be observed on the Lee curves. The electrode spacings (depths) at which subsurface changes take place are seldom clearly defined.

Seemingly minor changes in resistivity at increased electrode spacings on these plots may be representative of significant material differences masked by the overlying material (described earlier). These changes must be recognized and should be included in the subsurface interpretation when they appear systematically in a number of resistivity plots at a given site.

Moore (1944) introduced an empirical interpretation technique in which cumulative or summed apparent resistivity values for increasing "*a*" spacings are plotted on the ordinate versus the respective "*a*" spacings on the absissa, as illustrated in Module 7.3. Straight-line segments are fitted to the points and the intersections may indicate depths to layer interfaces when projected to the absissa. The method requires that the electrode increments are constant for a given plot and that they are small compared with layer thicknesses. Interpretation is improved when cumulative curves are compared with simple linear-data plots.

Although subject to considerable criticism over the years following its introduction, the cumulative plotting technique has proven to be a useful interpretation method when used in conjunction with subsurface control at a site. It often provides reasonable estimates of depths involving contrasting resistivity layers in shallow engineering applications (Telford et al., 1976). The plotting technique provides a graphical indication of changing rates of resistivity increase or decrease with changing depths of investigation.

Another empirical interpretation method, the Barnes layer method, has found increasing use in shallow engineering applications since its introduction (Barnes, 1952). This method was developed to minimize the masking effect on the apparent resistivity of the additional volumes being measured with increased electrode separations. This is accomplished by assuming a horizontally layered system wherein the individual layer resistances act as resistors in parallel with respect to the resistance of the total number of layers. This is shown by the equation:

$$1/R_n = 1/R_n - 1/R_{n-1} \tag{7.8}$$

Where:

$1/R_n$ = conductance of layer n

$1/R_n$ = conductance of mass from surface to bottom of layer n

$1/R_{n-1}$ = conductance of mass from surface to top of layer n

MODULE 7.3

Construction of a Moore cumulative resistivity graph.

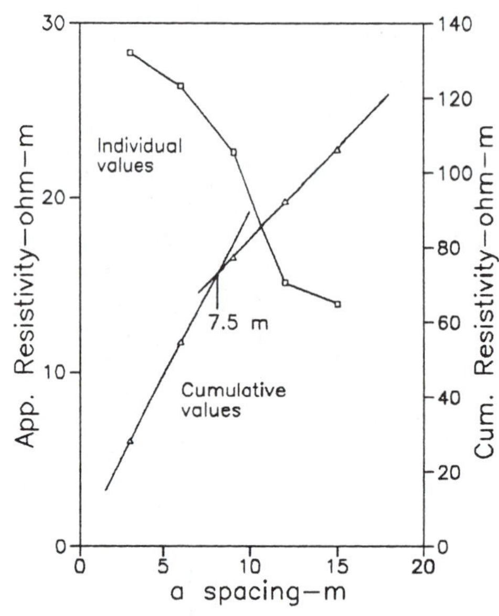

Site conditions
Sandy soil over shale bedrock. Site is same as that illustrated by Figure 7.43.

Wenner field data:

Spacing—m	*App. resis.—ohm-m*
3	28.3
6	26.4
9	22.6
12	15.1
15	13.9

Cumulative resistivity data:

Spacing—m	*App. resis.—ohm-m*
3	28.3
6	54.7
9	77.3
12	92.4
15	106.3

Data evaluation
The influence of the conductive shale under the less-conductive soil is apparent on the plot of individual resistivity values.

The depth to the conductive shale is interpreted from the change in slope seen on the plot of cumulative resistivity values. The depth of 7.5 m compares favorably with the 7.4-m depth obtained by the refraction seismic method.

Resistivity of a layer is obtained from $\rho = 2\pi a' R_n$, where a' is the layer thickness (see equation 7.5).

The layer thicknesses are arbitrarily chosen and their thicknesses and resultant resistivities are intended to serve only as guides and need bear no relation to actual thicknesses and resistivities. The method has produced satisfactory results, especially when combined with the cumulative interpretive approach (Malott, 1969; Franklin and McLean, 1973). Figure 7.47 illustrates the type of cross section that may be constructed from Barnes layer method data.

The only theoretically correct procedure for obtaining layer thicknesses and their resistivities is to match the field curves from vertical profiling to the theoretical curves for various numbers of layers, thicknesses, and relative resistivities. Curve fitting is applicable to data from both Wenner and Schlumberger electrode arrays and is the principal method used to inter-

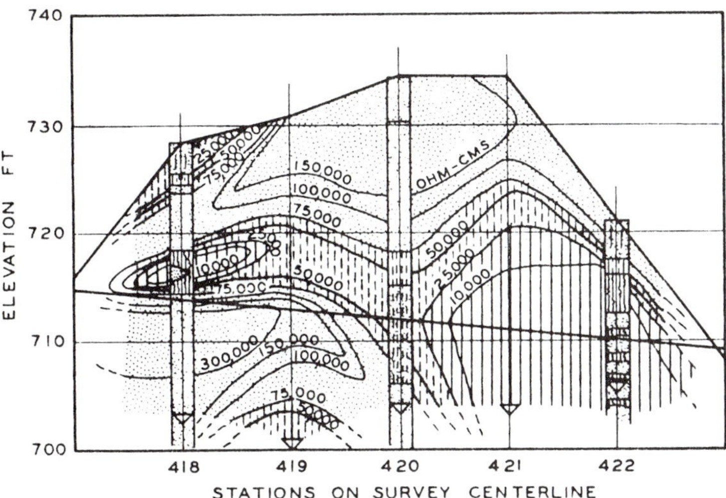

Figure 7.47 Resistivity-material cross section, using the Barnes layer method. (From Malott, 1967)

pret Schlumberger data. Master curves have been published for Wenner and Schlumberger data by Mooney and Wetzel (1956), Compagnie Générale de Géophysique (CGG) (1963), Orellana and Mooney (1966), and Orellana and Mooney (1972). Curve fitting requires layers that are free of lateral changes in resistivity.

Field data are plotted on log-log graph paper that has the same size and number of cycles as the master curves to be used. An example of the overlay of a field curve on a two-layer master curve is shown in Figure 7.48

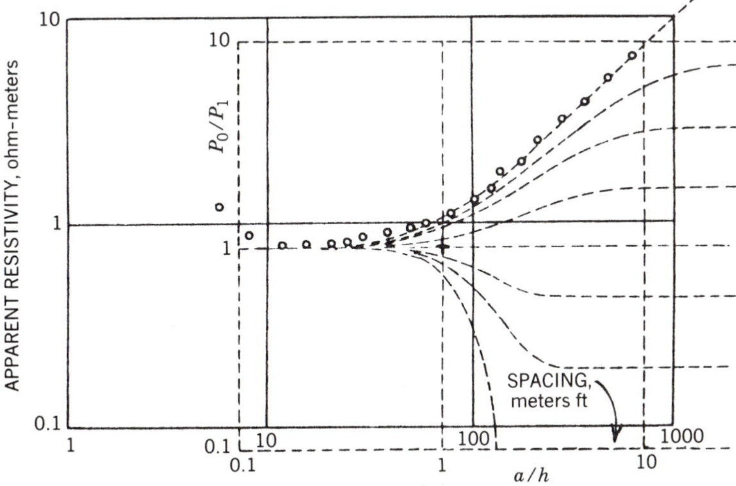

Figure 7.48 Example of the interpretation of a field curve by superposition with a set of two-layer resistivity curves. (Reprinted with permission from G. V. Keller and F. C. Frischknecht, Electrical Methods in Geophysical Prospecting. Copyright © 1966, Pergamon Press.)

from Keller and Frischknecht (1966). Complete and partial curve-matching procedures are too lengthy to reproduce here. Procedural instructions may be found in the publications cited earlier and in Van Nostrand and Cook (1966), Dobrin (1976), and Telford et al. (1976).

Engineering Applications. Although engineering applications of the resistivity method are numerous, success in all cases is dependent on correct evaluation of the probable subsurface electrical contrasts, availability of subsurface control, and selection of an interpretation technique that provides results compatible with the control data. In many cases, a trial test may be conducted prior to designing a field survey in order to determine the usefulness of the method. The presence of any buried conductors such as pipes and culverts must be known prior to a survey, as such features preclude the use of any electrical methods in the immediate vicinity of the conductor.

Of the several engineering applications of the resistivity method, depth determination to bedrock is the most common as illustrated by Module 7.3. Assuming a resistivity contrast between soil and underlying rock, the vertical and horizontal profile techniques employed depend on the need. Horizontal profiling that uses the Wenner electrode array with constant spacing is of special value in outlining irregular karst topography in which clay-rich soil overlies limestone or where water-filled voids occur. The karstic nature of the bedrock surface will be revealed as shown in Figure 7.49. The advantage of this application over exclusive dependence on drillholes is that numerous resistivity points may be obtained rapidly, thereby avoiding the possibility of missing some near-surface rock projections between drillholes.

The resistivity method has been used successfully to map subsurface limits of landslide masses by vertical and horizontal profiling. Landsliding often results in both the disruption of soil and rock materials above the failure surface and the development of an irregular contact along the failure surface. These are conducive to creation of electrical resistivity contrasts.

Conditions may cause accumulation of subsurface water at one point in a landslide mass and drainage at another. Where boring control or surface springs permit recognition of water influence on resistivity, zones of high water content and reduced stability may be isolated and drained for improved stability. These references provide case histories of value relative to use of the resistivity method for surveying landslide masses: Trantina, 1963; Takada, 1968; Moore, 1972; Brooke, 1973; Bogoslovsky and Ogilvy, 1977.

The position of the saturated zone in earth dams has been monitored by the resistivity method for use in seepage studies (Bogoslovsky and Ogilvy, 1970). Resistivity values of material above the water table have ranged from five to ten times those of the saturated zone in these studies. Amimoto and Nelson (1970) have reported on a novel resistivity application in a reservoir. Electrode and cable assemblies were floated and dragged over the reservoir bottom to locate the higher resistivity sand and gravel zones

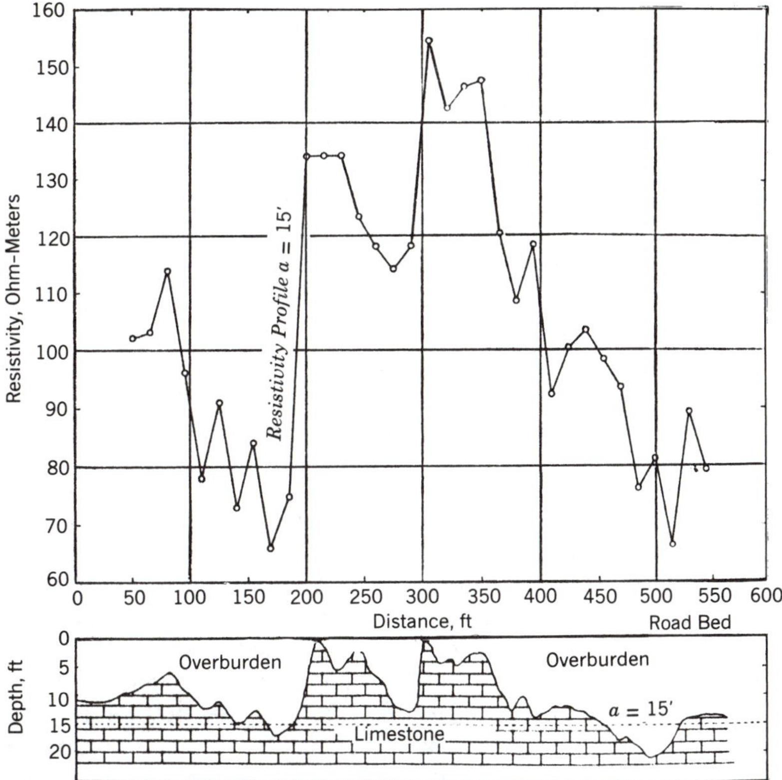

Figure 7.49 Horizontal electrical resistivity profile over sinkhole conditions on Pennsylvania Turnpike, Chester County, for a constant "a" spacing of 15 ft (4.6 m). (From Scharon, 1951. Reprinted with permission, from ASTM Spec. Tech. Publ. 122. Copyright ©, ASTM, 1916 Race Street, Philadelphia, PA 19103.)

through which water was leaking. Location of leakage areas permitted placement of an impermeable blanket on the bottom.

Rock quality has been mapped over a tunnel site in an area of normally high-resistivity igneous and metamorphic rocks (R. B. Johnson, 1970). Assessment of rock quality was based on the reduction in resistivity that occurs with increased porosity and moisture content in shear zones and hydrothermally altered rock. In similar fashion, Donaldson (1975) correlated resistivity, porosity, and rock strength. The ratio of constant short-spacing resistivity values both parallel and perpendicular to shale bedding in a tunnel was used by Schwarz (1972) to determine the occurrence of concretions, excessive gas, water seepage, fracturing, and major joints as tunneling progressed.

Other Methods

The methods described here do not constitute the only geophysical methods useful in civil engineering practice. Methods less-frequently used, but no less effective in special applications, are available. These include geophysical logging of the physical properties of boreholes, microgravity surveys of underground cavities, and investigation of shallow materials and material interfaces by electromagnetic (radar) techniques. Specialized texts and technical papers should be consulted for details and applications.

SUMMARY

Exploration is a body of methods for collecting facts useful in characterizing and defining conditions for solving problems addressed by an investigation.

Maps are a fundamental form of exploration. They provide specific information about the spatial distribution of factors or conditions. Another function of maps is to store and transmit information. Maps may be descriptive, showing factual information such as soil types or depths to the water table, or they may represent an interpretation of those facts for a specific need. Maps of slope stability or liquefaction potential are examples of interpretive maps.

Remote-sensing methods are useful to the engineering geologist in virtually all site-related investigations. Corridor surveys for transportation, railroads, pipelines, electric transmission lines, and water-distribution systems as well as inventories of construction materials and the mapping of terrain features, geologic structure, and rock and soil types all are aided by the utilization of available sensing systems. Unstable slopes and natural and humanmade subsidence are identifiable by selected combinations of sensors and enhancement techniques.

Aerial photographs (black-and-white, color, and color IR) remain the basic sources of remote-sensing data. The MSS systems provide additional information about lithology, soil types, geologic structure, moisture content, drainage patterns, and terrain. Side-looking airborne radar is an especially useful adjunct to air photos where knowledge of regional structural trends is of critical importance, as in power plant and tunnel siting. Thermal IR sensing helps to delineate landslides, sinkholes, and other surface and near-surface conditions influenced by soil moisture.

Subsurface exploration defines subsurface conditions that influence site selection, design, and performance of a project. Exploratory excavations should be planned and conducted so that they relate directly to project needs and enhance information gained from other sources. Face mapping of trenches and pits provides detailed data not available from other surface or subsurface exploration sources.

Several methods of borehole exploration are available for site investigations. Each has its advantages and disadvantages; method selection is dependent on site-investigation requirements, equipment availability, site accessibility, and cost. Both the representativeness and quality of the sampling of subsurface materials are as variable as the kinds of methods.

The geophysical methods most commonly used for subsurface investigations are varieties of the seismic and electrical methods. The refraction seismic method provides more accurate depth measurements if its subsurface requirements are met. Velocities are indicative of the *in situ* physical properties of discrete buried units without the averaging effect of overlying beds noted in resistivity data and with the reflection seismic method. The reflection seismic method is now available for shallow-site investigations, but equipment cost and computer requirements may limit its usefulness in engineering applications.

The resistivity method is a valuable one for identifying layered subsurface materials that have different electrical properties. It is ideally suited to cases that preclude the use of the refraction method because of velocity reversals even though the data are not representative of the physical properties of the individual buried units. Variations of the method may identify lateral material changes in the subsurface that may be of considerable value in site investigations.

REFERENCES

Abrams, M. J., 1980, Lithologic mapping: *in* Remote sensing in geology, John Wiley & Sons, New York, pp. 381–417.

Aikas, K., Loven, P., and Sakka, P., 1983, Determination of rock mass modulus of deformation by hammer seismograph: Bull. Intl. Assoc. Eng. Geol., No. 26–7, pp. 131–133.

Amimoto, P. Y., and Nelson, J. S., 1970, Surveying reservoir leakage using waterborne electrical resistivity method: Bull. Assoc. Eng. Geol., Vol. 7, No. 1/2, pp. 1–9.

Bailey, A. D., 1972, Rock types and seismic velocities versus rippability: Proc. 10th Ann. Eng. Geol. Soils Eng. Symp., Univ. of Idaho, Moscow, pp. 135–142.

Baoshan, Z., and Chopin, L., 1983, Shear wave velocity and geotechnical properties of tailings deposits: Bull. Intl. Assoc. Eng. Geol., No. 26–7, pp. 347–353.

Barbier, M. G., 1983, The Mini-Sosie method: Intl. Human Resources Devel. Corp., Boston, Mass., 90 pp.

Barnes, H. E., 1952, Soil investigation employing a new method of layer-value determination for earth resistivity interpretation: Hwy. Res. Bd. Bull. 65, pp. 26–36.

Barrett, E. C., and Curtis, L. F., 1976, Introduction to environmental remote sensing: Chapman & Hall, London, Eng., 336 pp.

Bhattacharya, P. K., and Patra, H. P., 1968, Direct current geoelectric sounding, principles and interpretations: Elsevier, Amsterdam, Neth., 135 pp.

Blair, M. L., and Spangle, W. E., 1979, Seismic safety and land-use planning—selected examples from California: U.S. Geol. Surv. Prof. Paper 941-B, 82 pp.

Bock, C. G., 1981, The interrelationship of geologic conditions with construction methods and costs, Washington Metro: Bull. Assoc. Eng. Geol., Vol. 18, pp. 187–194.

Bogoslovsky, V. A., and Ogilvy, A. A., 1970, Application of geophysical methods for studying the technical status of earth dams: Geophys. Prospect., Vol. 18, pp. 758–773.

———, 1977, Geophysical methods for the investigation of landslides: Geophys., Vol. 42, No. 3, pp. 562–571.

Bonilla, M. G., Alt, J. N., and Hodgen, L. D., 1978, Trenches across the 1906 trace of the San Andreas fault in northern San Mateo County, California: U.S. Geol. Surv. J. Res., Vol. 6, pp. 347–358.

Boyce, R. C., 1982, An overview of site investigations: Bull. Assoc. Eng. Geol., Vol. 19, pp. 167–171.

Brooke, J. P., 1973, Geophysical investigation of a landslide near San Jose, California: Geoexploration, Vol. 11, No. 2, pp. 61–73.

Bruce, I. G., Wightman, A., and Brown, F. R., Jr., 1983, Foundation dynamic properties of meta-sedimentary rocks from cross-hole seismic tests: Bull. Intl. Assoc. Eng. Geol., No. 26–7, pp. 201–206.

Carroll, R. D., Scott, J. H., and Lee, F. T., 1972, Seismic refraction studies: *in* Geological, geophysical and engineering investigations of the Loveland Basin landslide, Clear Creek County, Colorado, 1963–65, U.S. Geol. Surv. Prof. Paper 673C, pp. 17–19.

Chandler, P. B., 1975, Remote detection of transient thermal anomalies associated with the Portugene Bend landslides: Bull. Assoc. Eng. Geol., Vol. 12, pp. 227–232.

Clark, S. P., Jr. (ed.), 1966, Handbook of physical constants, rev. ed.: Geol. Soc. Am., Mem. 97, 587 pp.

Compagnie générale de géophysique (CGG), 1963, Master curves for electrical sounding, 2nd ed.: Eur. Assoc. Explor. Geophys., The Hague, Neth. 49 pp.

Cosma, C., 1983, Determination of rockmass quality by the crosshole seismic method: Bull. Intl. Assoc. Eng. Geol., No. 26–7, pp. 219–225.

Dechman, G. H., and Oudenhoven, M. S., 1976, Velocity-based method for slope failure detection: U.S. Bur. Mines, Rep. Invest. 8194, 19 pp.

Dedman, E. V., and Culver, J. L., 1972, Airborne microwave radiometer survey to detect subsurface voids: Hwy. Res. Rec., No. 421, pp. 66–70.

DeGraff, J. V., 1985, Using isopleth maps of landslide deposits as a tool in timber sale planning: Bull. Assoc. Eng. Geol., Vol. 23, pp. 445–453.

DeGraff, J. V., and Romesburg, H. C., 1980, Regional landslide-susceptibility assessment for wildlands management: a matrix approach, *in* Thresholds in geomorphology, Allen & Unwin, Boston, Mass. pp. 401–414.

Dellwig, L. F., and Burchell, C., 1972, Side-look radar: its uses and limitations as a reconnaissance tool: Hwy. Res. Rec., No. 421, pp. 3–13.

Dobrin, M. B., 1976, Introduction to geophysical prospecting, 3rd ed.: McGraw-Hill, New York, 630 pp.

Donaldson, P. R., 1975, Geotechnical parameters and their relationships to rock resistivity: Proc. 13th Ann. Eng. Geol. Soils Eng. Symp., Univ. of Idaho, Moscow, pp. 169–185.

Eardley, A. J., 1943, Aerial photographs and the distribution of construction materials: Proc. Hwy. Res. Bd., Vol. 23, pp. 557–569.

Estes, J. E., Mel, M. R., and Hooper, J. O., 1977, Measuring soil moisture with an airborne imaging passive microwave radiometer: Photogramm. Eng. & Remote Sensing, Vol. 43, No. 10, pp. 1,273–1,281

Franklin, A., and McLean, F., 1973, A test of the Barnes layer method: Bull. Assoc. Eng. Geol., Vol. 10, pp. 65–75.

Frost, R. E., and Woods, K. B., 1948, Airphoto patterns of soils of the western United States as applicable to airport engineering: Tech. Devel. Rep. No. 85, U.S. Dept. of Commerce, Civ. Aeronautics Admin., 76 pp.

Geological Society, 1977, The description of rock masses for engineering purposes: Geol. Soc. (London) Eng. Group Working Party, Q. J. Eng. Geol., Vol. 10, pp. 355–388.

Gillespie, A. R., 1980, Digital techniques of image enhancement: *in* Remote sensing in geology, John Wiley & Sons, New York, pp. 139–226.

Goodman, R. E., 1976, Methods of geological engineering in discontinuous rocks: West, St. Paul, Minn., 472 pp.

Hagedoorn, J. G., 1959, The plus-minus method of interpreting seismic refraction sections: Geophys. Prospect., Vol. 7, pp. 158-182.

Hardy, H. R., Jr., 1981, Applications of acoustic emission techniques to rock and rock structures: a state-of-the-art review: Am. Soc. Tes. Mater., Spec. Tech. Publ. 750, pp. 4-92.

Hasan, S. E., and West, T. R., 1982, Development of an environmental geology data base for land use planning: Bull. Assoc. Eng. Geol., Vol. 19, pp. 117-132.

Hatheway, A. W., 1982, Trench, shaft, and tunnel mapping: Bull. Assoc. Eng. Geol., Vol. 19, pp. 173-180.

Hatheway, A. W., and Leighton, F. B., 1979, Trenching as an exploratory method. *in* Geology in the siting of nuclear power plants: Reviews in Eng. Geol., Vol. 4, Geol. Soc. Am., Boulder, Colo., pp. 169-195.

Hawkins, L. V., 1961, The reciprocal method of routine shallow seismic refraction investigations: Geophys., Vol. 26, pp. 806-819.

Helley, E. J., LaJoie, K. R., Spangle, W. E., and Blair, M. L., 1979, Flatland deposits of the San Francisco Bay region, California—their geology and engineering properties, and their importance to comprehensive planning: U.S. Geol. Surv. Prof. Paper 943, 88 pp.

Hoar, R. J., and Stokoe II, K. H., 1978, Generation and measurement of shear waves in situ: Am. Soc. Tes. Mater., Spec. Tech. Publ. 654, pp. 3-29.

Hunter, J. A., Burns, R. A., Good, R. L., MacAuley, H. A., and Gagné, R. M., 1982, Optimum field techniques for bedrock reflection mapping with the multichannel engineering seismograph: *in* Current research, Pt. B, Geol. Surv. Can., Paper 82-1B, pp. 125-129.

Hunter, J. A., Pullan, S. E., Burns, R. A., Gagné, R. M., and Good, R. L., 1984, Shallow seismic reflection mapping of the overburden-bedrock interface with the engineering seismograph—some simple techniques: Geophys., Vol. 49, No. 8, pp. 1,381-1,385.

Janza, F. J. (ed.), 1975, Interaction mechanisms: *in* Manual of remote sensing, Vol. 1, Am. Soc. Photogramm., pp. 75-179.

Jenkins, D. S., Belcher, D. J., Gregg, L. E., and Woods, K. B., 1946, The origin, distribution and airphoto identification of United States soils: Tech. Devel. Rep. No. 52, and App. B, U.S. Dept. of Commerce, Civ. Aeronautics Admin., 202 pp.

Johnson, R. B., 1970, Refraction seismic and electrical resistivity studies of engineering geological problems at mountain tunnel sites: Proc. 8th Ann. Eng. Geol. Soils Eng. Symp., Idaho State Univ., Pocatello, pp. 1-20.

Johnson, S. R., Greenleaf, J. F., Ritman, E. L., Harris, L. D., and Rajagopolan, B., 1979, Three-dimensional analysis and display of rock and soil masses for site characterization: potential contributions from medical imaging techniques: *in* Site characterization and exploration, Am. Soc. Civ. Eng., New York, pp. 322-335.

Johnson, W. E., 1976, Seismic detection of water saturation in unconsolidated material: Proc. 14th Ann. Eng. Geol. Soils Eng. Symp., Boise State Univ., Idaho, pp. 221-231.

Keaton, J. R., 1982, Genesis-lithology-qualifier system of engineering geology mapping symbols: applications to terrain analysis for transportation systems, *in* Transportation Research Record 892, Transp. Res. Bd., Natl. Acad. Sci., Washington, D.C., pp. 69-74.

———, 1984, Genesis-Lithology-Qualifier (GLQ) system of engineering geology mapping symbols: Bull. Assoc. Eng. Geol., Vol. 21, pp. 355–364.

Keller, G. V., and Frischknecht, F. C., 1966, Electrical methods in geophysical prospecting: Pergamon, New York, 517 pp.

Kempton, J. P., and Cartwright, K., 1984, Three-dimensional geologic mapping: a basis for hydrogeologic and land-use evaluations: Bull. Assoc. Eng. Geol., Vol. 21, pp. 317–335.

Kennedy, B. A., and Niermeyer, K. E., 1971, Slope monitoring systems used in the prediction of a major slope failure at the Chuquicamata Mine, Chile: *in* Planning open pit mines, S. Afr. Inst. Min. and Metall., pp. 215–225.

King, K. W., Williams, R. A., and Johnson, R. B., 1986, A high-resolution seismic reflection investigation of shallow horizons at the Denver Federal Center, Lakewood, Colorado: U.S. Geol. Surv. Open-File Rep. OF 86–0448, 34 pp.

Knapp, R. W., and Steeples, D. W., 1986, High-resolution common-depth-point seismic reflection profiling: instrumentation: Geophys., Vol. 51, pp. 276–282.

Knill, J. L., 1970, The application of seismic methods in the prediction of grout take in rock: Conf. on In Situ Invest. in Soils and Rocks, Brit. Geotech. Soc., London, Eng., pp. 93–100.

———, 1974, Engineering geology related to dam foundations: Proc. 2nd Intl. Cong., Intl. Assoc. Eng. Geol., Vol. 1, pp. PC-1.1–PC-1.7.

Knill, J. L., and Price, D. G., 1972, Seismic evaluation of rock masses: Proc. 24th Intl. Geol. Congr., Montreal, Can., pp. 176–182.

Lacy, W. C., 1963, Quantitizing geological parameters for the prediction of stable slopes: Trans. Soc. Min. Eng., Am. Inst. Min. Metall. & Pet. Eng., Vol. 226, pp. 272–276.

Lane, R.G.T., 1964, Rock foundation diagnosis of mechanical properties and treatment: 8th Intl. Cong. Large Dams, Edinburgh, Scot., Vol. 1, pp. 141–165.

Leighton, F., and Steblay, B. J., 1977, Application of microseismics in coal mines: first conference on acoustic emissions/microseismic activity in geologic structures, Trans Tech Publ. Ser. on Rock and Soil Mech., pp. 205–229.

Lillesand, T. M., and Kiefer, R. W., 1979, Remote sensing and image interpretation: John Wiley & Sons, New York, 612 pp.

Lowe, D., 1980, Acquisition of remotely sensed data: *in* Remote sensing in geology, John Wiley & Sons, New York, pp. 48–90.

Lund, W. R., and Euge, K. M., 1984, Detailed inspection and geologic mapping of construction excavations at Palo Verde nuclear generating station, Arizona: Bull. Assoc. Eng. Geol., Vol. 21, pp. 179–189.

Lytle, R. J., 1979, Geophysical characterization using advanced data processing: *in* Site characterization and exploration, Am. Soc. Civ. Eng., New York, pp. 291–301.

MacDonald, H. C., 1980, Techniques and applications of imaging radars: *in* Remote sensing in geology, John Wiley & Sons, New York, pp. 297–336.

Malott, D. F., 1967, Shallow geophysical exploration by the Michigan Department of State Highways: Proc. 18th Ann. Hwy. Geol. Symp., Purdue Univ. Eng. Bull., Vol. 51, No. 4, pp. 104–134.

———, 1969, Uses of geophysics in subsurface surveying: Trans. Soc. Min. Eng., Am. Inst. Min. Metall. & Pet. Eng., Vol. 244, pp. 259–267.

Marcuson, W. F., and Curro, J. R., Jr., 1981, Field and laboratory determination of soil moduli: J. Geotech. Eng. Div., Proc. Am. Soc. Civ. Eng., Vol. 107, No. GT10, pp. 1,269–1,291.

McCauley, M. L., 1976, Microsonic detection of landslides: Transp. Res. Rec. 581, pp. 25–30.

McEldowney, R. C., and Pascucci, R. F., 1979, Application of remote-sensing data to nuclear power plant site investigations: *in* Reviews in Eng. Geol., Vol. 4, Geol. Soc. Am., pp. 121–139.

McKean, J. A., Johnson, R. B., and Maxwell, E. L., 1977, The application of color density enhancement of aerial photography to the study of slope stability: Proc. 15th Ann. Eng. Geol. Soils Eng. Symp., Pocatello, Idaho, pp. 199–216.

Miller, C. D., Mullineaux, D. R., Crandell, D. R., and Bailey, R. A., 1982, Potential hazards from future volcanic eruptions in the Long Valley–Mono Lake area, east-central California and southwest Nevada—a preliminary assessment: U.S. Geol. Surv. Circ. 877, 10 pp.

Mooney, H. M., 1974, Seismic shear waves in engineering: J. Geotech. Eng. Div., Proc. Am. Soc. Civ. Eng., Vol. 100, No. GT8, pp. 905–923.

———, 1976a, Handbook of engineering geophysics: Bison Instruments, Inc., Minneapolis, Minn.

———, 1976b, Shallow reflection seismology: H. M. Mooney, Dept. of Geology & Geophysics, Univ. of Minnesota, Minneapolis, 50 pp.

Mooney, H. M. and Wetzel, W. W., 1956, The potentials about a point electrode and apparent resistivity curves for a two-, three-, and four-layered earth: Univ. of Minnesota, Minneapolis, 146 pp. and reference curves.

Moore, R. W., 1944, An empirical method of interpretation of earth resistivity measurements: Am. Inst. Min. Metall. & Pet. Eng., Tech. Publ. 1743, 18 pp.

———, 1972, Electrical resistivity investigations: *in* Geological, geophysical and engineering investigations of the Loveland Basin landslide, Clear Creek County, Colorado, 1963–65, U.S. Geol. Surv. Prof. Paper 673B, pp. 11–15.

Murphy, V. J., and Rubin, D. I., 1974, Seismic survey investigations of landslides: Proc. 2nd Intl. Cong. Intl. Assoc. Eng. Geol., Sao Paulo, Braz., Paper V–26, Vol. 2, 3 pp.

NAVFAC, 1982, Design manual: soil mechanics: U.S. Dept. of Defense, NAVFAC DM–7.1, Dept. of the Navy, Washington, D. C., 360 pp.

Nilsen, T. H., and Brabb, E. E., 1975, Landslides, *in* Studies for seismic zonation of the San Francisco Bay region: U.S. Geol. Surv. Prof. Paper 941-A, pp. A75–A87.

Norman, J. W., and Watson, I., 1975, Detection of subsidence conditions by photogeology: Eng. Geol., Vol. 9, pp. 359–381.

Obert, L., and Duvall, W. I., 1942, Use of subaudible rock noises for the prediction of rock bursts, II: U.S. Bur. Mines, Rep. Invest. 3654, 22 pp.

———, 1957, Microseismic method of determining the stability of underground openings: U.S. Bur. Mines, Bull. 573, 18 pp.

Orellana, E., and Mooney, H. M., 1966, Master tables and curves for vertical electrical sounding over layered structures: Interciencia, Madrid, Spain, 125 pp., 68 plates.

———, 1972, Two- and three-layer master curves and auxiliary point diagrams for vertical electrical sounding using Wenner arrangement: Interciencia, Madrid, Spain, 43 sheets.

Paine, D. P., 1981, Aerial photography and image interpretation for resource management: John Wiley & Sons, New York, 571 pp.

Palmer, D., 1981, An introduction to the generalized reciprocal method of seismic refraction interpretation: Geophys., Vol. 46, pp. 1508–1518.

Price, D. G., Malone, A. W., and Knill, J. L., 1970, The application of seismic methods in the design of rock bolt systems: Intl. Cong. Intl. Assoc. Eng. Geol., Vol. 2, Sec. 6, pp. 740–752.

Pryor, W. T., 1964, Evaluation of aerial photography and mapping in highway development: Photogramm. Eng., Vol. 30, No. 1, pp. 111–123.

Real, C. R., Toppozada, T. R., and Parke, D. L., 1978, Earthquake epicenter map of California: Calif. Div. Mines and Geol. Map Sheet 39, Sacramento.

Redpath, B. B., 1973, Seismic refraction exploration for engineering site investigations: U.S. Army Eng. Waterw. Exp. Stn., Expl. Excav. Res. Lab., Tech. Rep. E–73–4, 55 pp.

Reeves, R. G. (ed.), 1975, Manual of remote sensing, Vols. 1 and 2: Am. Soc. Photogramm., Falls Church, Va., 2,144 pp.

Rengers, N., 1967, Terrestrial photogrammetry: a valuable tool for engineering geological purposes: Rock Mech. Eng. Geol., Vol. 5, pp. 150–154.

Rengers, N., and Soeters, R., 1980, Regional engineering geological mapping from aerial photographs: Bull. Intl. Assoc. Eng. Geol., No. 21, pp. 103–111.

Rib, H. T. (ed.), 1975, Engineering: regional inventories, corridor surveys, and site investigations: *in* Manual of remote sensing, Am. Soc. Photogramm., pp. 1,881–1,945.

Rib, H. T., and Liang, T., 1978, Recognition and identification: *in* Landslides, analysis and control, Transp. Res. Bd. Spec. Rep. 176, Natl. Acad. Sci., pp. 34–80.

Ross-Brown, D. M., and Atkinson, K. B., 1972, Terrestrial photogrammetry in open pits; Pt. 1—Description and use of the phototheodelite in mine surveying: Inst. Min. Metall., Trans. Sec. A, Vol. 81, p. 205.

Ross-Brown, D. M., Wickens, E. H., and Markland, J., 1973, Terrestrial photogrammetry in open pits; Pt. 2—An aid to geological mapping: Inst. Min. Metall., Trans., Sec. A, Vol. 82, pp. 115–130.

Ruhe, R. V., 1952, Topographic discontinuities of the Des Moines lobe: Am. J. Sci., Vol. 250, pp. 46–56.

Sabins, F. F., Jr., 1978, Remote sensing—principles and interpretation: W. H. Freeman, San Francisco, Calif., 426 pp.

Scharon, H. L., 1951, Electrical resistivity geophysical method as applied to engineering problems: ASTM, Spec. Tech. Publ. 122, pp. 104–114.

Schwarz, S. D., 1972, Geophysical measurements related to tunneling: Proc. No. Am. Rapid Excav. Tunnel. Conf., Vol. 1, pp. 195–208.

Sheriff, R. E., 1978, A first course in geophysical exploration: Intl. Human Resources Devel. Corp., Boston, Mass 313 pp.

Siegal, B. S., and Gillespie, A. R. (eds.), 1980, Remote sensing in geology: John Wiley & Sons, New York, 702 pp.

Siegal, B. S., and Goetz, A.F.H., 1977, Effect of vegetation on rock and soil type discrimination: Photogramm. Eng. & Remote Sensing, Vol. 43, No. 2, pp. 191–196.

Siegrist, A. T., and Schnetzler, C. C., 1980, Optimum bands for rock discrimination: Photogramm. Eng. & Remote Sensing, Vol. 46, No. 9, pp. 1,207–1,215.

Sjøgren, B., Øfthus, A., and Sandberg, J., 1979, Seismic classification of rock mass qualities: Geophys. Prospect., Vol. 27, pp. 409–442.

Soule, J. M., 1976, Geologic hazards in the Crested Butte–Gunnison area, Gunnison County, Colorado: Colo. Geol. Surv. Inf. Ser. 5, 34 pp.

———, 1980, Engineering geologic mapping and potential geologic hazards in Colorado: Bull. Intl. Assoc. Eng. Geol., No. 21, pp. 121–131.

Stallard, A. H., 1972, Use of remote sensors in highway engineering in Kansas: Hwy. Res. Rec. No. 421, pp. 50–57.

Stephansson, O., Lande, G., and Bodare, A., 1979, A seismic study of shallow jointed rocks: Intl. J. Rock Mech. Min. Sci. & Geomech. Abstr., Vol. 16, pp. 319–327.

Stokoe II, K. H., and Woods, R., 1972, In-situ shear wave velocity by cross-hole method: J. Soil Mech. Found. Eng. Div., Proc. Am. Soc. Civ. Eng., Vol. 98, No. SM5, pp. 443–460.

Takada, Y., 1968, A geophysical study of landslides: Bull. Disaster Prevention Res. Inst., Kyoto Univ., Japan, Vol. 18, Pt. 2, No. 137, pp. 37–58.

Tamaki, I., and Ohba, Y., 1971, Shallow reflection method and its use for landslide investigation: Proc. 4th Asian Reg. Conf. Soil Mech. Found. Eng., Bangkok, Thai., Vol. 1, pp. 227–233.

Telford, W. M., Geldart, L. P., Sheriff, R. E., and Keys, D. A., 1976, Applied geophysics: Cambridge Univ. Press, 859 pp.

Trantina, J. A., 1963, Investigation of landslides by seismic and electrical resistivity methods: Am. Soc. Tes. Mater. Spec. Tech. Pub. 322, pp. 120–133.

UNESCO, 1976, Engineering geological maps, a guide to their preparation: UNESCO Earth Sciences Ser. No. 15, UN Educational, Scientific, and Cultural Organization, Paris, France, 79 pp.

USBR, 1974, Earth manual, 2nd ed.: U.S. Bur. of Reclamation, Denver, Colo., 810 pp.

Van Nostrand, R. G., and Cook, K. L., 1966, Interpretation of resistivity data: U.S. Geol. Surv. Prof. Paper 499, 310 pp.

Varnes, D. J., 1974, The logic of geological maps, with reference to their interpretation and use of engineering purposes: U.S. Geol. Surv. Prof. Paper 837, 48 pp.

Viksne, A., 1976, Evaluation of in situ shear wave velocity measurement techniques: Report No. REC–ERC–76–6, U.S. Bur. Rec., Denver, Colo., 40 pp.

Wagner, T. W., 1972, Multispectral remote sensing of soil areas: A Kansas study: Hwy. Res. Rec. No. 421, pp. 71–77.

Warwick, D., Hartopp, P. G., and Viljoen, R. P., 1979, Application of the thermal infra-red linescanning technique to engineering geological mapping in South Africa: Q. J. Eng. Geol., Vol. 12, pp. 159–179.

Way, D. S., 1973, Terrain analysis, a guide to site selection using aerial photographic interpretation: Dowden, Hutchinson & Ross, Stroudsburg, Pa., 392 pp.

Wentworth, C. M., Ziony, J. I., and Buchanan, J. M., 1970, Preliminary geologic environmental map of the greater Los Angeles area, California: U.S. Geol. Survey, TID–25363, 41 pp., issued by U.S. Atomic Energy Comm., Oak Ridge, Tenn.

West, T. R., Mundy, S. A., and Moore, M. C., 1976, Evaluation of gravel deposits using remote sensing data, Wabash River Valley north of Terre Haute, Indiana: Proc. 27th Ann. Hwy. Geol. Symp., Orlando, Fla., pp. 199–214.

Wickens, E. H., and Barton, N. R., 1971, The application of photogrammetry to the stability of excavated rock slopes: Photogramm. Rec., Vol. 7, pp. 46–54.

Wieczorek, G. F., 1984, Preparing a detailed landslide-inventory map for hazard evaluation and reduction: Bull. Assoc. Eng. Geol., Vol. 21, pp. 337–342.

Williams, J.C.C., 1969, Simple photogrammetry: Academic Press, London, Eng., 211 pp.

Zall, L., and Michael, R., 1980, Space remote sensing systems and their application to engineering geology: Bull. Assoc. Eng. Geol., Vol. 17, pp. 101–152.

Zellmer, J. T., Roquemore, G. R., and Pannuto, B. J., 1984, Investigation and repair of tectonic and storm-related road damage near the Garlock fault, California: Bull. Assoc. Eng. Geol., Vol. 21, pp. 495–507.

8

Construction Uses of Rock

The use of rock as a construction material is varied, widespread, and of great economic importance—ranging from construction of highway and airport pavements, dams, buildings, and tunnel linings to railroad ballast, breakwater construction, riprap facings for earth dams, roofing materials, and in other less obvious, but nonetheless important, ways. Although here we restrict our discussion to the use of rock in construction, it should be noted that rock-products are used in the manufacture of portland cement, steel, aluminum, and alkalies as well as for municipal water filtration.

The term *mineral aggregate*, or aggregate, is used in this text for sand, gravel, and crushed rock or stone. Larger-size rock material is referred to by terms such as *riprap, armor stone, rockfill*, and *rubble*. Blast furnace slags and lightweight or artificially expanded aggregates are used in construction, but they are beyond the scope of this work.

Sands and gravels are classed as natural aggregates because they occur as products of natural processes. Other aggregates are produced from rock by quarrying and crushing or by the crushing of natural gravels and cobbles to obtain the desired sizes. Fine aggregates are those materials that pass a No. 4 (4.75-mm) sieve (ASTM, 1978). Coarse aggregates pass a 100-mm (4-in.) square-opening sieve and are retained on a No. 4 sieve. The sizes of selected sieve numbers for fine aggregates and grain sizes for coarse aggregates and larger materials are listed in Table 8.1.

In a general sense, aggregates and larger materials such as riprap are classed as sound if they perform satisfactorily in a given application. In addition, soundness may apply to the response of rock to a specified test utilizing sodium or magnesium sulfate. In such cases, to avoid confusion, it is preferable to refer to the sulfate soundness of the material.

There are many tests to determine the soundness of rock to be used for construction. However desirable it might be to be able to classify a mate-

Table 8.1 Selected Sizes for Larger Aggregates and Standard Sieve Numbers with Opening Sizes for Fine Aggregates

	Grain Size	
Sieve No.	mm	in.
—	1219.20	48.0
—	762.00	30.0
—	457.20	18.0
—	304.80	12.0
—	203.20	8.0
—	101.60	4.0
—	76.20	3.0
—	50.80	2.0
—	38.10	1.5
—	25.40	1.0
—	19.05	0.75
—	12.70	0.50
—	9.53	0.375
4	4.75	0.187
8	2.38	0.0937
10	2.00	0.0787
16	1.19	0.0469
30	0.595	0.0234
50	0.297	0.0117
100	0.149	0.0059
200	0.074	0.0029

rial's soundness, it is not possible to do so. The many uses of rock in construction preclude universal classification and uniform test specifications. In practice, the greatest number of applications of rock, related problems, and tests to ascertain rock suitability for a given application involve the coarse aggregates. With a few exceptions, coarse aggregates will be the size range dealt with during the remainder of this chapter. Construction applications will be an integral part of the description of aggregate properties, problems, and tests.

AGGREGATES

Regardless of how an aggregate is used, the geologic properties of the material will be basic to its performance and the selection of appropriate preconstruction testing. Rock type, mineralogy, and texture are several of the geologic factors that govern physical resistance to crushing, abrasion, volume change, and chemical decomposition—whether from normal weathering processes or from reaction to chemically reactive environments. A review of igneous rock composition and textures; metamorphic rock composition, texture, and structure; and the wide ranges of sedimentary rock types, compositions, and textures is recommended if the reader is not familiar with them.

Although somewhat dated, papers by Rhoades and Mielenz (1948) and Woods (1948) are classic treatments of the distribution, petrography, and mineralogic characteristics of mineral aggregates as they occur in the United States. In general, aggregates, whether natural or crushed rock, will be obtained from igneous, nonfoliated to weakly foliated metamorphic, and carbonate and well-cemented detrital sedimentary parent materials.

A survey of the desirable and undesirable, or deleterious, characteristics of aggregates in their many construction applications provides a basis for investigating aggregate properties, problems, and tests. We now address the combined geologic characteristics of aggregates and their performance in given applications. No attempt is made to prioritize our approach because of the differing engineering uses of aggregates, aggregate availability, climate, and economics.

Geologic and Performance Characteristics

Petrographic Analysis

An examination of the large- and small-scale characteristics of the rock (or rocks) used as aggregates is basic to any further testing or selection for use. The most obvious and essential first step is to identify the rock type or types involved. In many cases, microscopic examination is needed to make a proper assessment of potential problems and to select the pertinent tests. Some aggregate applications require a more-detailed examination of the insoluble residues that remain after solution in acid. X-ray diffraction and/or differential thermal analysis may be used in such cases. All of these procedures are grouped under the general heading of petrographic analysis.

Proper identification of rock type is possibly of greater importance when using mineral aggregates than in any other phase of engineering geology. Identification and understanding of the identification carries with it important factors such as mineral composition, grain or crystal size, textural characteristics, and internal structure. These singly or in combination may adversely affect the performance or in-service life of a structure as well as the safety of the public. Examples include the disfigurement of the exterior of concrete structures, structural failure of a concrete structure, short pavement life, and skid susceptibility of pavement surfaces in wet weather.

The importance of the use of specific geologic rock-type terminology has been demonstrated by Ramsay et al. (1974). In the United Kingdom, industry usage of the term *granite* includes granite pegmatite and gneiss. This has resulted in wide and artificial textural, grain-size, and structural ranges when compared to the restricted geologic definition of granite. In turn, some "granite" performance characteristics, such as crushing resistance, which are dependent on these rock properties, differ greatly with respect to the definition used.

The simple determination of whether an aggregate is a relatively soft carbonate or a hard silicate rock is essential from the standpoint of predicting abrasive wear. Grain or crystal size and texture also control such performance. Susceptibility of aggregates to chemical reactions in the highly alkaline environments of portland cement concrete is a function of the composition of the aggregate. The results of such reactions are examined microscopically. Resistance to structural failure of portland cement concrete in a climate where freezing and thawing cycles occur is related to aggregate mineralogy and porosity. An assessment of how large rock materials such as riprap will perform may be obtained from viewing natural exposures. Mineralogy and grain size are important factors that affect the initial and in-service skid resistance of asphaltic concrete (bituminous, blacktop, flexible) pavements. The application of petrography to aggregates is extensive and its use and associated petrographic techniques are noted throughout the remainder of this chapter. Mielenz (1966) has prepared a comprehensive summary of the reasons and methods for the petrographic examination of aggregates in general.

Alkali-Aggregate Reactions

The highly alkaline environment created when water is added to portland cement in making portland cement concrete may cause severe reactions with the coarse aggregate used. Sodium and potassium alkalies are normal, though variable, constituents in portland cement. Deleterious aggregate reactions may occur when a variety of siliceous minerals and rocks and certain dolomitic limestones are used as aggregate. Additional factors are the alkali type and content in the cement, the amount of water available, and the aggregate particle size.

All the reactions result in cracking of the hardened concrete paste from expansion of the aggregate particles (Figure 8.1). In addition, aggregate pieces may crack and the bond between the aggregate particle and the paste may be reduced by a silica-gel reaction at the particle-paste interface. The results of these reactions range from disfiguring (but not serious) surface popouts to structural failure of concrete structures. Well-illustrated papers by McConnell et al. (1950) and Mielenz (1962) are recommended for obtaining a visual appreciation of the reactions and their products. The various deleterious alkali reactions have been described and classified by Gillott (1975) as alkali-silica, alkali-silicate, and alkali-carbonate reactions.

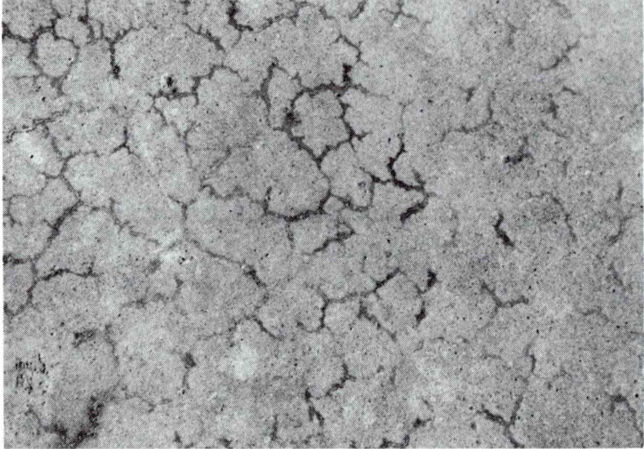

Figure 8.1 Map or pattern cracks in portland cement concrete from alkali-silica reactions.

Alkali-Silica Reactions. In the alkali-silica reaction, aggregate expansion results from the formation of a silica gel that continues to increase in volume as long as water is available. The reaction rate is dependent on the reactivity of the aggregate, abundance of reactive aggregate in the concrete aggregate, aggregate size, alkali content of the cement, temperature, and availability of water (Mielenz, 1962; Hansen, 1966; Gillott, 1975).

The highly reactive aggregates range from the finely crystalline, or microcrystalline, to the amorphous silicate minerals. These include chert, chalcedony, opal, cristobalite, tridymite, and cryptocrystalline quartz that occur as discrete aggregate particles or as inclusions within rock. Both volcanic rocks with glass and cryptocrystalline varieties of quartz and sedimentary rocks with reactive siliceous minerals of organic and diagenetic origin are the common sources of deleterious materials. The relative rates of expansion of chert-bearing mortar bars—the result of alkali-silica reaction to different alkali contents—are shown graphically by Figure 8.2. A factor that causes large variation in chert expansion or reactivity is porosity. The more porous cherts provide a greater surface area for alkali reactions to occur.

Alkali-Silicate Reactions. The low-grade metamorphic rock phyllite has been recognized as an expansive aggregate material in the alkaline portland cement environment for many years (McConnell et al., 1950; Mielenz, 1962). Today, it is known that some graywackes and argillites are also expansive. The presence of fine-grained quartz in these rocks has led to the conclusion that they expand solely from the alkali-silica reaction. However, although this may occur, expansion also has been shown to take place physically in certain of the phyllosilicate minerals found in these rock types (Gillott, 1975).

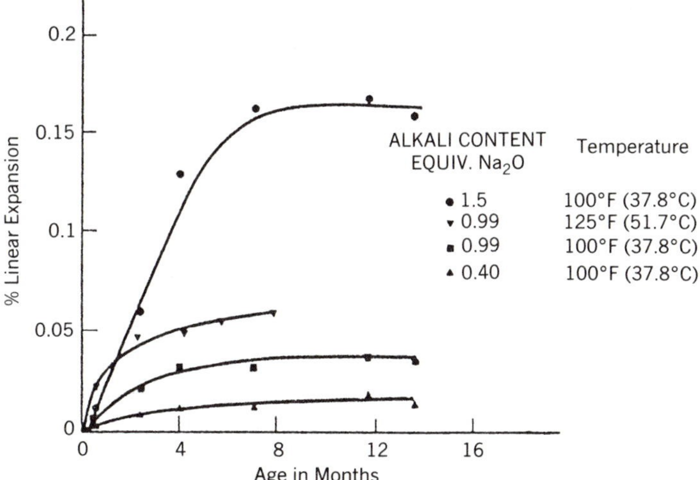

Figure 8.2 Effect of different alkali-content portland cements on expansion of chert-bearing mortar bars caused by alkali-silica reactions. (Reproduced by permission from Eng. Geol., Vol. 9, J. E. Gillott, Alkali-aggregate reactions in concrete, 1975, Elsevier Science Publishers, B. V.)

Several common phyllosilicate minerals are talc, chlorite, kaolinite, serpentine, and muscovite. They have been shown to swell, not unlike vermiculite, in an aqueous alkaline environment; this adds to the expansive characteristics of the aggregate pieces from other causes. Thus the term *alkali-silicate reaction* has been established for this more-complex expansive alkali reaction.

Alkali-Carbonate Reactions. Carbonate aggregates are not immune from alkali-reaction problems. Some argillaceous dolomitic limestones are expansive in portland cement concrete and are subject to the alkai-carbonate reaction. All such reactive aggregates are coarse in size and very fine grained, with small isolated rhombs of dolomite in a matrix of clay and fine-grained calcite (Hadley 1964; Gillott, 1975). One process that causes expansion and resultant breakage of the aggregate particle and surrounding cement paste is known as dedolomitization. It is the process by which dolomite is converted to calcite and brucite, resulting in volume increase. The process occurs in an alkaline environment by the following general reaction (Hadley, 1964):

$$\underset{\text{dolomite}}{CaMg(CO_3)_2} + \underset{\text{alkali}}{2MOH} \rightarrow \underset{\text{brucite}}{Mg(OH)_2} + \underset{\text{calcite}}{CaCO_3} + \underset{\text{carbonate}}{M_2CO_3} \tag{8.1}$$

where:

M represents potassium, sodium, or lithium

Until hardening of the paste occurs, the reaction is self-perpetuating, as the carbonate formed reacts with calcium hydroxide in the portland cement and water mix as follows:

$$M_2CO_3 + Ca(OH)_2 \rightarrow 2MOH + CaCO_3 \tag{8.2}$$

A number of factors influence the rate of dedolomitization, given the earlier stated aggregate characteristics. The greatest amount of reaction occurs when the following are combined: (1) approximately equal amounts of calcite and dolomite, (2) the finer the size of calcite grains, (3) the more alkaline the cement used, (4) increased aggregate size, and (5) when clay-sized material (acid insoluble residue) increases from 5% to 25% by rock weight (Hadley, 1964; Mather, 1974). The influence of the first three of these factors on carbonate aggregates is illustrated by Figures 8.3, 8.4, and 8.5.

The part played by the clay-sized insoluble residues in carbonate aggregates has been in question since the alkali-carbonate reaction was recognized. Hadley (1964) has concluded that this fine material (1) serves as an osmotic membrane that assists in water movement to the reaction sites; (2) keeps the dolomite crystals separated, thus providing greater surface areas for the reaction to occur; and (3) decreases the tensile strength of the rock; thus reducing its ability to withstand the expansive forces generated by the dedolomitization process. Gillott (1975) has accepted the dedolomitization

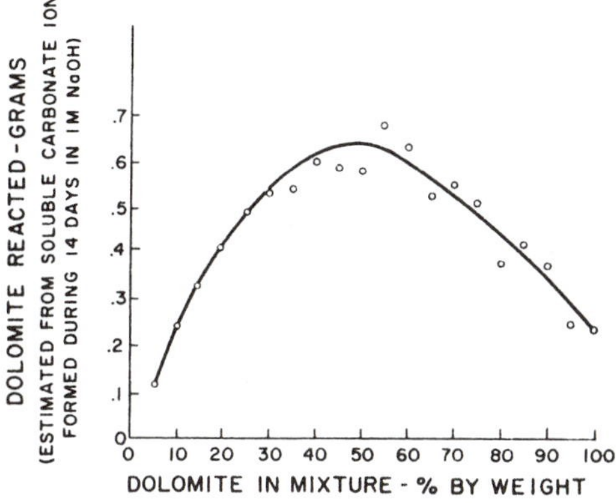

Figure 8.3 Reaction of dolomite in the dedolomitization process as a function of the dolomite-calcite ratio present in the aggregate. (From Hadley, 1964)

process as an integral part of the alkali-carbonate reaction but has argued that the process alone cannot explain the volumetric increases measured. He has viewed the role of dedolomitization as that of creating fractures in the cement paste through which moisture may enter the concrete.

The clay minerals present as clay-sized particles play an active role in expansion by means of generation of swelling pressures owing to moisture intake. Swelling pressures are also generated by surface hydration and sorption along the fractures. It is possible that some expansion or weakening is generated by the development of siliceous or positive reaction rims separating the aggregate and paste matrix (Lemish et al., 1958). The reaction rims are positive relative to the surrounding carbonate when the surface is etched with acid. Silicate migration from the aggregate insoluble residue and reaction with the newly formed brucite may be the source of both the rims and the expansion-generated cracks associated with them.

Alkali reactions with aggregates are not all deleterious. Limestone ag-

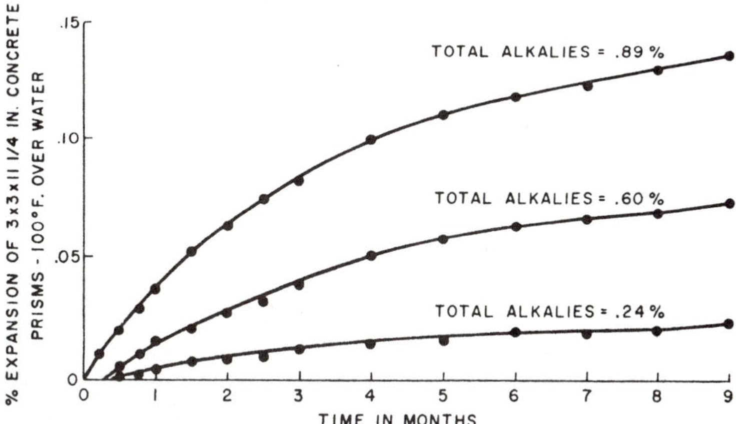

Figure 8.4 Influence of alkali content of cement on concrete expansion owing to alkali-carbonate reaction. Percentage expansion measured on 7.62 × 7.62 × 28.58-cm concrete prisms at 37.8°C over water. (From Hadley, 1964)

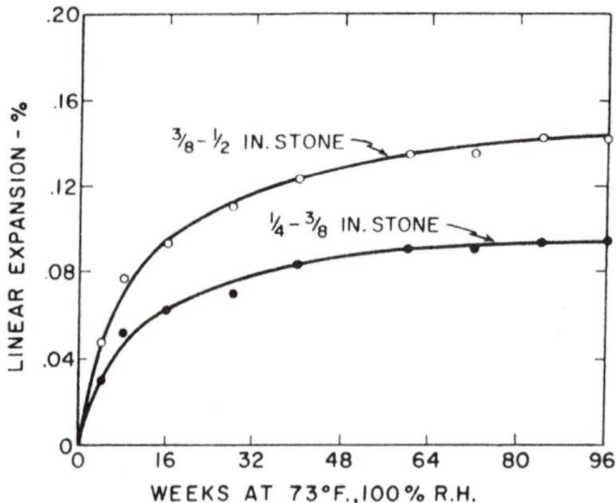

Figure 8.5 Effect of aggregate particle size on the alkali-carbonate reaction expressed as linear expansion of portland cement concrete. Percentage expansion for 0.64–0.95 cm and 0.95–1.27 cm particle sizes at 23°C (73°F) and 100% relative humidity. (From Hadley, 1964)

gregate long has been recognized as having an excellent performance record in portland cement concrete. Reaction rims may be seen serving as bonds at the aggregate-cement paste boundaries in many cases. When polished samples of concrete are etched with acid, the reaction rims dissolve more readily than the adjoining carbonate aggregate and paste. These negative reaction rims are the product of a beneficial alkali-aggregate reaction (Hadley, 1964; Mather, 1974).

Remedial Methods. The problem of alkali-reactive aggregates can be reduced by one or more methods. The method used will depend on economics and the engineering design requirements for the portland cement concrete. The availability of coarse aggregate to a construction site is of primary economic concern as haulage costs may require use of a nearby aggregate that is less than ideal. Economics may permit blending of higher-cost, less-reactive aggregate with the local less-satisfactory material. This procedure is called beneficiation. It may be possible in some cases to selectively produce aggregate from a nearby quarry to reduce the percentage of reaction-susceptible materials. When this is not possible, the strength requirements for the concrete may permit use of lower alkali-content cement.

When little can be done to alter the aggregate or alkali content of the cement, a pozzolanic cement may be used. A pozzolan is a fine-grained siliceous material that is added during the manufacture of portland cement. It may be volcanic ash (or tuff), siliceous shales, diatomaceous earth, fly ash, or similar material. The purpose of the fine pozzolanic additive is to react readily with the calcium hydroxide in the cement and water mix to reduce the alkali available for reaction with coarse aggregate. The product of the pozzolanic reactions contributes to the cementing properties of the mix. Other useful attributes of pozzolanic cement have been listed by Mielenz (1962). A summary of research on the complex interactions of pozzolanic material in the formation of portland cement concrete has been prepared by Gillott (1975).

Aggregate Degradation

The degradation of aggregates involves the breakdown of aggregate particles into smaller particles (including fines) by natural, construction, and in-service processes. A change in aggregate gradation may affect a variety of engineering uses of aggregate. When used as base-course and subbase-course materials to support rigid portland cement concrete and especially flexible asphaltic concrete pavements, densification will accompany a change in gradation. This results in loss of pavement support and ultimate pavement failure. The support provided railroad ties by ballast also can be reduced in similar fashion (Gaskin and Raymond, 1976).

The breakdown of aggregate by freezing and thawing and by wetting and drying are varieties of mechanical degradation. Another common form of mechanical degradation occurs during construction of pavement-supporting layers, or base courses, by crushing and abrasion. Following construction, degradation results from interparticle friction occurring within and under flexible asphaltic pavements.

The release of fine weathering and alteration products from some aggregates such as basalt causes coatings on aggregate particles and reduction in drainage that ultimately results in pavement failure. This variety of degradation is referred to as chemical degradation, which incorrectly implies chemical decomposition during and following construction. The fines released are the products of earlier chemical breakdown—for example, devitrification of natural glass particles—and the formation of clays from decomposition of the more unstable mineral components. In the United States, the area having the most problems of this type includes portions of Washington, Oregon, and Idaho where basalts that have these degradation characteristics are the most readily and economically available aggregates.

The definition of degradation does not state that the processes involved are detrimental. The change in gradation from construction may be a desirable means of obtaining a design density. However, degradation in general does have a negative impact before, during, and following construction. The emphasis, therefore, will be directed to the detrimental aspects of aggregate degradation and tests used to predict performance.

Freeze-Thaw Susceptibility and Testing. Response to cycles of freezing and thawing affects the soundness, or durability, of many aggregates. Some coarse aggregates will physically break down from being subjected to freezing and thawing cycles. The results may include a change in aggregate gradation as smaller sizes are formed, loss of adherence to asphaltic pavement materials as new surfaces are formed, popouts on portland cement concrete surfaces, and loss of strength in portland cement concrete as cracks form in the cement paste surrounding the frost-susceptible aggregate particles. The cracks may develop as map or pattern cracks (Figure 8.1) or as D-cracks parallel to edges and joints.

If fine materials are produced by freeze–thaw breakdown, supporting aggregate layers also undergo a reduction in drainage characteristics and may become susceptible to frost heaving. Retention of water may provide the moisture needed for continued freeze–thaw breakdown as well as further loss of support strength though friction reduction and material loss by the vertical motion (pumping) of pavement or ties under vehicle passage. Aggregate with a specified gradation stockpiled for later use may no longer have the required gradation following a season or two of multiple freeze–thaw cycles (West and Aughenbaugh, 1964).

In an early paper that dealt in part with the freeze–thaw susceptibility of certain aggregates, Rhoades and Mielenz (1948) recognized the combined factors of low permeability, a critical saturation level, and progressive freezing of the aggregate as being important. These factors continue to be of importance with some additions, refinement, and quantifying of threshold conditions. The aggregate materials subject to freeze–thaw breakdown are almost exclusively of sedimentary origin when occurring as fresh material. They include pure to argillaceous limestone and dolomite, shale, and chert (Mielenz, 1962; Stark, 1976). Strongly weathered, porous aggregate of most rock types may be frost susceptible (Dolar-Mantuani, 1983).

Effective porosity, a rock property related to permeability and used in reservoir engineering, has been combined with pore-size estimates to refine the requirement for low permeability. Effective porosity is a measure of just the interconnected pore space in a rock or mineral rather than total porosity. It is known that in addition to the stress directly applied by the freezing of pore water, aggregate particles and surrounding portland cement concrete undergo tensile failure as water is forced through minute passageways, or capillary channels, owing to the expansion caused by the progressive freezing of pore water (Dolch, 1966). Failure appears to occur as a function of the speed with which the water can move through the interconnected pores of low-permeability materials relative to the freezing rate and the distance it has to travel before hydraulic pressures are dissipated.

If an aggregate particle is small, the distance to the surface required for pressure relief to occur is short and failure may not occur. A maximum dimension of 1.27 cm (½ in.) for coarse aggregate has been found to be an optimum size for prevention of aggregate failure from freezing in susceptible materials (Dolch, 1966; Stark, 1976). The use of air-entrained concrete with many closely spaced, well-distributed small voids reduces the probability of portland cement concrete failure because pressures are dissipated in the voids, resulting in a minimum of fractures in the concrete.

In parts of the country where chert is a gravel constituent or is quarried with limestone, it creates freeze–thaw problems in addition to its alkali reactivity when used in portland cement concrete. Certain cherts fail from cycles of freezing and thawing and produce popouts when near the concrete surface. Structural failure, indicated by reductions in relative dynamic elastic moduli, occurs when the cherts are situated deeper within the concrete. The more porous cherts are the most frost susceptible of the varieties of chert. Lounsbury and Schuster (1964) found that concrete beams with greater than 6% chert having bulk-specific gravities less than

2.45 had significant deep-seated failure and surface deterioration from 300 freeze–thaw cycles. Similar results are illustrated by Figure 8.6, in which relative dynamic modulus E is defined by equation 8.3. Petrographic analysis indicated that chert porosity was, in part, the result of the solution of carbonate grains in the chert. In this study, from 6% to 12% shale in the concrete produced severe popout damage but no deep-seated failures within the beams.

Problem-causing lightweight, high-porosity chert may be removed by a heavy media separation process that is widely used in the economic minerals industry. A suspension of finely ground, high-specific-gravity material is used to separate the lower-specific-gravity chert from the aggregate. The use of iron-rich minerals and alloys permits required control of specific gravity and recovery of the heavy media suspension by magnetic recovery methods. The various techniques available to industry to improve aggregate quality prior to use have been described and summarized by Ames (1966) and Dolar-Mantuani (1976).

Moisture is necessary for all deleterious freeze–thaw occurrence in aggregate. A saturation level of 91.7% of the effective porosity of the aggregate either alone or with the surrounding concrete paste, if present, is considered to be the threshold, or critical saturation, for freeze–thaw failure to occur in aggregate pieces or concrete paste (Havens and Deen, 1977). The availability of moisture remains a problem throughout the service life of the concrete. Although moisture obviously is available at the

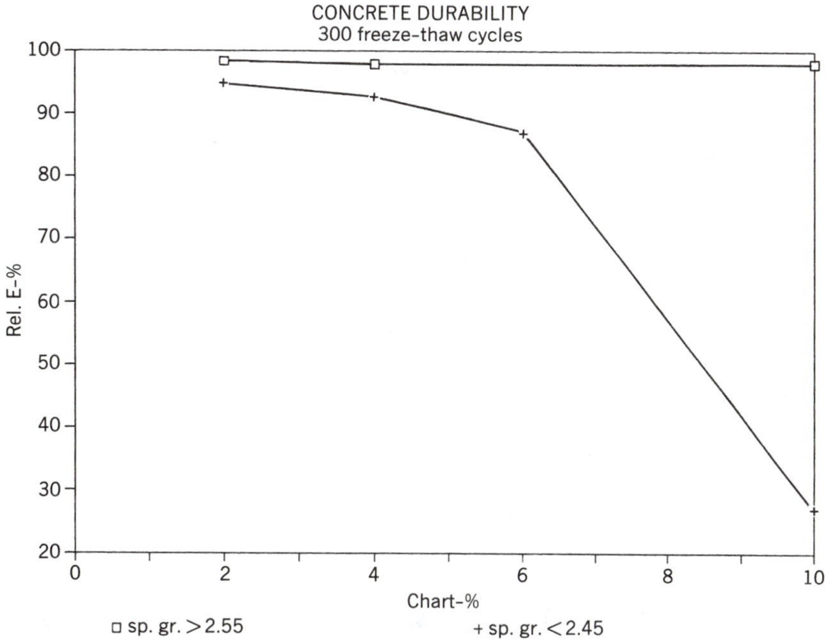

Figure 8.6 Influence of varying amounts of cherts with specific gravities greater than 2.55 and less than 2.45 on the relative dynamic elastic moduli of concrete beams after 300 freeze–thaw cycles. (Adapted from Schuster and McLaughlin, 1961)

time of construction, continued contact with moisture in the lower levels of pavement slabs or on exposed surfaces increases the probability of freeze–thaw deterioration owing to any of the subject aggregates. Cycles of wetting and drying that do not involve freezing and thawing generate swelling pressures in some of the frost-susceptible rocks. This may be a major contributor to failure of aggregate and portland cement concrete, explaining failures similar to freezing and thawing in frost-free areas (Hudec, 1978).

Testing coarse aggregates for susceptibility to freezing and thawing can be done on unconfined and confined samples. In unconfined testing, a sample of aggregate is sieved with a typical lower sieve size of 12.5 mm (½ in.). The number of aggregate particles in the sample is counted and the particles can be examined with a binocular microscope for cracks and other defects. The sample is then immersed in water for at least 24 hr to saturate the pores. There is considerable variation in testing procedures.

In summary, the samples are subjected to a number of freezing and thawing cycles. The number of cycles, rates of freezing and thawing, the freezing and thawing temperatures, and the freezing and thawing environments depend on the local or regional specifications that usually are based on experience. Freezing and thawing test environments include freezing in air and thawing in water as well as freezing and thawing in water. After the specified number of cycles has been reached, the sample is counted to determine breakage; it is also sieved. The difference in weight of the sample retained on the 12.5-mm sieve is used to determine the percentage loss in original weight. The particles are microscopically reexamined to determine any changes in cracks and other defects. Individual pieces can be identified by marker pen before testing for detailed analysis of crack development.

Confined freeze–thaw tests involve testing of an aggregate's performance in portland cement concrete. Figure 8.6 is an example of test results from a confined test of chert-bearing aggregates in concrete beams. Confined tests are concerned with the influence of an aggregate's freeze–thaw soundness on the dynamic elastic modulus of a concrete specimen; the assumption being that aggregate breakdown lowers the elastic properties of the concrete specimen. Concrete samples bearing the aggregate in question are subjected to approximately 300 rapidly repeated freeze–thaw cycles at specified temperature ranges and freeze–thaw rates. The relative dynamic elastic modulus, E (Figure 8.6), is obtained by the following equation:

$$\mathrm{E} = (n_1^2/n^2) \times 100 \tag{8.3}$$

Where:

n = fundamental transverse frequency of specimen before testing

n_1 = fundamental transverse frequency after testing

Details of freeze–thaw testing of aggregates and aggregate-bearing concrete may be obtained from the current Annual Book of ASTM Standards and from Dolar-Mantuani (1983). Freeze–thaw tests should not be considered as quantitative measures of the length of an aggregate's service because of the rigorous nature of the testing procedures.

The sulfate soundness, or sulfate, test has been used for many years as a substitute specifically for the more time-consuming freeze–thaw test of an aggregate's response to freeze–thaw cycles and for weathering resistance, or soundness, in general. In the context of soundness, an unsound aggregate is one that may undergo disruptive changes, as in failure during freezing or wetting and drying. It also may cause failure in pavements from expansion or breakdown in particle size. The sulfate soundness test is used to obtain an indication of aggregate soundness in this context. Whether it does so or not is the subject of continuing discussion (Bloem, 1966; Dolar-Mantuani, 1976).

In the sulfate test, a carefully graded and weighed aggregate sample is alternately immersed in a solution of either sodium or magnesium sulfate and oven dried for a specified number of cycles. The cumulative growth of salt crystals in pore spaces is thought to duplicate disruptive stresses in pores similar to those generated by freezing of water and exposure to other weathering processes. The soundness value is the percentage loss in weight of the original sample from material passing the smallest sieve used in the original gradation after five or ten cycles. Although the test is more severe than the freeze–thaw test, experience in Illinois has shown it to be a generally acceptable predictor of frost susceptibility. Aggregates having greater than 15% sulfate soundness loss tend to be frost susceptible (Harvey et al., 1974).

The ability to absorb water influences both the freeze–thaw and the sulfate tests of coarse aggregates. Kazi and Al-Mansour (1980b) have shown sulfate soundness percentage to vary linearly with absorption percentage. Particle size influences absorption values, so similar sizes of aggregate should be used for such tests. There is a trend toward including water-absorption percentage as a soundness indicator because of the influence of particle size, the difficulty of conducting freeze–thaw and sulfate soundness tests, and the differing degrees to which neither method is thought to be representative of natural processes (Hartley, 1974; Kazi and Al-Mansour, 1980b; Dolar-Mantuani, 1983).

The water-absorption characteristics of a coarse aggregate sample are determined by measuring the increase in sample weight owing to pore water, expressed as percentage of dry weight. Samples are immersed in water for approximately 24 hr, surface dried, and weighed. The sample is then oven dried, weighed, and the percentage of absorption relative to dry weight calculated. Quick, qualitative water-absorption evaluation of an aggregate can be obtained by timing the absorption of a drop of water placed on the surface of an aggregate particle. Relative absorption characteristics of the various rock types composing an aggregate sample can be determined rapidly by this procedure.

Construction-Related Degradation. The uses of aggregates in pavement construction will serve to illustrate this variety of degradation. Pavements, whether for highway or aircraft traffic, share certain features. The top layer in contact with vehicles is the running surface, or surface course. It may be constructed of portland cement concrete or of an asphaltic mixture. These are the rigid and flexible pavements, respectively, noted earlier. Some low-volume road surfaces may be constructed of a portland cement and soil mixture, of asphaltic material sprayed on an aggregate, or of an unprepared soil base. Although aggregate degradation plays a part in the construction and service life of all pavements, our attention will be directed toward those rigid and flexible pavements in which the surface course is a carefully designed and constructed discrete layer.

Most pavements are constructed on layers of aggregate called *base* and *subbase courses*. The purpose of a base course is to provide support for the pavement as well as drainage to prevent moisture-caused problems. Wheel loads and traffic density may require a second layer, or subbase course, beneath the base course of some flexible pavements for additional support. Both construction styles are shown in Figure 8.7.

Mechanical and "chemical" degradation have roles in both types of pavement construction. Degradation occurs in the base and subbase courses in both pavement types and in the surface courses of flexible pavements. The degradation that occurs in base and subbase courses during and following construction- and preconstruction-aggregate testing will be examined first.

Base-course and subbase-course aggregates must provide support and drainage (described earlier). Aggregate rock type(s), gradation, and construction methods are chosen to provide and maintain optimum conditions of support and drainage. Resistance to abrasion and crushing is of paramount importance in the selection of aggregate. The term *durability* has been applied by some to these attributes. Petrographic characteristics of aggregates influence abrasive and crushing durability. Silicate minerals typically are more resistant to abrasion than the softer carbonate minerals. Relative hardness of the different rock-forming minerals also influences abrasion. Therefore, igneous and the siliceous metamorphic rocks typically are more resistant to abrasion than limestone and dolomite; mica schist is a notable exception.

The texture of a rock is an additional and complex factor affecting abrasion and crushing. It includes crystal or grain size, relative orientation of crystals or grains, crystal or grain shape, porosity, crystalline versus grain-to-grain texture, and the kind and amount of cementing material if there is a grain-to-grain texture. Hartley (1974), Lees and Kennedy (1975), and Kazi and Al-Mansour (1980b) have addressed the many geological factors that affect the mechanical degradation of aggregates. Grain size consistently has been found to be directly proportional to the abrasion susceptibility of aggregates (West and Johnson, 1966; West et al., 1970; Kazi and Al-Mansour, 1980a).

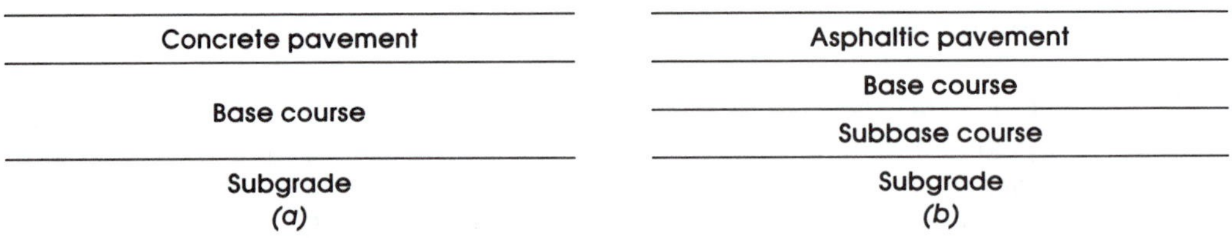

(a)	(b)
Concrete pavement	Asphaltic pavement
Base course	Base course
	Subbase course
Subgrade	Subgrade

Figure 8.7 Idealized cross sections of pavement construction. (*a*) Rigid pavement-portland cement-concrete. (*b*) Flexible pavement-asphaltic concrete.

Several tests are used singly or in combination to estimate the mechanical degradation characteristics of an aggregate. The freeze–thaw, sulfate, and absorption tests discussed earlier often are used to estimate degradation resulting from freezing and thawing, and wetting and drying. The Los Angeles (LA) abrasion test is commonly used in the United States to estimate aggregate-abrasion susceptibility. In this test, an oven-dried test sample is placed in a drum with an abrasive charge of steel spheres and rotated a specified number of revolutions. Following the test, the sample is removed and sieved according to the original sample gradation. The loss in weight in the form of fines is divided by the original weight and expressed as percentage loss, or %LA or LA loss. The fines generated are sieved according to specifications and may be tested for plasticity by using Atterberg limits or the sand equivalent test. Specific standards concerning the test gradations, procedures, and variations are found in the current ASTM Book of Standards and ISRM (1978). The LA test has undergone several variations by users for specific purposes. These include testing without steel spheres and the use of water in the drum (wet LA test).

Other tests that measure aggregate resistance to abrasion are in use. Among them are the Deval test and the Dorry test. Summaries of these tests have been prepared by the International Society of Rock Mechanics (ISRM, 1978), and comparisons between each and the Los Angeles abrasion test have been published by Woolf (1966) and Kazi and Al-Mansour (1980a). The latter have also experimented with the use of the "L"-type Schmidt hammer to predict LA%. The rationale for such experiments is the logical inverse relationship between uniaxial compressive strength and LA abrasion loss. The Schmidt hammer rebound number substitutes for compressive strength. The results of this work are shown in Figure 8.8. In the United Kingdom, aggregate impact and crushing tests have had greater usage than the LA abrasion test. A paper detailing these tests and analysis of results as they apply to mechanical aggregate degradation has been published by Dhir et al. (1971). Water-absorption and sulfate tests typically accompany the impact and crushing tests (Hartley, 1974).

The interrelationships among composition, texture factors, selected degradation tests, and in-service performance have been evaluated by West et al. (1970). The results of one comparison for carbonate aggregates in the form of a cluster diagram and selected partial correlation coefficients are shown in Figures 8.9 and 8.10. The LA abrasion loss percentage was chosen as the dependent variable because of its wide usage, thus providing a basis for comparison with other investigations. Examination of the simple and partial correlation coefficients greater than $+/-.20$ reveals predictable correlations. The low partial correlation coefficient for average grain size probably is the result of the uniform hardness of the essentially monomineralic composition of the samples. Crystalline igneous and metamorphic aggregates were tested and the data examined statistically. However, small sample numbers relative to the number of variables precluded any reliable correlations.

Factors other than petrographic characteristics influence aggregate performance in base- and subbase-course applications. The more important ones are aggregate shape, gradation, and the compactive effort used in

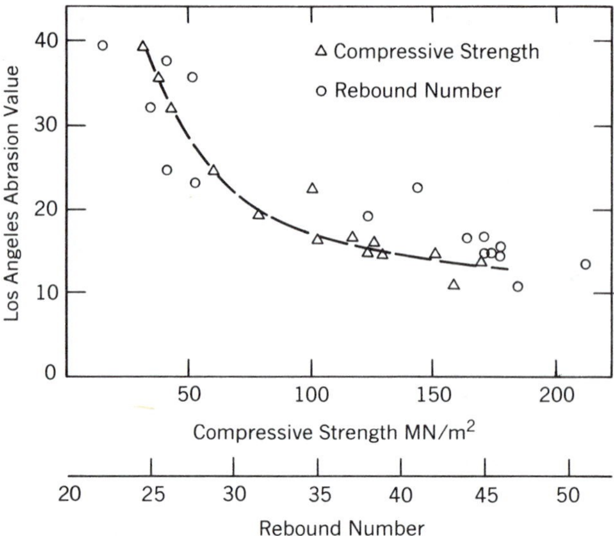

Figure 8.8 Relationship between LA abrasion value and compressive strength—some values estimated from Schmidt hammer-rebound number. (Reproduced by permission of the Geological Society from A. Kazi and Z. R. Al-Mansour, Quarterly Journal of Engineering Geology, Vol. 13, 1980.)

placing the aggregate courses. The role of shape on aggregate performance has been investigated in some detail by Dhir et al. (1971) and Ramsey et al. (1974) and summarized by Lees and Kennedy (1975). The detailed studies included four shape categories, which are cuboidal (irregular and angular), elongate, flaky, and flaky-elongate. Quantitative limits for each can be found in the paper by Ramsay et al. The descriptive nature of the class names is sufficient for this presentation. The abrading and crushing that takes place during construction affects the elongate, flaky, and flaky-elongate particles most. The tendency is toward developing subrounded cuboidal particles.

Gradation of an aggregate is a factor in maintaining the support required by overlying pavement and traffic. It must also provide for drainage. If excessive degradation and resulting changes in gradation occur, support and drainage will be reduced. In addition, frost-susceptible fines may be developed. Degradation in base courses during construction has been examined and reported on by Aughenbaugh et al. (1962). Three base-course gradations of carbonate aggregate were used: one size, open graded, and dense. Examples of cumulative curves for open-graded and dense-graded materials are shown in Figure 8.11. The differences in degradation or aggregate breakage for the three gradations in a 20.3-cm (8-in.) layer are illustrated by Figure 8.12. In this figure, it also can be seen that the greatest amount of degradation occurs as a result of the first trip, or pass, of compacting equipment over the aggregate layer. Breakdown of the aggregate was greatest in the upper 5 cm of the layers regardless of the number of compactor passes.

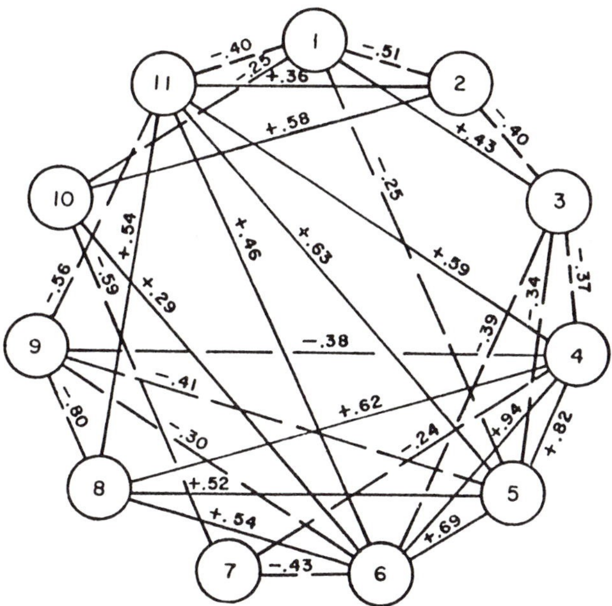

Figure 8.9 Cluster diagram of simple correlation coefficients for selected Indiana carbonate aggregates, using the following variables: (1) percentage loss by LA abrasion, (2) percentage insoluble residue, (3) percentage voids, (4) average grain size, (5) grain-size variation, (6) grain shape—highest values for most rounded particles, (7) grain interlock—highest numbers for best interlock, (8) specific gravity, (9) percentage absorption, (10) percentage loss by sodium sulfate test, and (11) degradation value—highest number for best performance. (From West and Johnson, 1966)

The dense gradation, which experienced the least degradation of the three gradations used by Aughenbaugh et al. (1962), has been shown to provide the best support, or greatest shear strength, by Jones et al. (1972). The optimum grading of ten gradations examined by Jones et al. is given in Table 8.2 along with the dense gradation used by Aughenbaugh (1963) for comparison.

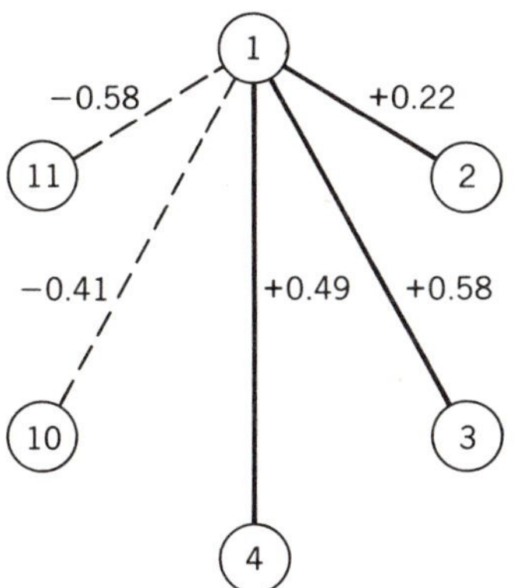

Figure 8.10 Cluster diagram of partial correlation coefficients greater than .2 for selected Indiana carbonate aggregates, using variables listed in Figure 9.9. (From West and Johnson, 1966)

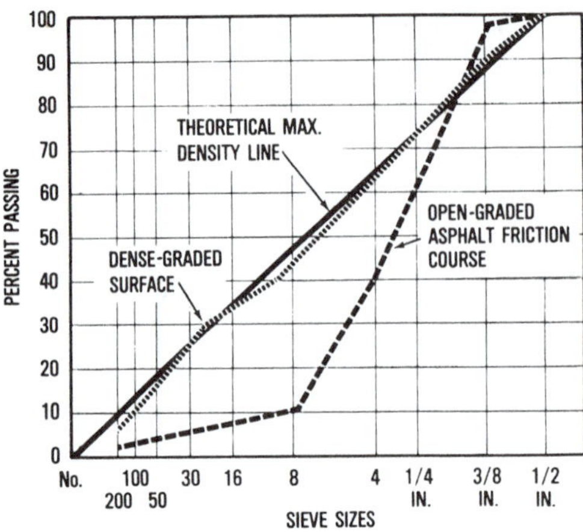

Figure 8.11 Cumulative curves illustrating open and dense gradations used in asphaltic pavement surface courses. See Table 9.1 for sieve sizes. (From NCHRP, 1978)

The type of compacting equipment also affects the amount of degradation in a base course during construction. Given the same aggregate rock type, gradation, and layer thickness, significant differences in breakage, or gradation, and generation of fines may be noted due to using different kinds of equipment. Figure 8.13 compares these factors for a steel-wheel roller, pneumatic, or rubber-tired, compactor, and vibratory compactor. A change in the amount of breakage need not be consistently proportional to the amount of fines generated among different rock types. The relationship is controlled by a combination of petrographic factors and weathering history of the aggregate.

Aggregate degradation assumes different roles in pavements or surface courses. In portland cement concrete, the gradation may change during the mixing process prior to laying the pavement. The proportion of cement to aggregate is altered, thus affecting the workability and ultimately the strength of the concrete (West and Aughenbaugh, 1964). Abrasion and crushing cease to be a problem after the concrete hardens. In flexible asphaltic concrete pavement, however, degradation occurring during and after placement of the surface course has been recognized since the 1930s as the cause of pavement failure. Many of the factors that govern degradation in base courses are active in this application. They include the geologic factors that control breakage and abrasion. Both breakage and generation

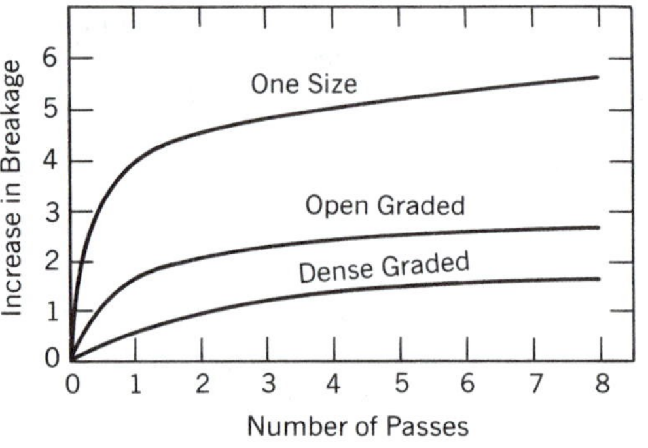

Figure 8.12 Effect of grading on carbonate aggregate breakage for different numbers of passes of compactor over base course. (Reprinted with permission from Trans. Soc. Min. Eng., AIME, Vol. 223, N. B. Aughenbaugh, R. B. Johnson, and E. J. Yoder. Factors influencing the breakdown of carbonate aggregates during field compaction, 1962.)

Table 8.2 Comparison of Dense Gradations Used for Optimum Base-Course Shear Strength and Degradation Characteristics

Jones et al. (1972)		Aughenbaugh (1963)	
Cumulative % Passing	Sieve Size	Cumulative % Passing	Sieve Size
100.0	2.00 in.	100.0	0.75 in.
94.0	1.50 in.	78.0	0.50 in.
80.0	0.75 in.	50.0	No. 4
58.5	0.375 in.	30.0	No. 10
42.5	No. 4	0.0	No. 40
15.5	No. 30		
5.0	No. 200		

of fines during construction increase the aggregate surfaces to be coated and break the bonds between the asphalt and aggregate particles, respectively. Similar actions occur in service under the cyclic wheel loadings from passing traffic. Uncoated cohesionless surfaces develop and the surface is said to ravel or gradually fail.

The influence of gradation, loads, repetition of loads, and aggregate type on in-service failure of asphaltic pavements was examined by Moavenzadeh and Goetz (1963). They used a gyratory testing machine, which duplicates magnitude and repetition of wheel loads on pavement. Three kinds of aggregate with LA values, ranging from high to intermediate to low, and three gradations, from open to dense, were used in asphalt mixtures. Gradation proved to be the most important factor with the densest gradation resulting in the least degradation. Gradation was of such overriding importance that a high-LA-loss aggregate degraded less in a dense-gradation-pavement mix than did a low-LA-loss material with an open grada-

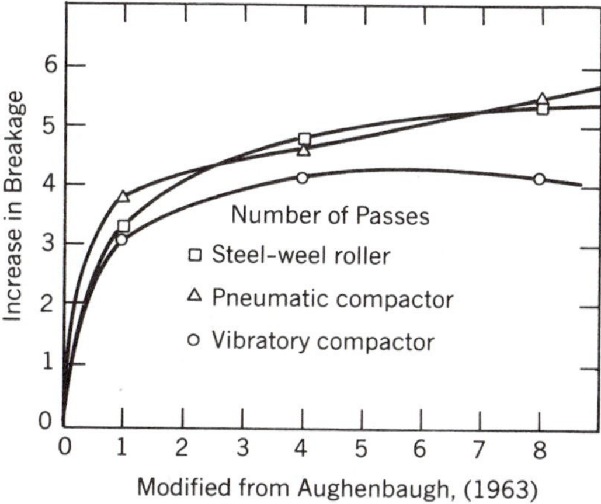

Figure 8.13 Effect of compactor type on aggregate breakage in base courses for different numbers of compactor passes. (From Aughenbaugh, 1963)

tion. Good crystal interlock and strong cementing material controlled degradation in an asphaltic pavement, just as it does in other degradation-causing situations. The magnitude of wheel loading affected degradation more than did the repetition of wheel loads. Increases in both resulted in increased degradation.

The deleterious influence of chemical weathering or alteration of rock used for coarse aggregate normally is detected by standard abrasion and crushing tests. However, exceptions occur. In the United States, the now "classic" area for such exceptions includes parts of Oregon, Washington, and Idaho where basalt is the most readily accessible rock type for use as aggregate (Erickson, 1960). The basalts in question pass the standard LA abrasion test but have failed in service in asphaltic pavements and their base courses. The problem has been found to be the release of swelling and nonswelling clays from a fresh, abrasion-resistant basalt matrix. This has resulted in reduction in asphalt bonding to aggregate, poor drainage, and loss of support. The clays or claylike alteration products result from devitrification of glass and/or alteration of silica-deficient minerals in the basalt. Day (1962) concluded that the various alteration products were present in the aggregate at the time of pavement or base-course construction, precluding any chemical decomposition during or after construction.

Petrographic examination of aggregates to identify the alteration products of devitrification of glass and the chemical weathering of minerals such as pyroxene is now considered to be essential before use of the Pacific Northwest basalts as road aggregate (Van Atta and Ludowise, 1976). Aggregate tests that involve a wet environment more conducive to release of clays than the dry-abrasive tests are now preferred. Several have been devised and have been described in the papers listed herein. The wet LA test is typical of such tests.

Skid-Resistant Pavements

The familiar highway signs warning that the pavement may be slippery when wet are indicative of an aggregate-related problem that occurs on pavement surfaces. Many factors are involved in the skid resistance of a pavement, including cumulative traffic, vehicle speed, and surface texture. Skidding and hydroplaning on a wet pavement present the most hazardous aspects of the problem. Related glare, splash, and spray also contribute to highway accidents. Selection of an aggregate that is resistant to polishing or smoothing from traffic and use of a surface-course gradation that will maintain a surface texture that permits water drainage beneath the tires at higher speeds are keys to the construction of a skid-resistant pavement. These and other aggregate-related factors in the construction and maintenance of skid-resistant pavements are summarized in: Kearney et al., 1972; Beaton, 1977; NCHRP, 1978; Henry and Dahir, 1979.

Portland cement concrete and asphaltic concrete pavements are both susceptible to smoothing and the development of a skid-susceptible pavement. The nature of asphaltic pavement-matrix materials results in greater problems. In portland cement concrete pavements, a combination of using a wear-resistant fine aggregate and mechanical surface texturing at time of construction serve to produce a longer-wearing, more skid-resistant pavement surface. Should in-service wear ultimately cause hazardous smoothing, the surface may be retextured by sawing shallow grooves in the surface at selected locations such as at traffic intersections and signals. In the asphaltic concrete pavement, skid resistance is more dependent on the physical and textural characteristics of the coarse aggregate exposed on the running surface. This section will address primarily the problems related to asphaltic pavements because of the greater magnitude of the skidding problem with these pavements.

Aggregate Roughness. As indicated previously, pavement surface texture is of major concern in the construction and maintenance of a skid-resistant pavement. Roughness and waviness as defined in chapter 4 (where they were used to describe discontinuity surfaces in rock) are utilized when dealing with pavement surfaces. The term *macrotexture* is analogous to waviness or first-order irregularities and *microtexture* is related to roughness or second-order irregularities. The term *asperity* is used in describing microtexture roughness features, just as with discontinuity surfaces. These surface textural features are illustrated by Figures 8.14 and 8.15. Macrotexture is obtained by use of coarse aggregate with an asphalt binder. It provides channels for escape of water from the tire-pavement interface. Microtexture refers to individual coarse aggregate surfaces and is dependent on aggregate textures and mineralogy. Both textures are necessary to prevent hydroplaning and skidding, respectively. Maintenance of optimum microtexture keeps the particles from smoothing or polishing.

Although shape of aggregate particles would seem to be a primary factor in skid resistance in asphaltic pavements, there is little concurrence as to the importance of shape because of the many variables introduced by composition and texture. The following references will serve to illustrate this for those requiring additional information: Goodwin, 1961; Huang and Ebrahimzadeh, 1973; Beaton, 1977; Henry and Dahir, 1979.

Aggregate Gradation. An open, coarse aggregate gradation in the running or friction surface of an asphaltic pavement has been found to minimize hydroplaning by permitting drainage while maintaining sufficient tire-pavement contact (Huang and Ebrahimzadeh, 1973; NCHRP, 1978; Ryell et al., 1979). Additional benefits are increased skid resistance at higher speeds (Figure 8.16); less glare, splash, and spray during wet

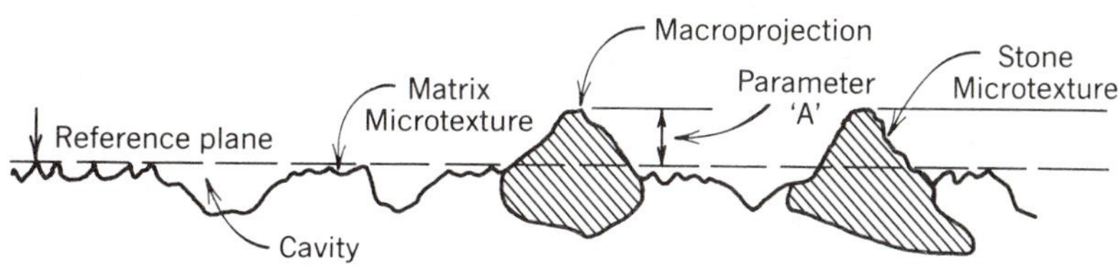

'A' – Height of macroprojections (mm)

Figure 8.14 Micro- and macrotextures relative to aggregate particles and surface roughness of asphaltic pavement. (From Ryell et al., 1979)

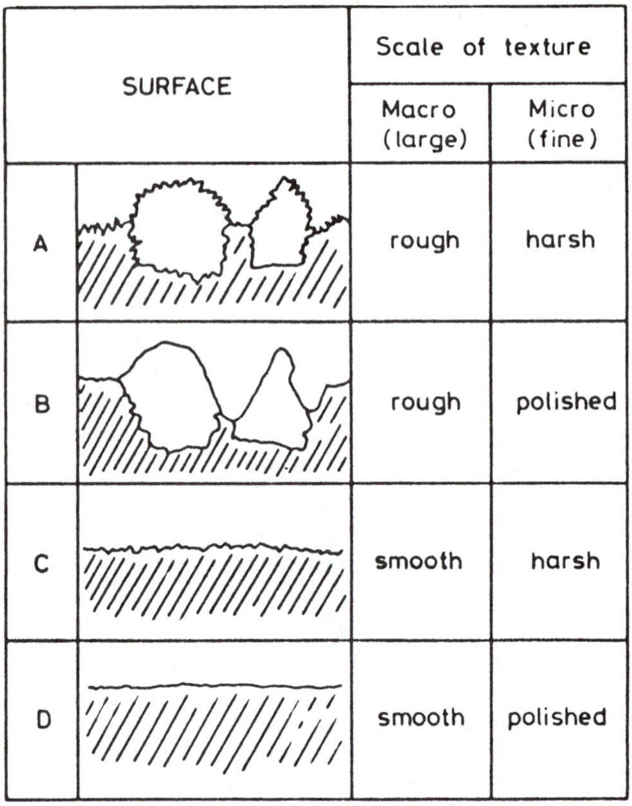

SURFACE		Scale of texture	
		Macro (large)	Micro (fine)
A		rough	harsh
B		rough	polished
C		smooth	harsh
D		smooth	polished

Figure 8.15 Texture scales used to describe an asphaltic pavement surface. (From Elsenaar et al., 1976)

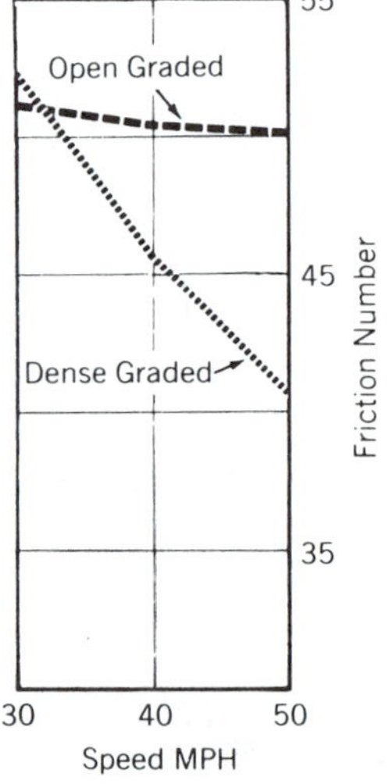

Figure 8.16 Influence of vehicle speed and aggregate gradation on skid resistance of open-graded and dense-graded friction on surface course of asphaltic-concrete pavements. (From NCHRP, 1978)

weather; and a smoother, quieter riding surface. The last benefit appears to be true in spite of earlier studies and continuing concern (Goodwin, 1961; NCHRP, 1978).

Several methods are in use to measure skid resistance in the laboratory and on pavements. They understandably generate test values that have different terminology, units, and meanings. The numerical representation of these units is not standardized. For some tests (Figure 8.17), the test number is inversely proportional to pavement skid resistance. In others (Figures 8.16 and 8.18), the number is directly proportional. For the sake of uniformity, higher numbers should be used to indicate better pavement resistance to skidding as in Figures 8.16 and 8.18.

Aggregate Composition and Texture. The maintenance of microtexture by coarse aggregate in asphaltic pavements is primarily a function of petrographic characteristics, given the same traffic conditions. Composition and grain size are the dominant petrographic factors. Concerning composition, it has been shown, quite logically, that the hard silicate minerals wear more slowly than the softer carbonates (Knill, 1960; Henry and Dahir, 1979). There may be variation in the wear rates of carbonates, however. The better wearing ones have coarse crystalline carbonate textures and higher proportions of acid-insoluble material, with the best having quartz particles as the insoluble material (Gray and Renninger, 1966; Kearney et al., 1972). Dahir and Meyer (1976) have shown that

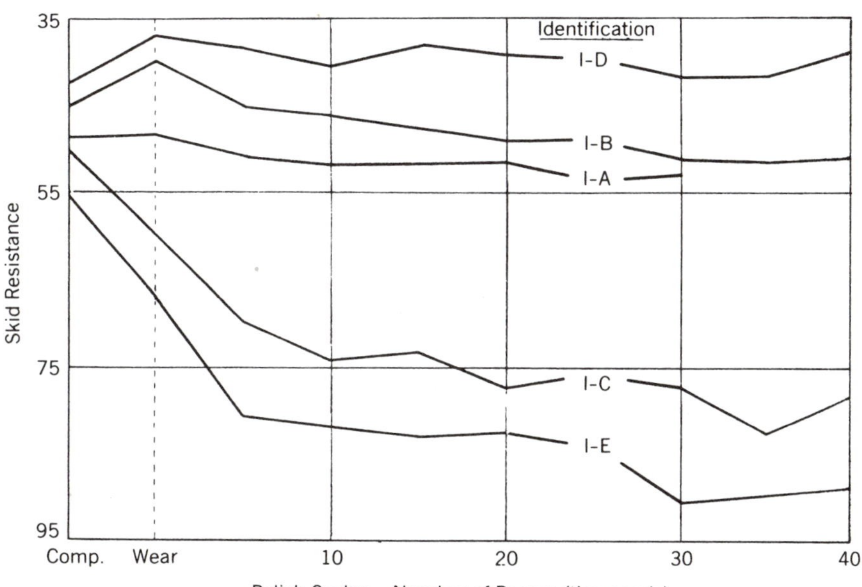

Figure 8.17 Polishing rate versus skid resistance of asphaltic pavement for carbonate aggregates having the following insoluble residues: (1–A) 42% sand-size of quartz, mica, and feldspar; (1–B) 36% of mostly mica; (1–C) 25%, of which 40% ≤0.05-mm diameter; (1–D) 48% sand-size quartz; (1–E) 9%, of which 2/3 <0.05-mm diameter. (From Gray and Renninger, 1966)

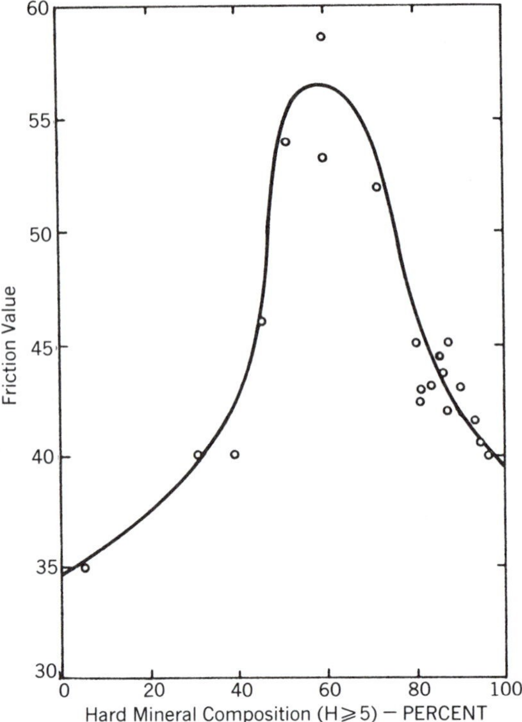

Figure 8.18 Optimum percentage of minerals of Moh's hardness greater than or equal to 5 in aggregate particles to achieve maximum skid resistance in asphaltic pavement. (From Dahir, 1978)

coarse-grained rocks require greater polishing effort than fine-grained rocks of uniform composition for any given composition.

Aggregate composition has an impact on maintenance of microtexture that surpasses those factors described to this point. Some aggregate particles are composed of minerals having different hardnesses. The differential wearing of such an aggregate particle maintains the desired microtexture far longer than does the uniform composition of another aggregate. For example, carbonate aggregates having quartz grains as insoluble residue perform better than those composed uniformly of carbonate.

The differential hardness between calcite and quartz has significant impact on the skid performance of the aggregates illustrated in Figure 8.17. Detrital sedimentary rocks (e.g., quartzose sandstone with soft carbonate cement) also retain a rough microtexture in service. A hard quartzite with uniform composition and grain size will polish more rapidly than such mixed composition aggregates. Dahir (1978) has reported that the minimum difference in Mohs hardness for aggregate minerals to maintain skid resistance should be 2. The optimum hard mineral content, given this minimum hardness differential, is in the 50% to 70% range (Figure 8.18). This figure also illustrates our earlier statement regarding aggregates of uniformly hard minerals. Comprehensive papers concerning skid-resistant properties of coarse aggregates in general and differential hardness in particular have been published by Knill (1960), Beaton (1977), and Dahir (1978).

Other Uses of Aggregates

To this point, attention has been restricted mainly to the use of aggregates in pavement. In addition to pavement construction, coarse aggregates are used as railroad ballast. Railroad ballast serves as support for ties and rails and provides for drainage in much the same way that coarse aggregate does in base and subbase courses under pavement.

Goldbeck (1948) has listed performance requirements of ballast, some of which are common to those of pavement courses. They are: (1) to give tie support and load distribution, (2) to restrict lateral tie movement owing to rail traffic and temperature changes, (3) to provide drainage around ties; (4) to provide nonrigid foundation for track subjected to large cyclic loads of varying intensity and frequency; (5) to resist freeze–thaw and other weathering degradation; and (6) to retard vegetative growth. All of these are affected by excessive breakdown of particle and/or shape of particles. Shape, as in the case of rounded aggregate particles, may create an unstable roadbed from the shear strength standpoint, thus affecting requirements 1, 2, and 4. Gradation must be considered also in terms of providing optimum support and drainage.

The petrographic requirements of ballast are similar to those of pavement aggregates. They include resistance to fracturing and abrasion; nature of the fines generated by degradation; grain size, texture, structure, amount, and kind of cementing material for detrital sedimentary rocks; and amount of prior chemical weathering. Among the tests found to provide measures of aggregate suitability as ballast are bulk specific gravity, LA abrasion, sulfate, crushing, freeze–thaw, and mineral hardness. Shape measurements are of special value in evaluating aggregate for the unique support needs of trackage. Representative measurements are sphericity, roundness, elongation, and flakiness (Gaskin and Raymond, 1976).

In the case of LA abrasion testing, the permissible amount of fines may vary, depending on whether the fines are nonplastic or plastic. Regardless of the plasticity of fines and potential drainage problems, a maximum LA% must be set to avoid loss of support as the gradation becomes more dense. This maximum ranges from 20% to 40% for U.S. and Canadian railroads. Similarly, maximum values of 5% sodium sulfate loss and 3% freeze–thaw loss are recommended along with a sphericity range of 0.55–0.70 and a minimum average Mohs hardness of 5.0.

RIPRAP AND OTHER LARGE ROCK MATERIALS

Large-size rock materials that range from coarse aggregate to volumes up to 1 m^3 have many construction applications. The larger sizes—called *riprap, armor stone, rockfill,* and *rubble*—are used where placement stability must be maintained, often in high wave-energy environments. Typical of high-energy environments are the reservoir or upstream sides of earth dams, stilling basins at dam-discharge sites, tailraces of power plants, canals, riverbanks, and a variety of shoreline and coastal protective installations. In each case, the rock prevents erosion from waves or running water. Figure 8.19 is a typical application of riprap, or armor stone.

Of the listed environments, those associated with shoreline and coastal-protective installations may impose unique requirements on the rock materials used. Jetties, causeways, and coastal-protection structures such as seawalls typically are subjected to higher wave energies as well as to marine environments with tides, differing salinities, and climatic extremes. The durability requirements involve mechanical and chemical weathering properties of the material. The many aspects of rock selection for such rigorous engineering applications and associated tests have been addressed by Fookes and Poole (1981).

Rockfill dams are constructed of compacted layers of well-graded rock, ranging from large sizes to fines. Compacted rockfill is designed to be the main structural element of dams. In addition to compacted rockfill, many rockfill dams have earth cores and concrete faces. Cooke (1984) has prepared a comprehensive summary of the construction of rockfill dams.

Rock suitable for the various applications noted must at the outset be sufficiently massive to satisfy the size requirements of the particular application. Thus, any closely jointed or thinly bedded rock must be rejected. When quarrying of suitable riprap is not possible or feasible for either geologic or economic reasons, talus and large boulders from stream beds or glacial till may be used either as is or crushed to size. Rounded boulders require some crushing as the high shear strength required for the various uses is a product of interlocking of angular pieces of the entire layer. The rock used must be durable enough to withstand processing, placement, and in-service conditions (Treasher, 1964). The latter include freezing and thawing, wetting and drying, temperature changes, and wave action.

Examination of the rock quality at a potential extraction site normally

Figure 8.19 Riprap facing of dike at Lake Granby water-storage facility, Colorado.

provides an indication of the rock's durability for the in-service elements. An exception would be in an arid or semiarid climate where the material would be subjected to atypical environments as in dams, canals, or similar moist environments. Further testing is similar to that conducted on aggregates, namely, petrographic, freeze–thaw, sulfate, absorption, LA abrasion, and bulk-specific gravity (Lienhart and Stransky, 1981; Summer and Johnson, 1982). In general, clay-bearing sedimentary rocks are unacceptable as riprap.

Riprap, or armor stone, used as facing material to prevent surface runoff erosion of the downstream faces of earth dams and embankments is smaller in size than that used on the reservoir, or upstream, side of dams. Still smaller material may be used as a drainage blanket under riprap to reduce the possibility of runoff erosion under the riprap cover. Rock used for these purposes is referred to as rockfill when used for construction of rockfill dams. The requirements are less rigorous than for riprap. The main requirement is that excessive breakdown will not occur on exposure to the elements. Optimum size, shape, and gradations are at least gravel size, equidimensional, angular, and well graded (USBR, 1974).

SUMMARY

Rock products have many specialized applications in construction. Mineral composition, texture, reaction to alkalis, susceptibility to freeze–thaw breakdown, and resistance to abrasion are among the properties used to evaluate a rock for a particular application. No one set of rock characteristics can be used as a criterion for usage of a rock as aggregate or large-size material (e.g., riprap) because of the many possible application and climatic variables. Neither can one test nor a constant set of tests be used to define a material's soundness or suitability as construction material. For instance, a test for aggregate chemical reactions in the alkaline environment of portland cement concrete may have no bearing on that aggregate's susceptibility to mechanical wear or freezing and thawing in other applications or its skid-resistant performance in asphaltic pavements.

An understanding of rock characteristics, available test procedures, engineering applications, construction methods, and in-service conditions is critical to the proper use of rock in construction.

REFERENCES

Ames, J. A., 1966, Limestone deposits vs. beneficiation: Ohio J. Sci., Vol. 66, No. 2, pp. 131–136.

ASTM, 1978, Book of ASTM Standards, Part 14, Concrete and Mineral Aggregates, Spec. C33, pp. 15–22.

Aughenbaugh, N. B., 1963, Degradation of base course aggregates during compaction: Ph.D. thesis, Purdue Univ., Lafayette, Ind. 190 pp.

Aughenbaugh, N. B., Johnson, R. B., and Yoder, E. J., 1962, Factors influencing the breakdown of carbonate aggregates during field compaction: Trans. Soc. Min. Eng., AIME, Vol. 223, pp. 402–406.

Beaton, J. L., 1977, Providing skid resistant pavements: Transp. Res. Rec. 622, pp. 39–50.

Bloem, D. L., 1966, Concrete aggregates: soundness and deleterious substances: Am. Soc. Tes. Mater, Spec. Tech. Publ. 169–A, pp. 497–512.

Cooke, J. B., 1984, Progress in rockfill dams: J. Geotech. Eng., Vol. 110, pp. 1,383–1,414.

Dahir, S. H., 1978, Petrographic insights into the susceptibility of aggregates to wear and polishing: Transp. Res. Rec. 695, pp. 20–27.

Dahir, S. H., and Meyer, W. E., 1976, Effects of abrasive size, polishing effort, and other variables on aggregate polishing: Transp. Res. Rec. 602, pp. 54–56.

Day, H. L., 1962, A progress report on studies of degrading basalt aggregate bases: Hwy. Res. Bd. Bull. 344, pp. 8–16.

Dhir, R. K., Ramsay, D. M., and Balfour, N., 1971, A study of the aggregate impact and crushing value tests: Inst. Hwy. Eng. J., Vol. 18, No. 11, pp. 17–27.

Dolar-Mantuani, L., 1976, Working with borderline aggregates: Am. Soc. Tes. Mater. Spec. Tech. Publ. 597, pp. 2–10.

———, 1983, Handbook of concrete aggregates: Noyes Publications, Park Ridge, N.J., 345 pp.

Dolch, W. L., 1966, Concrete aggregates: Porosity: Am. Soc. Tes. Mater. Spec. Tech. Publ. 164–A, pp. 443–461.

Elsenaar, P.M.W., Reichert, J., and Sauterey, R., 1976, Pavement characteristics and skid resistance: Transp. Res. Rec. 622, pp. 1–25.

Erickson, L. F., 1970, Degradation of aggregates used in base courses and bituminous surfacings: Hwy. Res. Bd. Circ. 416, 10 pp.

Fookes, P. G., and Poole, A. B., 1981, Some preliminary considerations on the selection and durability of rock and concrete materials for breakwaters and coastal protection works: Q. J. Eng. Geol., Vol. 14, pp. 97–128.

Gaskin, P. N., and Raymond, G. P., 1976, Contribution to selection of railroad ballast: Transp. Eng. J., Am. Soc. Civ. Eng. Proc., Vol. 102, No. TE2, pp. 377–394.

Gillott, J. E., 1975, Alkali-aggregate reactions in concrete: Eng. Geol., Vol. 9, pp. 303–326.

Goldbeck, A. T., 1948, Mineral aggregates for railroad ballast: Am. Soc. Tes. Mater., Spec. Tech. Publ. 83, pp. 197–204.

Goodwin, W. A., 1961, Evaluation of pavement aggregates for nonskid qualities: 12th Ann. Symp. on Geol. as Applied to Hwy. Eng., Bull. 24, Engr. Exp. Stn., Univ. of Tennessee, Knoxville, pp. 8–18.

Gray, J. E., and Renninger, F. A., 1966, The skid-resistant properties of carbonate aggregates: Hwy. Res. Rec. 120, pp. 18–34.

Hadley, D. W., 1964, Alkali reactivity of dolomitic carbonate rocks: Hwy. Res. Rec. 45, pp. 1–20.

Hansen, W. C., 1966, Concrete aggregates: Chemical reactions: Am. Soc. Tes. Mater., Spec. Tech. Publ. 169–A, pp. 487–496.

Hartley, A., 1974, A review of the geological factors influencing the mechanical properties of road surface aggregates: Q. J. Eng. Geol., Vol. 7, pp. 69–100.

Harvey, R. D., Fraser, G. S., and Baxter, J. W., 1974, Properties of carbonate rocks affecting soundness of aggregate: Ill. State Geol. Surv., Min. Notes 54, 19 pp.

Havens, J. H., and Deen, R. C., 1977, Possible explanation of concrete pop-outs: Transp. Res. Rec. 651, pp. 16–24.

Henry, J. J., and Dahir, S. H., 1979, Effects of textures and the aggregates that produce them on the performance of bituminous surfaces: Transp. Res. Rec. 712, pp. 44–50.

Huang, E. Y., and Ebrahimzadeh, T., 1973, Laboratory investigation of the effect of particle shape characteristics and gradation of aggregates on the skid resistance of asphalt surface mixtures: Am. Soc. Tes. Mater., Spec. Tech. Publ. 530, pp. 117–137.

Hudec, P. P., 1978, Standard engineering tests for aggregate: what do they actually measure?: *in* Decay and preservation of stone, Eng. Geol. Case Histories, No. 11, Geol. Soc. Am., pp. 3–6.

ISRM, 1978, Suggested methods for determining hardness and abrasiveness of rocks: Intl. Soc. Rock Mech. Comm. on Standardization of Laboratory and Field Tests, Intl. J. Rock Mech. Min. Sci. & Geomech. Abstr., Vol. 15, pp. 89–97.

Jones, T. R., Jr., Otten, E. L., Machemehl, C. A., Jr., and Carlton, T. A., 1972, Effect of changes in gradation on strength and unit weight of crushed stone base: Hwy. Res. Rec. 405, pp. 19–23.

Kazi, A., and Al-Mansour, Z. R., 1980a, Empirical relationship between Los Angeles abrasion and Schmidt hammer strength tests with application to aggregates around Jeddah: Q. J. Eng. Geol., Vol. 13, pp. 45–52.

———, 1980b, Influence of geological factors on abrasion and soundness characteristics of aggregates: Eng. Geol., Vol. 15, pp. 195–203.

Kearney, E. J., McAlpin, G. W., and Burnett, W. C., 1972, Development of specifications for skid-resistant asphalt concrete: Hwy. Res. Rec. 396, pp. 12–20.

Knill, D. C., 1960, Petrographic aspects of the polishing of natural roadstones: J. Appl. Chem., Vol. 10, pp. 28–35.

Lees, G., and Kennedy, C. K., 1975, Quality, shape and degradation of aggregates: Q. J. Eng. Geol., Vol. 8, pp. 193–209.

Lemish, J., Rush, F. E., and Hiltrop, C. L., 1958, Relationship of physical properties of some Iowa carbonate aggregates to durability of concrete: Hwy. Res. Bd. Bull. 196, pp. 1–16.

Lienhart, D. A., and Stransky, T. E., 1981, Evaluation of potential sources of riprap and armor stone—methods and considerations: Bull. Assoc. Eng. Geol., Vol. 18, pp. 323–332.

Lounsbury, R. W., and Schuster, R. L., 1964, Petrography applied to the detection of deleterious materials in aggregates: Proc. 15th Ann. Hwy. Geol. Symp., Mo. Div. Geol. Surv. and Water Resources, pp. 95–115.

Mather, B., 1974, Developments in specification and control: Transp. Res. Rec. 525, pp. 38–42.

McConnell, D., Mielenz, R. C., Holland, W. H., and Greene, K. T., 1950, Petrology of concrete affected by cement-aggregate reaction: *in* Application of geology to engineering practice, Berkey Vol., Geol. Soc. Am., pp. 225–250.

Mielenz, R. C., 1962, Petrography applied to portland-cement concrete: *in* Reviews in eng. geol., Vol. 1, Geol. Soc. Am., pp. 1–38.

———, 1966, Concrete aggregates: petrographic examination: Am. Soc. Tes. Mater., Spec. Tech. Publ. 169–A, pp. 381–403.

Moavenzadeh, F., and Goetz, W. H., 1963, Aggregate degradation in bituminous mixtures: Hwy. Res. Rec. 24, pp. 106–137.

NCHRP, 1978, Open-graded friction courses for highways: Synthesis of Hwy. Practice 49, Natl. Coop. Hwy. Res. Prog., Transp. Res. Bd., 50 pp.

Ramsay, D. M., Dhir, R. K., and Spence, I. M., 1974, The role of rock and clastic fabric in the physical performance of crushed rock aggregate: Eng. Geol., Vol. 9, pp. 267–285.

Rhoades, R., and Mielenz, R. C., 1948, Petrographic and mineralogic characteristics of aggregates: Am. Soc. Tes. Mater., Spec. Tech. Publ. 83, pp. 20–48.

Ryell, J., Corkill, J. T., and Musgrove, G. R., 1979, Skid resistance of bituminous-pavement test sections: Toronto by-pass project: Transp. Res. Rec. 712, pp. 51–61.

Schuster, R. L., and McLaughlin, J. A., 1961, A study of chert and shale gravel in concrete: Hwy. Res. Bd. Bull. 305, pp. 51–73.

Stark, D., 1976, Characteristics and utilization of coarse aggregates associated with D-cracking: Am. Soc. Tes. Mater., Spec. Tech. Publ. 597, pp. 45–58.

Summer, R. M., and Johnson, R. B., 1982, Rock durability evaluation procedure for riprap and diversion channel construction in coal mining areas: 1982 Symp. on Surface Mining Hydrology, Sedimentology and Reclamation, Univ. of Kentucky, Lexington, pp. 433–437.

Treasher, R. C., 1964, Geologic investigations for sources of large rubble: *in* Eng. Geol. Case Histories No. 5, Geol. Soc. Am., pp. 31–43.

USBR, 1974, Earth manual, 2nd ed.: U.S. Bur. of Reclamation, Denver, Colo., 810 pp.

Van Atta, R. O., and Ludowise, H., 1976, Causes of degradation in basaltic aggregates and durability testing: Proc. 14th Ann. Eng. Geol. Soils Eng. Symp., Boise State Univ., Idaho, pp. 241–254.

West, T. R., and Aughenbaugh, N. B., 1964, The role of aggregate degradation in highway construction: Proc. 15th Ann. Hwy. Geol. Symp., Mo. Div. Geol Survey and Water Resources, pp. 117–132.

West, T. R., and Johnson, R. B., 1966, Analysis of textural and physical factors contributing to the abrasion resistance of some Indiana carbonate aggregates: Proc. Ind. Acad. Sci., 1965, Vol. 75, pp. 153–162.

West, T. R., Johnson, R. B., and Smith, N. M., 1970, Tests for evaluating degradation of base course aggregates: Hwy. Res. Bd. Natl. Coop. Hwy. Res. Proj. Rep. 98, 92 pp.

Woods, K. B., 1948, Distribution of mineral aggregates: Am. Soc. Tes. Mater., Spec. Tech. Publ. 83, pp. 4–19.

Woolf, D. O., 1966, Concrete aggregates: toughness, hardness, abrasion, strength, and elastic properties: Am. Soc. Tes. Mater., Spec. Tech. Publ. 169–A, pp. 462–475.

9

Engineering Geology and Earth Processes

Engineering works and and earth processes interact in two distinct and important ways. First, natural processes influence how structural and other development occurs. The suitability of locations for development, their design, and cost are related to the effect of natural processes. Requiring a setback distance between buildings and the edge of sea cliffs reflects how the process of coastal erosion influences location of development. Second, structures and development can alter natural processes, creating changes that may be undesirable. Changes in stream channel-bed elevations and other channel characteristics below dams are an example of a natural process changed by development (Williams and Wolman, 1984). One of the primary responsibilities of an engineering geologist is assessing both ways in which natural processes may impact a specific project.

Engineering works that lack the benefit of geologic input are likely to cost more than necessary, function below their expected optimum, or fail altogether. This is not to say information on geologic factors is the only important consideration in the siting, design, and construction of engineering works. Unfortunately, examples are far too common in which lack of geologic information is at the root of problems encountered during development. Robinson and Spieker (1978) emphasize this point by quoting a statement made 350 years ago by the English philosopher Francis Bacon, "Nature to be commanded must be obeyed!" There is an abundance of benefits for engineering works to be derived from a full understanding of geologic processes—thus, geologic data cannot be safely ignored!

SAFETY, RISK, AND GEOLOGIC FORECASTING

Forecasting, or predicting, the interaction of engineering works with earth processes is necessary for safety and reliability. Lowrance (1976) points out that being safe, or free of risk, is a matter of degree rather than being absolute. An element of risk is involved in engineering works, as it is with most human endeavors. The point of forecasting is to ensure that the risk is at an acceptable level. This involves two activities. One is the measurement of risk. The other is judging whether that degree of risk is acceptable. The former is an objective process; the latter is a more subjective and often personal activity. Measuring risk for geologic processes is based on knowledge and principles. Acceptability of that measured risk is generally dictated by social, political, and economic forces. Therefore, determining acceptable risk requires the engineering geologist to participate in formulating public policy. Failure to be a partner to public-policy formulation would be irresponsible and limit opportunities for reducing geologic hazards in development (Lundgren, 1976; Lucchitta et al., 1981; Fiske, 1984; Petak, 1984).

Table 9.1 provides definitions of risk, vulnerability, and natural hazard. The engineering geologist is usually charged with defining natural hazard, that is, earth process, for use in establishing risk. The definition of natural hazard requires some statement of probability of occurrence. This prediction of future events of a given magnitude usually depends on basic geologic studies. Because these studies usually relate present and past conditions in an effort to identify some empirical relationship, they are more suitable for inferring than predicting future events (Kitts, 1976). The engineering geologist typically is forced to make the most of existing information to predict expected risk (Lundgren, 1976).

Long-term prediction can be a source of conflict between the engineering geologist and the engineer concerned with addressing landslide and other earth processes that influence a project (Cotecchia, 1978). The engineering geologist will define the essential conditions and characteristics of the process in largely qualitative terms. The engineer will often view this with some skepticism, wishing to have a quantitatively defined model. The engineering geologist can overcome much of this skepticism by interpreting the geologic information in a manner that will highlight any relevant relationships, empirical or otherwise, that may be pertinent.

Table 9.1 Terms Used in Risk Assessment and Their Definitions

Term	Definition
Natural hazard (H)	means the probability of occurrence within a specified period of time and within a given area of a potentially damaging phenomenon.
Vulnerability (V)	means the degree of loss to a given element or set of elements at risk (see below) resulting from the occurrence of a natural phenomenon of a given magnitude. It is expressed on a scale from 0 (no damage) to 1 (total loss).
Elements at risk (E)	means the population, properties, economic activities, including public services, etc., at risk in a given area.
Specific risk (R_s)	means the expected degree of loss due to a particular natural phenomenon. It may be expressed by the product of $H \times V$.

Source: From D. J. Varnes, Landslide Hazard Zonation: A Review of Principles and Practice. Copyright © UNESCO 1984. Reproduced by permission of UNESCO.

One important empirical relationship needed in defining hazard is the probability of a process occurring. Calculating probability requires knowing the recurrence interval between events of a certain magnitude. This is based on the record of past occurrence. The elements and formula to use in finding recurrence intervals and probability are well described in Fohn's (1978) paper on defining snow-avalanche hazard. Historical records may be too short to serve as a reliable basis for recurrence. In the case of seismic hazard, a relationship between frequency of occurrence and earthquake magnitude is sometimes used. This permits measurement of frequently occurring microearthquakes to forecast the recurrence of infrequent, large-magnitude ones (Figure 9.1). Minimum ages based on datable features such as ash layers and paleosols displaced by fault movement are another means of determining recurrence. Records of past landslide movement and other geologic processes may be compiled from tree-ring data (Shroder, 1980; DeGraff and Agard, 1984). The reliability of hazard histories reconstructed by these means is limited. A prediction made under these circumstances should: (1) identify the time frame for which the prediction is made, (2) use current understanding of the processes involved and geologic principles to estimate frequency and magnitude of future events, and (3) reflect a conservatism appropriate for the degree of uncertainty associated with the prediction.

Once the hazard is defined, vulnerability can be determined for the proposed or existing project. As noted earlier, risk is the product of the hazard and vulnerability. Specific remedies can be proposed to reduce or eliminate risk. These remedies result in different levels of risk that are then judged as to acceptability (i.e., how safe they are) in the decision-making process for the project. Remedies or measures specific to reducing the risk associated with different processes will now be discussed. These measures

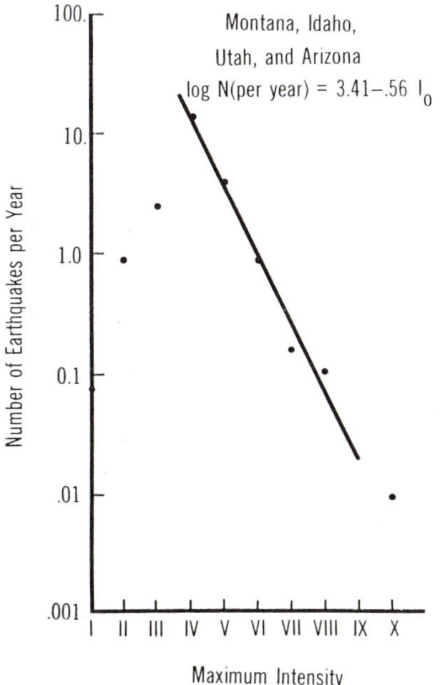

Figure 9.1 A typical relationship of earthquake frequency to intensity. (From Algermissen and Perkins, 1972)

generally involve: (1) controlling where and what types of development may occur, (2) engineering-design changes, and (3) distributing the losses that may result (Hays and Shearer, 1981; Olshansky and Rogers, 1987). Kockelman (1985, 1986) provides examples for accomplishing hazard reduction for landslides and earthquakes (Table 9.2). Similar specific actions can also be suggested for reducing the hazard of other earth processes on development.

Table 9.2 Actions That Can Mitigate Effects of Earthquakes and Landslides

	Actions	
Techniques	Earthquakes[1]	Landslides[2]
Discouraging new development in hazardous areas by	• Adopting seismic-safety or alternate-land-use plans • Developing public-facility and utility service-area policies • Disclosing the hazards to potential buyers • Enacting presidential and gubernatorial executive orders • Informing and educating the public • Posting warnings of potential hazards	• Disclosing the hazard to real-estate buyers • Posting warnings of potential hazards • Adopting utility and public-facility service-area policies • Informing and educating the public • Making a public record of hazards
Removing or converting existing unsafe development through	• Acquiring or exchanging hazardous properties • Clearing and redeveloping blighted areas before an earthquake • Discontinuing uses that do not conform to the zoning ordinance • Reconstructing damaged areas after an earthquake • Removing unsafe structures	• Acquiring or exchanging hazardous properties • Discontinuing nonconforming uses • Reconstructing damaged areas after landslides • Removing unsafe structures • Clearing and redeveloping blighted areas before landslides
Providing financial incentives or disincentives by	• Adopting landing polices that reflect risk of loss • Clarifying the legal liability of property owners • Conditioning federal and state financial assistance • Making public capital improvements in safe areas • Providing tax credits or lower assessments to property owners • Requiring nonsubsidized insurance related to level of hazard	• Conditioning federal and state financial assistance • Clarifying the legal liability of property owners • Adopting lending policies that reflect risk of loss • Requiring insurance related to level of hazard • Providing tax credits or lower assessments to property owners
Regulating new development in hazardous areas by	• Creating special hazard-reduction zones and regulations • Enacting subdivision ordinances • Placing moratoriums on rebuilding • Regulating building setbacks from known hazardous areas	• Enacting grading ordinances • Adopting hillside-development regulations • Amending land-use zoning districts and regulations • Enacting sanitary ordinances • Creating special hazard-reduc-

(continued)

Table 9.2 *(continued)*

Techniques	Actions: Earthquakes[1]	Actions: Landslides[2]
	• Requiring appropriate land-use zoning districts and regulations	tion zones and regulations • Enacting subdivision ordinances • Placing moratoriums on rebuilding
Protecting existing developments through	• Creating improvement districts that assess costs to beneficiaries • Operating monitoring, warning, and evacuating systems • Securing building contents and nonstructural components • Stabilizing potential earthquake-triggered landslides • Strengthening or retrofitting unreinforced masonry buildings	• Controlling landslides and slumps • Controlling mudflows and debris flows • Controlling rockfalls • Creating improvement districts that assess costs to beneficiaries • Operating monitoring, warning, and evacuating systems
Ensuring the construction of earthquake-resistant structures by	• Adopting or enforcing modern building codes • Conducting appropriate engineering, geologic, and seismologic studies • Investigating and evaluating risk of a proposed site, structure, or use • Repairing, strengthening, or reconstructing after an earthquake • Testing and strengthening or replacing critical facilities	

Sources: (1) From Kockelman, 1985; (2) From William J. Kockelman, Some Techniques for Reducing Landslide Hazards, *Bulletin of the Association of Engineering Geologists*, Vol. 23, No. 1, February 1986, pp. 29–52. Used by permission.

EARTHQUAKE-INDUCED PROCESSES

Earthquake-induced processes represent a major concern to engineering works in many parts of the United States and the world. Figure 9.2 illustrates that damage from earthquakes is not limited to California or Alaska. Earthquakes are measured by two different factors. One is magnitude, a measure of the energy released. This measurement is made from records of earthquake shock waves at various seismograph stations (Figure 9.3). The most commonly used scale for magnitude is the Richter scale, or Gutenberg Richter scale (as it is sometimes referenced). The other measure is of intensity. This is based on the effect of an earthquake on people and structures. Table 9.3 is the Modified Mercalli scale, which is typically used for measuring intensity. Intensity has the most direct relationship to the concerns of the engineering geologist defining seismic hazard. Figure 9.4 is an isoseismal map for the 1971 San Fernando, California, earthquake. This was a 6.4 magnitude (moderate) earthquake (Blair and Spangle, 1979).

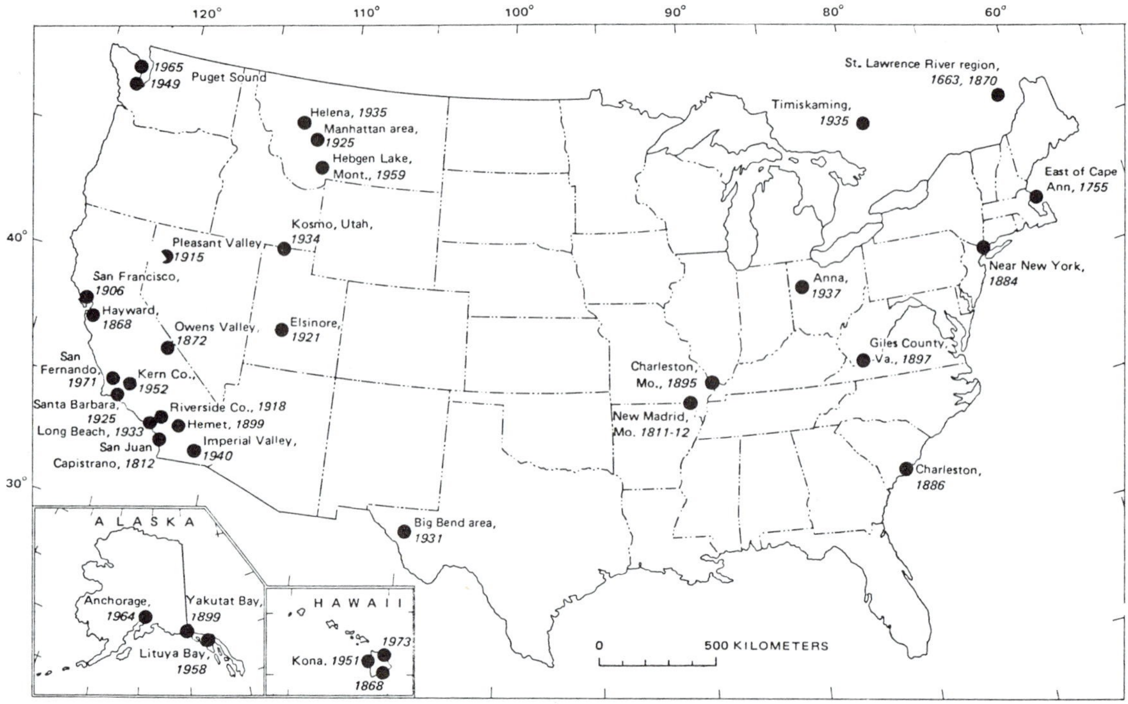

Figure 9.2 The location of historic destructive earthquakes in the United States. (From Hays, 1980)

Earthquake intensity clearly has importance beyond the immediate area of the earthquake. Figure 9.5 relates acceleration to the different levels in the Modified Mercalli scale. This provides some indication of g-values—ground motion expressed in terms of gravitational acceleration—appropriate to engineering calculations.

There are four types of earthquake-induced processes: (1) surface rupture, (2) ground shaking, (3) ground failure, and (4) tsunami and seiche occurrence. Each type has a significance to safe design of structures and is a concern for engineering geologists defining hazard levels.

Surface rupture is the actual displacement and cracking of the ground surface along a fault trace. Figure 9.6 suggests the degree of importance to engineering works that this process may represent. Displacement beneath a building that exceeds 1 or 2 in. can have a catastrophic effect (Nichols and Buchanan-Banks, 1974). Surface rupture is confined to a narrow zone along an active fault. Rupture may happen rapidly during an earthquake or it may not occur at all. In general, the potential for a great earthquake and associated greater amount of displacement increases the longer the fault trace (Nichols and Buchanan-Banks, 1974). The effect of surface rupture on structures is not only limited to rupture occurring during a major earthquake. Some active faults undergo imperceptibly slow movement, termed *fault creep.* Although a serious problem for structures, the cumulative effect of this long-term displacement, unlike rupture during an earthquake, is not catastrophic.

Ground shaking is the actual trembling or jerking motion produced by an earthquake. It causes widespread damage and is one of the more difficult seismic effects to quantify and predict (Nichols and Buchanan-Banks, 1974). Structural damage accounted for a large part of the $500 million worth of damage caused by the 1971 earthquake in San Fernando, California. The degree of damage varies with the wave length and duration of shaking, the nature of underlying materials, and the character of structures. For the same earthquake, the effect of ground shaking can be several times greater at sites with thick, water-saturated soil than at those on competent bedrock (Figure 9.7). Buildings have a fundamental period that is roughly equivalent to the number of their stories (Figure 9.8). A resonance, or amplification, of ground shaking develops where the building and underly-

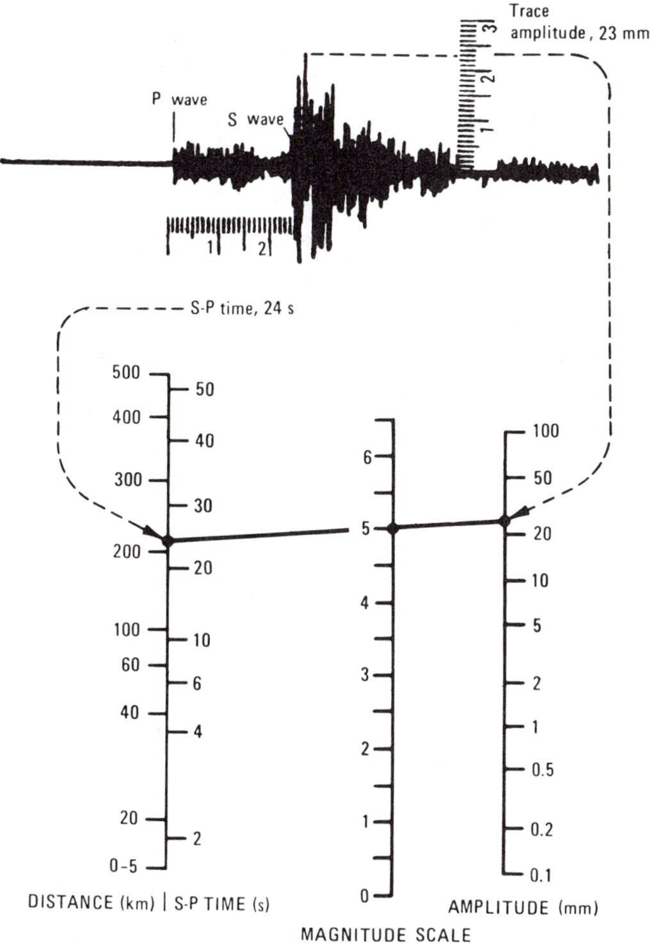

Figure 9.3 Diagram of how Richter magnitude of an earthquake is determined. The maximum amplitude of the seismogram and the difference in arrival time between the S- and P-waves is used on the nomogram to read the magnitude value. (From Hays, 1980)

Table 9.3 Descriptions of the 12 Levels of Earthquake Intensity on the Modified Mercalli Scale

I. Not felt
II. Felt by persons at rest, on upper floors, or favorably placed.
III. Felt indoors. Hanging objects swing. Vibration like passing of light trucks. Duration estimated. May not be recognized as an earthquake.
IV. Hanging objects swing. Vibration like passing of heavy trucks; or sensation of a jolt like a heavy ball striking the walls. Standing automobiles rock. Windows, dishes, doors rattle. Wooden walls and frame may creak.
V. Felt outdoors; direction estimated. Sleepers wakened. Liquids disturbed, some spilled. Small unstable objects displaced or upset. Doors swing. Shutters, pictures move. Pendulum clocks stop, start, change rate.
VI. Felt by all. Many frightened and run outdoors. Persons walk unsteadily. Windows, dishes, glassware broken. Knickknacks, books, etc., off shelves. Pictures off walls. Furniture moved or overturned. Weak plaster and masonry D[1] cracked.
VII. Difficult to stand. Noticed by drivers of automobiles. Hanging objects quiver. Furniture broken. Weak chimneys broken at roof line. Damage to masonry D, including cracks; fall of plaster, loose bricks, stones, tiles, and unbraced parapets. Small slides and caving in along sand or gravel banks. Large bells ring.
VIII. Steering of automobiles affected. Damage to masonry C; partial collapse. Some damage to masonry B; none to masonry A. Fall of stucco and some masonry walls. Twisting, fall of chimneys, factory stacks, monuments, towers, elevated tanks. Frame houses moved on foundations if not bolted down; loose panel walls thrown out. Decayed piling broken off. Branches broken from trees. Changes in flow or temperature of springs and wells. Cracks in wet ground and on steep slopes.
IX. General panic. Masonry D destroyed; masonry C heavily damaged, sometimes with complete collapse; masonry B seriously damaged. General damage to foundations. Frame structures, if not bolted, shifted off foundations. Frames racked. Serious damage to reservoirs. Underground pipes broken. Conspicuous cracks in ground and liquefaction.
X. Most masonry and frame structures destroyed with their foundations. Some well-built wooden structures and bridges destroyed. Serious damage to dams, dikes, embankments. Large landslides. Water thrown on banks of canals, rivers, lakes, etc. Sand and mud shifted horizontally on beaches and flat land. Rails bent slightly.
XI. Rails bent greatly. Underground pipelines completely out of service.
XII. Damage nearly total. Large rock masses displaced. Lines of sight and level distorted. Objects thrown in the air.

[1] Masonry A: Good workmanship and mortar, reinforced designed to resist lateral force.
Masonry B: Good workmanship and mortar, reinforced.
Masonry C: Good workmanship and mortar, unreinforced.
Masonry D: Poor workmanship and mortar and weak materials, like adobe.

Source: From Blair and Spangle, 1979.

ing soil have a similar fundamental period and causes more-extensive damage. Buildings with differing fundamental periods will sway differently. Where they are in close proximity, this difference may cause buildings swaying with differing motion to impact and damage each other.

Ground failure because of ground acceleration from an earthquake produces landslides, ground cracking, subsidence, and differential settlement. Hillslopes that might remain stable under a static load—the weight

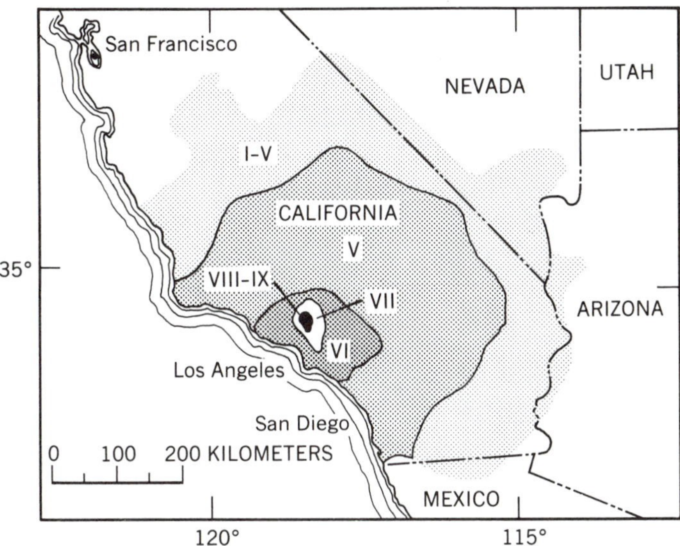

Figure 9.4 An isoseismal map showing the extent of intensity for the 1971 San Fernando, California, earthquake. (From Hays, 1980)

of overlying material—can fail under a dynamic load, the stress from earthquake motion. Slopes composed of unconsolidated soft sediments or surficial deposits having steep slopes, high seasonal groundwater levels, and shallow-rooted or sparse vegetation, and slopes composed of rock containing many open discontinuities are especially prone to landsliding during an earthquake. Four earthquakes with magnitudes about 6.0 occurred in the vicinity of Mammoth Lakes, California, in May 1980. Several hundred rockfalls and rockslides were triggered in the surrounding Sierra Nevada as a consequence of earthquake motion (Figure 9.9). Although property losses were minimal due to the undeveloped condition of the mountains, two hikers in Yosemite National Park were fatally injured by falling rock (Harp et al., 1984). Liquefaction is another mechanism of ground failure during an earthquake. Ground motion transforms loose water-saturated granular material to a liquid state (Figure 9.10). This occurs because the cyclic stress of succeeding waves of ground motion causes pore-water pressure to build up. This rapid increase in pore-water pressure nearly reduces the effective stress of the soil mass to zero, a state where it has the least resistance to applied stress. Otherwise solid level ground becomes unable to bear the weight of overlying structures. Building and structural foundations sink into the liquified materials, causing tilting and related damage (Figure 9.11*a*). Differential settlement can cause damage to buildings as some of the soil settles more than other parts underlying the building foundation (Figure 9.11*b*). Liquified soil may flow on very low slopes. This special type of landslide is called a lateral spread failure (Figure 9.11*c*). For example, the Juvenile Hall failure in the 1971 San Fernando earthquake involved a mile-long area on a 2.5% slope (Nichols and Buchanan-Banks, 1974).

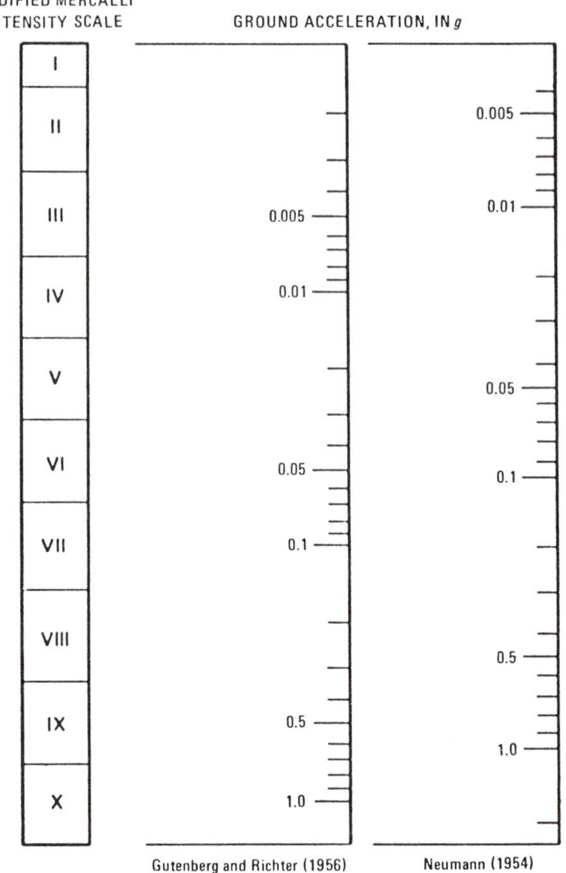

Figure 9.5 Diagram showing proposed relations of intensity to ground acceleration. (From Hays, 1980)

Figure 9.6 Rupture along the Lost River fault caused by the Borah Peak earthquake on October 28, 1983. Vertical displacement at this point on the scarp is about 4 m. This 6.9-magnitude event was centered about 110 miles northwest of Pocatello, Idaho. (Photo courtesy of Earl P. Olson.)

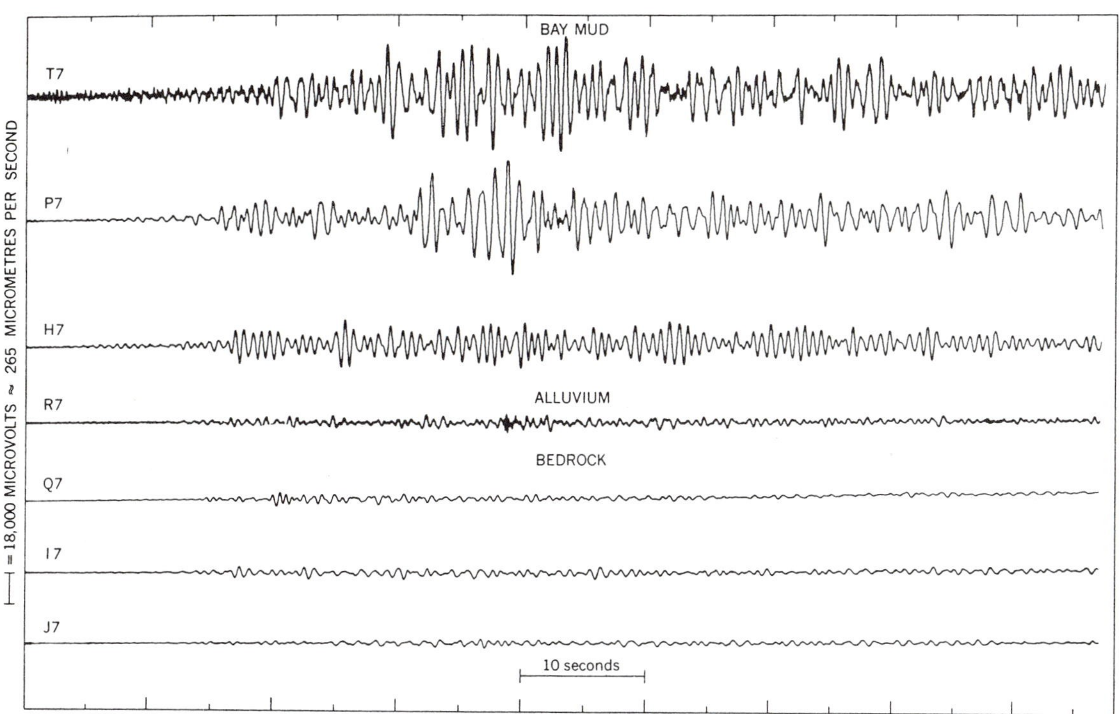

Figure 9.7 Recordings of horizontal ground motions at San Francisco, California, generated by two nuclear explosions. These recordings show the difference in ground motions for different underlying materials. (From Borcherdt et al., 1975)

Table 9.4 summarizes the different forms of liquefaction and the kinds of structures most likely to be damaged by them.

Tsunamis and seiches are similar effects that occur in bodies of water. *Tsunami* is a Japanese term for large ocean waves generated by submarine earthquakes. They are sometimes mislabeled, tidal waves. Rapid displacement on an undersea fault can cause a wave capable of traveling thousands of miles from the earthquake epicenter. On the deep ocean, the tsunami height typically is 1 ft. The wave height increases as it reaches shallower water near shorelines (Hays, 1981b). Tsunamis tens-of-feet high spread more than 1,500 mi from their point of origin during the 1964 Alaskan earthquake (Nichols and Buchanan-Banks, 1974). In distant Crescent City, California, on the coast near the border with Oregon, the tsunami from this earthquake caused $7 million worth of damage. Of the several waves reaching Crescent City, the third wave was especially devastating. It ran inland a distance of 500 m, carrying logs with velocities as high as 10-m-per-second through the streets. The wave height was about 6 m above mean low water and inundated 30 city blocks (Bolt et al., 1975). Seiche or earthquake-generated standing waves are not quite as destructive as tsunamis. They occur in enclosed or restricted bodies of water and consist of a standing wave that oscillates across the surface. Sieches generally are of low amplitude (less than 1 ft). However, in shallow areas, the wave may run up to 20–30 ft. This can destroy shoreline facilities and damage reservoirs and

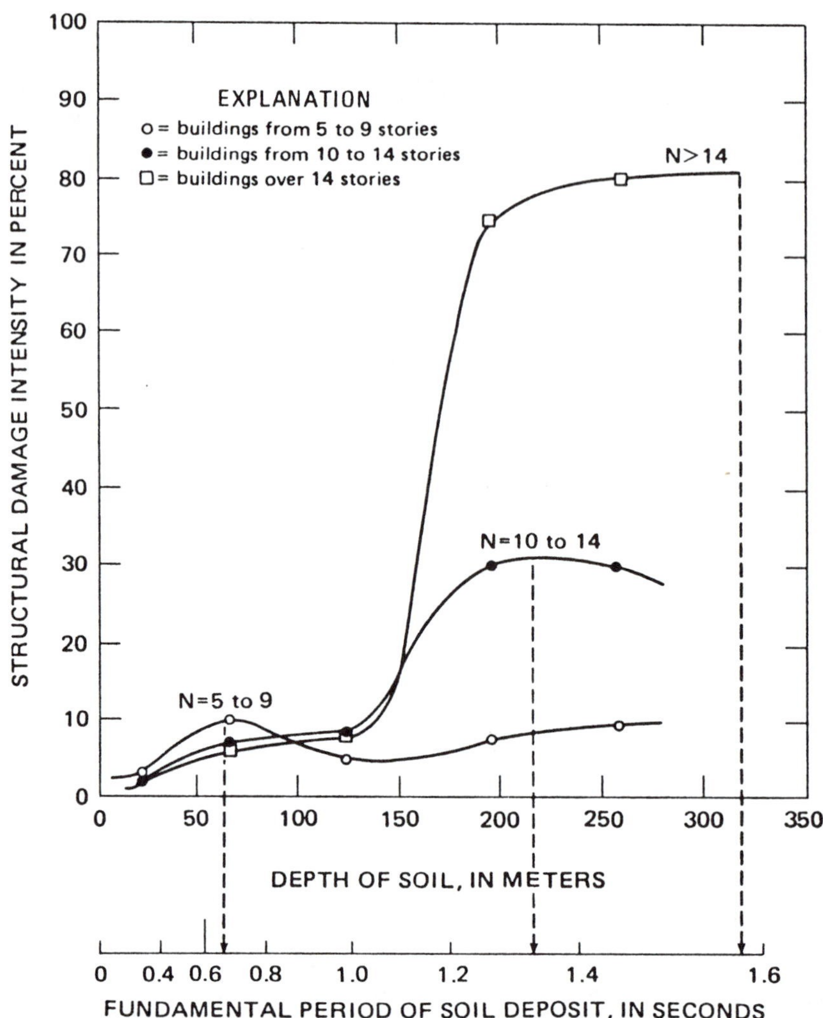

Figure 9.8 Graph showing the expected structural damage for buildings of different height on soil with varying depth and fundamental period; N is equal to the number of stories in the building. (Reprinted with permission from Am. Soc. Civ. Eng. Proc., J. Soil Mech. and Found. Eng. Div., Vol. 98, H. B. Seed, R. V. Whitman, H. Dezfulian, R. Dobry, and I. M. Idriss, Soil Conditions and Building Damage in 1967 Caracas Earthquake, 1972.)

dams by causing overtopping. The fundamental period of oscillation for a sieche is controlled by the depth and size of the body of water. Several standing waves with different periods may be possible for complicated bodies of water. Of the several standing waves possible for San Francisco Bay, the most important has a period of 39 min (Bolt et al., 1975).

Evaluating Earthquake-Induced Processes

The preceding discussions show that evaluating earthquake effects is not a simple task. Such an evaluation depends both on the effects that are of concern and the type of project they might influence. Even when a specific

Figure 9.9 Rockfalls and rockslides in the vicinity of Convict Lake and Mount Baldwin near the crest of the Sierra Nevada, California. These landslides were generated by the May 1980 earthquake sequence centered at Mammoth Lakes, California. (From Harp et al., 1984)

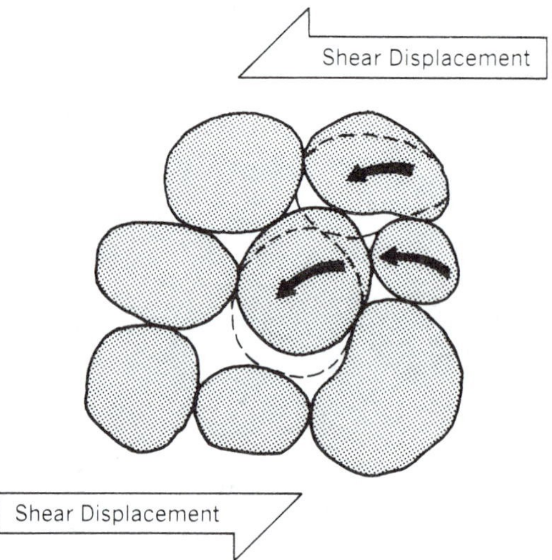

Figure 9.10 An illustration of the process of liquifaction. Horizontal ground motion from an earthquake represented by the large arrows distorts a water-saturated mass of sand grains. Small arrows show how looser packed grains collapse, transferring stress from grain-to-grain contacts to water in void space. Pore-water pressures build until the mass behaves like a liquid. (From Youd and Keefer, 1981)

effect such as ground motion is under study, multiple approaches are useful in the absence of a standard procedure for evaluating ground motion (Krinitzsky and Marcuson, 1983). An engineering geologist can always expect to deal with questions of earthquake location and magnitude. In some instances, this will involve determining the location of faults and estimating the magnitude of past earthquakes. Establishing how an earthquake of a particular magnitude might affect an area or project is the other way that questions of location and magnitude arise.

Determining the location of faults and estimating their activity level is a typical task of the engineering geologist. This requires using standard geologic mapping techniques and methods. Aerial photography and other remote-sensing techniques are employed to detect faults. Cluff et al. (1972) note some of the traditional and nontraditional techniques applied to investigating faults and their past seismicity (Table 9.5). Ground-level investigation is an important component of most stages of fault studies. Figure 9.12 is an example of some ground features that may indicate strike-slip faulting. The presence of faults does not necessarily mean that earthquakes may occur. It is important to distinguish faults likely to cause an earthquake from those that are unlikely. This is part of determining the degree of hazard for risk assessment. Hays (1980) notes that key factors for assessing whether a fault will produce an earthquake are: (1) fault length, (2) magnitude and nature of displacement, (3) geologic history of displace-

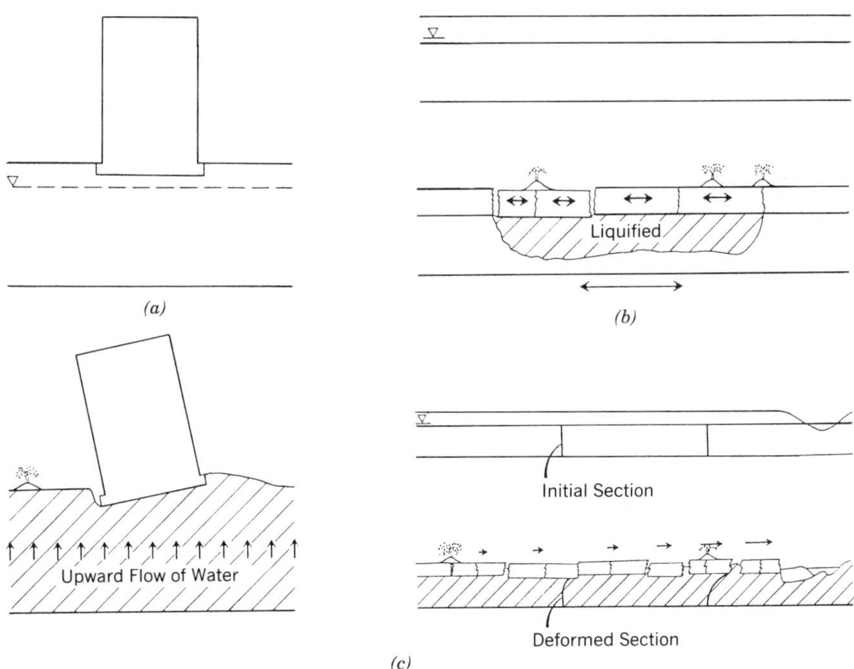

Figure 9.11 Diagram of three types of damaging effects of liquefaction. (*a*) soil able to bear weight of building prior to an earthquake (above) fills as liquefaction weakens the soil and reduces support (below). (*b*) Normally solid soil (above) can crack, oscillate, and form sand boils and volcanos owing to liquefaction of soil below the water table (below). (*c*) Liquefaction within a soil on a low slope (above) can cause movement, called *lateral spreading* (below). (From Youd, 1984)

Table 9.4 Examples of Structures Most Likely to Be Affected by Different Types of Liquefaction Effects

Types of Structural Instability	Structures Most Often Affected
Loss of foundation-bearing capacity	Buried and surface structures
Slope instability slides	Structures built on or at the base of the slope
	Dam embankments and foundations
Movement of liquefied soil adjacent to topographic depressions	Bridge piers
	Railway lines
	Highways
	Utility lines
Lateral spreading on horizontal ground	Structures, especially those with slabs on grade
	Utility lines
	Highways
	Railways
Excess structural buoyancy caused by high subsurface pore pressure	Buried tanks
	Utility poles
Formation of sinkholes from sand blows	Structures built on grade
Increase of lateral stress in liquefied soil	Retaining walls
	Port structures

Source: From National Research Council, 1985a.

Table 9.5 Listing Fault-Investigation Techniques and the Stage of Investigation in Which They Are Most Often Employed

		Investigative Stage			
Investigation	Description	Regional Fault Activity	Site Fault Activity	Regional Seismicity	Site Response
Literature Review	Published and unpublished data.	x	x	x	x
Aerial Photos	Photointerpretation of stereopairs, taken different years, different scales.	x	x		x
Geologic Reconnaissance	Field check of features seen in aerial photographs, etc.	x	x		x
Aerial Reconnaissance	Spot check of broad geologic features not obvious on ground.	x	x		
Low Sun-Angle Photos	Faults, landslides accentuated by shadows.	x	x		
Field Mapping	Detail needed depends on purpose and information otherwise obtained.	x	x		x
Seismic Refraction	Surface and downhole refraction to determine depth, continuity, and properties of subsurface materials.		x		x
Resistivity	Determine water level differences, clay-rich zones.		x		x
Magnetic	Locate iron mineralization concentrations, sometimes associated with faulting.		x		
Geodetic Nets	Determine relative fault displacement in active creep areas.	x			
Borings	Spot observations and testing of materials, water level determination.		x		x
Trenching	Directly observe subsurface materials for evidence of displacement.		x		
Age Dating	Determine age of most recent fault offset.	x	x		
Seismologic Records	Statistical studies, earthquake recurrence projections, concentrations of activity.	x	x	x	
Historic Accounts	Preinstrumental records of major earthquakes and surface effects in project area.	x		x	x

Source: From Cluff et al., 1972.

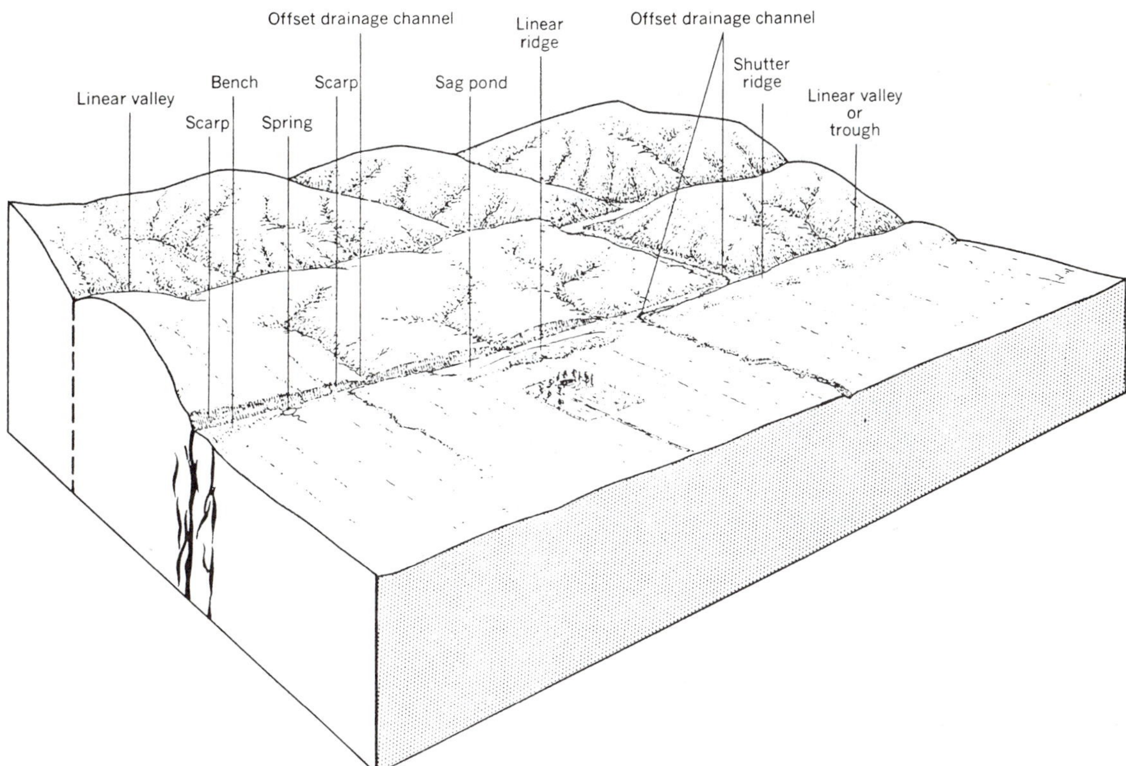

Figure 9.12 Block diagram showing some of the landform features indicating recent movement on a strike-slip fault. (From Wesson et al., 1975)

ments, including the age of recent movements, and (4) the relationship of the fault to regional tectonics. Fault mapping provides data such as fault length and relationship to regional tectonics. Trenching of faults permits confirmation of fault location and detection of evidence for estimating the magnitude and frequency of past movement. Figure 9.13 shows how the vertical displacement of sedimentary contacts in relation to the age of deposits yielded the rate of movement of the previous 3000 years (Clark et al., 1972). Microearthquake monitoring and examination of seismic records provide seismic activity information to correlate with these geologic studies. Determining whether a fault is active or potentially active generally requires several lines of evidence. Cluff et al. (1972) show how these lines of evidence can be applied to classify fault activity (Table 9.6).

Even where fault rupture may pose no concern at a project site, it may be necessary to estimate the ground motion likely to be experienced at a location. This hypothetical, or design, earthquake is used to assess the response of the engineering structure, to provide for a design that has an adequate factor of safety, or to use for investigating the response of physical conditions at the location. As Blair and Spangle (1979) note, the design earthquake is usually one that is at or close to the maximum magnitude that may be reasonably expected. This is sometimes referred to as the

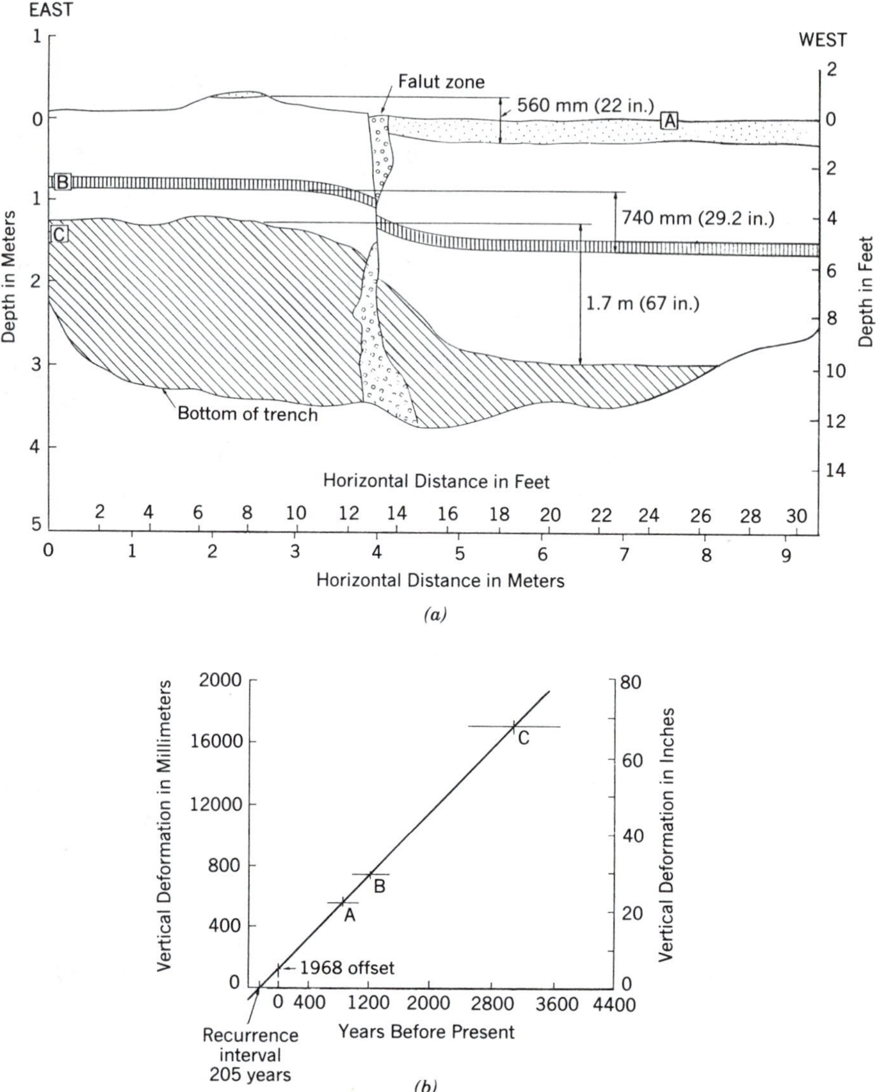

Figure 9.13 (a) Face map of trench showing the displacement of sediment by the Coyote Creek fault, southern California. (b) Radiometric dating of the deposits (points A, B, C) combined with measured vertical displacement permit recurrence interval for past fault movement to be determined. (Adapted by Wesson et al., 1975, from Clark et al., 1972.)

maximum credible earthquake. All faults within a reasonable distance around the area of interest need to be considered. The maximum credible earthquake on a more distant fault may cause greater damage than a maximum credible earthquake on a closer fault (Blair and Spangle, 1979). Recent work suggests that magnitude, and frequency relationships used to estimate recurrence of large landslides may not be appropriate for all faults (Schwartz and Coppersmith, 1984). Algermissen and Perkins (1972) proposed a method for estimating the ground motion that might reasonably be

Table 9.6 Descriptions of Historical, Geological, and Seismological Evidence Indicating Activity State of a Fault

	Active	Potentially Active	Uncertain Activity	Inactive
Historical	Surface faulting and associated strong earthquakes. Tectonic fault creep or geodetic indications of movement.	No reliable report of historical surface faulting.		No historical activity.
Geological	Generally young deposits have been displaced or cut by faulting. Fresh geomorphic features characteristic of active fault zones present along fault trace. Physical groundwater barriers in geologically young deposits.	Geomorphic features characteristic of active fault zones subdued, eroded, and discontinuous. Faults are not known to cut or displace the most recent alluvial deposits but may be found in older alluvial deposits. Water barrier may be found in older materials. Geologic setting in which the geomorphic relation to active or potentially active faults suggests similar levels of activity.	Available information does not satisfy enough criteria to establish fault activity. If the fault is near the site, additional studies are necessary.	Geomorphic features characteristic of active fault zones are not present, and geologic evidence is available to indicate that the fault has not moved in the recent past.
Seismological	Earthquake epicenters are assigned to individual faults with a high degree of confidence.	Alinement of some earthquake epicenters along fault trace but locations have a low degree of confidence.		Not recognized as a source of earthquakes.

Source: From Cluff et al., 1972.

expected at a site. Initially, records of past earthquakes are used to relate earthquake frequency to maximum Modified Mercalli intensity in order to identify earthquake source areas. Five source areas for Utah and Arizona are described in Algermissen and Perkins' example. Maximum intensities in each source area were converted to magnitude and accelerations associated with earthquakes of each magnitude level were calculated. Acceleration attentuation relations by Schnabel and Seed (1972) were included in these calculations. By using the expected occurrences of accelerations in each source area, a probabilistic estimate of the maximum acceleration likely to occur for a 50-year return period was made (Figure 9.14). Choos-

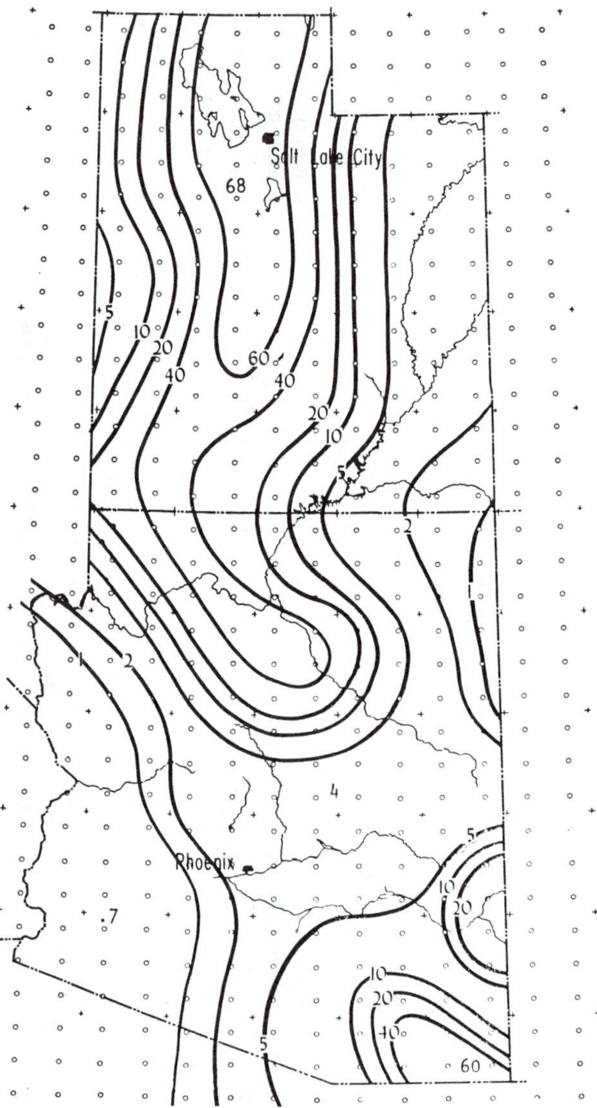

Figure 9.14 Map of Utah and Arizona showing the maximum acceleration (expressed in percentage of gravity) with a 90% chance of its occurrence within a 50-year period. (From Algermissen and Perkins, 1972)

ing a design earthquake magnitude or peak acceleration is a subjective judgment and should depend on the kind of project, the stage in the decision-making process, and the consequences of its failure (Krinitzsky and Marcuson, 1983). It should be clear that we have only introduced some of the means for evaluating earthquake-induced processes. Blair and Spangle (1979), Hays (1980), Krinitzsky and Marcuson (1983), and current literature describe a variety of methods and procedures for assessing fault location, determining expected magnitude, and selecting design earthquake motions. Because this topic is being actively researched by both the geologic and engineering communities, practitioners should be aware that new methods and knowledge are changing how future evaluation is done.

Mitigating the Effects of Earthquake-Induced Processes

Avoiding areas subject to earthquake-induced processes and engineering design are the two principal ways of mitigating potentially damaging effects (Bonilla, 1981; Hays, 1981a; Hays 1981b; Youd and Keefer, 1981). The examples herein give an overview of how these approaches can be applied to different types of earthquake-induced processes.

Avoiding areas subject to one or more earthquake-induced processes involves hazard zonation or mapping. Mapping areas of potential surface rupture is one of the most effective means of mitigating this process. In California, mapping potential surface rupture is mandated by the Alquist-Priolo Special Studies Zones Act. Figure 9.15 is an example of a Special Studies Zones map along the San Andreas fault system. This zonation permits local government to establish setbacks, or areas along the line of potential rupture, where construction of buildings will not be allowed. Figure 9.16 is an example of setbacks showing how the setback distance varies with the degree of uncertainty associated with fault trace location.

Ground motion can be zoned on the basis of past landslide damage. The seismic risk map of the United States produced by Algermissen (1969) used the distribution of damaging earthquakes and their effect as defined by the Modified Mercalli intensity scale to define four zones, ranging from no expected damage to anticipated major damage (Figure 9.17). The record of past earthquakes can also be used to estimate expected effective peak ground acceleration. Peak acceleration is a variable useful in engineering evaluations of structural response to earthquake motions. The magnitude of past earthquakes along with generalized curves on attenuation of generated motion for different regions can be combined to provide estimates of expected peak ground acceleration. The basic procedure is described in Algermissen and Perkins (1972). Figure 9.18 shows the resulting map with contours of expected maximum peak acceleration.

Zonation of areas for potential ground failure requires a good understanding of the surficial materials in an area. Wieczorek et al. (1985) have mapped slope-stability zones for San Mateo County, California. Slope-stability maps usually depend on static slope-stability analysis. This is a shortcoming in predicting relative hazard areas during the motion of an earthquake. The San Mateo County map is based on a dynamic analysis

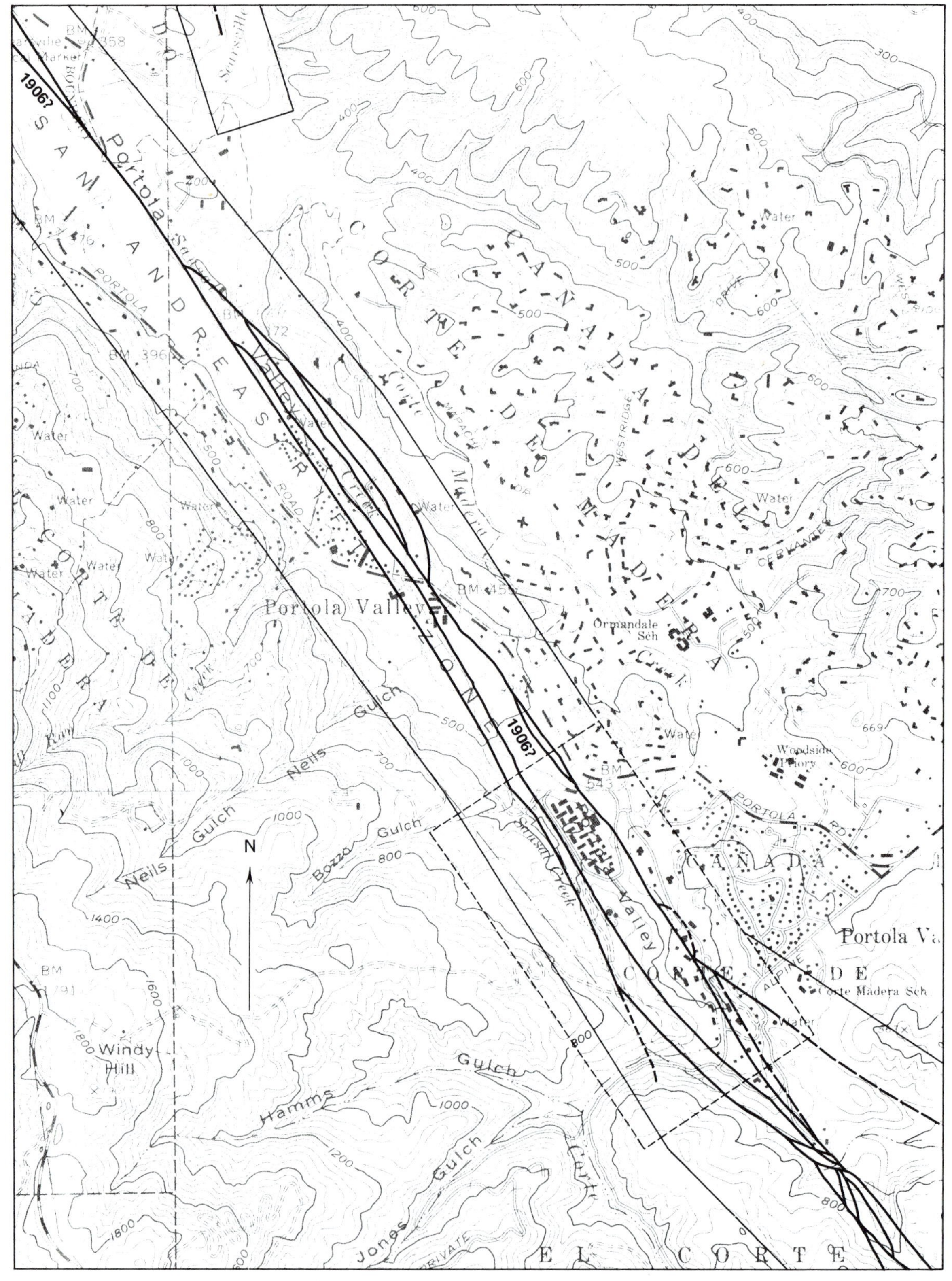

Figure 9.15 Part of the Special Studies Zones map for San Mateo County, California, showing fault traces. The rectangular area outlined by dashed line is shown at a larger scale in Figure 9.16. (From Blair and Spangle, 1979)

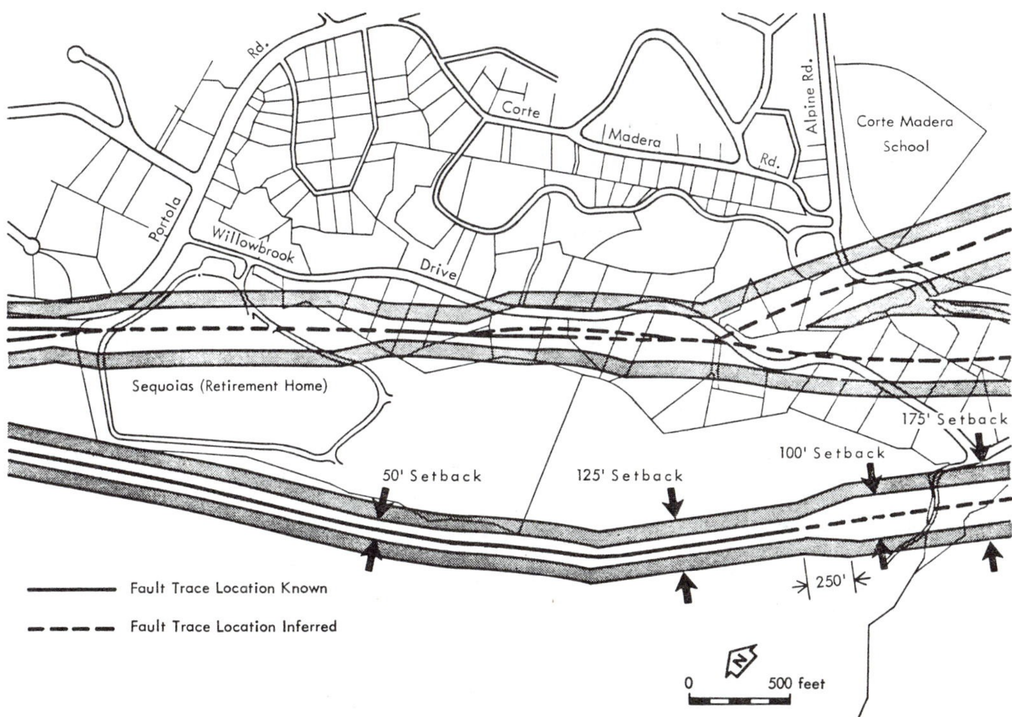

Figure 9.16 Map illustrating the use of setback distances from known and inferred fault traces. Setbacks avoid locating buildings across locations with likely fault rupture during a future earthquake. (From Mader et al., 1972)

technique developed by Newmark (1965) that permits evaluation of ground motion as a factor in defining slope stability. Like most hazard zonation maps, the slope-stability map is appropriate for some purposes, not for others. This is an important point to remember in using maps of hazard zones; Table 9.7 defines these uses and limitations. Liquefaction, an equally important form of ground failure, requires detailed information of surficial materials and expected ground motion. Blair and Spangle (1979) describe efforts in Santa Clara County, California, to address liquefaction. Figure 9.19 shows their map, which delineates zones of different types of liquefaction. These zones are based on the likely response of underlying surficial material under both seismic and aseismic (static) loading. Table 9.8 describes each zone and the expected surface effect. Various land and building uses are considered appropriate to each zone by local governments in Santa Clara County. Restriction of land and building uses to certain zones permits avoidance of severe damage from liquefaction (Table 9.9).

Nichols and Buchanan-Banks (1974) note that zonation for tsunamis and seiches is hampered by extremely limited historical data and the absence of theoretical knowledge. Bolt et al. (1975) describe an empirical relationship for relating earthquake magnitude to tsunami magnitude and expected maximum run-up at the shoreline (Table 9.10). This applies to shallow-focus submarine earthquakes and the actual run-up will be greatly

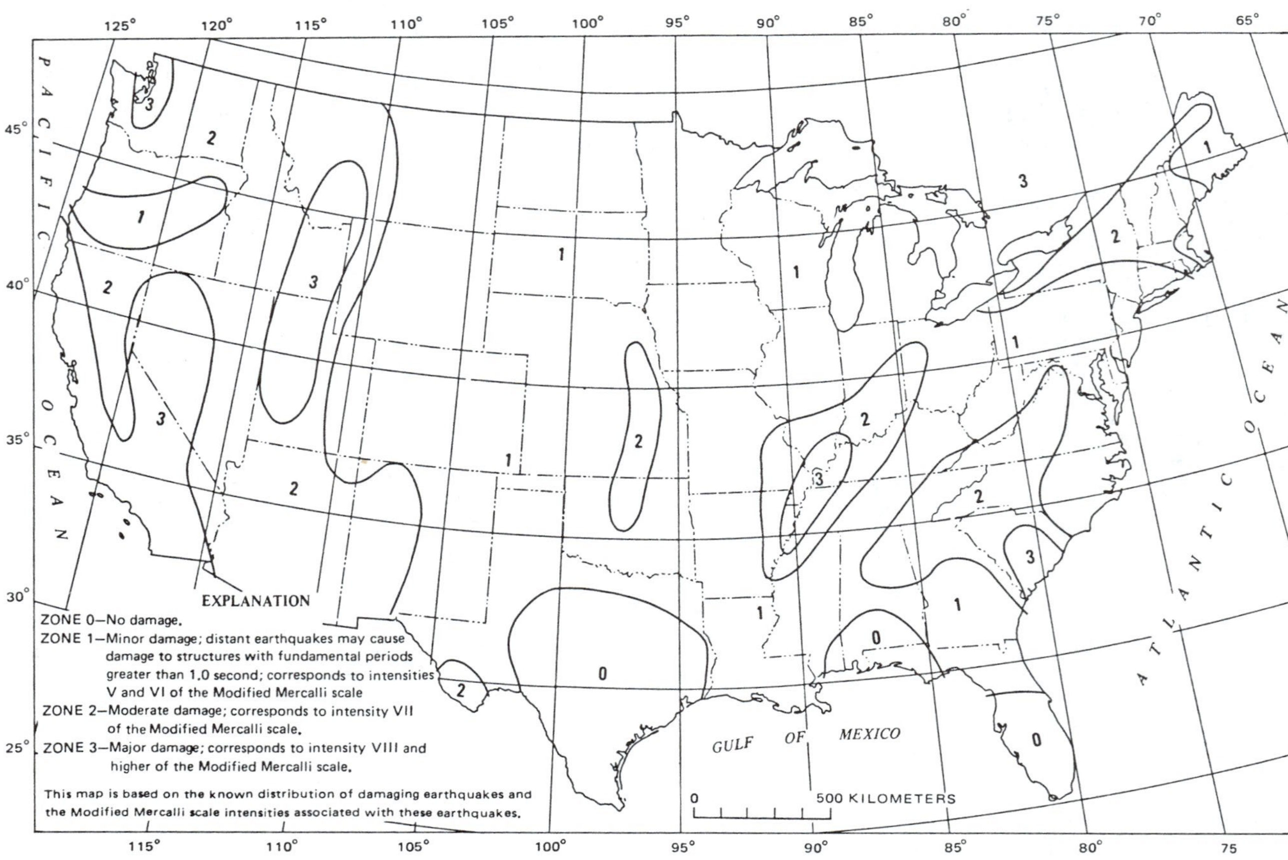

Figure 9.17 Seismic risk zones for the United States based on the distribution of damaging earthquakes and Modified Mercalli intensity scale. (From Algermissen, 1969)

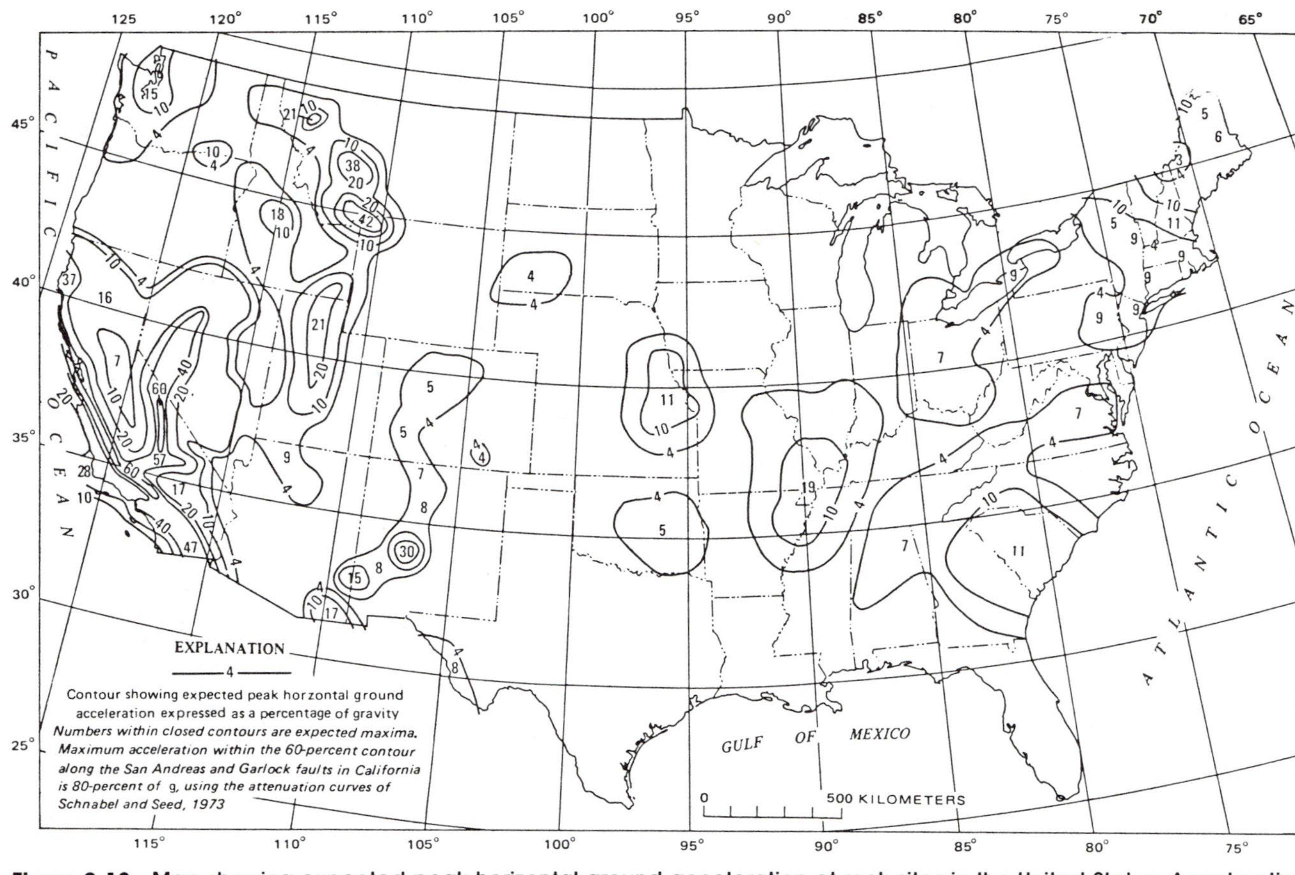

Figure 9.18 Map showing expected peak horizontal ground acceleration at rock sites in the United States. Acceleration is expressed as the percentage of gravity. Contoured values have a 10% chance of being exceeded in a 50-year period. (From Algermissen and Perkins, 1976)

Table 9.7 Uses and Limitations Appropriate to a Zonation Map of Earthquake-Induced Effects

Map Appropriate for Following Uses	Map Inadequate, Not Recommended for Following Uses
1. A comparative guide to seismically induced landslide susceptibility of different areas in San Mateo County.	1. A basis of determine the absolute risk of seismically induced landslides at any specific site.
2. One of several elements in regional land-use planning on the county level (e.g., transportation corridors, open space, etc.).	2. The sole justification for zoning or rezoning of any specific site or parcel.
3. Input for preliminary routing of lifeline or transportation corridors that could be seriously damaged by seismic landsliding.	3. Detailed design or routing of lifeline or transportation corridors.
4. To prepare emergency response plans for a future earthquake—e.g., selecting emergency transportation routes, planning for lifeline disruptions, etc.	4. Specific site planning to reduce hazards.
5. Guide to estimate losses from seismic landsliding in a future earthquake, on a countywide basis.	5. To set or modify earthquake insurance rates.

Source: From Wieczorek et al., 1985.

influenced by the near-shore water depths and coastal configuration. This is clearly a hazard needing more research.

Engineering design is another means for mitigating earthquake-induced effects. There is little that design can do to prevent damage from surface rupture on faults or ground failure. However, certain lifeline facilities such as oil, gas, and water pipelines and electrical and telephone transmission lines can be equipped with means for limiting damage. For example, automatically closing valves can be placed in oil pipelines on either side of an area of possible fault rupture. This limits any spilled oil to that contained within the short span actually broken by the fault movement.

Engineering design can do far more in reducing damage from ground motion. The Unified Building Code includes a formula that uses the seismic zones from Algermissen's 1969 map in calculating the force to be resisted by the building design. This illustrates how the mapping produced by the engineering geologist becomes part of the engineer's design effort. This engineering design applies to both new and existing construction. Existing construction incapable of withstanding an earthquake can be a serious problem in mitigating earthquake-induced hazard. An estimated eight thousand unreinforced masonry buildings that are likely to partially or completely collapse under a moderate earthquake are found in the city of Los Angeles (Kockelman and Campbell, 1983). Eighty percent are commercial and industrial buildings where about 70,000 people work. Another 137,000 people live in residential apartments and hotels that

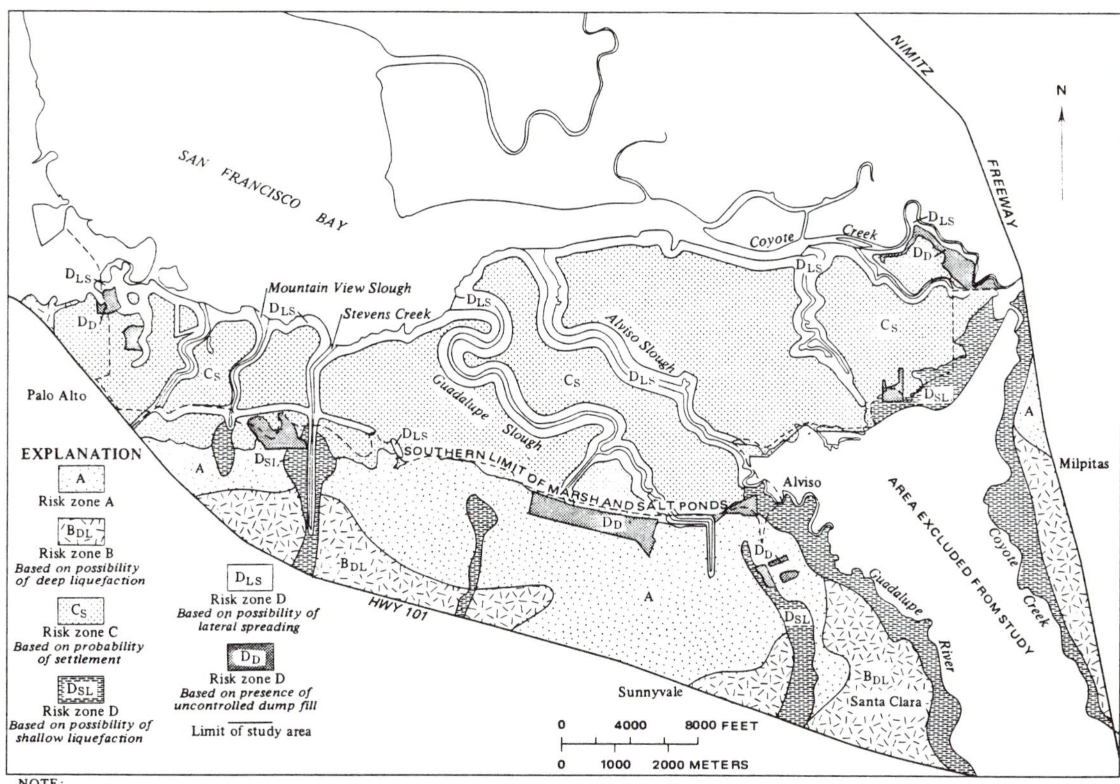

NOTE:
The delineation of risk zones is based on limited subsurface information. Risk zone designations could change as more detailed subsurface information becomes available.

Figure 9.19 Liquefaction risk zones for part of Santa Clara County, California. (From Blair and Spangle, 1979)

Table 9.8 Description of Surface Effect and Subsurface Cause Expected in the Mapped Risk Zones Shown in Figure 9.19

Risk Zone	Surface Effect	Subsurface Cause
A	Little risk of settlement or ground failure	
B_{DL}	Significant settlement	Liquefaction of confined granular layer in alluvium (seismic loading)
C_S	Moderate to substantial settlement and/or differential settlement	Consolidation of bay mud or soft clay (static loading)
D_D	Substantial settlement and/or differential settlement	Consolidation of uncontrolled dump fill or sanitary land fill (static loading)
D_{SL}	Failure of ground surface	Liquefaction of granular surface layer (seismic loading)
D_{LS}	do	Lateral spreading toward free face (seismic loading)

Source: From Blair and Spangle, 1979.

Table 9.9 Categories of Land and Building Uses and Their Permitted Risk Zones for Siting.[a]

Land and Building Uses	Risk Zones A	B	C	D
Group A Buildings				
Hospitals and nursing homes	x			
Auditoriums and theatres	x			
Schools	x			
Transportation and airport	x			
Public and private office	x			
Major utility	x			
Group B Buildings				
Residential — multiple units	x	x		
Residential — 1 and 2 family	x	x		
Small commercial	x	x		
Small public	x	x		
Small schools — one story	x	x		
Utilities	x	x		
Group C Buildings				
"Industrial park" commercial	x	x	x	
Light and heavy industry	x	x	x	
Small public, if mandatory	x	x	x	
Airport maintenance	x	x	x	
Group D Buildings				
Water-oriented industry	x	x	x	
Wharves and docks	x	x	x	
Warehouses	x	x	x	
Group D Open Space				
Agriculture, marinas, public and private open spaces, marshlands and saltponds, and small appurtenant buildings	x	x	x	x

[a] Based on the effects described in Table 9.8.

Source: From Blair and Spangle, 1979.

Table 9.10 Tsunami Magnitude and Expected Maximum Run-up Distance for Different Magnitude, Shallow-Focus, Submarine Earthquakes

Earthquake Magnitude	Tsunami Magnitude	Maximum Run-up (m)
6	Slight	
6.5	−1	0.5 0.75
7	0	1 1.5
7.5	1	2 3
8	2	4 6
8.25	3	8 12

Source: Reprinted with permission from Geological Hazards, B. A. Bolt, W. L. Horn, G. A. Macdonald, and R. F. Scott, 1975, Springer-Verlag.

account for 14% of such buildings. Strengthening these buildings requires the same type of information of ground motion that new construction requires. Figure 9.20 provides three examples of engineering methods for reinforcing existing homes to resist ground motion.

VOLCANIC PROCESSES

Thousands of volcanoes worldwide have erupted with destructive consequences over the last few thousand years. In historic time, more than five hundred volcanoes have been active (Scott, 1984). Much of the land threatened by volcanic processes is found around the margin of the Pacific Ocean. This includes hazardous areas in Alaska, Hawaii, the Pacific Northwest, and California (Figure 9.21). Other regions with a significant concentration of volcanic activity are the Caribbean and Mediterranean.

Many examples of the destructive effect of volcanic eruptions are found in historic accounts. The 1902 eruption of Mount Pelée on the Caribbean island of Martinique is often cited. It completely destroyed the city of

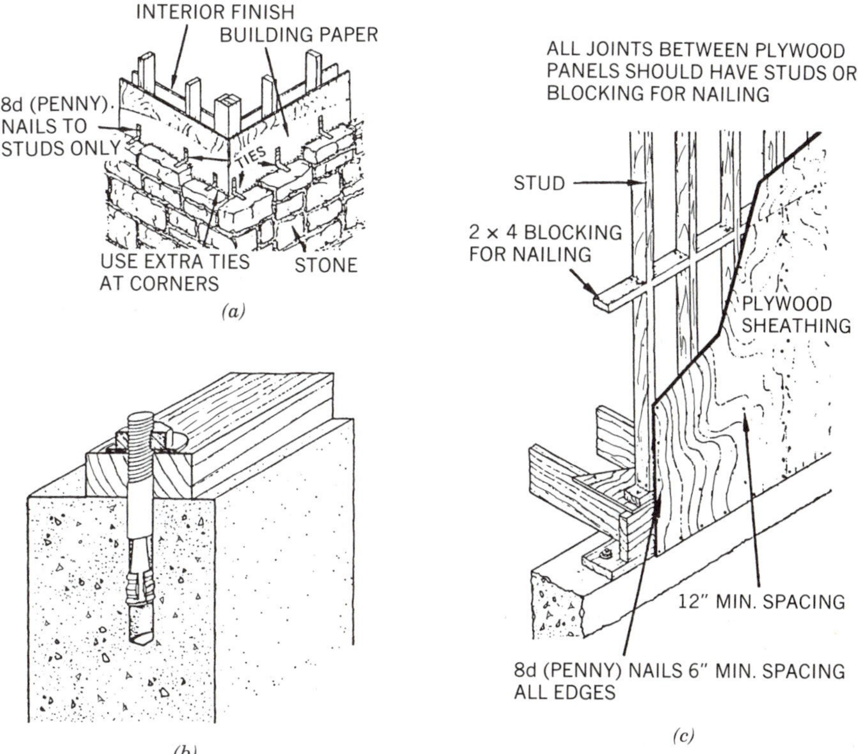

Figure 9.20 Examples of home-construction design features to reduce damage during an earthquake. (*a*) Diagram displays anchoring device for masonry veneer. (*b*) Shows a means of anchoring a wood-frame structure to an existing foundation with expansion bolts. (c) An illustration of plywood bracing to strengthen walls. (From Yanev, 1977)

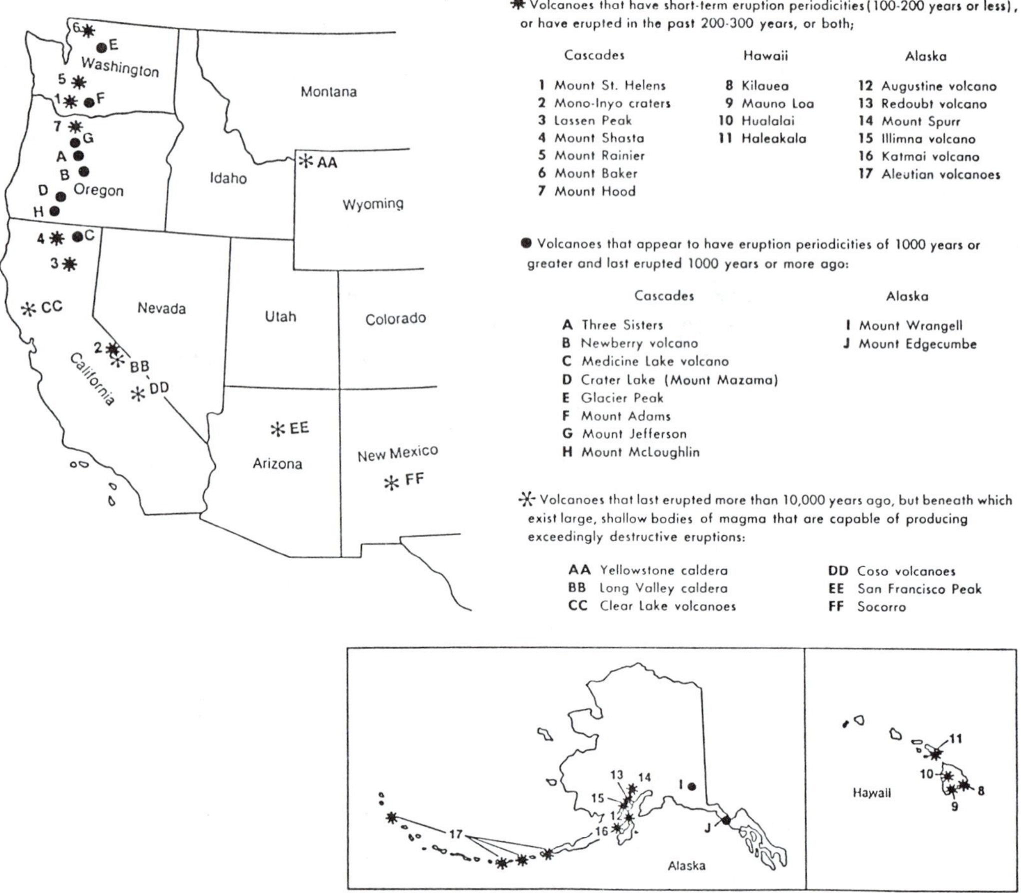

Figure 9.21 Location of potentially hazardous volcanoes in the United States—to be revised based on studies in progress or planned by the U. S. Geological Survey. (From Scott, 1984)

Saint-Pierre, killing nearly thirty thousand inhabitants (Bolt et al., 1975). Despite this awful toll, Indonesia, Japan, and the Philippines account for about 50% of the reported fatalities from historic volcanic eruptions (Scott, 1984). The eruption of Tambora volcano in Indonesia on April 10–11, 1815, is regarded by many as the greatest eruption in historic time. Directly and indirectly through famine, tsunamis, and disease, it resulted in the death of more than ninety thousand persons in the central Indonesian region (Self et al., 1984).

Volcanoes are locations where molten rock, or magma, issues from deep within the earth. This may take the form of an explosive or nonexplosive eruption. The explosiveness of the magma varies with the viscosity and water content. High viscosity produces greater explosiveness. Water turning to steam within magma also increases explosiveness. The explosiveness of the magma and other characteristics of an eruption influence the volcanic processes that occur.

The specific types of volcanic processes that may need to be addressed by an engineering geologist vary with the type of volcanic activity at a location and the distance a site may be from the eruptive center. Table 9.11 contains a detailed description of the three classes of volcanic processes and their effects (Mullineaux, 1981). Baker (1979) and Scott (1984) both provide detailed information on the characteristics of different volcanic processes. This information serves as a guide to the engineering geologist who must address volcanic-process issues for a particular project. The engineering geologist is generally interested in the expected effects of future volcanic eruptions rather than prediction of the eruptions themselves. Prediction requires monitoring of premonitory activity to serve as a basis for reducing loss of life (Crandell and Mullineaux, 1975). In addition to volcanic processes, the engineering geologist must keep in mind that earthquakes, floods, and landslides may be induced (Mullineaux, 1981). These secondary processes may prove equally important to a project.

Schuster (1983) provides instructive examples of the effect of some volcanic processes on engineering works. He notes that the 1980 eruptions of Mount Saint Helens caused over $350 million damage to engineering works in the Pacific Northwest. This damage is attributable to (1) the directed blast, (2) associated debris avalanches, (3) mudflows, and (4) ashfall. The directed blast on May 18, 1980, totally destroyed everything within a radius of 13 km of the Mount Saint Helens crater. This was a consequence of the force of the moving pyroclastic cloud and its associated temperature in excess of 350° +/−50° C. The debris avalanches and mudflows obliterated or buried buildings, roads, bridges, and water-supply systems, and sewage-disposal systems—often at significant distances from the eruptive center of Mount Saint Helens. Table 9.12 shows the amount of highways, roads, and railroads destroyed or damaged. Downstream civil works and operations were threatened or impaired by the debris avalanche and mudflow effects. This included the forming of impoundments in valley bottoms dammed by debris-avalanche deposits and sediment generated by the mudflow. Sedimentation in the Columbia River blocked navigation and required several months of dredging to restore a navigable channel (Schuster, 1983). Ashfall from the May 18, 1980, eruption extended across eastern Washington, northern Idaho, and western Montana (Figure 9.22). It required removal and disposal to restore function to highways, streets, roads, airports, and storm-drain systems. Entry into sanitary-sewage systems resulted in plugging that left them inoperable for various periods. Schuster (1983) notes that the direct cost of damage to the Yakima, Washington, sewage-disposal system is estimated to exceed $1 million. Subsequent eruptions caused significant volumes of ash to fall. Ash removal and disposal was required in these cases, too.

Evaluating Volcanic Processes

The behavior of a volcano is often highly individualistic (Baker, 1979). Recent studies in the Long Valley–Mono Lake area of California illustrate the importance of relating the individualistic behavior of a particular volcanic area to hazard evaluation. Miller et al. (1982) note that concern

Table 9.11 Description of Main Types of Volcanic Hazards and Their Effects

	Lava Flows	Hot Avalanches, Mudflows, and Floods	Volcanic Ash (Tephra) and Gases
Origin and characteristics	Result from nonexplosive eruptions of molten lava. Flows are erupted slowly and move relatively slowly; usually no faster than a person can walk.	Hot avalanches can be caused directly by eruption of fragments of molten or hot solid rock; mudflows and floods commonly result from eruption of hot material onto snow and ice and eruptive displacement of crater lakes. Mudflows also commonly caused by avalanches of unstable rock from volcano. Hot avalanches and mudflows commonly occur suddenly and move rapidly, at tens of miles per hour.	Produced by explosion or high-speed expulsion of vertical to low-angle columns or lateral blasts of fragments and gas into the air, materials can then be carried great distances by wind. Gases alone may issue nonexplosively from vents. Commonly produced suddenly and move away from vents at speeds of tens of miles per hour.
Location	Flows are restricted to areas downslope from vents; most reach distances of less than 6 miles. Distribution is controlled by topography. Flows occur repeatedly at central-vent volcanoes, but successive eruptions may affect different flanks. Elsewhere, flows occur at widely scattered sites, mostly within volcanic "fields."	Distribution nearly completely controlled by topography. Beyond volcano flanks, effects of these events are confined mostly to floors of valleys and basins that head on volcanoes. Large snow-covered volcanoes and those that erupt explosively are principal sources of these hazards.	Distribution controlled by wind directions and speeds, and all areas toward which wind blows from potentially active volcanoes are susceptible. Zones around volcanoes are defined in terms of whether they have been repeatedly and explosively active in the last 10,000 years.
Size of area affected by single event	Most lava flows cover no more than a few square miles. Relatively large and rare flows probably would cover only hundreds of square miles.	Deposits generally cover a few square miles to a few hundreds of square miles. Mudflows and floods may extend downvalley from volcanoes many tens of miles.	An eruption of "very large" volume could affect tens of thousands of square miles, spread over several States. Even an eruption of "moderate" volume could significantly affect thousands of square miles.

Effects	Land and objects in affected areas subject to burial, and generally they cause total destruction of areas they cover. Those that extend into areas of snow, may melt it and cause potentially dangerous and destructive floods and mudflows. May start fires.	Land and objects subject to burning, burial, dislodgement, impact damage, and inundation by water.	Land and objects near an erupting vent subject to blast effects, burial, and infiltration by abrasive rock particles, accompanied by corrosive gases, into structures and equipment. Blanketing and infiltration effects can reach hundreds of miles downwind. Odor, "haze," and acid effects may reach even farther.
Predictability of location of areas endangered by future eruptions	Relatively predictable near large, central-vent volcanoes. Elsewhere, only general locations predictable.	Relatively predictable, because most originate at central-vent volcanoes and are restricted to flanks of volcanoes and valleys leading from them.	Moderately predictable. Voluminous ash originates mostly at central-vent volcanoes; its distribution depends mainly on winds. Can be carried in any direction; probability of dispersal in various directions can be judged from wind records.
Frequency in conterminous United States as a whole	Probably one to several small flows per century that individually cover less than 10 square miles. Flows that cover tens to hundreds of square miles probably occur at an average rate of about once every 1,000 years. (In Hawaii, eruption of many flows per decade would be expected.)	Probably one to several events per century caused directly by eruptions. Probably only about one event per 1,000 years caused directly by eruption at "relatively inactive" volcanoes.	Probably one to a few eruptions of "small" volume every 100 years. Eruption of "large" volume may occur about once every 1,000 to 5,000 years. Eruption of "very large" volume, probably no more than once every 10,000 years.
Degree of risk in affected area	To people, low. To property, high.	Moderate to high for both people and property near erupting volcano. Risk relatively high to people because of possible sudden origin and high speeds. Risk decreases gradually downvalley and more abruptly with increasing height above valley floor.	Moderate risk to both people and property near erupting volcano; decreases gradually downwind to very low.

Source: From Mullineaux, 1981.

Table 9.12 The Lengths of Roads, Highways, and Railroads Destroyed or Damaged by the Eruption of Mount St. Helens, Washington

	Length (km)
Public highways and roads	
State of Washington	48
U.S. Forest Service	38
Cowlitz County	16
Private roads	198
Total highways and roads	300
Private railway	27

Source: From Robert L. Schuster, Engineering Aspects of the 1980 Mount St. Helens Eruptions, *Bulletin of the Association of Engineering Geologists*, Vol. 20, Number 2, May 1983, pp. 125–144. Used by permission.

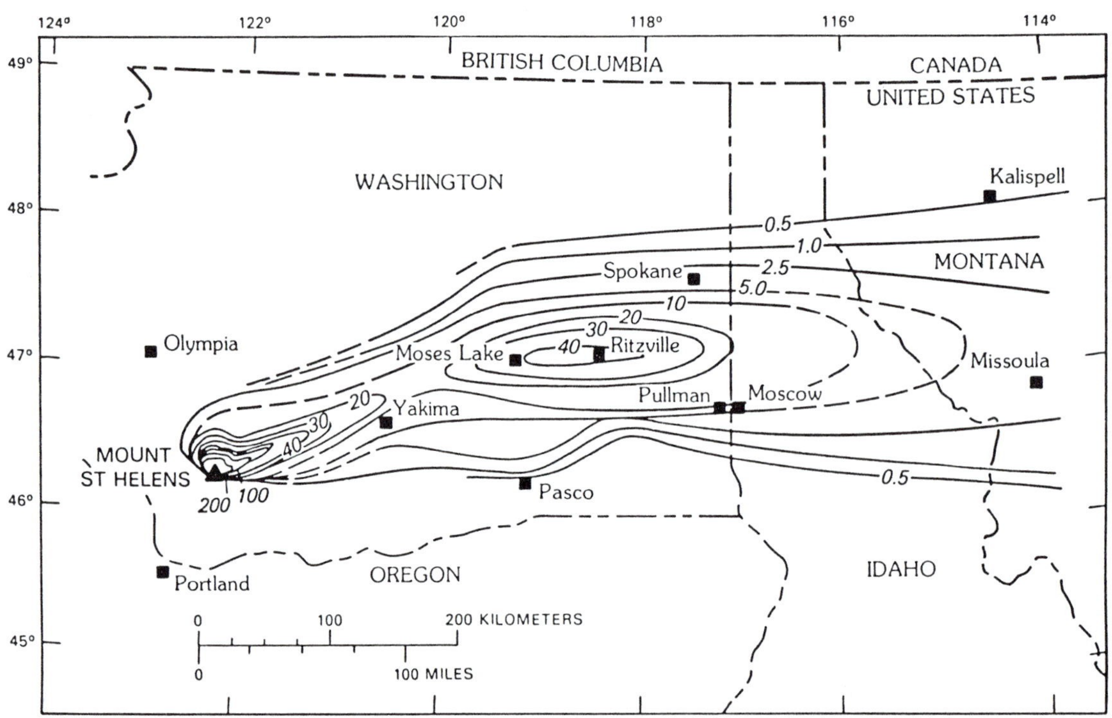

Figure 9.22 An isopach map showing the ashfall across eastern Washington, northern Idaho, and western Montana resulting from the May 18, 1980, eruption of Mount Saint Helens. The lines represent accumulation in mm. (Source: From Robert L. Schuster, Engineering Aspects of the 1980 Mount St. Helens Eruptions, *Bulletin of the Association of Engineering Geologists*, Vol. 20, No. 2, May 1983, pp. 125–144. Used by permission.

about eruptions in the foreseeable future arose in the aftermath of the May 1980 Mammoth Lake earthquake sequence. Concern centered on the Long Valley region, an old caldera. Four Richter magnitude 6.0 earthquakes shook the southern part of the caldera near Mammoth Lakes, California, within a 48-hr period. Subsequently, earthquake swarms were noted repeatedly occurring underneath a site near Casa Diablo Hot Springs (Figure 9.23). The depth of these swarms seemed to have become shallower over time. Increased fumarolic activity was noticed in this same site in 1982. Measurement along U.S. Highway 395 across Long Valley showed uplift in the western part of the caldera. The overall interpretation of this evidence indicated that magma beneath the caldera had moved upward at the time of the 1980 earthquakes, thus causing bulging of the caldera floor and opening fractures at depth in the southern part of the caldera. This allowed a tongue of magma to move toward the surface beneath the Casa Diablo Hot Springs epicentral site.

Miller (1985) published the results of stratigraphic studies of Holocene eruptions in the Inyo volcanic chain that related to the possible volcanic hazard in Long Valley Caldera. The stratigraphic sequence and timing of events occurring on the Inyo volcanic chain 550–650 years ago indicate

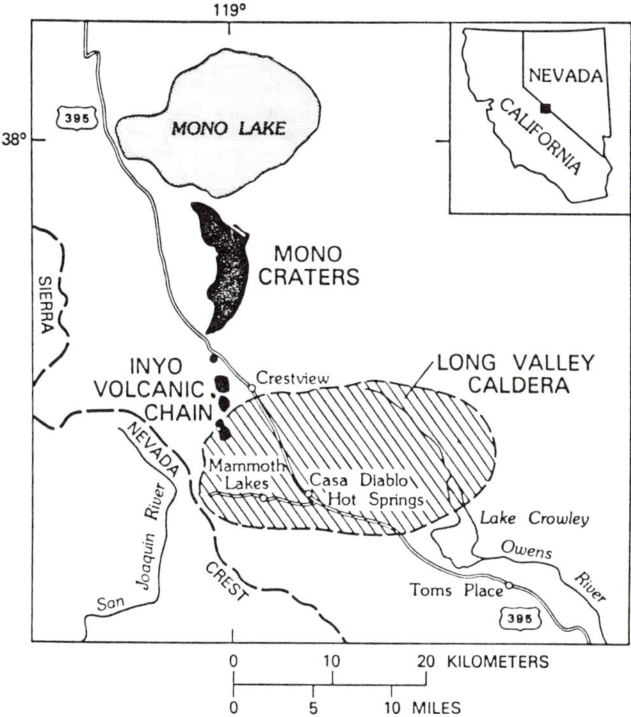

Figure 9.23 Location of Casa Diablo Hot Springs and volcanic features in the Long Valley–Mono Lake area. (Reprinted with permission from Geology, Vol. 13, C. D. Miller, Holocene Eruptions at the Inyo Volcanic Chain, California, 1985. Published by the Geological Society of America.)

that a dikelike conduit produced surface faulting and explosive magmatic and phreatic eruptions at about the same time at many of the vents. It is reasonable to expect that future eruption on the possible dike intrusion near Casa Diablo Hot Springs, should it occur, would likely follow a sequence of events like the 550–650-year-old sequence in the Inyo volcanic chain. These findings also point out another important aspect of volcanic hazard evaluation. Past activity in an area can provide valuable insight about the nature of volcanic hazard in the long-term (see the volcanic-hazard zonation for Long Valley–Mono Lake region, Figure 7.10). However, it rarely provides a basis for evaluating volcanic hazard in the short-term (Baker, 1979). Factors such as the nature of the magma and other physical characteristics will control the specific size and nature of any future eruptions in Long Valley despite the detailed knowledge of the 550–650-year-old events on the Inyo volcanic chain (Miller, 1985).

As with other processes, the engineering geologist must evaluate the degree of hazard to permit assessment of risk from volcanic processes (Fournier d'Albe, 1979). The engineering geologist is usually addressing the long-term hazard, but it is clear that this evaluation must be tied closely to the behavior of the volcano under study. Crandell and Mullineaux (1975) describe the basic technique and rationale that serves as the basis for appraising volcanic processes. They note that this evaluation should identify the types of volcanic processes that may possibly occur, offer some approximation of the likelihood of future events, and delineate the areas likely to be affected by different processes in a future event. Genetic, stratigraphic, chronologic, and cartographic studies are necessary to fully define the hazard from volcanic processes (Crandell and Mullineaux, 1975). Genetic studies involve the determination of the origin of volcanic deposits produced by past eruptions. The similar appearance of some deposits with distinctly different modes of origin and transportation makes this a difficult task. Table 9.13 shows some characteristics that have proven useful in distinguishing the origin and transportation of coarse deposits of volcanic rock. Stratigraphy sorts out the sequence of volcanic processes over time. Crandell and Mullineaux (1975) determined composite sequences or assemblages for various events. This requires a decision on an interval of time appropriate to the history of the volcano under study. Table 9.14 is an example of results from a stratigraphic study. It is important to know not only the relative occurrence of volcanic processes that is gained from stratigraphic study, but also the actual chronology of past events. Tephrachronology, lichenometry, tree-ring dating, and radiocarbon dating are all techniques that may prove useful in this effort to establish the chronology of past volcanic events.

Genetic, stratigraphic, and chronologic studies identify the behavior of a particular volcano. Cartographic studies that show the areal distribution of different products of volcanic processes indicate the areas likely to be impacted in future eruptions. Crandell and Mullineaux (1975) note that too many unknowns exist about future eruptions to assign hazard ratings that are truly specific. However, these areas can reasonably be expected to have some hazard. Figure 9.24 is representative of hazard zonation from this type of evaluation. More detailed versions of this map would provide input to land-use planning.

Table 9.13 Features Useful in Distinguishing Volcanic and Nonvolcanic Deposits as Part of the Genetic Phase of a Volcanic-Hazards Evaluation

	Genesis of deposits			
Feature	Hot Pyroclastic Flow	Lahar	Avalanche	Glacier
Variety of rock types	Monolithologic	Monolithologic or heterolithologic	Typically monolithologic	Typically heterolithologic
Shape of clasts	Typically angular and subangular	Typically angular and subangular	Typically angular	Typically subrounded; may have facets, striations, snubbed ends
Evidence of heat	Generally present	Rarely present	Absent	Absent
Distribution within valley	Generally limited to valley floors or other depressions; may be considerably higher on obstacles in path of flow	Generally limited to valley floors or other depressions; are not appreciably higher on obstacles in path	Extend directly downslope from source; may be many tens of m higher on obstacles in path	Limited only by thickness of glacier
Length	Generally less than 15 km	May be many tens of km	Generally only a few km	Limited only by length of glacier
Topography of deposits	Generally form flat-topped fill in valley	Generally form flat-topped fill in valley; may veneer valley sides up to maximum height reached by lahar; mounds of rock debris may be present	Generally very irregular	Generally irregular
Nature of margins in flat areas	Typically digitate	Typically digitate	May be digitate or smooth	Generally smooth
Textural gradation	Vertical change in clast size not common	Clast size generally decreases upward	No vertical change in clast size	No vertical gradation in clast size

Source: Reprinted with permission from Environmental Geology, Vol. 1, D. R. Crandall and D. R. Mullineaux, Technique and Rationale of Volcanic-Hazards Appraisals in the Cascade Range, Northwestern United States, 1975, Springer-Verlag.

Table 9.14 Generalized Stratigraphy of Mount St. Helens, Washington, Based on Stratigraphic Studies

Tephra Deposit and Age (in years)	SE Side	SW Side	N Side
	Lahars	Pyroclastic flows	Lahars
	Lava flows		Lava flows
W (450)	Tephra	Tephra	Tephra
			Lava flows
		Lahars	Pyroclastic flows, lahars, and tephra
B (1500–2500)	Lahars and tephra	Tephra	
		Lava flow(s)	Lava flow(s)
P (2500–3000)	Pyroclastic flows, lahars, and tephra	Tephra	Pyroclastic flows, lahars, and tephra
Y (3000–4000)	Pyroclastic flows, lahars, and tephra	Tephra	Tephra

Source: Reprinted with permission from Environmental Geology, Vol. 1, D. R. Crandall and D. R. Mullineaux, Technique and Rationale of Volcanic-Hazards Appraisals in the Cascade Range, Northwestern United States, 1975, Springer-Verlag.

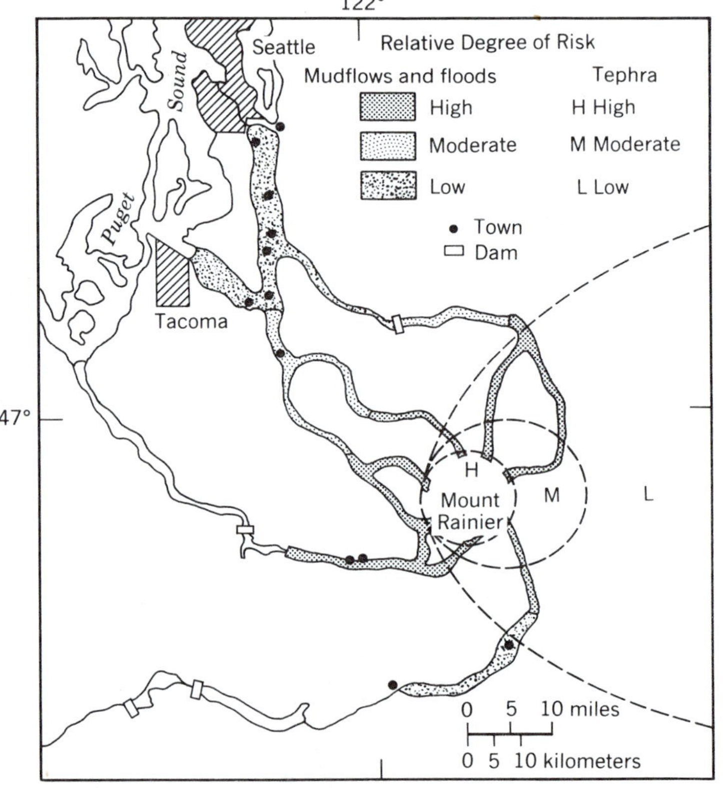

Figure 9.24 Relative degrees of potential volcanic hazard from tephra, mudflows, and floods in the vicinity of Mount Rainier, Washington. (Reprinted with permission from Environmental Geology, Vol. 1, D. R. Crandall and D. R. Mullineaux, Technique and Rationale of Volcanic-Hazards Appraisals in the Cascade Range, Northwestern United States, 1975, Springer-Verlag.)

Mitigating the Effects of Volcanic Processes

Land-use regulation that prevents concentrating residences and critical facilities in high-hazard areas is one of the most effective forms of mitigation. Damage from volcanic bombs and ash accumulation can be limited by building design, too. In some instances, protective measures can reduce potential damage from lahars (hot avalanches and mudflows) and lava flows. However, building design and protective devices are best applied to existing facilities that may be difficult to relocate to safer areas or in areas with a low degree of risk.

Crandell and Mullineaux (1975) note that maps showing relative hazard in the vicinity of a volcano permit land-use planning. Decisions on appropriate uses of land can be made with emphasis on public safety and limiting property losses. As Booth (1979) notes, this is especially important in areas where volcanoes are inactive for extended periods of time such as the Cascades in the northwestern United States and on the Canary Islands. These extended periods allow shifts in land use that may place many people and considerable property at risk unless land-use planning is carefully applied to hazard areas. In addition to aiding land-use planning, these maps are useful in identifying secure evacuation routes and areas.

Suryo and Clarke (1985) provide a detailed example of the application of volcanic-hazard zonation to mitigate anticipated effects. As noted earlier, Indonesia is one of three countries that have historically suffered the greatest loss of life from volcanic eruptions. Two classes of hazard-zonation maps are currently being used. Detailed study and analysis is the basis for zonation around the six most active and potentially destructive volcanoes. Three zones are defined (Figure 9.25). The forbidden zone is the area likely to be affected by nuées ardentes and is recommended for permanent abandonment. Areas around the summit that are unlikely to be affected by nuées ardentes but which may be affected by bombs or blocks are called the first danger zone. Second danger zones are valleys or areas near the summit where inundation by lahars is a possibility. This zone is subdivided to show where escape is possible (alert zone) and where it is not (abandoned zone). Active volcanoes defined as class A with less available data have a preliminary hazard zonation consisting of two zones. The danger zone is the area to be abandoned immediately should increased activity occur. People are warned and evacuated as needed in the alert zone.

After 64 years of inactivity, Galunggung volcano erupted in 1982–83, providing some indication of the value of preliminary hazard zonation based on guidelines established by the Volcanological Survey of Indonesia. Figure 9.26 shows that the danger from nuées ardentes and lahars was fairly accurately represented for this recent event. Mitigating the effects of lahars are further addressed by engineering structures. Check and consolidation dams, dikes, and pockets are employed to trap and store lahar material, limit their spread downslope, and reduce their destructive force. Sabo dams (Ikeya, 1976) designed to retain sand and gravel rather than water are a main feature of these lahar-control efforts. Figure 9.27 shows the distribution of planned or completed control structures on Mount Merapi. The engineering control is incorporated with efforts to improve the lahar-warning network. Figure 9.28 shows the mitigation measures at

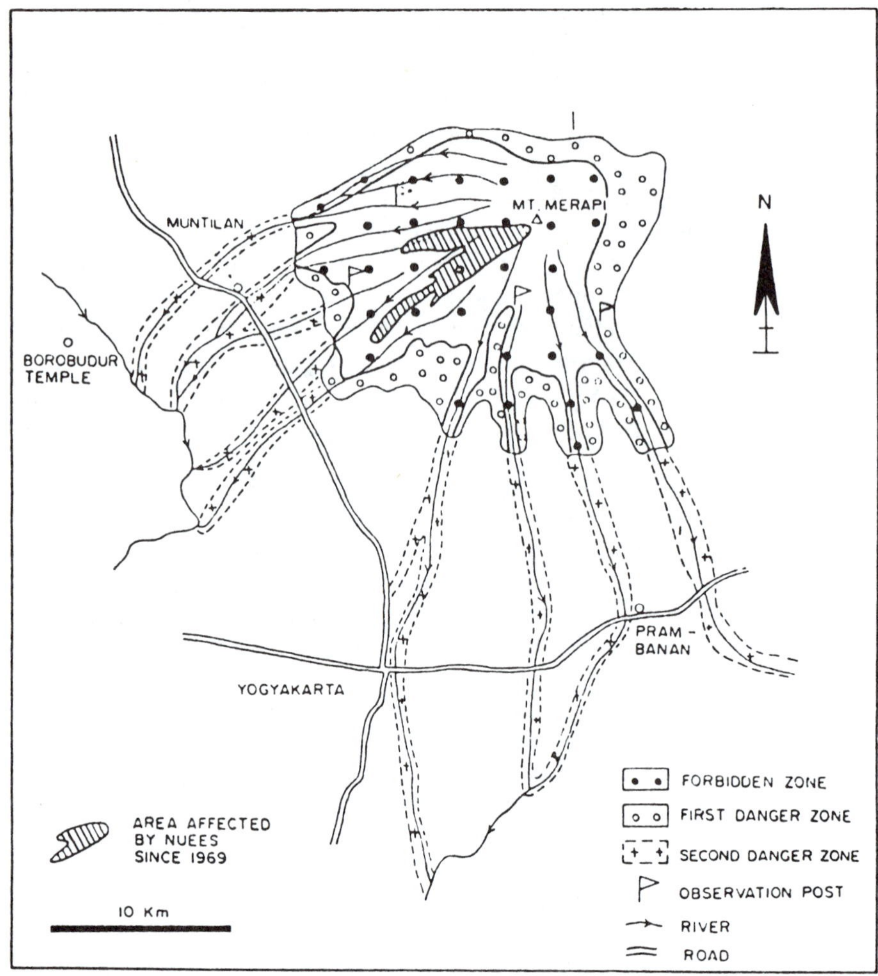

Figure 9.25 Location of the three hazard zones for the Mount Marapi volcano, Indonesia. (Reproduced by permission of the Geological Society from T. Suryo and M.C.G. Clarke, Quarterly Journal of Engineering Geology, Vol. 18, 1985.)

Galunggung, including the location of monitoring stations intended to improve the warning system. Although the lahar pocket on the upper Ciloseh drainage was unable to fully prevent lahar overflow into parts of the city of Tasikmalaya, it did appear to reduce expected damage. The overall effectiveness of these engineering structures for lahar mitigation will be clearer with additional performance data in the future.

Blong (1981) surveyed literature to determine the effect of volcanic bombs and ash accumulation on buildings. This effort provides a basis for aiding the design of buildings to limit damage from this type of volcanic process. Most bombs are capable of perforating or damaging plywood, asphalt, and cedar shingles as well as roofing tiles. Steel sheeting provides a greater degree of protection from all but the larger bombs. Increasing their thickness only marginally improves their performance. Information on the effects of snow loads on roofs was extrapolated to indicate the effect of ash accumulation. The greater structural strength of pitched roofs makes them less likely to collapse under the weight of accumulated ash. Gables, chimneys, or other projections cause drifting of ash that increases accumulation and a greater chance of collapse. Blong (1981) notes that these conclusions represent only an initial analysis. Specific data from volcanic events and testing is needed to develop more precise guidelines for building design.

Protective measures can be applied to lava flows as well as lahars. The objective is to direct lava away from critical structures or valuable property. One protective measure is to apply water to cool the advancing lava flow. In 1973, 47 barge-mounted pumps conveyed seawater at a rate 1 m^3/sec onto a flow encroaching on the harbor of Vestmannaeyjar, the main town on the Icelandic island of Heimaey (Lambert, 1974). This effort not only prevented the harbor from becoming completely blocked, it also improved the sheltering of the entrance. Bolt et al. (1975) describe another measure, the early efforts to use explosives, specifically bombs, to divert lava flows in Hawaii in 1935 and 1942. Diversion of lava flows from Kilauea, Hawaii, by use of earthen barriers was attempted in 1955 and

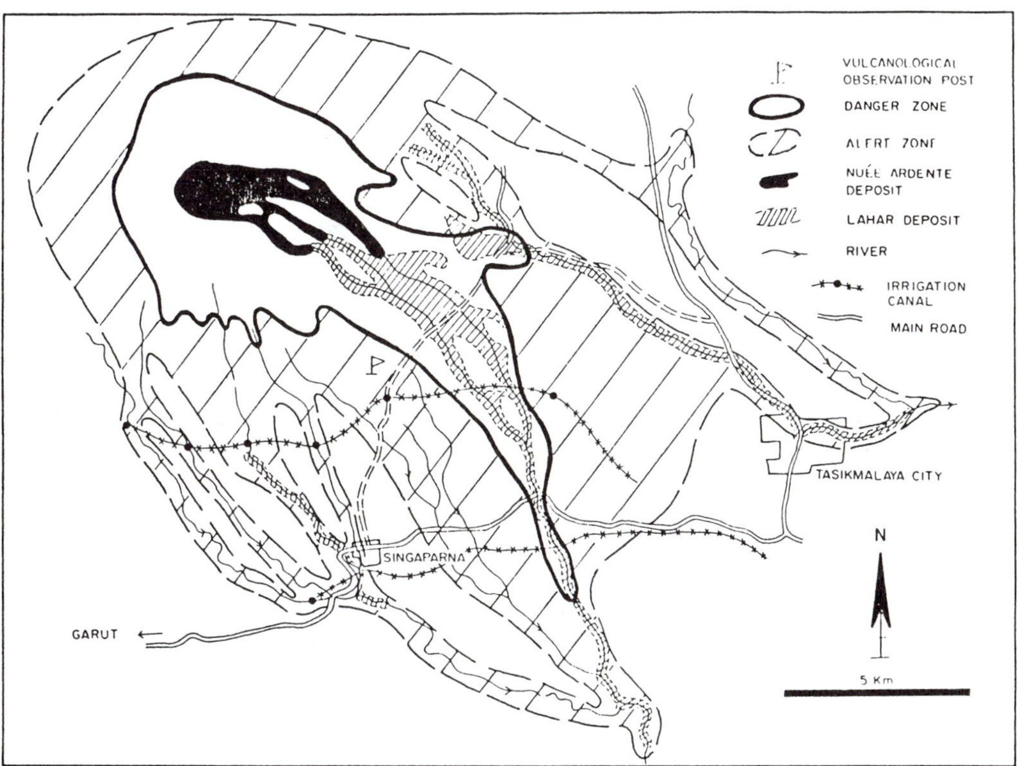

Figure 9.26 Two-zone preliminary volcanic-hazard map for Galunggung volcano, Indonesia. Note that most of the nuée ardente and lahar activity in the 1982–83 eruption falls within the danger zone. (Reproduced by permission of the Geological Society from T. Suryo and M.C.G. Clarke, Quarterly Journal of Engineering Geology, Vol. 18, 1985.)

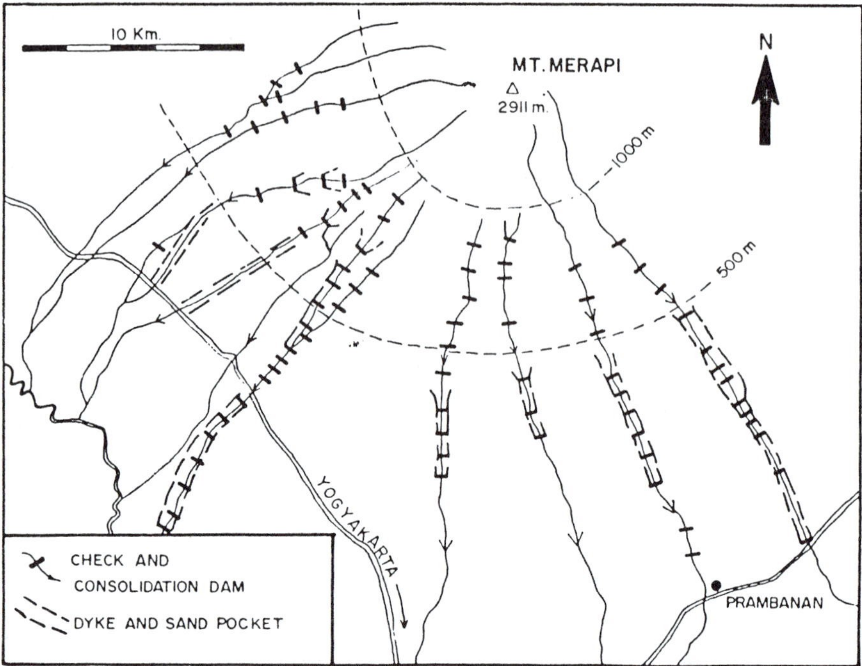

Figure 9.27 Location of structures planned for control of lahars and sediment at Mount Marapi, Indonesia. (Reproduced by permission of the Geological Society from T. Suryo and M.C.G. Clarke, Quarterly Journal of Engineering Geology, Vol. 18, 1975.)

1960. Little success was achieved by aerial bombing or barriers (Lockwood and Romano, 1985).

Explosives and barriers proved more successful in diverting lava at Mount Etna, Sicily, during the 1983 eruption (Lockwood, and Romano, 1985). Beginning on March 28, 1983, Etna yielded flows responsible for destroying tourist facilities, summer homes, and threatening property in Rifugio Sapienza and Monte Vetore. Figure 9.29 shows the location of features and where protective measures were applied. An experimental use of explosives attempted to divert lava into an artificial channel. This was intended to reduce flow toward Sapienza to enable construction of a barrier there. Completion of 60 drillholes for explosives was rendered impossible owing to unexpected problems. Cooling of the holes to permit safe placement of the explosives also cooled the lava-channel wall. The resulting constriction caused lava to overflow the channel and bury some of the drillholes. By using an air-gun device, 400 kg of explosives were injected and detonated in the remaining holes. Although the flow diverted into the artificial channel ceased within two days, it did reduce flow toward Sapienza by two thirds, but for unexpected reasons. The explosives blew cooled rock into the active channel, thus causing multiple blockages. This resulted in the channel becoming roofed over as a lava tube, which luckily altered the direction of the lava flow into the nearby artificial channel.

Earthen barriers constructed at Monte Vetore and Sapienza required a major effort by men and machines in a nearly round-the-clock effort. The first barrier at Monte Vetore exceeded 150,000 m^3 and was over 500-m long. After being overtopped on May 27, 1983, a second barrier was started 100 m farther west. Diversion by the second barrier was aided by continued, but impaired, functioning of the first barrier. The barrier at Sapienza was continuously raised as the lava flowed directly against it. It reached a volume of 120,000 m^3 and a length of over 300 m. It was partially overtopped (Figure 9.30) on May 29, 1983. Fortunately, the main part of the flow remained diverted and a 2-m increase in the barrier height from cooled lava atop the barrier resulted. The work at Mt. Etna demonstrated that lava flows can be diverted by earthen barriers, and it provided an example for comparing the costs of such an effort to the value of property saved.

LANDSLIDE PROCESSES

A landslide involves vertical and horizontal movement of soil, rock, or some combination of the two under the influence of gravity (Schuster, 1978). These slope movements are usually categorized as being either: (1)

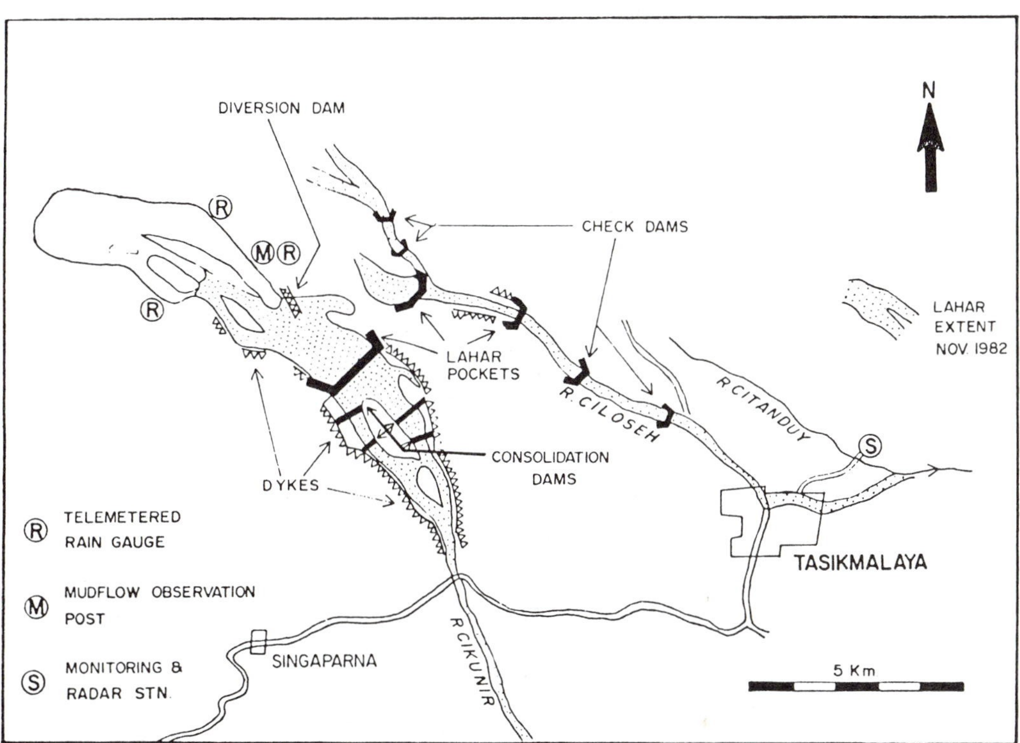

Figure 9.28 Lahar-mitigation measures at Galunggung volcano, Indonesia. Check dams, lahar pockets, and dykes are used to trap and divert debris. Note that observation stations for the early warning system are part of the measures taken. (Reproduced by permission of the Geological Society from T. Suryo and M.C.G. Clarke, Quarterly Journal of Engineering Geology, Vol. 18, 1985.)

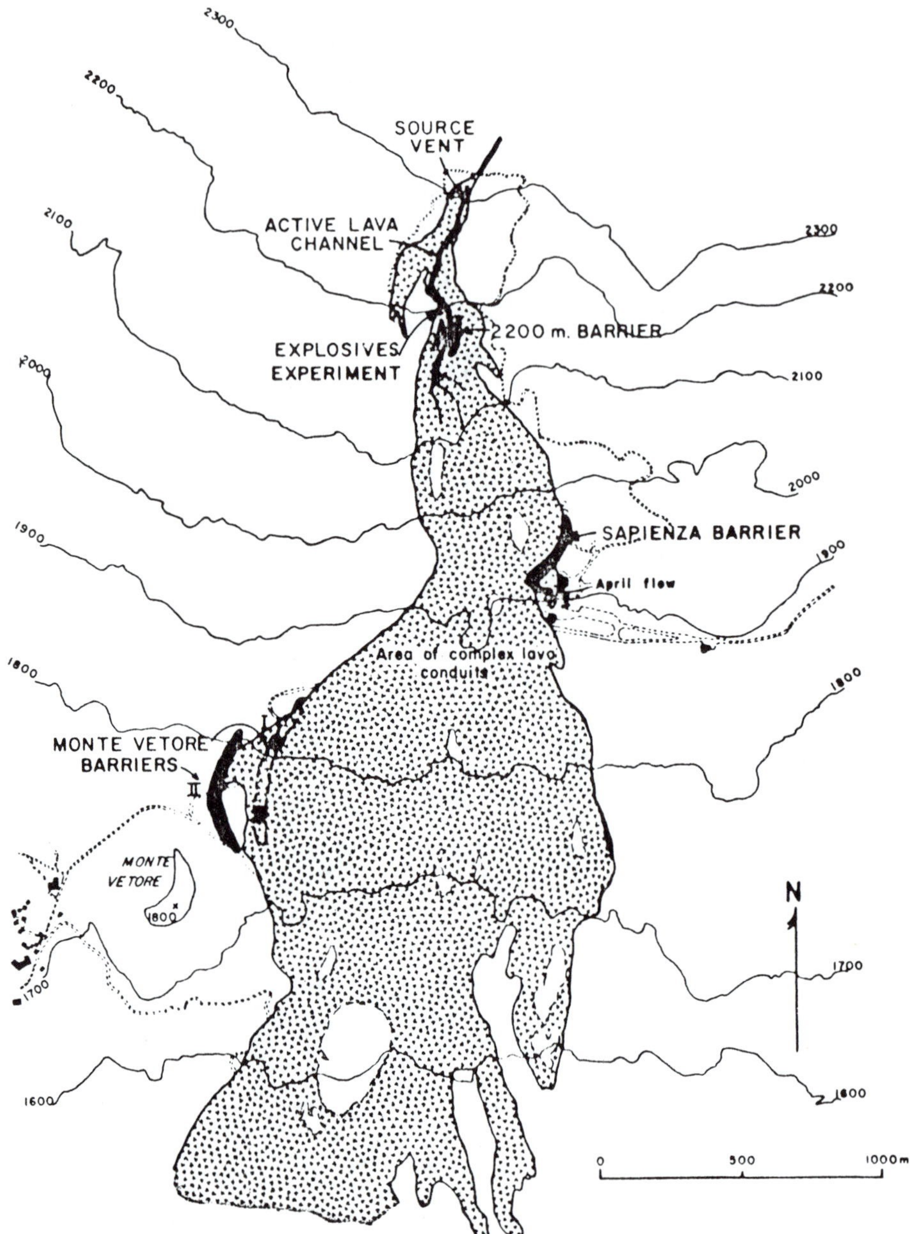

Figure 9.29 Location of lava-flow-control efforts and protected facilities during the 1983 eruption of Mount Etna, Sicily. (From Lockwood and Romano, 1985)

falls, (2) topples, (3) slides, (4) spreads, or (5) flows. Figure 9.31 illustrates falls, slides, and flows, the three most common types of movement. The terms *slope movement* and *slope failure* as well as the term *landslide* when used in a generic sense are generally equivalent in meaning. One of the most widely used classifications of slope movements is that of Varnes (1978). Another accepted classification of landslides is that of Coates

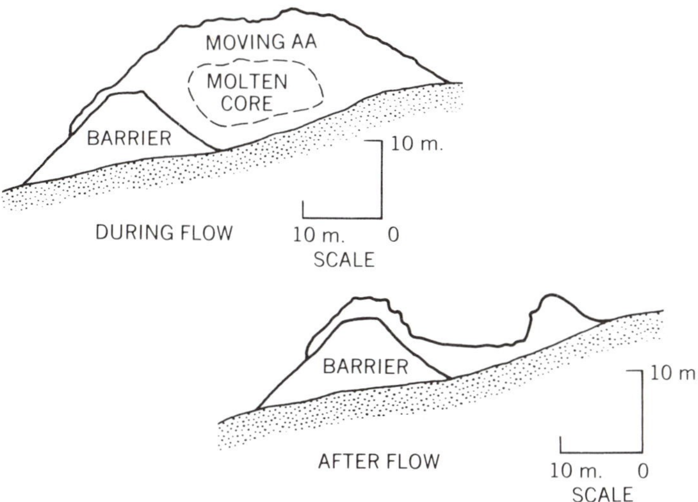

Figure 9.30 Cross section of the earthen barrier constructed to protect La Sapienza on Mount Etna, Sicily. During the flow the barrier was partly overtopped. The cooled lava added 3–4 m to the height of the barrier. (From Lockwood and Romano, 1985)

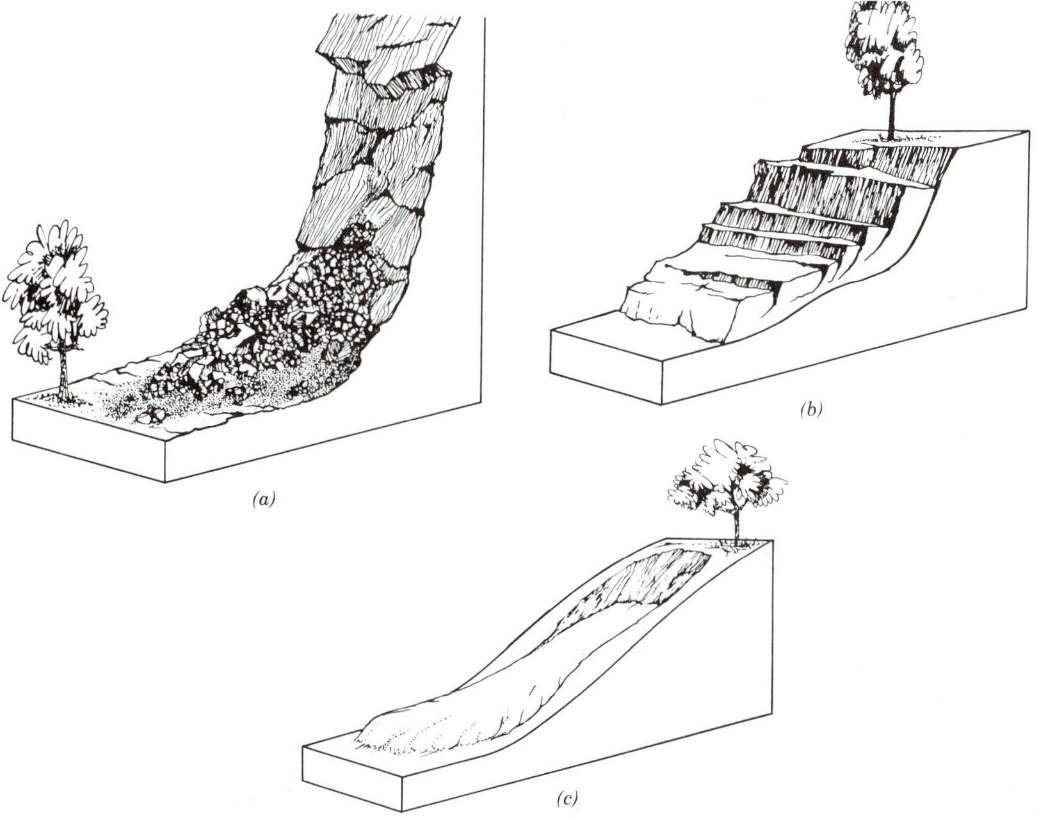

Figure 9.31 Block diagrams showing three of the more-common types of slope movement: (*a*) A fall. (*b*) A slide. (*c*) A flow. (From Nilsen and Wright, 1979)

(1977). Both use the type of movement and nature of the materials being displaced as the basis for their respective classifications (Figure 9.32). Classification provides a means for comparing landslide phenomena to determine underlying causes and to identify common characteristics. For these reasons, engineering geologists should attempt to adhere to a recognized classification system. As our understanding of different types of landslides increases, the ability to reduce their hazard becomes greater. A

TYPE OF MOVEMENT			TYPE OF MATERIAL		
			BEDROCK	ENGINEERING SOILS	
				Predominantly coarse	Predominantly fine
FALLS			Rock fall	Debris fall	Earth fall
TOPPLES			Rock topple	Debris topple	Earth topple
SLIDES	ROTATIONAL	FEW UNITS	Rock slump	Debris slump	Earth slump
	TRANSLATIONAL	FEW UNITS	Rock block slide	Debris block slide	Earth block slide
		MANY UNITS	Rock slide	Debris slide	Earth slide
LATERAL SPREADS			Rock spread	Debris spread	Earth spread
FLOWS			Rock flow (deep creep)	Debris flow	Earth flow
				(soil creep)	
COMPLEX			Combination of two or more principal types of movement		

(a)

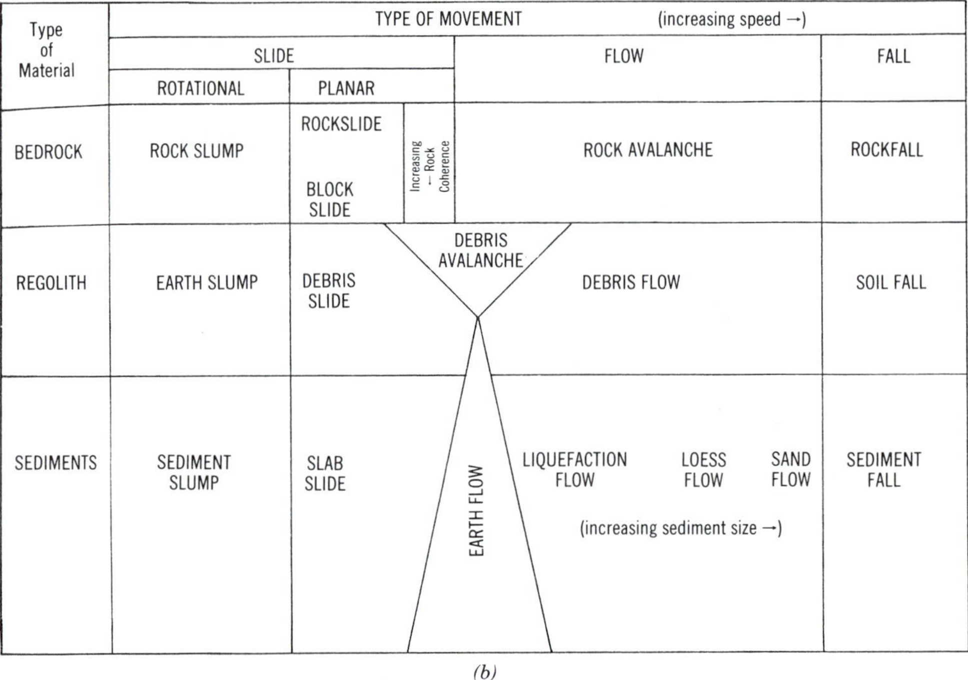

(b)

Figure 9.32 Two of the commonly used classifications of slope movements. (*a*) Abbreviated version of Varnes's classification. (*b*) Coates's classification. ((*a*) From Varnes, 1978. (*b*) Reprinted with permission from Reviews in Engineering Geology, Vol. 3, D. R. Coates, Landslide Perspectives, 1977. Published by the Geological Society of America.)

number of studies in recent time have advanced this understanding. For example, Keefer and Johnson (1983) conducted studies leading to insights on the morphology, mobilization, and movement of earthflows. Similarly, Costa (1984) provides a comprehensive examination of all aspects of debris flows.

Landslides are caused by either the decreased ability of a slope to resist gravitational influence, an increased effectiveness of gravity acting on the slope, or a combination of these two elements. Some causes are inherent or basic conditions (Varnes, 1984): the nature of the bedrock or soil present, topography, vegetation, and climate. For example, changing the type of vegetation covering a slope can increase soil moisture. At Sheep Creek watershed in central Utah, changing cover from aspen to grass increased soil moisture, as demonstrated by higher baseflows to the stream. Within a year, this greater moisture content produced a 300% increase in visible landslide activity (DeGraff, 1979). Other causes are due to more transitory conditions that produce changes, either increasing stress or decreasing material strength. Among these conditions are vibration from earthquake motion, fluctuating groundwater level, and the reshaping of slopes by man. DeGraff and Cunningham (1982) documented conditions along a 5.1-km section of road subject to frequent landsliding in west-central Utah. Relocating the road on the canyon slope had resulted in excavation that coincided with an inherently weak zone in the stratigraphic sequence. Changing the slope at this susceptible location initiated the current landslide problems. Varnes (1984) and Sidle et al. (1985) offer in-depth explanations of these causes, with supporting studies documenting their influence on slope movement.

Landslides cause damage in a number of ways. One of the most obvious is by tilting, shearing, and pulling apart of rock and soil. This results in damage to structures constructed of rock or soil and to those made of nonearth materials placed on a foundation of soil or rock. Road pavement warped and cracked by landslide movement is no longer suitable for use (Figure 9.33). Likewise, few buildings remain fully functional and safe after their foundation is sheared by landslide movement. Impact forces, ranging from individual boulders to a debris avalanche mass, shatter or displace structures and structural elements. Debris from landslides can bury or partially bury facilities and impair their use. The blocking of roads by landslide debris is an obvious example (Figure 9.34). Damming of streams by landslides can result in later catastrophic flooding, another form of damage arising from slope movement (Schuster and Costa, 1986).

Landslides inflict economic losses and casualties in many areas and on a frequent basis. Too often, their coincidence with other natural disasters such as earthquakes, floods, or volcanic eruptions tends to mask their true impact. Annual economic losses from landslides are estimated at $1 billion or more in the United States, Japan, Italy, and India (Schuster and Fleming, 1986). Despite incomplete information, it is known that significant landslide losses are sustained in China, Nepal, Indonesia, the Soviet Union, and South America. Annual loss of life from landslides in the United States is estimated to be 25 – 50 people (National Research Council, 1985b). These losses are capable of being reduced. Fleming and Taylor

Figure 9.33 Cracked and warped pavement on Stump Springs Road, a two-lane 14-ft wide road used for timber harvest and recreational vehicle traffic in the Sierra Nevada, California. Damage resulted from movement on a large rockslide above the road in 1983.

(1980) note that reducing landslide damage requires a better grasp of actual losses by government officials and the public. Incomplete or unavailable records coupled with the absence of centralized collection of damage information hamper efforts to accurately represent these losses.

Evaluating Landslide Processes

Assessing hazard from landslides usually takes the form of either a regional appraisal or a detailed site study. Regional appraisals typically range from mapping the landslide features of an area to producing a zonation of

Figure 9.34 Structural damage and partial burial of a building resulting from a 1986 debris slide. This represents loss of the only school building in a small village on Dominica, West Indies.

hazard based on causative factors. Detailed site studies usually attempt to quantify the forces responsible for the landslides or to ascertain the existing forces acting on an unfailed slope or soil mass.

Varnes (1985) and Coates (1977) include a number of good examples of regional appraisals. Figure 9.35 is part of map showing the landslide features around Montespertoli near Florence, Italy. Inventories of existing landslides and their features commonly form the basis for hazard zonation. These inventories serve as a basis for identifying inherent or basic conditions favoring landslide occurrence. DeGraff (1978) used a landslide inventory to examine how different bedrock, slope inclinations, and slope orientations influenced landslide occurrence in two parts of Utah. This analysis provided a general statement on the conditions associated with the highest degree of hazard. In contrast, an inventory of landslides in West Virginia used an analysis of factors influencing past landslides to distinguish slide-prone areas from relatively stable ground (Lessing et al., 1976) (Figure 9.36). A similar inventory effort by Brabb et al. (1972) resulted in defining seven categories of relative landslide susceptibility (Figure 9.37). These are only a few examples of the many innovative mapping approaches available for use in regional appraisals (Brabb, 1984). Regional appraisals do not always involve areal zonation of landslide hazard potential. Other regional appraisals deal with landslide potential for specific types of projects. Determining the potential for inducing landslides at a reservoir is a good example. Schuster (1979) describes a number of examples in which fluctuating reservoir levels have initiated failure in the reservoir rim.

Evaluating landslide hazard for sites requires a close coordination between geology and engineering (Leighton, 1976). The geologist is concerned with determining: Where is the landslide problem? Why is the landslide here? What are the geologic conditions the engineer needs to know for design? In some instances, the geologic conditions may suggest that the landslide hazard is less severe than initially thought. Leighton (1976) illustrates this point with the case history of the San Juan Creek Landslide in Orange County, California. Property there appeared unsuitable for extensive development because of a recognized old landslide. Subsequent geologic investigation revealed that the geomorphic evolution of the slide area had effectively buttressed the old landslide. Consequently, the degree of landslide hazard was less than initially assumed.

Often, evaluation of landslide hazard at sites relies on some form of slope-stability analysis. Figure 9.38 shows a stability analysis applied to designing stabilization of a landslide. Most methods of slope-stability analysis are based on theoretical models developed to ensure the stability of engineered embankments where the moisture content, material characteristics, and slopes vary within narrow, defined limits. For this reason, the engineering geologist must realize that the inherent heterogeneity of natural slopes presents a problem in using these methods (Sidle et al., 1985). Morgenstern and Sangrey (1978) and NAVFAC (1982) provide detailed description of the principal slope-stability methods.

A practical example of slope stability analysis is provided by Hamel and Spencer (1984). They note that active landslides were initiated during construction of the reservoir outlet on the Fort Peck Dam in Montana.

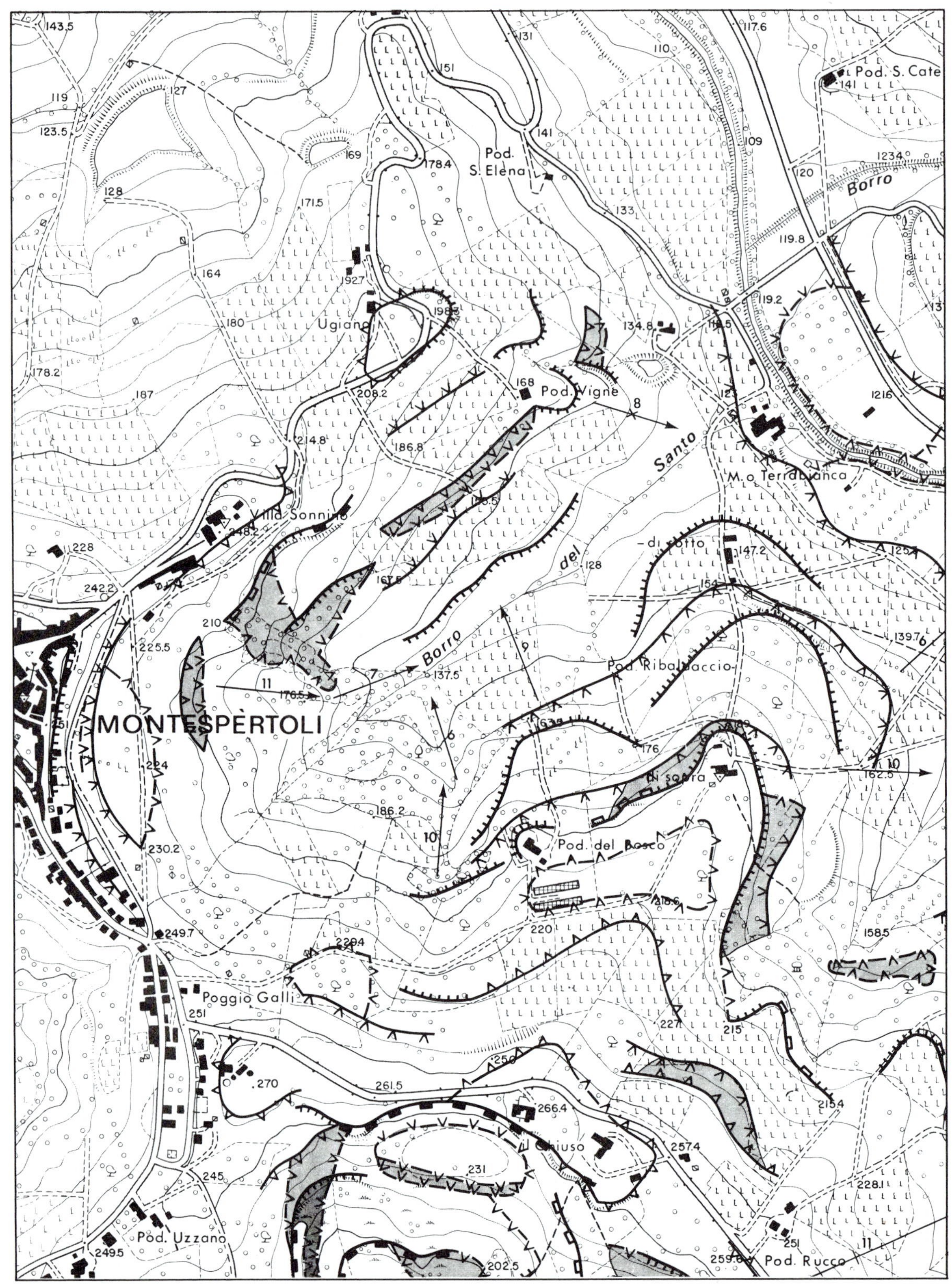

Figure 9.35 Part of a detailed landslide inventory map for the area around Montespertoli near Florence, Italy. Symbols show the location of scarps of various heights, depressions, and other landslide-related morphology. (From Canuti et al., 1982)

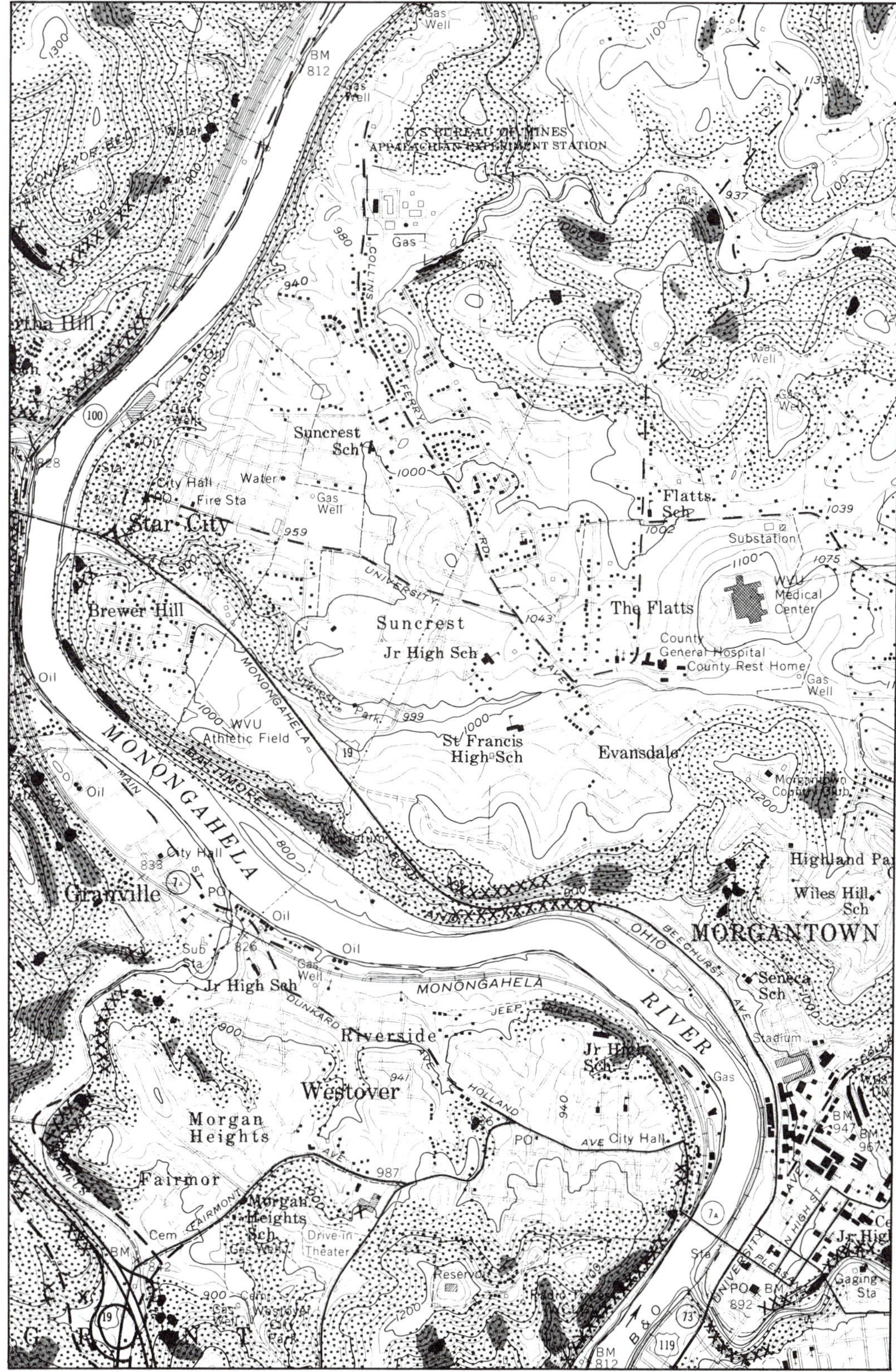

Figure 9.36 Part of a landslide-susceptibility map showing the area around Morgantown, West Virginia. Zones with differing degrees of susceptibility are based on mapping of existing landslides and factors contributing to slope instability. (From Lessing et al., 1976)

Areas least susceptible to landsliding. Very few small landslides have formed in these areas. Formation of large landslides is possible but unlikely, except during earthquakes. Slopes generally less than 15%, but may include small areas of steep slopes that could have higher susceptibility. Includes some areas with 30% to more than 70% slopes that seem to be underlain by stable rock units. Additional slope stability problems; some of the areas may be more susceptible to landsliding if they are overlain by thick deposits of soil, slopewash, or ravine fill. Rockfalls may also occur on steep slopes. Also includes areas along creeks, rivers, sloughs, and lakes that may fail by landsliding during earthquakes. If area is adjacent to area with higher susceptibility, a landslide may encroach into the area, or the area may fail if a landslide undercuts it, such as the flat area adjacent to sea cliffs.

Low susceptibility to landsliding. Several small landslides have formed in these areas and several of these have caused extensive damage to homes and roads. A few large landslides may occur. Slopes vary from 5–15% for unstable rock units to more than 70% for rock units that seem to be stable. The statements about additional slope-stability problems mentioned in I above also apply in this category.

Moderate susceptibility to landsliding. Many small landslides have formed in these areas and some of these have caused extensive damage to homes and roads. Some large landslides likely. Slopes generally greater than 30% but includes some slopes 15–30% in areas underlain by unstable rock units. See I for additional slope stability problems.

Moderately high susceptibility to landsliding. Slopes all greater than 30%. These areas are mostly in undeveloped parts of the county. Several large landslides likely. See I for additional slope-stability problems.

High susceptibility to landsliding. Slopes all greater than 30%. Many large and small landslides may form. These areas are mostly in undeveloped parts of the county. See I for additional slope-stability problems.

Very high susceptibility to landsliding. Slopes all greater than 30%. Development of many large and small landslides is likely. Slopes all greater than 30%. The areas are mainly in undeveloped parts of the County. See I for additional slope stability problems.

Highest susceptibility to landsliding. Consists of landslide and possible landslide deposits. No small landslide deposits are shown. Some of these areas may be relatively stable and suitable for development, whereas others are active and causing damage to roads, houses, and other cultural features.

Definitions: Large landslide—more than 500 ft in maximum dimension
Small landslide—50 to 500 ft in maximum dimension

Figure 9.37 A map explanation showing how factors contributing to landslide movement are used in defining landslide-susceptibility zonation. (From Brabb et al., 1972)

This movement continued from 1934 to 1974. At that time, 1.6×10^6 yd^3 of material were removed to stabilize the slope. By applying the Morgenstern–Price (1965) method of stability analysis to past landslides, Hamel and Spencer were able to calculate shear-strength parameters for the colluvium. It is interesting to note that field residual strength was

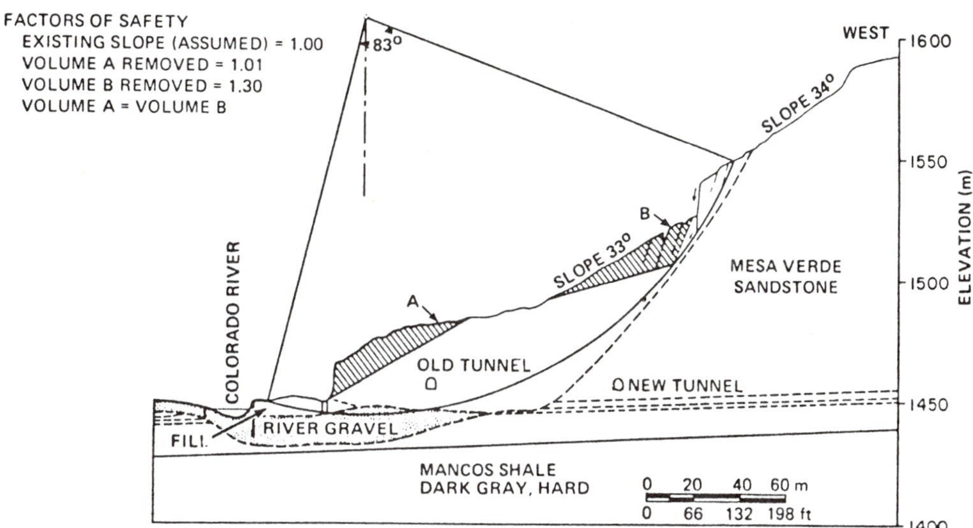

Figure 9.38 An example of stability analysis applied to landslide evaluation is provided by the Cameo slide in the Colorado River valley. Analysis showed that removing volume B would reduce forces driving the landslide more than removing volume A would. (From Gedney and Weber, 1978)

computed to be significantly greater than values obtained from laboratory testing of soil samples. Whether differences in the values used in stability-analysis methods arise from laboratory versus field testing or natural variation in the material, these differences can cause the computed stability factor to be under- or overestimated.

Mitigating the Effects of Landslide Processes

Over large areas, mitigation usually takes the form of land-use planning and grading ordinances. It is possible to use zonation of relative landslide hazard to avoid placing residences, commercial buildings, or critical facilities in higher hazard areas. Figure 9.39 is a part of the landslide hazard map for the Caribbean nation of Saint Lucia. This map was compiled using the regional appraisal approach of DeGraff and Romesburg (1980). Planning for economic development employs this hazard map to avoid higher-hazard areas in siting facilities. Grading ordinances are an effort to recognize the degree of hazard present and provide ways to ensure its reduction.

Fleming et al. (1979) document the reduction in landslide hazard resulting from efforts in the Los Angeles, California, region. Originally, hillside development was not required to attempt mitigation of landslide hazard. Then, in 1952 a grading ordinance that required soils engineering was established. Dissatisfaction with the results led to a more-restrictive grading code in 1962, one that required greater geologic input to design and construction. Table 9.15 compares damage sustained in the storms of 1969. It is clear from the differences among damage figures that regulation ensuring engineering-geologic input succeeded in reducing landslide hazard.

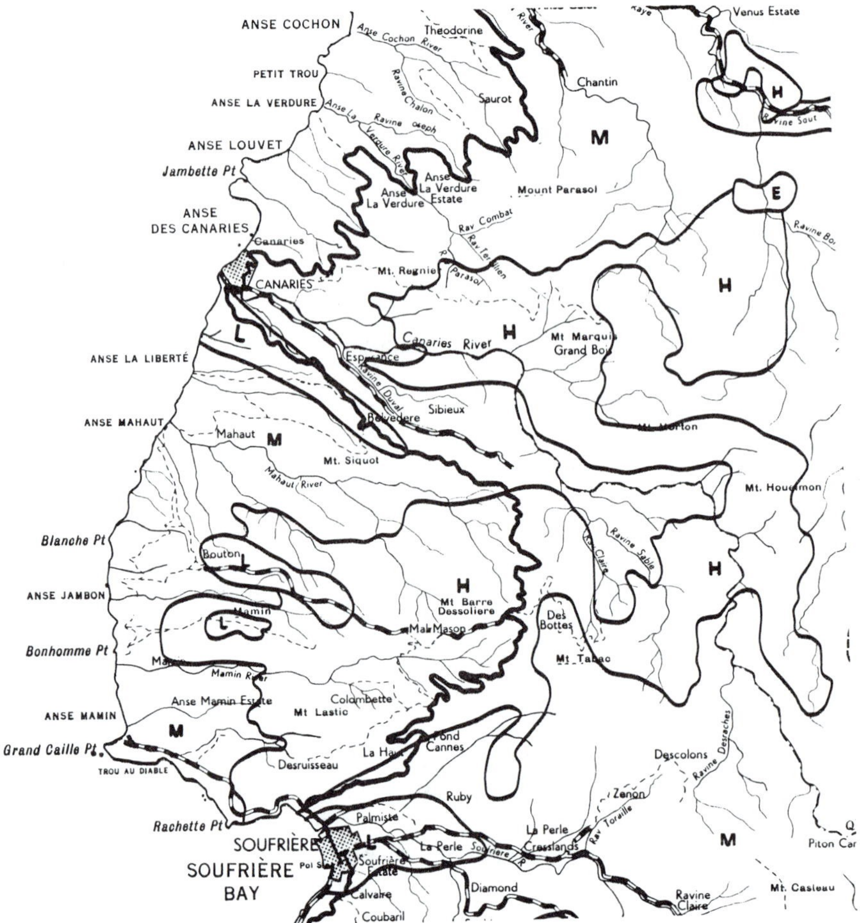

Figure 9.39 Part of a regional appraisal of landslide hazards for Saint Lucia, West Indies. The hazard zones are based on analysis of existing landslides and factors found to influence their occurrence. (From Organization of American States, 1986)

Table 9.15 Comparison of Damage Sustained in Hillside Areas of Los Angeles, California, During the Storms of 1969

	Sites Developed Prior to 1952	Sites Developed 1952–62	Sites Developed 1963–69
Number of sites constructed	10,000	27,000	11,000
Total damage	$3,300,000	$2,767,000	$182,400
Average damage per site	$300	$100	$7
Percentage of sites damaged	10.4	1.3	0.15

Source: Adapted from Slosson, 1969. From R. W. Fleming, D. J. Varnes, and R. L. Schuster, Landslide Hazards and Their Reduction. Reprinted by permission of the *Journal of the American Planning Association*, Vol. 45, 1979.

Table 9.16 Brief Summary of Different Slope-Design Procedures for Avoiding Slope Movement in Construction of Soil Slopes

Category	Procedure	Best Application	Limitation	Remarks
Avoid problem	Relocate highway	As an alternative anywhere	Has none if studied during planning phase; has large cost if location is selected and design is complete; also has large cost if reconstruction is required	Detailed studies of proposed relocation should ensure improved conditions
	Completely or partially remove unstable materials	Where small volumes of excavation are involved and where poor soils are encountered at shallow depths	May be costly to control excavation; may not be best alternative for large slides; may not be feasible because of right-of-way requirements	Analytical studies must be performed; depth of excavation must be sufficient to ensure firm support
	Bridge	At sidehill locations with shallow-depth soil movements	May be costly and not provide adequate support capacity for lateral thrust	Analysis must be performed for anticipated loadings as well as structural capability to restrain landslide mass
Reduce driving forces	Change line or grade	During preliminary design phase of project	Will affect sections of roadway adjacent to slide area	
	Drain surface	In any design scheme; must also be part of any remedial design	Will only correct surface infiltration or seepage due to surface infiltration	Slope vegetation should be considered in all cases
	Drain subsurface	On any slope where lowering of water table will effect or aid slope stability	Cannot be used effectively when sliding mass is impervious	Stability analysis should include consideration of seepage forces
	Reduce weight	At any existing or potential slide	Requires lightweight materials that are costly and may be unavailable; may have excavation waste that creates problems; requires consideration of availability of right-of-way	Stability analysis must be performed to ensure proper use and placement area of lightweight materials

(*continued*)

Table 9.16 *(continued)*

Category	Procedure	Best Application	Limitation	Remarks
Increase resisting forces	Drain subsurface	At any slide where water table is above shear plane	Requires experienced personnel to install and ensure effective operation	
	Use buttress and counterweight fills	At an existing slide, in combination with other methods	May not be effective on deep-seated slides; must be founded on a firm base	
	Install piles	To prevent movement or strain before excavation	Will not stand large strains; must penetrate well below sliding surface	Stability analysis is required to determine soil-pile force system for safe design
	Install anchors	Where rights-of-way adjacent to highway are limited	Involves depth control based on ability of foundation soils to resist shear forces from anchor tension	Study must be made of in situ soil shear strength; economics of method is function of anchor depth and frequency
	Treat chemically	Where sliding surface is well defined and soil reacts positively to treatment	May be reversible action; has not had long-term effectiveness evaluated	Laboratory study of soil-chemical treatment must precede field installation
	Use electro-osmosis	To relieve excess pore pressures at desirable construction rate	Requires constant direct current power supply and maintenance	
	Treat thermally	To reduce sensitivity of clay soils to action of water	Requires expensive and carefully designed system to artificially dry out subsoils	Methods are experimental and costly

Source: From Gedney and Weber, 1978.

Table 9.17 Possible Instability Occurrences in Natural Slopes and the Geologic-Engineering Considerations in Their Mitigation

Failure of Thin Wedge, Position Influenced by Tension Cracks

Failure at Relatively Shallow Toe Circles

Low Groundwater

High Groundwater

(1) Slope in coarse-grained soil with some cohesion

With low groundwater, failure occurs on shallow, straight, or slightly curved surface. Presence of a tension crack at the top of the slope influences failure location with high groundwater, failure occurs on the relatively shallow toe circle whose position is determined primarily by ground elevation.

Analyze with effective stresses using strengths C′ and ϕ' from CD tests. Pore pressure is governed by seepage condition. Internal pore pressures and external water pressures must be included.

Stable Slope Angle = Effective Friction Angle

Stable Slope Angle = 1/2 Effective Friction Angle

Low Groundwater

High Groundwater

(2) Slope in coarse-grained, cohesionless soil

Stability depends primarily on groundwater conditions. With low groundwater, failures occur as surface sloughing until slope angle flattens to friction angle. With high groundwater, stable slope is approximately ½ friction angle.

Analyze with effective stresses using strength ϕ'. Slight cohesion appearing in test envelope is ignored. Special consideration must be given to possible flow slides in loose, saturated fine sands.

Location of Failure Depends on Variation of Shear Strength with Depth

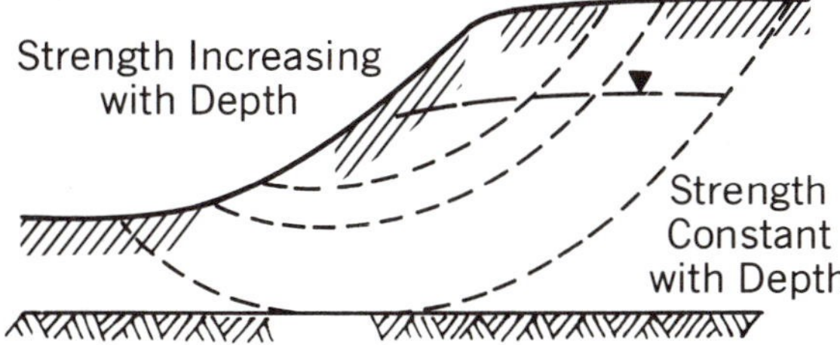

(3) Slope in normally consolidated or slightly preconsolidated clay

Failure occurs on circular arcs whose position is governed by theory. Position of groundwater table does not influence stability unless its fluctuation changes strength of the clay or acts in tension cracks.

Analyze with total stresses, zoning cross section for different values of shear strengths, determine shear strength from unconfined compression test, unconsolidated undrained triaxial test, or vane shear.

Location of Failure Depends on Relative Strength and Orientation of Layers

Strata of Low Strength

(4) Slope in stratified soil profile

Location of failure plane is controlled by relative strength and orientation of strata. Failure surface is combination of active and passive wedges with central sliding block chosen to conform to stratification.

Analyze with effective stress using C′ and ϕ' for fine-grained strata and ϕ' for cohesionless material.

(continued)

Table 9.17 *(continued)*

Bowl-shaped Area of Low Slope (9% to 11%) Bounded at Top by Old Scarp Failure Surface of Low Curvature that is a Portion of an Old Shear Surface. (5) Depth creep movements in old slide mass	Strength of old slide mass decreases with magnitude of movement that has occurred previously. Most dangerous situation is in stiff, overconsolidated clay that is softened, fractured, or slickensided in the failure zone.

Source: From NAVFAC, 1982.

Site-specific solutions often call for an engineered design. Maintaining a stable slope or stabilizing a failed one is a common circumstance in road construction. Gedney and Weber (1978) describe a variety of means of achieving an appropriate design. Table 9.16 summarizes some commonly used procedures.

It is important for the geologist to be aware of both the geologic and the engineering considerations for designing mitigation. This ensures that the proper information will be gathered. Table 9.17 gives some examples for natural slopes. An example of the value of this perspective is remedial work on Stump Springs Road, a major route used to haul harvested timber in the Sierra Nevada. In 1982 and 1983, 17 slope movements occurred along a 23-km section of this road (DeGraff et al., 1984). Site-specific investigation identified the importance of groundwater in inducing most of these failures. As a consequence, repair design included drainage to avoid groundwater concentration. Retaining walls such as the reinforced earth and concrete crib-style structures in Fig. 9.40 were designed to permit free groundwater drainage.

Figure 9.40 A concrete crib-style retaining wall stabilizing a small landslide above a road.

Hungr et al. (1984) provide another example of site-specific mitigation of landslide hazard. They performed quantitative analyses of debris torrents to aid in the design of remedial measures. The especially troublesome question of delineating the area likely to be overrun by debris was addressed. Velocity and flow-depth calculations permit the design of road bridges capable of allowing debris torrents to pass beneath. Hungr et al. also note that clearance should consider the rapidly moving particles projecting from the surface of the torrent. For example, an additional safety clearance of 3 m was applied to the height of debris-torrent surfaces expected to pass under bridges on the Squamish Highway in British Columbia, Canada. This approach also permits determining the expected magnitude, discharge, depth, and velocity of a future torrent in order to evaluate the design of deflection walls or embankments.

SUBSIDENCE

Subsidence is displacement of the ground surface vertically over a broad region or at localized areas. It may be either a gradual lowering or a collapse. This can have a costly effect on facilities and structures over a subsiding area. It is estimated that damage from subsidence in the United States annually amounts to several tens of millions of dollars (Lee and Nichols, 1981).

Subsidence results from a number of different mechanisms (Sowers, 1976; Lee and Nichols, 1981). It can occur as a consequence of natural processes. The dissolving of limestone, salt, or other soluble materials creates underground openings that may collapse. Collapse may also occur in the roofs of lava tubes in areas underlain by volcanic rock. Withdrawal of fluids from subsurface reservoirs can create human-induced subsidence. This type of subsidence has resulted from extracting oil, gas, and groundwater. Underground mining is another mechanism for creating subsidence by creating subsurface openings.

Natural solution of rock leading to collapse of the overlying surface can be a rather spectacular form of subsidence. A single sinkhole 324-ft wide and 100-ft deep was formed by collapse on May 8–9, 1981, in Winter Park, Florida. It destroyed a house, several cars, streets, parts of neighboring buildings, and the city swimming pool, causing losses estimated to exceed $2 million (Lee and Nichols, 1981). Another example is the sudden collapse of both the east- and westbound lanes of Interstate Highway 4 northeast of Lakeland, Florida, on June 22, 1975 (Sowers, 1976).

Limestone is not the only soluble rock type that is subject to this natural process. Its importance stems from the fact that it underlies extensive areas of Florida and many other states. Figure 9.41 illustrates the natural development of sinkholes in a limestone area. Groundwater levels and movement control this process. When surface activities alter the natural groundwater system, it can accelerate this natural process. In Alabama, nearly four thousand induced sinkholes that have occurred since 1900 are the product of either a decline in the water table or related to construction (Newton, 1984). Groundwater withdrawal for surface uses or dewatering of quarries and mines causes a general lowering of the water table. Figure 9.42

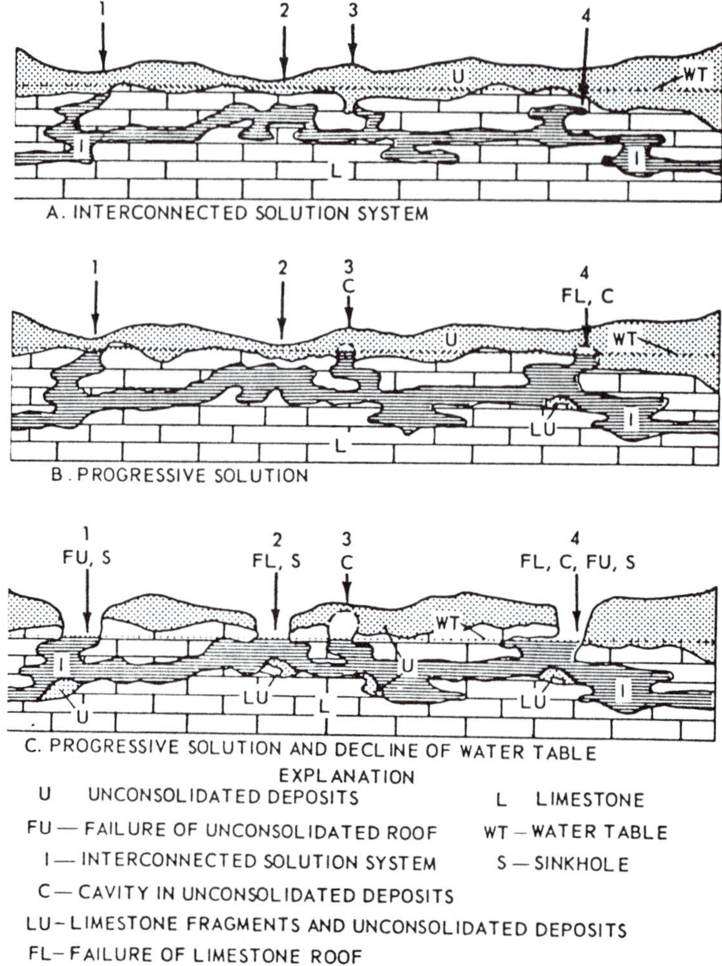

Figure 9.41 Illustrates the progressive development of sinkholes under natural conditions. (From Newton, 1976)

shows the changes leading to increased collapse that can result. Construction activities are a less-common cause of collapse. Subsidence because of construction can result from loading the ground surface over a cavity or from the diversion of surface water, thus changing the groundwater system and increasing sinkhole development (Newton, 1976).

Subsidence caused by underground mining results in severe economic losses in some areas. It is estimated that damage amounting to $30 million annually results from subsidence over abandoned coal mines (Lee and Nichols, 1981). In some areas such as Centralia, Pennsylvania, coal seams burning out of control increase the area at risk from subsidence. Figure 9.43 illustrates the extent of coal fields and subsidence in the western United States. Pennsylvania and other Appalachian states have experienced similar subsidence problems.

Underground mining, notably coal mining, creates subsurface openings. Rock layers bridging these voids may fracture and collapse into the open-

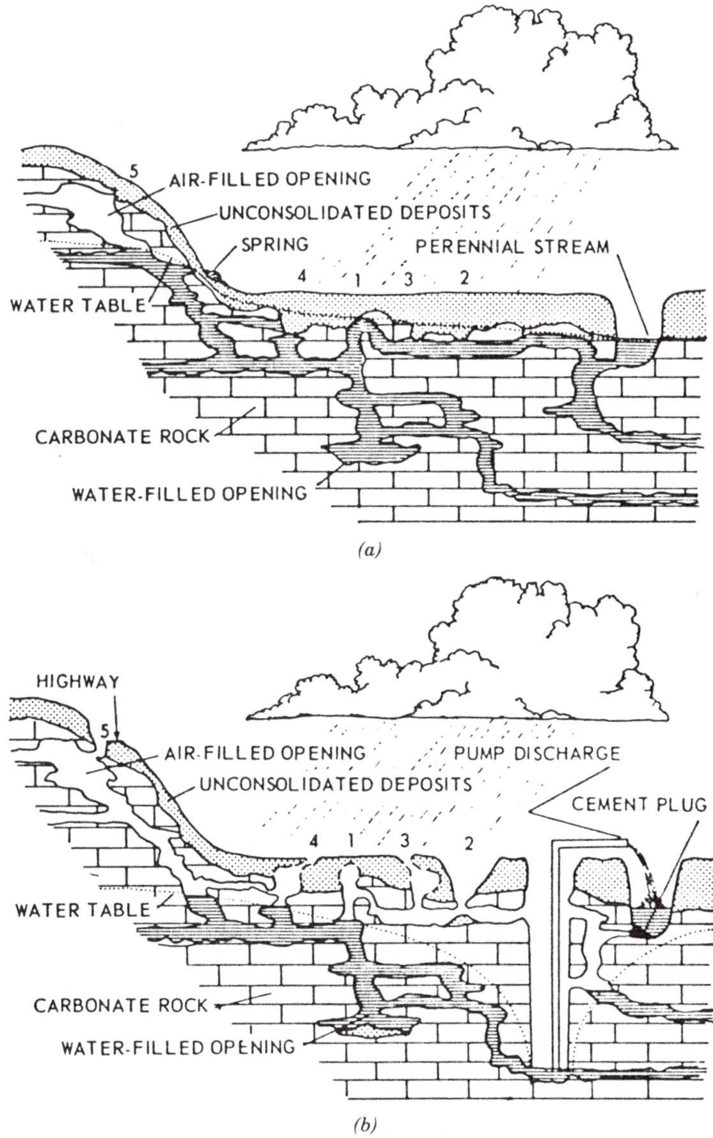

Figure 9.42 Two cross sections of geologic and hydrologic conditions in a solution-prone area. (*a*) The natural system. (*b*) Changes resulting from water development. (From Newton, 1976)

ing, with resultant lowering of the ground surface (Figure 9.44). Subsidence causes differential settlement, with the greatest amount near the center of the opening. Associated with this differential settlement are ground cracks (Figure 9.45). The extent and size of cracks will change until subsidence is complete in an area (DeGraff and Romesburg, 1981).

Withdrawal of oil, gas, and water has produced subsidence that has resulted in extensive losses in Arizona, California, and along the Gulf coast of Texas. A notable example is the failure of the Baldwin Hills Reservoir in Los Angeles, California. James (1968) notes that periodic level observa-

Figure 9.43 The location of major coal fields, subsidence areas, and reported underground fires in the western United States. (Reprinted with permission from Reviews in Engineering Geology, Vol. 6, R. E. Gray and R. W. Bruhn, Coal Mine Subsidence—Eastern United States, 1984. Published by the Geological Society of America.)

tions in Baldwin Hills detected subsidence amounting to 9.7 ft from 1917 to 1963. Development of ground cracks in the area adjoining the southeastern part of the reservoir was further evidence of deformation. On December 14, 1963, a crack opening along a minor fault split the reservoir floor from the abutment of the main dam. Rapidly escaping water enlarged this passage to the point that the dam was breached. Property losses exceeded $15 million and five deaths resulted from subsequent flooding. Although the subsidence cannot be wholly attributed to production from the Inglewood oil field, the subsurface relationship of this oil field to local structures suggests a contributing role at a minimum (James, 1968).

Subsidence causes damage in several ways. The most obvious is tilting, cracking, and shearing of structures where subsidence produces differential settlement. Large-scale collapse can completely destroy some structures. Destruction results when water-containment structures such as reservoirs and canals are breached. Subsidence causes damage by impairing the function of some surface facilities. It can create low points in pipelines and

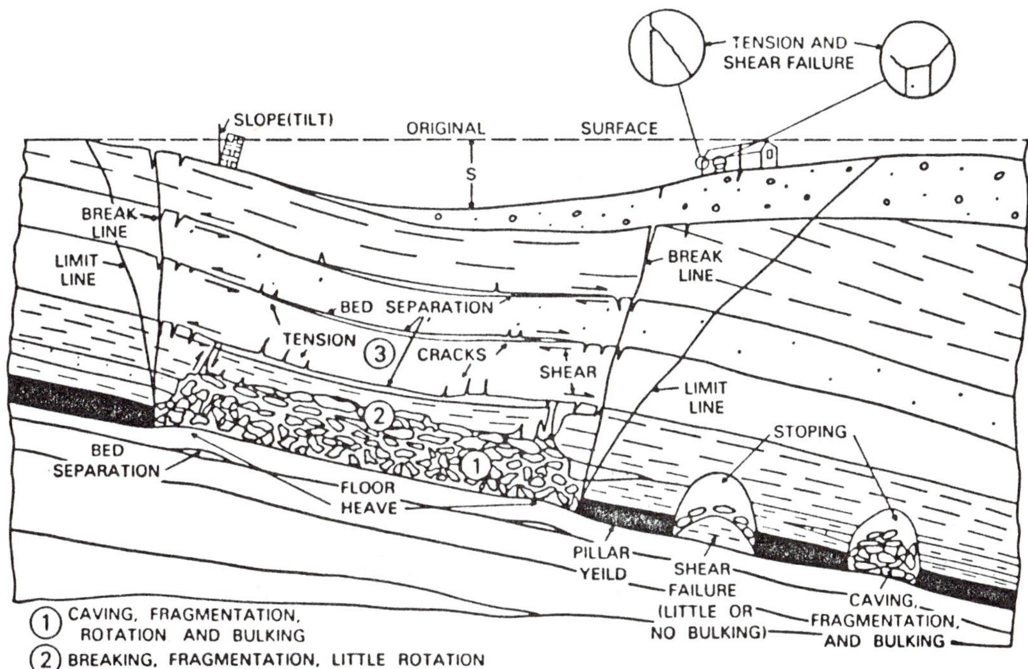

Figure 9.44 Schematic representation of subsurface conditions before and after subsidence at a mine. (Reprinted with permission from Reviews in Engineering Geology, Vol. 6, C. R. Dunrud, Coal Mine Subsidence—Western United States, 1984. Published by the Geological Society of America.)

Figure 9.45 Ground cracking because of subsidence on the Wasatch Plateau in central Utah. The ground surface is almost 700 m above a mine where a single seam of coal 7-ft thick is being removed.

alter the alignment of microwave transmission stations. A more-subtle consequence of subsidence is ground lowering that makes more land subject to flooding. Lee and Nichols (1981) note that Baytown, Seabrook, and other Texas coastal towns experienced as much as 3 ft of subsidence because of fluid withdrawal. These towns are now more vulnerable to flooding from storm surges and hurricanes.

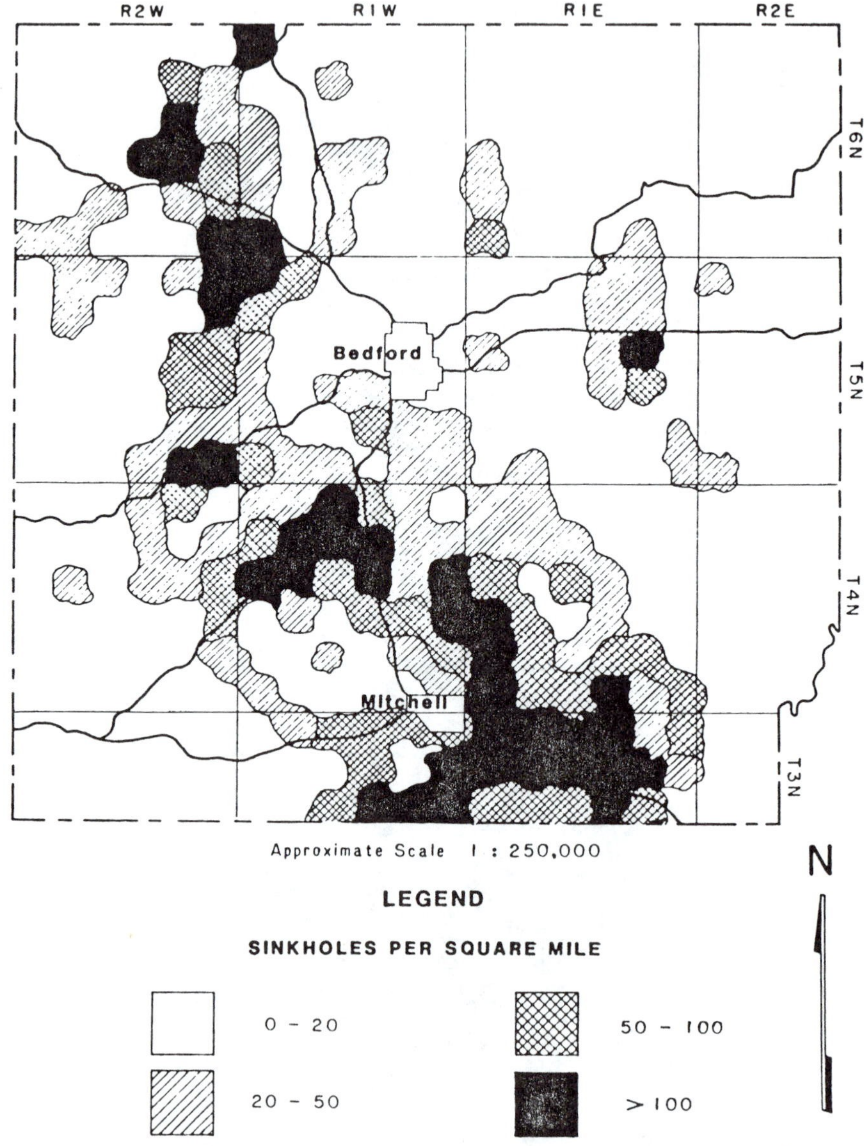

Figure 9.46 Sinkhole-density map for Lawrence County, Indiana, indicates localities especially prone to subsidence from limestone dissolution. (From Adams and Lovell, 1984)

Evaluating Subsidence Processes

It should be clear from our discussion that evaluation of subsidence processes depends on a variety of methods. Evaluating possible underground openings will require a very different approach than estimating subsidence from fluid withdrawal.

The potential for solution-caused subsidence depends on the presence of limestone or other soluble rock types. Examining the natural subsidence occurring where these rocks are present serves as an initial indicator. For example, a sinkhole-density map compiled by means of aerial photography is a useful measure of relative collapse potential (Adams and Lovell, 1984) (Figure 9.46). Similarly, the potential for subsidence from underground mining exists only where mining is active or was conducted in the past. Historical records or the detailed surface surveys normally conducted by many present-day mines provide a basis for evaluating subsidence potential. The extent of subsidence depends on factors such as: (1) thickness of the mined coal, (2) mine geometry and mining methods, and (3) thickness, lithology, structure, and hydrology of the bedrock and surficial material in the mining area (Dunrud, 1984).

Detecting solution cavities or abandoned mine openings is most reliably done with drilling. However, expense limits such drilling over large areas. Fountain (1976) noted that earth-resistivity surveys can detect openings at depths up to 25 m. Gravity surveys are marginally successful; only large openings near the surface are detectable. Subsurface radar techniques proved too unreliable for practical use.

Helm (1984) describes two field-based techniques for subsidence prediction where fluid is being withdrawn. One is the depth-porosity method, and the other is the aquitard-drainage method. Estimating subsidence in an area where fluid withdrawal is being initiated is best done with the depth-porosity method. For the more complicated situation in which subsidence is already active, the aquitard-drainage method is recommended. Holzer (1984) explains that ground failures in areas subsiding owing to fluid withdrawal can be predicted. These ground failures range from tension cracks to surface faults. Areas subject to ground failure are found in many parts of the Southwest (Figure 9.47). Prediction of ground failure in areas where deformation is underway requires monitoring of surface conditions for signs of failure. In areas not yet subject to subsidence, prediction requires determining the particular subsurface conditions conducive to failure. Larson and Péwé (1986) provide a useful example of the value of this approach. Their study of ground fissures in a subsiding area of Phoenix, Arizona, identified three geologic settings with potential for ground failure (Figure 9.48). This establishes some of the types of subsurface conditions to look for in areas not yet undergoing subsidence.

Mitigating the Effects of Subsidence Processes

Controlling land use to avoid large-scale changes in the regional water table is one way to avoid subsidence in areas underlain by soluble rock. Avoiding withdrawal overdraft from compressible groundwater aquifers is

Figure 9.47 Location of subsidence-induced ground failures because of groundwater withdrawals in the western United States. (Reprinted with permission from Reviews in Engineering Geology, Vol. 6, T. L. Holzer, Ground Failure Induced by Ground-Water Withdrawal from Unconsolidated Sediment, 1984. Published by the Geological Society of America.)

equally effective in avoiding subsidence. Reservoirs from which oil, gas, or geothermal fluid is being withdrawn can be reinjected with water to compensate for the lost fluids. In some instances, subsidence over mines need not result in structural damage. This requires knowing the specific factors that influence subsidence at that locality and conducting mining in a manner that permits a general lowering of the entire area in which a structure is situated. This minimizes the differential settlement responsible for most of the distress to structures.

Structures damaged or impaired by subsidence can be restored in many cases. Kemmerly (1984) describes corrective procedures for sinkhole collapse. As shown in Figure 9.49, this involves sealing the cleaned-out sinkhole, restoring the ground surface, ensuring that surface water to the site is minimized, and promoting groundwater flow down gradient from the repair location. Royster (1984) notes that some problems with solution-related subsidence are human related. This is especially true where natural sinkholes are used for drainage. Diverted water can increase the groundwater gradient in areas, leading to greater subsidence, as was the case at a site on Tennessee Highway 76. Resolution of the problem required extensive subsurface investigation. Figure 9.50 shows the situation at the time of subsidence and the change to surface-water drainage necessary to cure it.

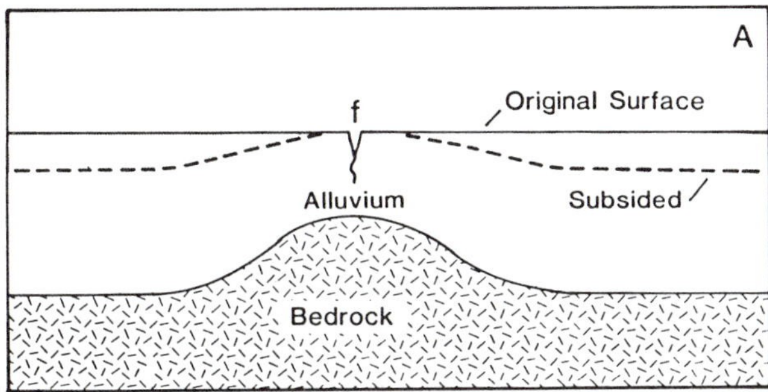

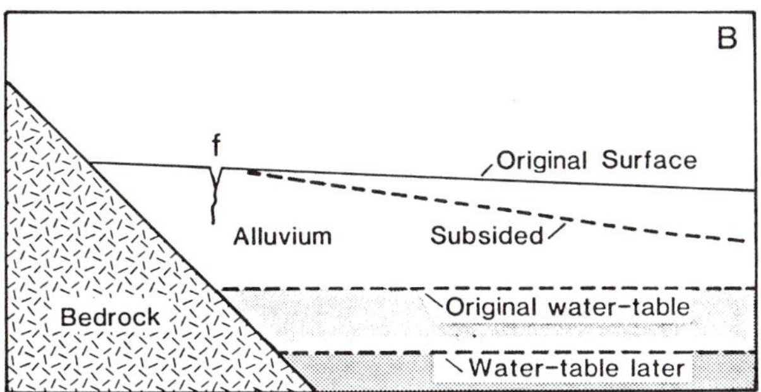

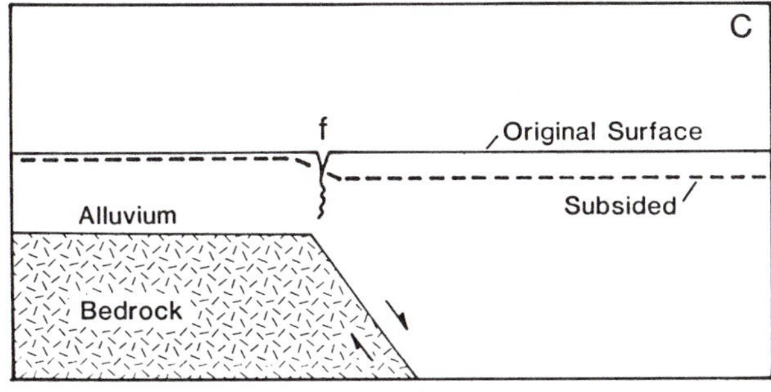

Figure 9.48 Three geologic conditions found associated with ground fissures in Phoenix, Arizona. (Source: From Michael K. Larson and Troy L. Péwé, Origin of Land Subsidence and Earth Fissuring, Northeast Phoenix, Arizona, *Bulletin of the Association of Engineering Geologists*, Vol. 23, No. 2, May 1986, pp. 139–165. Used by permission.

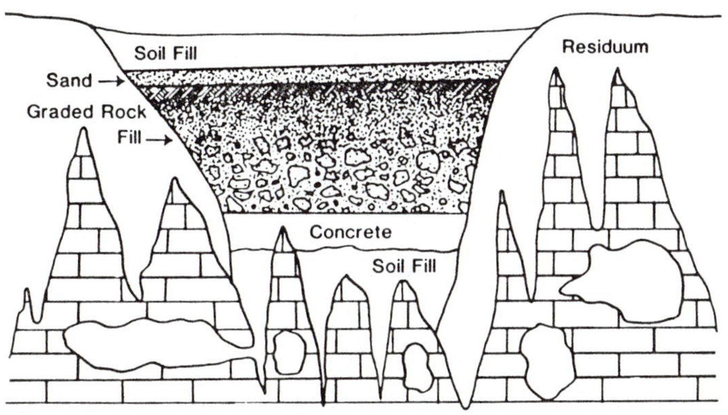

Figure 9.49 Final repair of a sinkhole collapse. (From Kemmerly, 1984)

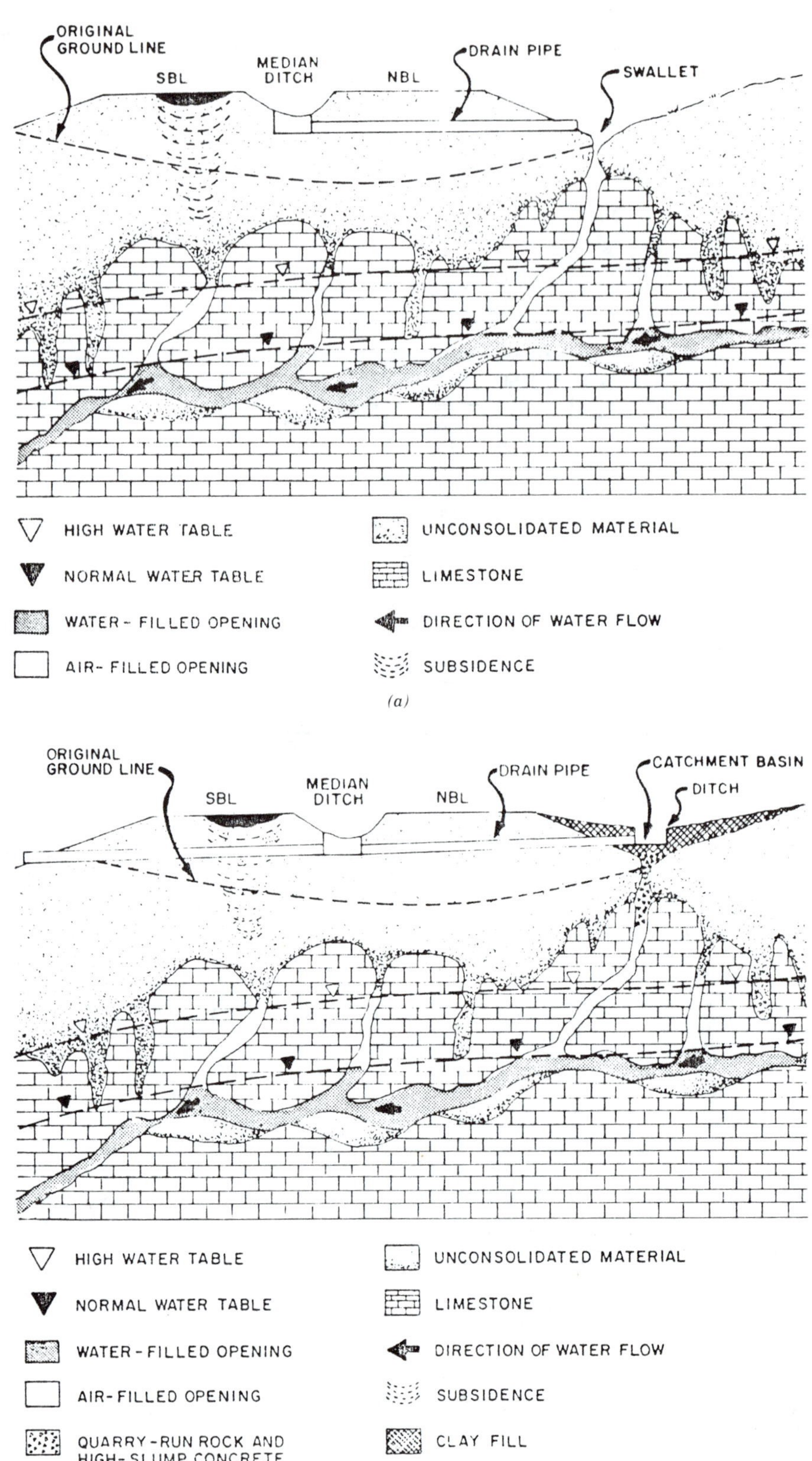

Figure 9.50 (*a*) Cross sections showing postsubsidence conditions at Tennessee Highway 76. (*b*) TN-76's subsequent repair. (lower diagram). (From Royster, 1984)

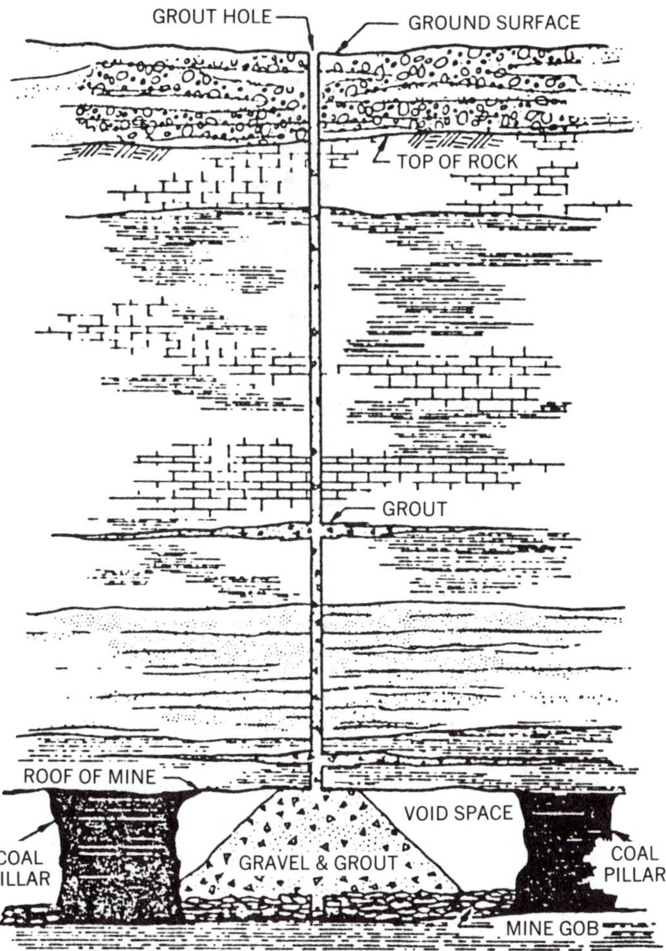

Figure 9.51 Shows one of the grout columns used in controlling subsidence at an electric substation in Pennsylvania. (From Gray et al., 1976)

Subsidence control over mined areas generally takes the form of either providing selective support for the structure or filling the underground space to halt further subsidence. Gray et al. (1976) describe efforts to control subsidence affecting an electric substation. Both approaches to dealing with subsidence were used. Selective support involved placing drilled piers and piling seated into rock below the base of the mined coal seam. Fly ash was injected through drillholes to fill some underground openings. In 15 selected locations, grout columns were constructed (Figure 9.51).

EXPANSIVE SOILS

Expansive soil is soil or soft rock that shrinks or swells as its moisture content changes. Montmorillonite or other clay minerals of the smectite group will almost always be present. Expansive soils are found in the western, central, and southeastern United States (Figure 9.52). There are two major sources for expansive soils in the United States (Schuster, 1981). Volcanic rocks, ash, and glass are one source. The other is sedimentary rocks that contain clay minerals. Expansive soils are found in other parts of the world where similar source rocks exist.

Despite being virtually unknown to the general public, expansive soils are responsible for major economic losses. Annual losses in the United States from this process are estimated to be about $6 billion (National Research Council, 1985b). About 10% of the new homes constructed annually in the United States are subjected to significant damage during their useful lives by expansive soils. An additional 60% sustain minor damage (Jones and Holtz, 1973). Figure 9.53 illustrates the modes of failure that can be expected in a typical frame house founded on expansive soil.

Shrinking and swelling of soil and soft rock requires two conditions to be satisfied before it occurs. First, the soil or rock must have the potential for volume change. This is dependent on the type and amount of clay minerals present. The potential is much higher with smectites than with other clay minerals such as illites and kaolinites. For example, intercrystalline expansion of pure sodium montmorillonite can amount to as much as twenty times its dry volume (Mielenz and King, 1955). Unless the soil is pure smectite, which seldom occurs, the percentage of clay present in the soil will also control swelling potential. Second, water must be both available and mobile enough to reach the expansive soil. In semiarid environments, home development can increase available water through discharge from roof gutters and downspouts. Reducing the forest cover in wetter areas may have the same effect.

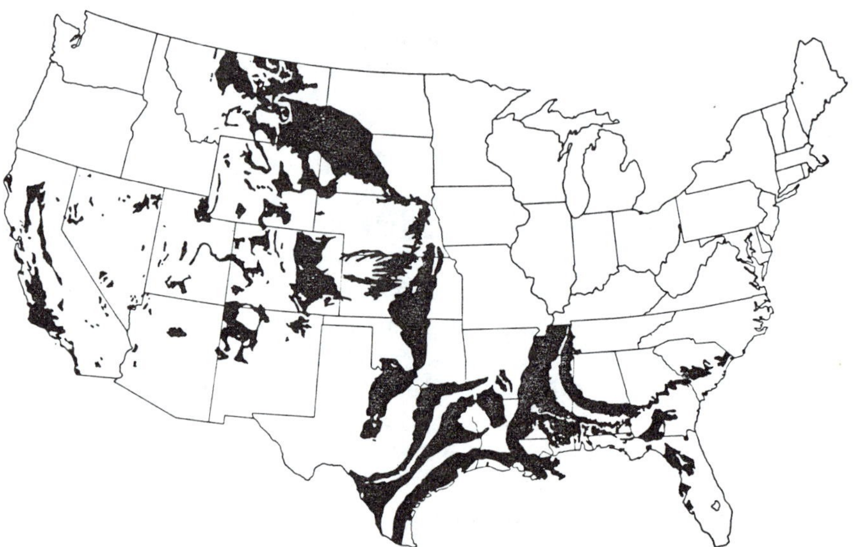

Figure 9.52 The distribution within the United States of soils subject to high volume changes. (From Witczak, 1972)

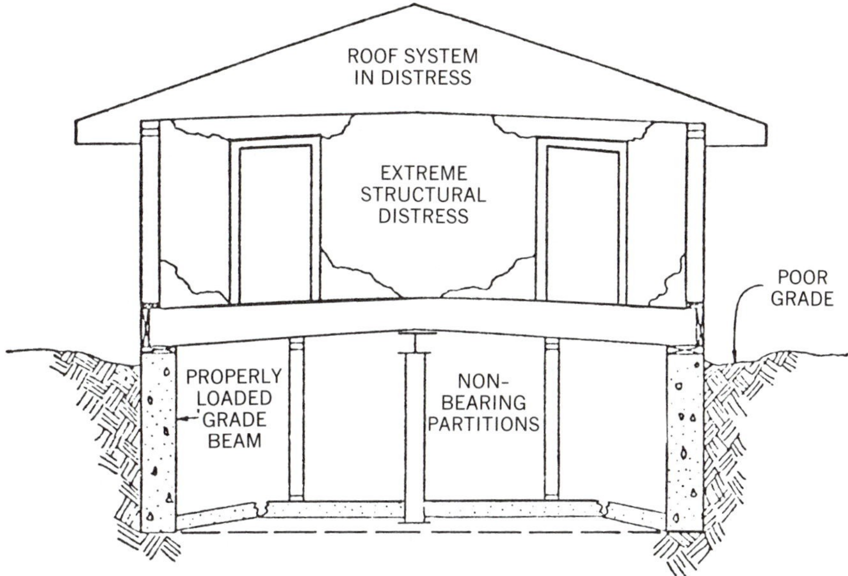

Figure 9.53 Typical major house damage owing to swelling of foundation soil. (From Holtz and Hart, 1978)

The nature of the soil mass will also influence the degree of shrinking and swelling that will take place (Schuster, 1981). A greater volume change will be associated with a thick layer of expansive soil than with a thin one. The nearer the expansive soil is to the ground surface, the greater the likelihood of shrinking and swelling. This stems from the reduced confining pressure to be overcome before volume change occurs.

Evaluating Expansive Soils

Obviously, expansive soils are likely to be present in the United States anywhere either of the two parent materials are found. Surface exposures of soil and soft rock containing montmorillonite frequently display a characteristic "popcorn" appearance (Figure 9.54). Another source of general information is the swelling characteristics noted for various soils mapped in the National Cooperative Soil Surveys. These agricultural soil classifications contain laboratory measurement of shrink–swell characteristics. Often, they include information on the nature of clay minerals present.

Various laboratory methods are available for identifying expansive soils. These tests either involve a mineralogical identification of clay minerals likely to produce high volume changes or are direct evaluations of volume-change characteristics or of physical properties closely allied with volume changes (Schuster, 1981). X-ray diffraction, differential thermal analysis, and microscopic examination are three methods for determining clay–mineral content. Common methods for identifying expansive soils on the basis of volume-change characteristics or related physical properties are:

Figure 9.54 Example of "popcorn" appearance developed on expansive soils, Slumgullion Landslide, Lake City, Colorado.

(1) the free-swell test, (2) the Atterberg limits and colloid-content determination, and (3) direct measurement of volume change.

Mitigating the Effects of Expansive Soils

The best mitigation for the effects of expansive soils is to avoid founding structures on them. However, their extensive occurrence in some areas can render this approach infeasible. This means that other techniques must be applied.

One method is to remove the expansive soil and replace it with nonexpansive soil. Where the entire expansive soil thickness can be removed, it can completely prevent occurrence of the process. However, the thickness may be too great to permit complete removal. The solution then requires determining the thickness to remove and backfill with nonexpansive soil. The thickness of nonexpansive soil must be sufficient to resist uplift on the underlying expansive soil completely or, at least, enough to avoid significant damage. The Mohawk and Melton canals in Arizona crossed areas with expansive soils. Removal of expansive soils and replacement with nonexpansive sand-gravel soil was used to avoid damage to the canal lining (Holtz, 1959).

Foundation treatments may prove effective in preventing the expansive-soil process. Applying a confining load is one type of foundation treatment. This usually involves placing a blanket or embankment of nonexpansive soil over the expansive soil. The surcharge resists the uplift of the underlying expansive soil. It is especially effective as a foundation treatment for large buildings that add to the load.

Other foundation treatments tend to limit the effect of the expansive soil. Cellular foundations and granular mat "cushions" placed between the soil and building foundation dissipate the effect of expanding soil. The effect of expansive soil can also be limited by placing reinforced concrete piers below the depth of the expansive soil. The reinforcing steel in these piers is connected into the steel in the concrete foundation of the houses. With small structures, a stiff, reinforced concrete slab will permit the structure to float on uplift pressures.

Chemical stabilization of expansive soil is another method (Schuster, 1981). It attempts to modify the ionic character of the soil-and-water combination, minimizing or preventing, swelling. The most commonly used chemical is hydrated lime, Ca(OH). The strong positively charged calcium ions in the lime replace the weaker ions such as sodium on the surface of clay particles. This reduces base-exchange capacity of the clay particles and results in lower volume-change potential. Portland cement is another chemical used in stabilizing expansive soils. It seems to have two separate effects. The lime within the cement acts in the same way as hydrated lime. In additional, the hardened cement matrix in the soil physically resists movement.

Isolating water from expansive soils is another approach to mitigation. Techniques for accomplishing isolation depend on whether the water is surface water percolating downward or groundwater moving upward. With surface water, ditches and pipes can be used to keep water away from sensitive areas underlain by expansive soil. Sand and gravel have been used to form breaks in capillary continuity when groundwater is moving upward. Enveloping masses of expansive soil in impermeable membranes is another means for isolating water from expansive soil.

Steinberg (1985) reported on the use of deep vertical geomembranes to control expansive soil effects in highways. A zone of expansive clay activity is present in the upper 6–8 ft in the San Antonio, Texas, area. Minimizing fluctuations in moisture within this zone would perhaps reduce destructive movement on road subgrades. Geomembranes placed to a depth of 8 ft serve as vertical barriers to moisture movement through the soil. The geomembranes help maintain a uniform moisture regime and thereby seem to limit shrink and swell. Figure 9.55 shows the pretreatment and

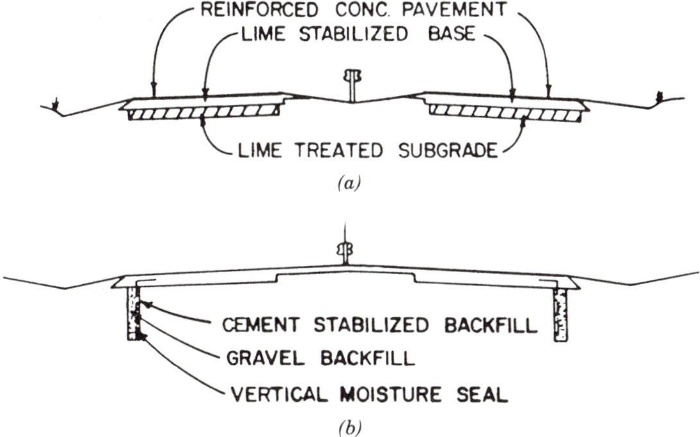

Figure 9.55 (*a*) Interstate 37 section near San Antonio, Texas, prior to rehabilitation. (*b*) Section after rehabilitation with geomembranes. (lower diagram). (From Steinberg, 1985)

posttreatment sections of Interstate 37, one of four highways involved in this test.

SHORELINE PROCESSES

At the shoreline, a dynamic interaction occurs between solid land and moving water. Contact between land and water may occur at a rocky bluff, tidal mudflat, sandy beach, or barrier island. Each landform involves its own set of conditions that are important to engineering works. Waves, currents, and changing water levels are the means by which moving water affects these landforms.

The dynamic interaction occurring at the shoreline is of more than academic interest. Griggs and Johnson (1983) note the losses along the California coast owing to devastating storms in 1978 and 1983. These storms resulted in damages of over \$18 million and \$100 million, respectively. Three thousand homes were damaged with 27 ocean-front residences totally destroyed. Twelve businesses were destroyed with damage done to nine hundred others. Griggs and Johnson point out that mild climatic conditions experienced between 1946 and 1976 are being replaced by more frequent and intense storm activity. Rapid and extensive development along the California coast took place during the milder climatic period. As a consequence, development in areas previously thought safe from storm action sustained losses during recent storms.

Significant economic consequences can result from shoreline processes less dramatic than major storms. Hoyt (1981) describes beach processes and sand removal that have affected stabilized inlets along the coast of Delaware. The jetties used to stabilize these heavily used inlets interrupt the transport of sand along the shore: resulting beach erosion threatens near-shore structures. Artificial replenishment to maintain the beach cost over \$2 million at North Indian River inlet between 1957 and 1978. Hoyt notes the continuing high cost for maintaining this inlet will require consideration of its value and who should pay for its maintenance.

Without major intervention by human actions, there is a long-term balance of forces along a shoreline. When this balance is altered, the system attempts to return to the earlier balance or achieve a new balance. The response induced by structures is a good example of how this balance can be altered. Figure 9.56 shows a shoreline after construction of a series of groins. The groins trap sand being transported by currents along the shoreline. However, the loss of sediment on one side of the groin results in beach erosion on the other side in order to replenish the diminished sand supply. The net effect is to change the shape of the shoreline rather than greatly increase the amount of beach present. Where such efforts have increased the beach along a stretch of shoreline, a new balance was achieved by the obvious loss of beach farther down the coast.

Although leaving shorelines in a natural state might be most desirable, it is not always practical. Many coastal areas were altered decades ago. The need for navigable channels, harbors, and other facilities also means engineering works will be part of many shorelines. The successful design and

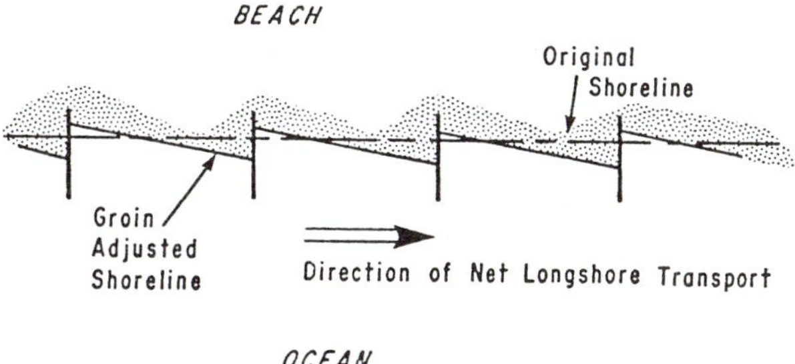

Figure 9.56 The original shoreline in balance with natural forces is forced to seek a new balance because of the alteration of sediment transport by a groin field. (From U.S. Army Coastal Engineering Research Center, 1984)

operation of these works in this dynamic environment require geologic information permitting the least amount of alteration to the long-term balance of shoreline processes. To do otherwise is to resort to brute-force engineering that has a poor track record in this environment. Table 9.18 describes the "success" of beach-protection measures for a location in northern Monterey Bay on the California coast. The costly nature of resisting natural shoreline processes suggests a need to be selective in efforts to alter shorelines and the ways to accomplish this. The engineering geologist plays a key role by providing critical information for deciding such issues.

Not all human-induced changes in shoreline processes result from efforts directed at controlling shoreline processes. Table 9.19 states ways human activities can induce changes in shoreline processes. Subsidence because of oil and gas withdrawal from the Wilmington oil field at Long Beach, California, lowered parts of the harbor as much as 27 ft (Lee and Nichols, 1981; Martin and Serdenjecti, 1984). Seawalls, dikes, and other structures had to be constructed to protect structures along the shoreline. This is clearly a human-induced change unrelated to coastal engineering activities. One item noted in Table 9.19 is reduced sediment supply to the littoral zone. This can result from actions other than efforts to control shoreline erosion and maintain navigable channels. An extensive development of reservoirs on a river system traps sediment and reduces the amount contributed to the littoral zone. The likely consequence is increased shoreline erosion.

Processes at the Shoreline

Shoreline processes are examined here with reference to marine coasts; marine shorelines hold much in common with lake and river shorelines (Figure 9.57). A comparison of sea cliff, lake bluff, and riverbank erosion illustrates the similarities between marine and nonmarine shorelines. Galster (1987) describes sea cliff development in a broad examination of coastal engineering geology along the Pacific Northwest coast. In simplistic

Table 9.18 Damage to Structures, Especially Those Installed for Protection from Waves, at Seacliff Beach, Northern Monterey Bay, California

Date of Storm	Damage Description	Direction/Type of Storm
Feb. 14–16, 1927	Concrete seawall at Seacliff Beach destroyed.	"heavy southwester"
Dec. 9–10, 1931	Timber bulkhead at Seacliff destroyed.	"southwest wind waves"
Dec. 23–29, 1931	Concession building and bathing pavilion at Seacliff wrecked.	"winds first from southwest then northwest"
Dec. 26–27, 1940	Crux of local weather problem at Seacliff. Logs up to 10 feet tossed onto road, houses damaged, 80 feet of state park lost, two sections of bulkhead ripped out.	
Jan. 8–13, 1941	At Seacliff Beach, about one half of a timber bulkhead destroyed. Beach eroded to bedrock.	"waves and swell from southwest"
Feb. 11–13, 1941	Residents in Seacliff cut off by slides.	
Feb. 9–10, 1960	Camping sites destroyed, restroom nearly destroyed.	"southerly and westerly storms"
Feb. 11–15, 1976	High waves washed completely over new seawall, carrying debris back to cliff. Portions of seawall undercut and caved in.	"southerly gale"
Jan. 8–9, 1978	Seawall overtopped and logs and debris scattered across parking and camping areas. Extensive damage to seawall.	"storm from southwest"
Feb. 1980	$1.1 million in damage at Seacliff. Storm destroyed entire lower beach portion of park, taking roads, parking lots for 324 cars and a 2672-foot seawall.	"southwest"
Jan. 28–30, 1983	$740,000 in damage. 2800 feet of new seawall damaged, 700 feet totally destroyed; eleven RV sites destroyed, restroom heavily damaged, logs and debris washed back to cliff.	"waves from southwest"

Source: From Griggs and Johnston, 1983.

terms, cliff retreat by mass movement is influenced by shoreline processes that remove the material that accumulates at the cliff base. This maintains the cliff face at an unstable or quasi-stable angle. Galster also notes that exposed material in the cliff and groundwater conditions are major controls on mass-movement activity. High, steep-faced bluffs, similar in appearance to sea cliffs, can be found along the shoreline of the Great Lakes. As with sea cliffs, waves and currents remove the products of mass movement from the base of the bluffs. Thus retreat is dependent on the nature of the bluff material and the seepage effects from groundwater movement

Table 9.19 Ways Development Can Change the Natural Balance of Forces Along Shorelines

Man-induced
Land subsidence from removal of subsurface resources
Interruption of material in transport
Reduction of sediment supply to the littoral zone
Concentration of wave energy on beaches
Increase water-level variation
Change natural coastal protection
Removal of material from the beach

Source: Modified from U.S. Army Coastal Engineering Research Center, 1984.

Figure 9.57 Shoreline views illustrate the similarities of lake bluff and sea cliff retreat. (*a*) Chimney Bluffs, a dissected drumlin on the shores of Lake Ontario. (*b*) Point Multrie on the Atlantic coast of Dominica, West Indies.

that influences mass movement (Edil and Vallejo, 1980). Large rivers such as the Ohio and Mississippi are bordered in many locations by high, steep-faced banks. Rivers erode debris from bank failures, removing it from the base of the banks. Hamel (1983) notes the importance of the materials in the bank and groundwater action to the mass-movement activity that causes bank erosion. Although important differences exist between the shoreline processes operating at seacoasts, lake shores, and riverbanks, the general principles applicable to engineering geology in a marine coastal area are often usable on nonmarine shorelines.

The processes operating along a shore involve erosion, transportation, or deposition of material. Most shoreline processes result from moving ocean water (Komar, 1983; U.S. Army Coastal Engineering Research Center, 1984). Wind acting on the surface of the ocean generates waves. These waves vary greatly in size and are called oscillatory waves. Figure 9.58 illustrates some of the terminology of waves. *Wavelength* and similar terms are used in relationships developed to evaluate shoreline processes and their effects. Waves reaching shore may be steepened by the winds of local or even distant storms. Shoreline erosion results from the action of waves.

Currents are another feature of shore processes (U.S. Army Coastal Engineering Research Center, 1984). Material is often transported along shorelines by currents. The discharge of rivers into the sea creates some of these currents. Surface currents in the oceans are generated by wind patterns. Waves usually approach the shore at an angle (Figure 9.59). This creates a longshore, or littoral, current, that is, a current in shallow water parallel to the shore.

Change in the level of water along a shore influences where waves and currents act on the shoreline (Komar, 1983; U.S. Army Coastal Engineering Research Center, 1984). A storm surge is a good example. Such a surge is created by the heavy winds of a major disturbance, producing currents that pile water against the shore. This temporarily raises water levels along a shoreline. As a consequence, waves act upon parts of the shore that would usually not be subject to this action (Figure 9.60). Storm surges can produce flooding of low-lying coastal areas and dramatic erosion along a shoreline. Another cause of water level is earthquake-generated waves

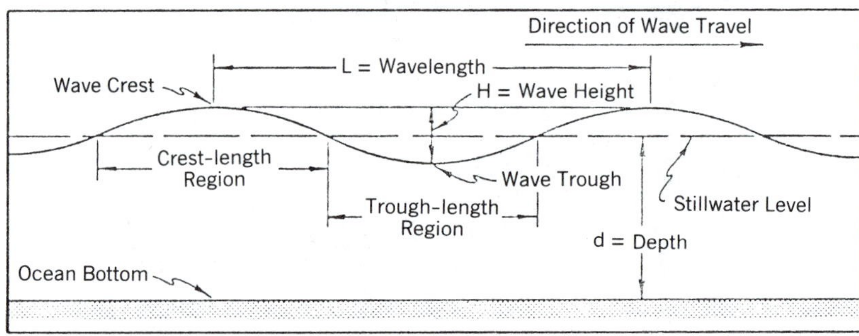

Figure 9.58 A diagram defining terms related to oscillatory waves. (From U.S. Army Coastal Engineering Research Center, 1984)

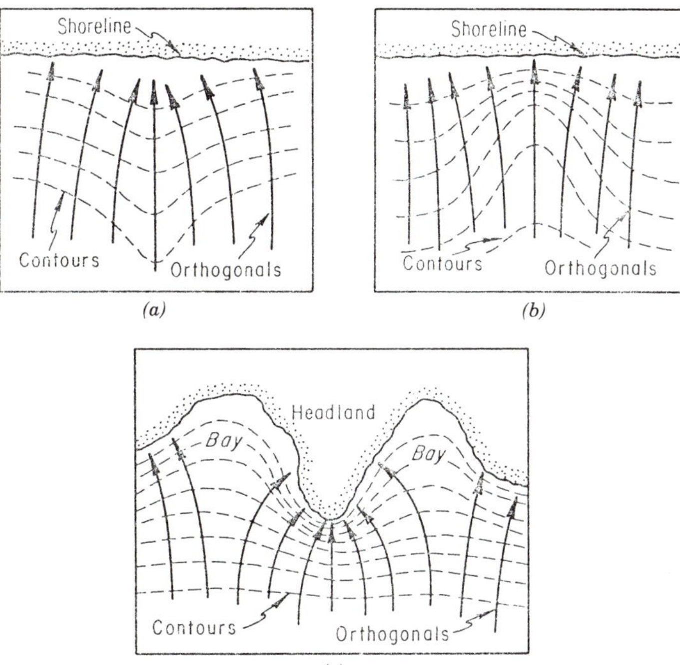

Figure 9.59 The diagrams (*a*) and (*b*) show how near-shore bathymetry can influence the direction waves approach a shoreline. Diagram (c) shows how the features on an irregular shoreline can have much the same effect. (From U.S. Army Coastal Engineering Research Center, 1984)

called *tsunamis.* (These seismic-induced waves were addressed earlier). In addition, daily changes in water level result from tides. Tides cause the water surface to rise and fall, principally in response to the gravitational attraction of the moon.

Storm surges, tsunamis, and tides represent short-term changes in water level. Water level along a shoreline can result from long-term changes in ocean levels. Shorelines worldwide record past changes in sea level owing to Pleistocene glaciation and other geologic changes. Pilkey (1981) notes that erosion along barrier islands and other sandy beaches on the East and Gulf coasts are responding to a rising sea level. For some engineering works, planning for the effects of this long-term change in water level may be as important as addressing storm surge or tidal effects.

All of these processes must be considered in understanding the geologic conditions along a particular shore. Thus, water-level changes at the shoreline depend on storm waves, daily tides, spring and neap tides, storm surges, and sea level changes operating at a particular point in time. Dolan and Lins (1985) report that normal wave heights for the Outer Banks area of the North Carolina coast average from 2–3 ft. The combined effect of tide, storm surge, and waves during the March 7, 1962, storm produced waves over 30-ft high and caused millions of dollars of damage along the mid-Atlantic coast. Likewise, the 1978 and 1983 storm damage along the

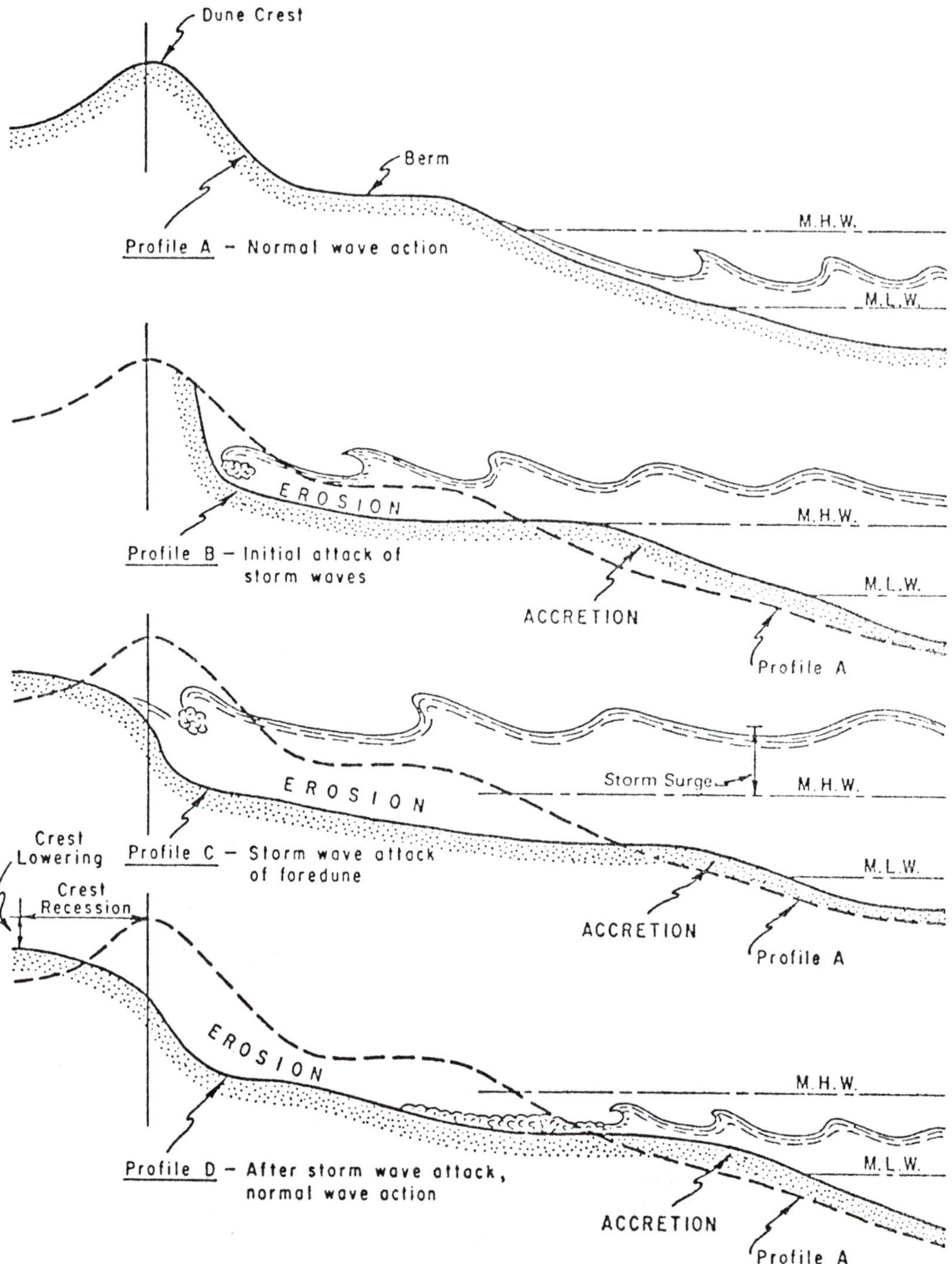

Figure 9.60 Profiles of a beach showing how storm waves erode a shoreline. M.H.W. = mean high water; M.L.W. = mean low water. (From U.S. Army Coastal Engineering Research Center, 1984)

California coast described earlier involved the simultaneous occurrence of high tides, storm surges, and storm-generated waves (Griggs and Johnson, 1983).

Evaluating Shoreline Processes

Engineering geologists need to consider the geologic setting in which the shoreline occurs in order to fully understand the processes in operation. Williams (1987) points up critical differences in geologic characteristics important to engineering along the shores of the East, Gulf, and West coasts. Low, broad plains extend inland from the shores of the East and Gulf coasts. In contrast, the tectonically active west coast generally consists of rugged terrain extending inland. Differences in offshore bathymetry exist between these coasts. A wide continental shelf with generally subdued canyons is found offshore from the East coast. The West coast, especially along the California portion, has a relatively narrow continental shelf with numerous deep submarine canyons heading close to shore. Williams (1987) notes that the differences among the East, Gulf, and West coasts extend to marine conditions that influence coastal processes. The East and Gulf coasts are struck by large cyclonic storms that approach from the southeast during late summer and fall. Most major storms impacting the West coast arrive from the northwest, many of them occurring during the winter. This general pattern is punctuated with an occasional cyclonic event from the southwest. Coastal currents modifying the climatic conditions along coasts also differ. The East coast is affected by the warm waters of the Gulf Stream moving northward from the Caribbean. Major currents have little effect on the Gulf coast. The colder water of the California Current flows southward along the West coast.

While a broad perspective of geologic factors affecting a coastal area is useful, a more detailed examination of shorelines is generally needed to adequately address engineering problems. Figure 9.61 is a diagram representing most of the features and processes operating at a shoreline. Few shorelines include all of the elements represented. It is important to observe that shorelines are dynamic areas where a balance exists between energy supplied by the moving ocean water and resistance from the materials along the land's edge. A detailed understanding of the materials along a particular shoreline is critical to evaluating the ways in which this balance is achieved or maintained. Equally critical is information on the processes acting on the materials exposed at the shoreline. The barrier islands off the North Carolina coast illustrate this point. Dolan and Lins (1985) explain that these islands are steadily migrating landward in response to rising sea level. The shoreline of the Outer Banks has moved toward the mainland at the rate of 3–5 ft per year for more than 100 years. Efforts to stabilize barrier-island shorelines ignores this long-term trend. Figure 9.62 compares cross sections of a natural and stabilized barrier island. Natural barrier islands are adapted to shoreline processes, especially extreme storms. Wave energy during storm surges is dissipated across the wide berm and other low-lying features. Resulting shoreline migration leads to new dune growth from the net gain in material from surges washing over

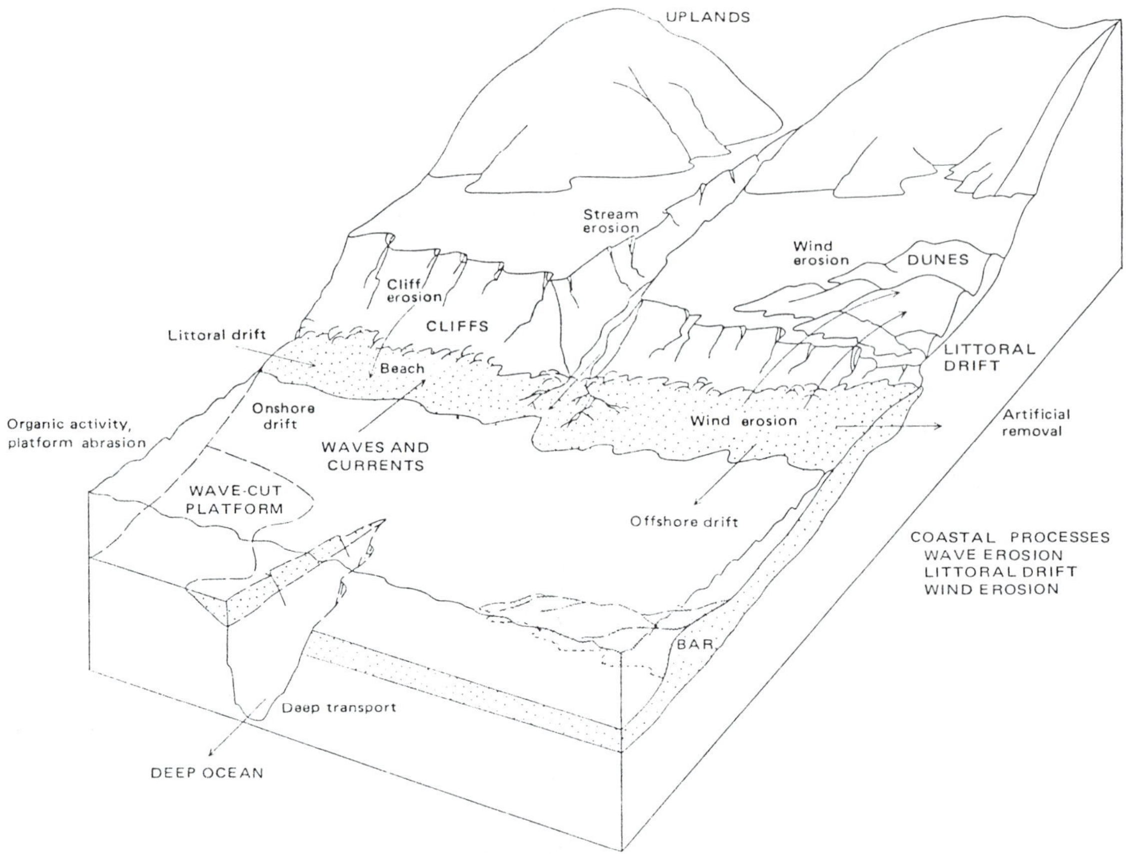

Figure 9.61 Block diagram showing the many processes and features that may be present along a shoreline. (From Helley et al., 1979)

the island and from maintenance of landform elements (Dolan, 1973). In contrast, a stabilized barrier island resists storm surge along the human-stabilized dunes. This obstructing feature on a stabilized island alters energy dissipation and material movement, causing loss of the beach. Shoreline migration still occurs, but it is accompanied with loss of landform features.

A difficulty with assessing shoreline processes to satisfy engineering needs is how much of the shoreline should be examined. Hayes et al. (1973) provide insight on choosing an appropriate stretch of shoreline to examine. Their efforts to investigate form and process in the coastal zone benefited from a zonal approach. Zonal studies conducted at the reconnaissance level start with identification of a single physiographic unit. Their example is a zonal study of the southern Gulf of the St. Lawrence off New Brunswick, Canada. Differences in shoreline morphology or major processes leads to subdividing the major zone. This permits detailed study at a few selected locations to be used to characterize major parts of the zone. Hayes et al. (1973) detail how their Gulf of St. Lawrence zone was subdivided into 10 physiographic elements.

It is often difficult to see how the large-area studies typical of the zonal approach can yield useful information for engineering efforts. It would seem that only site-specific studies would have value. A good example of the utility of large-area studies is provided by Griggs (1987). He describes the recognition along the California coast of shoreline elements called *littoral cells*. These cells consist of a source of sediment carried by littoral drift, the shoreline segment along which drift occurs, and a sediment sink such as a submarine canyon where the sediment is eventually deposited. Griggs examined the need for harbor dredging in relation to littoral cells. Marinas or harbors between or nearer the upcoast end of littoral cells tended to require little dredging owing to insignificant littoral drift. In contrast, harbors located in the middle reaches or downcoast ends of littoral cells required extensive dredging to maintain their usefulness. Griggs concludes that the harbor location within a littoral cell and the annual littoral-drift volume is more critical than the structural measures applied at the harbor entrance in reducing the need for dredging. In addition to providing insight into coastal processes, large-area assessments may permit one to discern long-term trends useful in improving engineering design.

Assessment techniques in engineering-geologic investigations along shorelines include both mapping and sampling. These techniques must be applied to both the land- and the water-related processes along the shoreline. Williams and Bedrossian (1978) describe efforts for characterizing the land along the California coast. Although resembling traditional bedrock mapping, the resulting product emphasizes conditions important to coastal development. Table 9.20 describes how areas subject to tsunami inundation, landsliding, and flooding are designated.

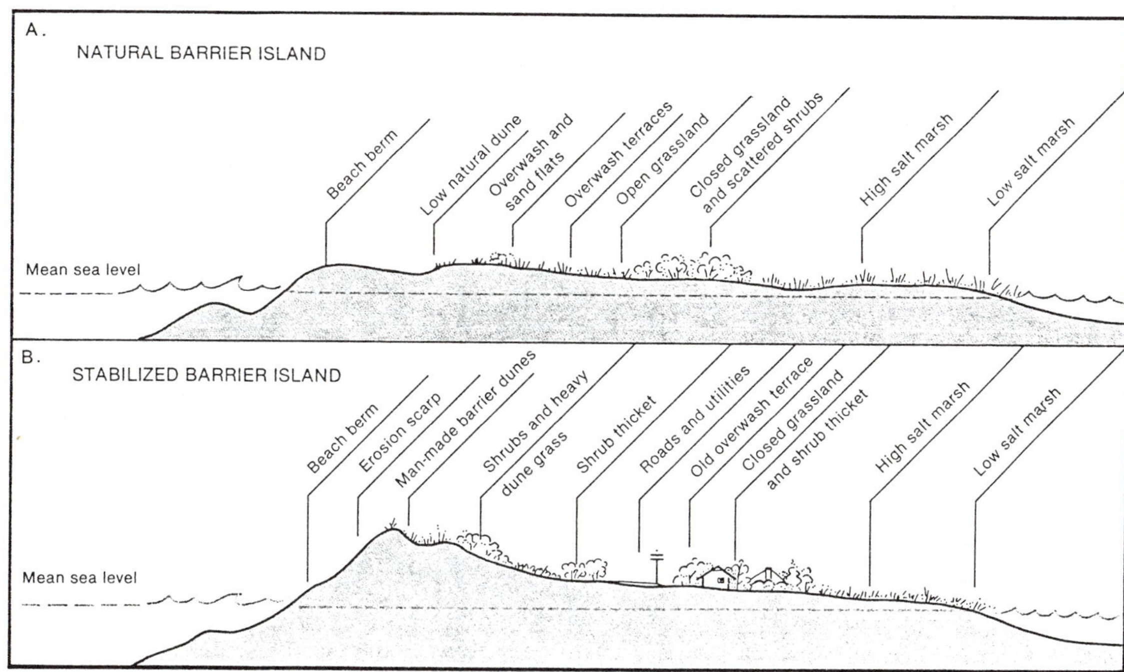

Figure 9.62 Cross section depicts natural and stabilized barrier islands, showing the differences that lead to varying erosional response to storms. (From Dolan and Lins, 1985)

Table 9.20 Descriptions of Geologic Features Important to Coastal-zone Mapping and Hazard Identification[a]

Symbol[b]	Geologic Factor	Criteria for Area Designation	Characteristics	Examples of Possible Mitigative Measures
	Landslides (known and potential)	Although not indicated on map, all slopes have the potential for failure depending on the slope angle and underlying materials.	Slides of variable depth and size; generally involve colluvium, soils, and vegetation on steep, water-saturated slopes.	Avoidance, drainage control, special construction techniques, selective removal of slide.
B	Flooding	Area designated by the U.S. Geological Survey as being subject to a 100-year flood.	Low-lying areas adjacent to major streams.	No structures permitted in these areas unless special construction techniques, levees, or dams are utilized.
C	Settlement and/or liquefaction	Saturated clay-free sand and silt layers within 50 feet of surface have potential for liquefaction; all are subject to settlement.	Generally, problems of differential settlement and liquefaction during seismic events occur on younger unconsolidated geologic materials.	Avoidance of units having potential for settlement or liquefaction, special construction techniques.
D	Susceptibility to earthquake shaking D1—characterized by relatively strong, low-frequency shaking of long duration. D2—characterized by intermediate-frequency shaking of less duration than that of D1. D3—characterized by high-frequency shaking of relatively short duration.	D1—unconsolidated, water-saturated sand, gravel, and clay. D2—semiconsolidated sand, gravel, and clay. D3—bedrock.	Reaction of structures on various units dependent on many variables, including but not limited to: size of earthquake, distance and direction of epicenter, depth of local water table, type of structure and foundation.	Site evaluation prior to selection of construction techniques, particularly foundation selection; select engineering analysis procedures, prior to construction, that are appropriate to building size, occupancy rate, and specific site conditions.

(continued)

Table 9.20 (continued)

Symbol[b]	Geologic Factor	Criteria for Area Designation	Characteristics	Examples of Possible Mitigative Measures
E	Tsunami inundation	Coastal areas lower than 25 feet above mean sea level.	Degree of tsunami flooding is dependent on many variables, including location of tsunami-generating event, direction of wave arrival, nature of offshore and onshore topography, phase of tide, direction and velocity of wind, volume of freshwater runoff.	Avoidance of low-lying coastal areas, establishment of warning system, selective construction of breakwaters, levees.
⚒ ⚔	Loss of mineral resource areas	Areas are considered mineral resource areas if there is evidence that they contain deposits that are or have been the source of mineral products.	Mineral resources of areas appear to be limited to construction materials of low unit cost. Proximity to market is important, as cost of transportation is a significant portion of total cost.	Within general plans, include provisions for designating mineral resources areas; avoid inappropriate zoning which might lead to incompatible neighboring uses; develop reclamation plans for depleted mineral deposits.
G	Fault rupture	Area designated by the California Division of Mines and Geology as being within an Alquist-Priolo Special Studies Zone, July 1, 1974.	Zone contains faults deemed sufficiently active and well defined as to constitute a potential hazard to structures from surface rupturing or fault creep.	Avoid construction across traces of potentially active faults; detailed studies required to accurately determine traces of such faults.

[a] Geologic factors: Schooner Gulch to Gualala River, Mendocino County, California.

[b] Areas that are subject to the impact of several geologic factors are identified with more than one symbol.

Source: From Williams and Bedrossian, 1977.

Mapping of changes in the shoreline position may yield valuable information. Dolan (1973) points out that the changes in barrier island shorelines and overwash zones seem best analyzed by using repetitive aerial photography. This was one of several techniques used to determine the natural shoreline erosion rate near the Cape Hatteras lighthouse (Lisle and Dolan, 1984). A 7.5-m-per-year natural erosion rate was determined for the 41-km shoreline where the lighthouse is located. Projecting this rate to the year 2005, indicates that the shoreline will migrate to a point 90 m west of the present lighthouse. Unless measures are taken, the lighthouse will be destroyed.

Mapping and sampling of shoreline processes to provide data comparable to the land along a shoreline can prove a difficult task. Pethick (1984) reports that it is no simple task to measure sediment movement under 6-ft-high breakers. Often a surrogate variable that is more readily measured can be employed.

Such a surrogate variable might be obtained by making profiles across beaches such as those used by Fox and Davis (1978) in their study of seasonal variation in beach erosion and sedimentation. Their work permitted estimation of sediment movement in the near-shore environment along part of the Oregon coast. By conducting repetitive cross-section surveys in conjunction with sampling the materials present, they showed how beach erosion responded to major storms and computed an estimate of the volume of material transported.

Mitigating the Effects of Shoreline Processes

Mitigation measures applied to shorelines are usually intended either to improve navigation or to reduce shoreline erosion (Komar, 1976). As noted earlier, the balance achieved along a natural shoreline is a desirable state. However, past alteration of the shoreline may require mitigation of erosion, material transport, or deposition. It is instructive to note that many places now requiring mitigation efforts were altered by past attempts to limit shoreline erosion. Dolan and Lins (1985) note tens of millions of dollars have been expended in the last 20 years in efforts to stabilize and protect beachfront property along the mid-Atlantic coast. Restoring the typical 50-ft-beachfront lot in the Outer Banks of North Carolina alone was estimated in 1972 to cost nearly $20,000. Annual maintenance would require an additional $1,000 to $2,000 each year to ensure stability. Such a large investment indicates erosion-control projects are necessarily limited to important shoreline areas.

The engineering geologist provides information useful to two design criteria for engineering-mitigation works along shorelines. Information is needed to ensure that a structure is designed to withstand extreme conditions without being rendered inoperable. This is sometimes referred to as a structural-stability criterion (U.S. Army Coastal Engineering Research Center, 1984). Data are also required to ensure that the structure achieves the purpose for which it is designed. In other words, a structure placed to prevent waves from eroding the foundation of another shoreline structure is expected to accomplish that purpose. This is sometimes called the functional-performance criterion (U.S. Army Coastal Engineering Center, 1984).

Mitigation measures along shorelines usually attempt to counter erosion, transportation, or deposition. Erosion that may remove a desirable beach or threaten shoreline structures may require mitigation by structures designed to limit wave attack or replenish beach sand. Transportation of materials can remove shoreline fills or offshore sources of sand for natural beach nourishment. Such material movement would need mitigation through means that inhibit transport by currents. Deposition can render channels and harbors unnavigible and block inlets leading to the sea. Removal by dredging is the principal mitigation for deposition.

Structures threatened with direct wave attack can either be removed or protected. Ideally, shoreline areas vulnerable to wave action can be identified and restricted to nonstructural uses. A setback requirement for eroding sea cliffs is a form of removal. This avoids the threat posed to structures by wave action that undermines cliff stability. Geologic information is needed to determine the erodibility of material exposed in the sea cliffs along coastal segments. This provides a basis for setback distances from the cliff edge that permits reasonable use of coastal property while limiting potential losses. Atwater (1978) describes the development of setback and regulation of bluff-top development in San Mateo County, California. Depending on bedrock resistance, historic rates of bluff erosion, and exposure to wave action, bluff tops may be designated as excluding structures, permitting structures only where geotechnical study demonstrates stability, or permitting structures with no special study of stability.

Where existing structures cannot be removed, mitigating direct attack by waves commonly relies on construction of seawalls, bulkheads, and revetments. These structures differ mainly in purpose (U.S. Army Coastal Engineering Research Center, 1984). Seawalls are designed to resist the full force of waves. They are often used to protect structures located on the shoreline. Figure 9.63*a* shows a rubble-mound seawall. Bulkheads serve mainly to retain fill material placed at or near the shoreline. These structures are generally designed to resist moderate wave action (Figure 9.63*b*). The lightest structure is the revetment. Revetments are used to protect shorelines from erosion by currents or mild wave action (Figure 9.63*c*). A variety of materials are used in constructing seawalls, bulkheads, and revetments, as can be seen from the examples given in Figure 9.63.

Several geologic factors require consideration when selecting such structural measures. An obvious factor is the exposure to wave action. This bears on the choice of seawall, bulkhead, or revetment. Foundation conditions influence the type of structure. Of prime concern is whether the foundation material is suitable for the proposed structure. One that requires driving structural elements into the foundation material is a poor choice for a bedrock-floored site. Another consideration is whether the foundation material will resist scour. Some types of structures may require additional design elements if scour should prove likely. Finally, the availability of materials, often aggregate, is a selection criterion.

Groins and jetties are structures that extend from the shore into the sea, inhibiting transport of material by littoral drift. Groins are intended to

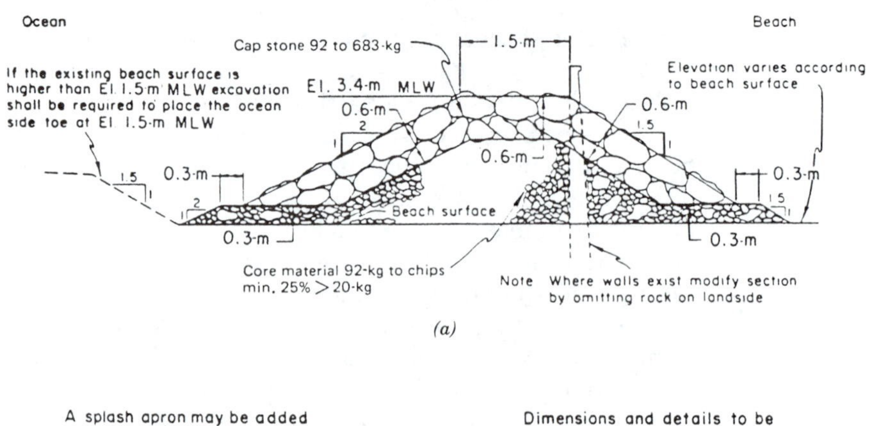

(a)

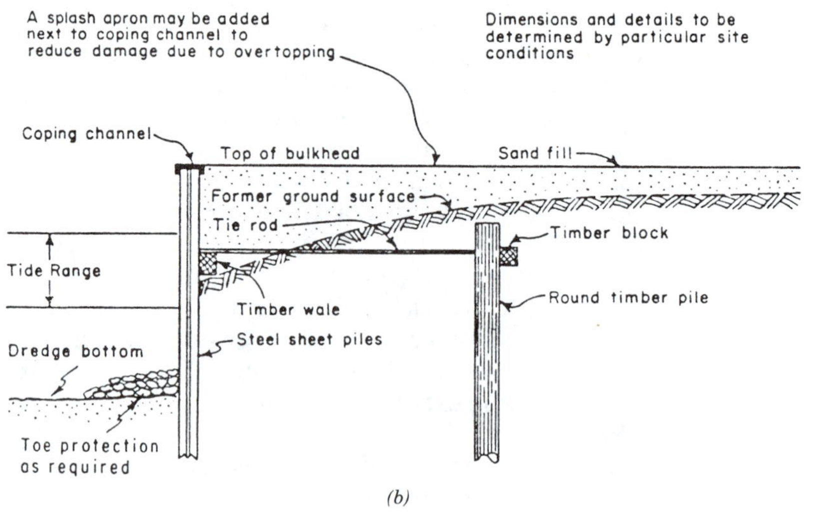

(b)

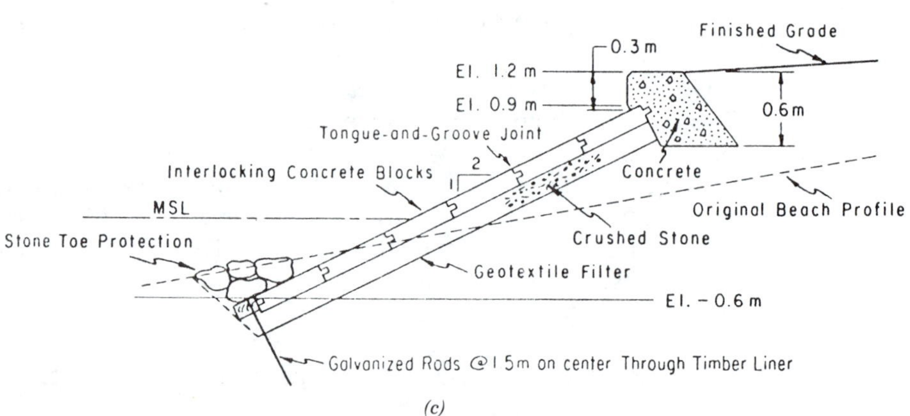

(c)

Figure 9.63 Examples of different structures used to inhibit the effects of waves at the shoreline. (*a*) Rubble-mound seawall. (*b*) Sheetpile bulkhead. and (*c*) Concrete revetment. (From U.S. Army Coastal Engineering Research Center, 1984)

protect the shoreline by trapping material from the littoral drift in order to create a protective beach. An idealized example of a groin-adjusted shoreline is shown in Figure 9.64. Jetties are located at the entrances of harbors or river mouths. Their purpose is to prevent littoral drift from accumulating and to channel tidal or river flow through the inlet. Figure 9.65 shows the placement of jetties at Ponce de Leon inlet, just south of Daytona Beach, Florida.

Like seawalls, bulkheads, and revetments, geologic considerations for constructing groins and jetties include foundation conditions and aggregate availability. In additional, the effectiveness of groins depends on the magnitude and direction of the littoral current. The engineering geologist needs to provide information to determine how far shoreward groins should be founded to avoid flanking and how far seaward to achieve the intended interruption of littoral drift. The length, height, and spacing of the groins depend on this information in conjunction with other data such as expected wave heights. Similar information on littoral drift is needed to determine the effectiveness of jetties in maintaining inlets. Information on shoreline migration is another geologic factor important to planning jetties.

Deposition along the shoreline can sometimes be handled through by-passing rather than dredging. This approach not only maintains the channel or coastal inlet, it avoids interrupting littoral drift, which may cause shoreline erosion. Figure 9.66 shows the arrangement of a sand by-passing operation at Rudee Inlet, Virginia. This inlet was made navigable in 1952 by constructing two jetties and dredging the channel. However, littoral drift immediately began to fill the channel and erosion took place on beaches downdrift. The by-pass station illustrated was installed in the early 1970s to permit maintenance of the channel without downdrift erosion. This system successfully passes 70,000–120,000 yd^3 of material per year past Rudee Inlet (U.S. Army Coastal Engineering Research Center, 1984).

To maintain navigable waterways and to protect beaches and shoreline

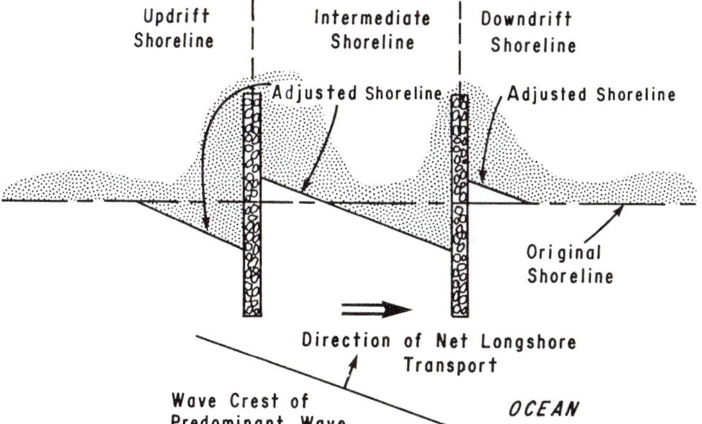

Figure 9.64 Diagram showing how groins may be employed to inhibit the effects of littoral drift. (From U.S. Army Coastal Engineering Research Center, 1984)

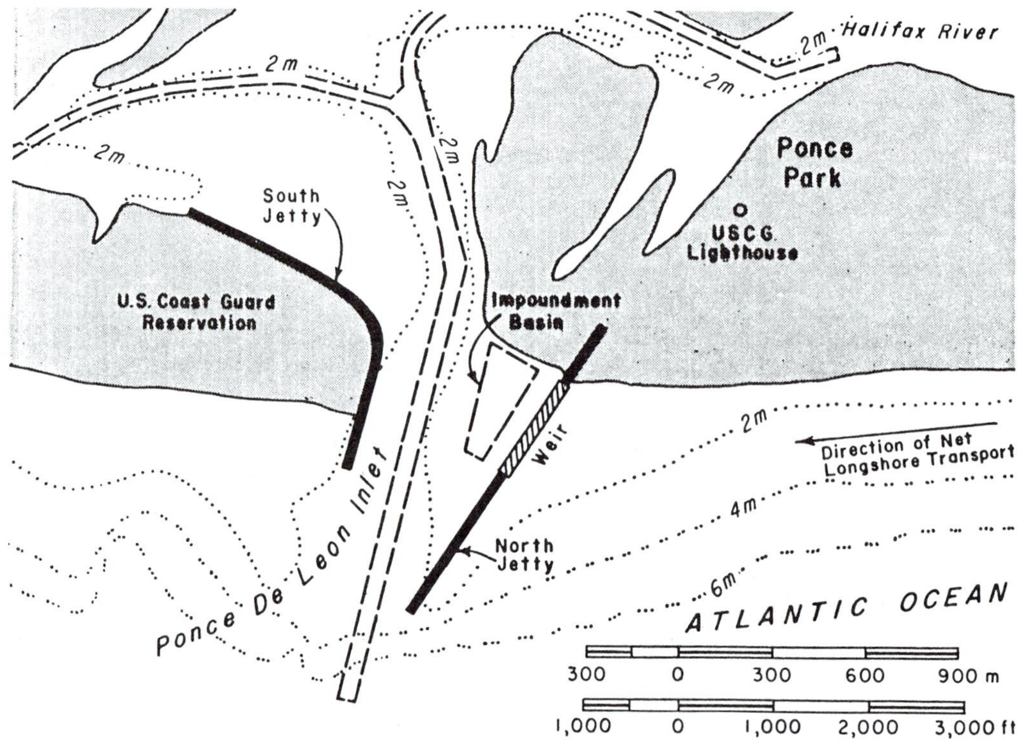

Figure 9.65 The application of jetties to stabilize an outlet illustrated by the Ponce de Leon inlet south of Daytona Beach, Florida. (From U.S. Army Coastal Engineering Research Center, 1984)

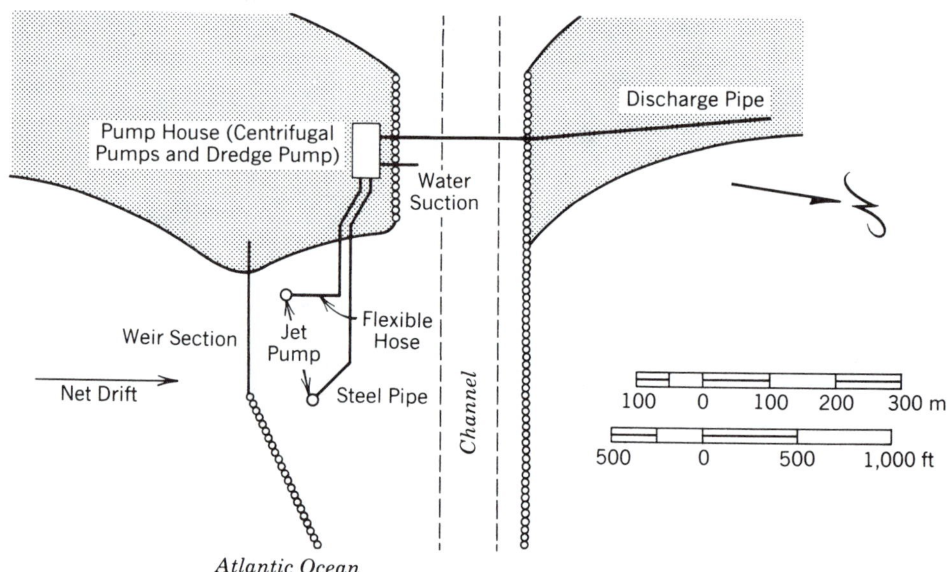

Figure 9.66 Features associated with the sand bypassing operation that maintains the channel through Rudee Inlet. (From U.S. Army Coastal Engineering Research Center, 1984)

structures, geologic data are needed on both the materials and landforms along the shoreline and on the action of water-related processes. This information is necessary to ensure mitigation measures are properly designed to survive in this dynamic environment and that they will achieve their intended purpose. It should be remembered that each coastal site hosts differing conditions. To simply transfer mitigation that worked at one site to another is to court a costly failure. It is more prudent to treat each site as unique when planning mitigation of shoreline processes (Pilkey, et al., 1978).

SUMMARY

The geologist develops information on earth processes that influences how development occurs and how it can avoid inducing undesirable changes in earth processes. Where this input is lacking, engineering works are likely to cost more than necessary, function below expected optimum, or even fail.

Prediction is an important part of assessing earth processes for engineering needs. This requires judgment as well as facts. Mapping, sampling, and other means of data collection can be applied to subsidence, landslide, shoreline, and the other processes discussed. In most situations, this collected information needs to be interpreted, which requires judgment based on a clear understanding of both the process and the engineering need being addressed.

Each of the earth processes described herein can pose a problem to engineering works. This ranges from the dramatic effects of earthquake-caused ground motion to the subtle stresses of swelling soils. Successfully avoiding major problems will always require accommodating rather than ignoring these natural forces.

REFERENCES

Adams, F. T., and Lovell, C. W., 1984, Mapping and prediction of limestone bedrock problems: Transp. Res. Rec., Vol. 978, pp. 1–5.

Algermissen, S. T., 1969, Seismic risk studies in the United States: Proc. 4th World Conf. on Earthquake Eng., Santiago, Chile, Vol. 1, pp. 19–27.

Algermissen, S. T., and Perkins, D. M., 1972, A technique for seismic zoning: general considerations and parameters: Proc. Intl. Conf. on Microzonation, Seattle, Wash., Vol. 2, pp. 865–878.

———, 1976, A probabilistic estimate of maximum acceleration in rock in the contiguous United States: U.S. Geol. Surv. Open-File Rep. 76–416. 45 pp.

Atwater, B., 1978, Central San Mateo County, California: land-use controls arising from erosion of seacliffs, landsliding, and fault movement, *in* Nature to be commanded: U.S. Geol. Surv. Prof. Paper 950, pp. 11–19.

Baker, P. E., 1979, Geological aspects of volcano prediction: J. Geol. Soc. Lond., Vol. 136, pp. 341–345.

Blair, M. L., and Spangle, W. E., 1979, Seismic safety and land-use planning—selected examples from California: U.S. Geol. Surv. Prof. Paper 941-B, 82 pp.

Blong, R. J., 1981, Some effects of tephra falls on buildings, *in* Tephra studies: D. Reidel, Dordrecht, Neth., pp. 405–420.

Bolt, B. A., Horn, W. L., Macdonald, G. A., and Scott, R. F., 1975, Geological hazards: Springer-Verlag, New York, 328 pp.

Bonilla, M. G., 1981, Surface faulting, *in* Facing geologic and hydrologic hazards, U.S. Geol. Surv. Prof. Paper 1240-B, pp. 16–23.

Booth, B., 1979, Assessing volcanic risk: J. Geol. Soc. Lond., Vol. 136, pp. 331–340.

Borcherdt, R. D., Joyner, W. B., Warrick, R. F., and Gibbs, J. F., 1975, Response of local geologic units to ground shaking, *in* Studies for seismic zonation of the San Francisco Bay region: U.S. Geol. Surv. Prof. Paper 941–A, pp. 52–67.

Brabb, E. E., 1984, Innovative approaches to landslide hazard and risk mapping, *in* Proc. 4th Intl. Symp. on Landslides, Toronto, Can., Vol. 1, pp. 307–324.

Brabb, E. E., Pampeyan, E. H., and Bonilla, M. G., 1972, Landslide susceptibility in San Mateo County, California: U.S. Geol. Surv. Misc. Field Studies Map MF-360.

Canuti, P., Focardi, P., Garzonio, C. A., Rodolfi, G., and Vannocci, P., 1982, Stabilità dei versanti nell'area rappresentativa di Montespertoli (Firenze): Consiglio Nazionale delle Richerche, scale 1 : 10,000.

Clark, M. M., Grantz, A., and Meyer, R., 1972, Holocene activity of the Coyote Creek fault as recorded in the sediments of Lake Cahuilla, in the Borrego Mountain earthquake of April 9, 1986: U.S. Geol. Surv. Prof. Paper 787, pp. 1,112–1,130.

Cluff, L. S., Hansen, W. R., Taylor, C. L., Weaver, K. D., Brogan, G. E., Idriss, I. M., McClure, F. E., and Blayney, J. A., 1972, Site evaluation in seismically active regions—an interdisciplinary team approach: Proc. Intl. Conf. on Microzonation, Seattle, Wash., Vol. 2., pp. 161–168.

Coates, D. R., 1977, Landslide perspectives, *in* Geol. Soc. Am., Reviews in Eng. Geol., Landslides, Vol. 3, pp. 3–28.

Costa, J. E., 1984, Physical geomorphology of debris flows, *in* Developments and applications of geomorphology, Springer-Verlag, Berlin, W. Ger., pp. 268–317.

Cotecchia, V., 1978, Systematic reconnaissance mapping and registration of slope movements: Bull. Intl. Assoc. Eng. Geol., No. 17, pp. 5–37.

Crandell, D. R., and Mullineaux, D. R., 1975, Technique and rationale of volcanic-hazards appraisals in the Cascade Range, northwestern United States: Environ. Geol., Vol. 1, pp. 23–32.

DeGraff, J. V., 1978, Regional landslide evaluation: two Utah examples: Environ. Geol., Vol. 2, pp. 203–214.

———, 1979, Initiation of shallow mass movement by vegetative-type conversion: Geology, Vol. 7, pp. 426–429.

DeGraff, J. V., and Agard, S. S., 1984, Using tree-ring analysis to define geologic hazards for natural resources management: Environ. Geol. & Water Sci., Vol. 6, pp. 147–155.

DeGraff, J. V., and Cunningham, C. G., 1982, Highway-related landslides in mountainous volcanic terrain: an example from west-central Utah: Bull. Assoc. Eng. Geol., Vol. 19, pp. 319–325.

DeGraff, J. V., McKean, J., Watanabe, P. E., and McCaffery, W. F., 1984, Landslide activity and groundwater conditions: insights from a road in the central Sierra Nevada, California: Transp. Res. Rec., Vol. 965, pp. 32–37.

DeGraff, J. V., and Romesburg, H. C., 1980, Regional landslide-susceptibility assessment for wildlands management: a matrix approach, *in* Thresholds in geomorphology, Allen & Unwin, Boston, Mass., pp. 401–414.

———, 1981, Subsidence crack closure: rate, magnitude, and sequence: Bull. Intl. Assoc. Eng. Geol., No. 23, pp. 123–127.

Dolan, R., 1973, Barrier islands: natural and controlled, *in* Coastal geomorphology, State Univ New York, Binghamton, pp. 263–278.

Dolan, R., and Lins, H., 1985, The outer banks of North Carolina: U.S. Geol. Surv. Prof. Paper 1177-B, 47 pp.

Dunrud, C. R., 1984, Coal mine subsidence—western United States, *in* Geol. Soc. Am., Reviews in Eng. Geol., Vol. 6, pp. 151–194.

Edil, T. B., and Vallejo, L. E., 1980, Mechanics of coastal landslides and the influence of slope parameters: Eng. Geol., Vol. 16, pp. 83–96.

Fiske, R. S., 1984, Volcanologists, journalists, and the concerned local public: a tale of two crises in the eastern Caribbean, *in* Explosive volcanism: inception, evolution and hazards: National Academy Press, Washington, D.C., pp. 170–176.

Fleming, R. W., and Taylor, F. A., 1980, Estimating the costs of landslide damage in the United States: U.S. Geol. Surv. Circ. 832, 21 pp.

Fleming, R. W., Varnes, D. J., and Schuster, R. L., 1979, Landslide hazards and their reduction: J. Am. Planning Assoc., Vol. 45, pp. 428–439.

Fohn, P.M.B., 1978, Avalanche frequency and risk estimation in forest sites: Proc. IUFRO Intl. Seminar, Mountain Forests and Avalanches, Davos, Switz, pp. 241–254.

Fountain, L. S., 1976, Subsurface cavity detection: field evaluation of radar, gravity, and earth resistivity methods: Transp. Res. Rec. 612, pp. 38–46.

Fournier d'Albe, E. M., 1979, Objectives of volcanic monitoring and prediction: J. Geol. Soc. Lond., Vol. 136, pp. 321–326.

Fox, W. T., and Davis, R. A., Jr., 1978, Seasonal variation in beach erosion and sedimentation on the Oregon coast: Geol. Soc. Am. Bull., Vol. 89, pp. 1,541–1,549.

Galster, R. W. 1987, A survey of coastal engineering geology in the Pacific Northwest: Bull. Assoc. Eng. Geol., Vol. 24, pp. 161–198.

Gedney, D. S., and Weber, W. G., Jr., 1978, Design and construction of soil slopes, *in* Landslides, analysis and control, Transp. Res. Bd. Spec. Rep. 176, pp. 172–191.

Gray, R. E., and Bruhn, R. W., 1984, Coal mine subsidence—eastern United States, *in* Geol. Soc. Am., Reviews in Eng. Geol., Vol 6, pp. 123–149.

Gray, R. E., Salver, H. A., and Gamble, J. C., 1976, Subsidence control for structures above abandoned coal mines: Transp. Res. Rec., Vol. 612, pp. 17–24.

Griggs, G. B., 1987, Littoral cells and harbor dredging along the California coast: Environ. Geol. Water Sci., Vol. 10, pp. 7–20.

Griggs, G. B., and Johnston, R. E., 1983, Impact of 1983 storms on the coastline of northern Monterey Bay: Calif. Geol., Vol. 36, pp. 163–174.

Gutenburg, B., and Richter, C. F., 1956, Earthquake magnitude, intensity, energy, and acceleration: Seismol. Soc. Am. Bull., Vol. 46, pp. 105–145.

Hamel, J. V., 1983, Geotechnical perspective on river bank instability, *in* Frontiers of hydraulic engineering, ASCE Hydraulics Div., Specialty Conf. Proc., Cambridge, Mass., pp. 212–217.

Hamel, J. V., and Spencer, G. S., 1984, Powerhouse slope behavior, Fort Peck

Dam, Montana, *in* Proc. Intl. Conf. on Case Histories in Geotech. Eng., Vol 2, pp. 541–551.

Harp, E. L., Tanaka, K., Sarmiento, J., and Keefer, D. K., 1984, Landslides from the May 25–27, 1980, Mammoth Lakes, California, earthquake sequence: U.S. Geol. Surv. Misc. Inv. Ser. Map I–1612.

Hays, W. W., 1980, Procedures for estimating earthquake ground motions: U.S. Geol. Surv. Prof. Paper 1114, 77 pp.

———, 1981a, Ground shaking, *in* Facing geologic and hydrologic hazards: U.S. Geol. Surv. Prof. Paper 1240–B, pp. 6–15.

———, 1981b, Tsunamis, *in* Facing geologic and hydrologic hazards: U.S. Geol. Surv. Prof. Paper 1240–B, pp. 32–38.

Hayes, M. O., Owens, E. H., Hubbard, D. K., and Abele, R. W., 1973, The investigation of form and processes in the coastal zone, *in* Coastal geomorphology, State Univ. of New York, Binghamton, pp. 11–41.

Hays, W. W., and Shearer, C. F., 1981, Suggestions for improving decision making to face geologic and hydrologic hazards, *in* Facing geologic and hydrologic hazards: U.S. Geol. Surv. Prof. Paper 1240–B, pp. 103–108.

Helley, E. J., LaJoie, K. R., Spangle, W. E., and Blair, M. L., 1979, Flatland deposits of the San Francisco Bay region, California—their geology and engineering properties, and their importance to comprehensive planning: U.S. Geol. Surv. Prof. Paper 943, 88 pp.

Helm, D. C., 1984, Field-based computational techniques for predicting subsidence due to fluid withdrawal, in Geol. Soc. Am., Reviews in Eng. Geol., Vol. 6, pp. 1–22.

Holtz, W. G., 1959, Expansive clays—properties and problems, *in* Theoretical and practical treatment of expansive soils, Colo. Sch. Mines Q., Vol. 54, No. 4, pp. 89–117.

Holtz, W. G., and Hart, S. S., 1978, Home construction on shrinking and swelling soils: Colo. Geol. Surv. Spec. Publ. 11, 18 pp.

Holzer, T. L., 1984, Ground failure induced by ground-water withdrawal from unconsolidated sediment, *in* Geol. Soc. Am, Reviews in Eng. Geol., Vol. 6, pp. 67–105.

Hoyt, W. H., 1981, Beach processes and sand removal downdrift of stabilized inlets: a case study of Indian River and Roosevelt inlets, coastal Delaware: Northeastern Geol., Vol. 3, pp. 259–267.

Hungr, O., Morgan, G. C., and Kellerhals, R., 1984, Quantitative analysis of debris torrent hazards for design of remedial measures: Can. Geotech. J., Vol. 21, pp. 663–677.

Ikeya, H., 1976, Introduction to Sabo works: Japan Sabo Assoc., Tokyo, 168 pp.

James, L. B., 1968, Failure of Baldwin Hills reservoir, Los Angeles, California, in Geol. Soc. Am., Eng. Geol. Case Histories, Vol 6, pp. 1–11.

Jones, D. E., Jr., and Holtz, W. G., 1973, Expansive soils—the hidden disaster: Civ. Eng., Vol. 43, pp. 49–51.

Keefer, D. K., and Johnson, A. M., 1983, Earth flows: morphology, mobilization and movement: U.S. Geol. Surv. Prof. Paper 1264, 56 pp.

Kemmerly, P. R., 1984, Corrective procedures for sinkhole collapse on the Western Highland Rim, Tennessee: Transp. Res. Rec., Vol. 978, pp. 12–18.

Kitts, D. B., 1976, Certainty and uncertainty in geology: Am. J. Sci., Vol. 276, pp. 29–46.

Kockelman, W. J., 1985, Reducing losses from earthquakes through personal preparedness: Earthquake Inf. Bull., Vol. 17, pp. 50–59.

———, 1986, Some techniques for reducing landslide hazards: Bull. Assoc. Eng. Geol., Vol. 23, pp. 29–52.

Kockelman, W. J., and Campbell, C. C., 1983, Two examples of earthquake-hazard reduction in southern California: Earthquake Inf. Bull., Vol. 15, pp. 216–225.

Komar, P. D., 1976, Beach processes and sedimentation: Prentice-Hall, Englewood Cliffs, N.J., 429 pp.

———, 1983, Beach processes and erosion—an introduction, *in* CRC handbook of coastal processes and erosion, CRC Press, Boca Raton, Fla., pp. 1–20.

Krinitzsky, E. L., and Marcuson III, W. F., 1983, Principles for selecting earthquake motions in engineering design: Bull. Assoc. Eng. Geol., Vol. 20, pp. 253–265.

Lambert, M. B., 1974, The mighty volcano drama in Iceland: Can. Geogr. J., Vol. 89, pp. 4–10.

Larson, M. K., and Péwé, T. L., 1986, Origin of land subsidence and earth fissuring, northeast Phoenix, Arizona: Bull. Assoc. Eng. Geol., Vol. 23, pp. 139–165.

Lee, S., and Nichols, D. R., 1981, Subsidence, *in* Facing geologic and hydrologic hazards: U.S. Geol. Surv. Prof. Paper 1240–B, pp. 73–86.

Leighton, F. B., 1976, Geomorphology and engineering control of landslides, *in* Geomorphology and engineering, Dowden, Hutchinson & Ross, New York, pp. 273–287.

Lessing, P., Kulander, B. R., Wilson, B. D., Dean, S. L., and Woodring, S. M., 1976, West Virginia landslides and slide-prone areas: W. Va. Geol. & Econ. Surv., Environ. Geol. Bull. 15, 64 pp.

Lisle, L. D., and Dolan, R., 1984, Coastal erosion and the Cape Hatteras lighthouse: Environ. Geol. Water Sci., vol. 6, pp. 141–146.

Lockwood, J. P., and Romano, R., 1985, Diversion of lava during the 1983 eruption of Mount Etna: Earthquake Inf. Bull., Vol. 17, pp. 124–133.

Lowrance, W. W., 1976, Of acceptable risk: William Kaufmann, Inc., Los Altos, Calif., 180 pp.

Lucchitta, I., Schleicher, D., and Cheney, P., 1981, Of pride and prejudice: the importance of being earnest about environmental impact statements: Geology, Vol. 9, pp. 590–591.

Lundgren, L., 1976, Geology and public policy: Geology, Vol. 5, pp. 657–660.

Mader, G. G., Danehy, E. A., Cummings, J. C., and Dickinson, W. R., 1972, Land use restrictions along the San Andreas fault in Portola Valley, California: Proc. Intl. Conf. Microzonation, Seattle, Wash., Vol. 2, pp. 845–857.

Martin, J. C., and Serdengecti, S., 1984, Subsidence over oil and gas fields, *in* Geol. Soc. Am., Reviews in Eng. Geol., Vol. 6, pp. 23–34.

Mielenz, R. C., and King, M. E., 1955, Physical-chemical properties and engineering performance of clays, *in* Clays and clay technology, Calif. Div. Mines and Geol. Bull. 169, pp. 196–254.

Miller, C. D., 1985, Holocene eruptions at the Inyo volcanic chain, California: implications for possible eruptions in Long Valley caldera: Geology, Vol. 13, pp. 14–17.

Miller, C. D., Mullineaux, D. R., Crandell, D. R., and Bailey, R. A., 1982, Potential hazards from future volcanic eruption in the Long Valley–Mono Lake area, east-central California and southwest Nevada—a preliminary assessment: U.S. Geol. Surv. Circ. 877, 10 pp.

Morgenstern, N. R., and Price, V. E., 1965, The analysis of the stability of general slip surfaces: Geotechnique, Vol. 15, pp. 79–93.

Morgenstern, N. R., and Sangrey, D. A., 1978, Methods of stability analysis, *in* Landslides, analysis and control, Transp. Res. Bd. Spec. Rep. 176, pp 155–171.

Mullineaux, D. R., 1981, Hazards from volcanic eruptions, *in* Facing geologic and hydrologic hazards: U.S. Geol. Surv. Prof. Paper 1240–B, pp. 86–101.

National Research Council, 1985a, Liquefaction of soils during earthquakes: National Academy Press, Washington, D.C., 240 pp.

———, 1985b, Reducing losses from landsliding in the United States: National Academy Press, Washington, D.C., 41 pp.

NAVFAC, 1982, Design manual: soil mechanics: U.S. Dept. of Defense, NAVFAC DM–7.1, Dept. of the Navy, Washington, D.C., 360 pp.

Neumann, N. M., 1954, Earthquake intensity and related ground motion: Washington Univ. Press, Seattle, 40 pp.

Newmark, N. M., 1965, Effects of earthquakes on dams and embankments: Geotechnique, Vol. 15, pp. 139–160.

Newton, J. G., 1976, Induced and natural sinkholes in Alabama—a continuing problem along highway corridors: Transp. Res. Rec., Vol. 612, pp. 9–16.

———, 1984, Sinkholes resulting from ground-water withdrawals in carbonate terranes—an overview: Geol. Soc. Am., Reviews in Eng. Geol., Vol. 6, pp. 195–202.

Nichols, D. R., and Buchanan-Banks, J. M., 1974, Seismic hazards and land-use planning: U.S. Geol. Surv. Circ. 690, 33 pp.

Nilsen, T. H., and Wright, R. H., 1979, Relative slope stability of the San Francisco Bay region, in Relative slope stability and land-use planning in the San Francisco Bay region, California, U.S. Geol. Surv. Prof. Paper 944, pp. 16–55.

Olshansky, R. B., and Rogers, J. D., 1987, Unstable ground: landslide policy in the United States: Ecology Law Q., Vol. 13, pp. 939–1,006.

Organization of American States, 1986, Map of landslide hazards, St. Lucia, West Indies: Natural Resources Evaluation Map, OAS, Washington, D.C.

Petak, W. J., 1984, Geologic hazard reduction: the professional's responsibility: Bull. Assoc. Eng. Geol., Vol. 21, pp. 449–458.

Pethick, J., 1984, An introduction to coastal geomorphology: Edward Arnold, London Eng., 260 pp.

Pilkey, O. H., 1981, Geologists, engineers, and a rising sea level: Northeastern Geol., Vol. 3, pp. 150–158.

Pilkey, O. H., Jr., Neal, W. J., and Pilkey, O. H., Sr., 1978, From Currituck to Calabash, *in* Living with North Carolina's barrier islands, N.C. Sci. and Tech. Res. Ct., 370 pp.

Robinson, G. D., and Spieker, A. M., 1978, Nature to be commanded: U.S. Geol. Surv. Prof. Paper 950, 95 pp.

Royster, D. L., 1984, Use of sinkholes for drainage: Transp. Res. Rec., Vol. 978, pp. 18–25.

Schnabel, P. B., and Seed, H. B., 1972, Accelerations in rock for earthquakes in the western United States: Seismol. Soc. Am. Bull., Vol. 62, pp. 501–516.

Schuster, R. L., 1978, Introduction, *in* Landslides, analysis and control, Transp. Res. Bd. Spec. Rep. 176, pp. 1–10.

———, 1979, Reservoir-induced landslides: Bull. Intl. Assoc. Eng. Geol., No. 20, pp. 8–15.

———, 1981, Expansive soils, *in* Facing geologic and hydrologic hazards: U.S. Geol. Surv. Prof. Paper 1240–B, pp. 66–73.

———, 1983, Engineering aspects of the 1980 Mount St. Helens eruptions: Bull. Assoc. Eng. Geol., Vol. 20, pp. 125–144.

Schuster, R. L., and Costa, J. E., 1986. A perspective on landslide dams, *in* Landslide dams: processes, risk, and mitigation, Am. Soc. Civ. Eng., Geotech. Spec. Publ. No. 3, pp. 1–20.

Schuster, R. L., and Fleming, R. W., 1986, Economic losses and fatalities due to landslides: Bull. Assoc. Eng. Geol., Vol. 23, pp. 11–28.

Schwartz, D. P., and Coppersmith, K. J., 1984, Fault behavior and characteristic earthquakes: examples from the Wasatch and San Andreas fault zones: J. Geophys. Res., Vol. 89, pp. 5,681–5,698.

Scott, W. E., 1984, Hazardous volcanic events and assessments of long-term volcanic hazards, *in* U.S. Geol. Surv. Open-File Rep. 84–760, pp. 447–498.

Seed, H. B., Whitman, R. V., Dezfulian, H., Dobry, R., and Idriss, I. M., 1972, Soil conditions and building damage in 1967 Caracas earthquake: Am. Soc. Civ. Eng. Proc., J. Soil Mech. and Found. Div., Vol. 98, pp. 787–806.

Self, S., Rampino, M. R., Newton, M. S., and Wolff, J. A., 1984, Volcanological study of the great Tambora eruption of 1815: Geology, Vol. 12, pp. 659–663.

Shroder, J. F., Jr., 1980, Dendrogeomorphology: review and new techniques of tree-ring dating: Prog. Phys. Geogr., Vol. 4, pp. 161–188.

Sidle, R. C., Pearce, A. J., and O'Loughlin, C. L., 1985, Hillslope stability and land use: Am. Geophys. Union, Water Resources Monograph Ser. No. 11, 140 pp.

Slosson, J. E., 1969, The role of engineering geology in urban planning, *in* Governor's conf. on Environ. Geol., Colo. Geol. Surv. Spec. Publ. 1, pp. 8–15.

Sowers, G. F., 1976, Mechanisms of subsidence due to underground openings: Transp. Res. Rec., Vol. 612, pp. 2–8.

Steinberg, M. L., 1985, Controlling expansive soil destructiveness by deep vertical geomembranes on four highways: Transp. Res. Rec., Vol. 1032, pp. 48–53.

Suryo, I., and Clarke, M.C.G., 1985, The occurrence and mitigation of volcanic hazards in Indonesia as exemplified at the Mount Merapi, Mount Kelut, and Mount Galunggung volcanoes: Q. J. Eng. Geol., Vol 18, pp. 79–98.

U.S. Army Coastal Engineering Research Center, 1984. Shore protection manual: U.S. Army Corps of Eng., Vicksburg, Miss., Vols. 1 and 2.

Varnes, D. J., 1978, Slope movement types and processes, *in* Landslides, analysis and control, Transp. Res. Bd. Spec. Rep. 176, pp. 11–33.

———, 1984, Landslide hazard zonation: a review of principles and practice: Natural Hazards Ser. No. 3, UNESCO, Paris, France, 63 pp.

Wesson, R. L., Helley, E. J., LaJoie, K. R., and Wentworth, C. M., 1975, Faults and future earthquakes, *in* Studies for seismic zonation of the San Francisco Bay region: U.S. Geol. Surv. Prof. Paper 941–A, pp. 5–30.

Wieczorek, G. F., Wilson, R. C., and Harp, E. L., 1985, Map showing slope stability during earthquakes in San Mateo County, California: U.S. Geol. Surv. Misc. Inv. Ser. Map I–1257–E.

Williams, G. P., and Wolman, M. G., 1984, Downstream effects of dams on alluvial rivers: U.S. Geol. Surv. Prof. Paper 1286, 83 pp.

Williams, J. W., 1987, Similarities and differences of engineering geological factors in the marine coastal areas of the continental United States: Bull. Assoc. Eng. Geol., Vol. 24, pp. 153–160.

Williams, J. W., and Bedrossian, T. L., 1977, Coastal zone geology near Gualala, California: Calif. Geol., Vol. 30, pp. 27–34.

———, 1978, Geologic mapping for coastal zone planning in California—background and examples: Environ. Geol., Vol. 2, pp. 151–163.

Witczak, M. .W., 1972, Relationships between physiographic units and highway design factors: Natl. Coop. Hwy. Res. Program Report 132, Hwy. Res. Board, Washington, DC, 161 pgs.

Yanev, P., 1977, Earthquake safety for homes: Cal. Geol., Vol. 30, pp. 272–277.

Youd, T. L., 1984, Geologic effects—liquefaction and associated ground failure, *in* U.S. Geol. Surv. Open-File Rep. 84–760, pp. 231–238.

Youd, T. L., and Keefer, D. K., 1981, Earthquake-induced ground failures. *in* Facing geologic and hydrologic hazards—earth science considerations: U.S. Geol. Surv. Prof. Paper 1240–B, pp. 23–31.

Appendix

	Unit Conversions		
	English	Standard International (SI)	cgs
Force	1 lbf	4.448 N	0.4536 kgf
	0.2248 lbf	1.0 N	10^5 dynes
Stress,	1 lbf/in.^{2}	6895 N/m^2	0.0703 kg/cm^2
Pressure	14.503×10^{-5} lbf/in.^{2}	1 N/m^2	10 dynes/cm^2
	14.223 lbf/in.^{2}	0.9807 N/m^2	1 kg/cm^2
Density	1 lbm/ft^3	16.019 kg/m^3	16.019 kg/m^3
	62.428 lbm/ft^3	1 kg/m^3	10^{-3} g/cm^3

lbf/in.^{2} = psi; to convert to N/m^2, divide by 145×10^{-6}
N/m^2 = Pascal (Pa); to convert to psi, multiply by 145×10^{-6}
1 kiloPascal = 1 kPa = 1,000 Pa
1 megaPascal = 1 mPa = 1,000,000 Pa
1 gigaPascal = 1 gPa = 1,000,000,000 Pa

Index